IN MEMORIAM

GEORGII · ARTVRI · PLIMPTON

VIRI

DOCTISSIMI · ILLVSTRISSIMI

ARTIVM MAGISTRI · LEGVM DOCTORIS

LIBRORVM VETVSTIORVM AMATORIS

SCRIPTORVM AMICI · SOCII ERVDITORVM

LITTERARVM FAVTORIS

RARA ARITHMETICA

PLATE I. FROM A MANUSCRIPT OF BOETHIUS, C. 1294, SHOWING FIGURATE NUMBERS

RARA ARITHMETICA

¶ A CATALOGVE OF THE ARITHMETICS WRITTEN BEFORE THE YEAR MDCI WITH A DESCRIPTION OF THOSE IN THE LIBRARY OF GEORGE ARTHVR PLIMPTON OF NEW YORK

BY DAVID EVGENE SMITH OF TEACHERS COLLEGE COLVMBIA VNIVERSITY

CHELSEA PVBLISHING COMPANY
NEW YORK ANNO MDCCCCLXX

FOURTH EDITION,

INCLUDING A. DE MORGAN'S 'ARITHMETICAL BOOKS'

LIBRARY OF CONGRESS CATALOG CARD 74-113148

INTERNATIONAL STANDARD BOOK NO. 8284-0192-6

PRINTED IN THE UNITED STATES OF AMERICA

PREFACE TO THE FOURTH EDITION

David Eugene Smith's *Rara Arithmetica* was first published, at Boston and London, in 1908. A second edition (not called such) was issued in 1910, which consisted of the first edition, textually unaltered, together with a few pages of addenda. Several decades later, in 1939, there was published *Addenda to Rara Arithmetica*; it was issued both as a separate publication and bound in one volume with the second (1910) edition of the *Rara*; and the two constituted, in effect, a third edition.

A much earlier work on early arithmetics, and one to which reference is made many times in the *Rara*, is Augustus De Morgan's *Arithmetical Books*, published at London in 1847. It is rather more informal than the *Rara* and does not confine itself, as the latter does, to works published before the year 1601. Indeed, the very last work it discusses is the fifth edition of De Morgan's own book on Arithmetic, published in 1846, and De Morgan's comment on his own book is quite characteristic (see page 675 of the present text).

The four writings just described have been put together to form the present, fourth edition of David Eugene Smith's *Rara Arithmetica*. The main text of the first edition, the first set of Addenda, the second set of Addenda, and De Morgan's *Arithmetical Books* are represented, respectively, by pages 1-494, 495-498, 499-548, and 551-703 of the present volume. To this has been added a new Index.

Of course, some changes and rearrangements have been necessary. Primarily, there has been inserted in the main text cross-references to the various Addenda. These references have been indicated by superior sans-serif numbers, usually placed at the end of the relevant paragraph, and the reader is referred to the pages indicated by these numbers. In a few instances, the originally published addendum indicated a correction to the text rather than an addition; in each such instance, the correction has been made in the present text and either the original addendum has been omitted entirely from the present text or it has been supplied with a dagger (†). The book by De Morgan has been included in its entirety without change or abridgement.

In the Index to the present work, where the version of a name as given in De Morgan's book or in the Addenda differs from that given in the original *Rara*, preference has been given to the latter version. It may be noted that the original 'List of Names' of De Morgan's book has been retained and constitutes pages 686-703 of the present text.

T. C.

PREFACE

One of the first and most important questions for the student of mathematical history is that relating to the available sources of information. In the fields of higher mathematics scholars have been more or less successful in bringing together these sources, and in listing them in bibliographies; but in that humbler field in which primitive mathematics first found root, only a few bibliophiles have sought to preserve the original material, and no one has seriously attempted to catalogue it. Libri, it is true, brought together two large libraries of Rara Arithmetica, but he was neither a true book-lover nor a true scholar, for he gathered his treasures purposely to see them dispersed, his commercial spirit scattering at random what should have been kept intact for the use of scholars. Prince Boncompagni, the most learned of all collectors in this domain, lived to see an unappreciative city ignore his offer to make his magnificent library permanent, and at his death it was scattered abroad, as had been the lesser ones of Kloss and De Morgan. 498 The third great collection of early textbooks which has been made in recent years is that of Mr. Plimpton. Of the libraries of arithmetics printed before the opening of the seventeenth century his is the largest that has ever been brought together, not excepting Boncompagni's, and it may well be doubted if another so large will again be collected by one man. De Morgan was able to examine, in the British Museum and elsewhere, less than a hundred arithmetics written before 1601, including all editions; but Mr. Plimpton has more than three hundred, a number somewhat in excess of that reached by Boncompagni. Indeed there are few arithmetics of much importance that are not found, in one edition or another, in his library.

The writers of these early printed books, not all themselves of the centuries under consideration, were by no means obscure men. Among them was Boethius, whom Gibbon called "the last of the Romans whom Cato or Tully could have acknowledged for their countryman." In the list are the names of Cassiodorus and Capella, who at least represented what there was of culture in their day, and Isidorus, the learned Bishop of Seville. There are also the names of Archimedes, who deemed it a worthy labor to improve the number system of the Greeks; of Euclid, whose contributions were by no means confined to geometry; of Nicomachus and Iamblichus, who represent the declining Hellenic civilization, and of Psellus, who was a witness of its final decay. There, too, are the names of the Venerable Bede, of Sacrobosco, and of Bradwardin, all of whom testify to the culture of mediæval England; of that great Renaissance compiler, Paciuolo; of Tartaglia and Cardan, who helped to make the modern algebra, and of such scholars as Ramus, Melanchthon, and Bishop Tonstall. Worthy as such a list may be, it is rendered none the less so by the names of Widman, Köbel, Borghi, Riese, and Gemma Frisius, mere arithmeticians though they were, for few who read their works can fail to recognize that they powerfully influenced education, not only in their own time but for generations after they had passed away.

In view of the fact that the fifteenth and sixteenth centuries constituted the formative period in the history of printed arithmetics, I have felt it a not unpleasant duty to catalogue the volumes in Mr. Plimpton's library that belong to this period and subject, including such later editions as it may contain, — to give a brief statement of their contents, and to supplement this work by a list of other arithmetics published before 1601. That a complete bibliography is impossible is evident to any one who considers the subject. It is a simple matter to consult the few lists of early mathematical works, and to trace the names thus secured through such catalogues as those of the British Museum and the Bibliothèque nationale (unfortunately only just begun), and through bibliographies like those of Graesse and Hain and Coppinger. It is also an easy matter to examine the masterly work of Riccardi, the less accurate list of Murhard, the catalogues of Libri, and numerous other works of a similar nature. But it is manifestly impossible to read all of the published catalogues, almost invariably arranged only by authors. Therefore many extant books will necessarily remain undiscovered, and it is probable that some will always elude the eye of the special bibliographer. Such is the work that has been done in preparing this volume, and such is the feeling of insufficiency of achievement that remains. There is, however, a satisfaction in knowing that the bibliography is based in large measure upon an examination of the books themselves in various libraries, and that the secondary sources are of recognized authority. Over five hundred and fifty different works are mentioned, or, including the various editions, nearly twelve hundred books in all. Of the five hundred and fifty books, about four hundred and fifty are, strictly speaking, arithmetics. Of all arithmetics known to have been printed in the sixteenth century, and to have been important enough to have two or more editions, Mr. Plimpton's library lacks less than twenty-five. Of those which were published but once, some are known only by name, while the rest are mostly 'abachetti' or 'Rechenbüchleins'—mere primers of a few pages and of no importance.

It must also be borne in mind that it was inevitable that certain arithmetics of the sixteenth century should have perished utterly, leaving not even a record of their existence. Their very commonness often caused their destruction, a law of which the unique surviving copy of more than seventy thousand New England Primers of the Franklin-Hall press is a lonely witness in our country.

It is inevitable that there should be errors in such a list. Titles have been included when mentioned in even one standard bibliography, although they cannot be found in others, it being practically impossible for one person to verify every item. It is hoped, however, that a foundation has been laid upon which others may build, eliminating or otherwise as the case may be.

Of other works in this field, little need be said. De Morgan's *Arithmetical Books* is still one of our best single sources, although sixty years have elapsed since it first appeared. De Morgan, however, mentions altogether about one hundred and twenty editions of works originally appearing before 1601, against nearly twelve hundred listed here. Unger and Tropffke are both scholarly writers, but their bibliographies are almost exclusively German. Sterner was less of a student, and his list is correspondently less valuable. Historians like Cantor and Zeuthen have given this particular period only nominal attention, save as to a few great arithmeticians, ignoring those contributions which set forth the real work of the people's schools. The titles of the works of such writers have not been given, since any reader of a bibliography like this will know the more general histories.

Mr. Plimpton's library has also a number of valuable manuscripts on arithmetic. Since these are not available for students generally, although of great value in themselves, they have been placed after the printed works instead of being inserted in chronological sequence. A study of our numeral system has justified the inclusion of books which, if printed, would hardly have place. Manuscripts written before the forms of numerals were fixed are often valuable in tracing their development, even though the books themselves are not arithmetical. Only those have been catalogued which bear with some directness upon arithmetic, and which were written before the year 1601, although numerous others, in many respects as valuable, and including several interesting works on the calendar, the sphere, and astrology, are in the library.

One difficulty attendant upon a labor of this kind is to determine what printed books to include. The number might easily have been increased by listing, as De Morgan occasionally did, works that are not at all arithmetical — Peletier's algebra for example. It has been thought better to draw the line in general more closely than he did, and to depart from genuine arithmetics only in the case of the works that discuss at least some questions relating to the science or art of numbers. No effort has been made to add to the supplementary lists books that are not purely arithmetical, such as treatises on the ancient measures, although those that are in Mr. Plimpton's library, and are of value in the study of the history of arithmetic, have been included.

The arrangement is chronological by first editions, but the Index allows for alphabetical and geographical reference. Although the nature of the work is usually discussed very briefly, an examination of the Index will show that a fairly complete history of Renaissance arithmetic has been included in the work — a history which I hope to present in other form in the future. The uncertainty in the use of such symbols as 4° and 8° has led to the measuring of the page and text. The page varies owing to the binder's work, and the text is not uniform page for page, but this plan seems the most satisfactory one for giving the size of the book. The centimeter has been taken as a unit of measure, since all English and American readers of a work like this will be familiar with it, while our popular units would be unknown to most others. These measures, like the number of lines to a page, of course vary considerably in the same book. The statement 'There were no other editions' is to be understood to mean that I have found no others that were printed before 1601. The illustrations have, in general, been selected with a view to bibliographical needs, although many have a marked historical interest.

In copying the titles it has been the intention to follow the original as closely as possible, without attempting to imitate particular forms of type or to use capitals except as initials. In the cases of misspelled words, omitted capitals, and peculiar punctuation, the errors have been copied as faithfully as possible. At the same time mistakes must have been made in transcribing, although it is hoped that they are not of a serious nature.

DAVID EUGENE SMITH

PREFACE TO THE ADDENDA

It is now a third of a century since the *Rara Arithmetica* was published, and in that space of time there have been many discoveries in the field of bibliography with which it was concerned. Mr. Plimpton passed away in 1936, and to the last he continued to add to the several branches of his large library — arithmetics printed before 1601, medieval manuscripts relating chiefly to mathematics, books and manuscripts concerned with history, religion, geography, calligraphy, and belles lettres in various languages. Some years before his death he had in mind a list of addenda to the *Rara Arithmetica* which should include the acquisitions made by the library during the preceding period of thirty years. In this he also proposed to add such information relating to other works and editions as had come to his attention through recent catalogues of important public or private libraries or those of prominent and reliable dealers.

Mr. Plimpton was well aware of the value of adding the names of publishers and the value of brief descriptions of such editions as were not in his own library, including complete collations, the names of the libraries consulted, and the present owners in case of permanent collections, but this would require an amount of space that would prohibit wide circulation. As the *Rara Arithmetica* stood it had served and still serves the original purpose of its publication. Today there are few prominent dealers' lists of early literature that fail to refer to the *Rara Arithmetica* as an aid to scholars in the history of this branch of elementary mathematics.

Since his death the question has arisen how his wishes could best be carried out so as to meet the needs of the many libraries which have copies of the *Rara Arithmetica.* After careful consideration it has seemed best to prepare a list of addenda which could readily be used in connection with such copies.

To assure the success of this undertaking the firm of Ginn and Company has undertaken the publication of this addendum, looking upon it as a memorial to Mr. Plimpton, a man who for a number of years had been the senior member of the firm. The Library staff of Columbia University has willingly joined in the work of extending the bibliography to include the later items pertaining to the collection. In this way Columbia University has recognized in some degree its gratitude to a man who had long been interested in its Library and connected with its Barnard College. His extensive collection, presented to the University, is now placed in a room of noble proportions in the Low Memorial Library, remodeled in accord with his wishes as expressed a short time before his death.

The editor of both the *Rara* and the *Addenda* wishes to express his thanks to Samuel Anthon Ives for his work in compiling a list of such arithmetics, printed before 1601, as were added to the library before Mr. Plimpton's death, the list having been begun by the owner's request. Special mention should also be made of the assistance rendered by Miss Bertha Margaret Frick, Curator of the Plimpton collection. Without such assistance it would have been almost impossible to select the material for the completion of this work.

It is hardly necessary to say that the list of addenda is by no means complete. Such a list in any line can hardly claim to be perfect. Unique publications are possible, and limited editions even of arithmetics are not infrequently found. In the case of the present work, however, this list of addenda, together with that given in the *Rara Arithmetica* itself, will be found to be fairly complete and fairly accurate.

For the convenience of readers each paragraph has a prominent reference to the related page in the original *Rara Arithmetica*, referred to as *Rara*. Such well-known bibliographies as those of Hain-Copinger, Proctor, and Dr. Klebs's recent *Incunabula Scientifica*, together with such well-known catalogues as those of Tregaskis, Goldschmidt, Sotheran, Quaritch, Stechert, Rosenthal, and Maggs, have been of great assistance in the compilation of the addenda. In general the items containing extended descriptions refer to books in the Plimpton Library.

DAVID EUGENE SMITH

LIST OF PLATES

ABBREVIATIONS

b., from the bottom of the page
c., *circa*, about
cm., centimeters
ed. pr., *editio princeps*, first edition
f., ff., folio, folios
fol., 4°, 8°, . . ., folio, quarto, octavo, . . .
ib., *ibidem*, the same place
l., ll., line, lines
p., pp., page, pages
r., recto, the first page of a leaf
v., verso, the second page of a leaf
s. a., *sine anno*, without date of publication
s. l., *sine loco*, without place of publication
s. l. a., without place or date of publication
/, or, to, and, as in 1471/2. Often used in the original as the equivalent of a comma.
/ /, the end of a line of print
[ ], uncertain items — names, dates, titles
Rara, the *Rara Arithmetica*

The modern form of *s* is used instead of the long one frequently found in title pages, and the umlaut as in ä instead of the older forms. The symbol ∴ is used instead of various medieval symbols for such terminations as *ium* or *orum*. The mark — over a letter is used to denote a missing letter as usual in early printing.

The spelling *Bullettino* is used for *Bulletino* in Italian works.

---

† the change indicated has already been made in the text.

PART I
PRINTED BOOKS

PRINTED BOOKS

ANONYMOUS. Ed. pr. 1478. Treviso, 1478.

Title. 'Incommincia vna practica molto bona et vtile // a ciaſchaduno chi vuole vxare larte dela mercha-//dantia. chiamata vulgarmente larte de labbacho.' (F. 1, r. See Fig. 1.)

Colophon. 'A Triuiſo : : A di .io. Decēbꝛ : : .i478.' (F. 62, r. See Fig. 4.)

Description. 4°, 14.6 × 20.5 cm., the text being 7.3 × 12.8 cm. 62 ff. unnumb., 32 ll. Treviso, 1478.

Editions. There was no other edition.

So far as known this is the first practical arithmetic to appear in print, for Albert of Saxony's Tractatus (p. 9) and the Ars Numerandi (p. 23), even if earlier, and the Etymologies of Isidorus (p. 8), are not, strictly speaking, of this class.

The author of the book is unknown, but from the opening lines it seems that he was a teacher of arithmetic in Treviso. The printer is also unknown, although it was probably one Manzolo, or Manzolino. The history of the work has been carefully studied by Boncompagni, his results appearing in the *Atti dell' Accademia Pontificia de' Nuovi Lincei*, vol. 16. This particular copy was in the Pinelli collection, and was sold on February 6, 1790, to a Mr. Wodhull. It afterward found its way into the library of Brayton Ives, Esq., of New York, and at the sale of that library was acquired by Mr. Plimpton.

The work is commercial in character, the fundamental processes being taken up in the common order, and these being followed by the rule of three. A curious feature not at all common in early arithmetics is the rule of two. The practical applications are chiefly included under the rule of three and partnership. There is also a brief treatment of the calendar, for Church purposes. The book is lacking in applications to exchange, and probably on this account it did not appeal to the merchant class sufficiently to warrant a second edition.

In considering these early works it is necessary to understand the four types of arithmetics which the Renaissance inherited from the Middle Ages. These types are as follows:

1. *The theoretical books.* These works were based chiefly upon

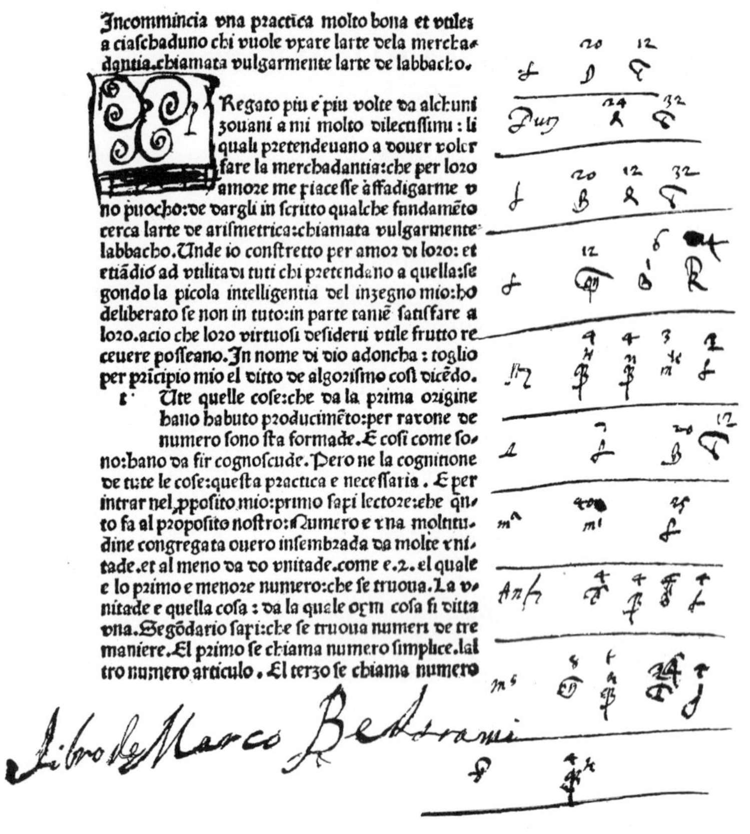

Incommincia vna practica molto bona et vtile a ciascaduno chi vuole vxare larte dela merchadantia. chiamata vulgarmente larte de labbacho.

Regato piu e piu volte da alchuni zouani a mi molto dilectissimi: li quali pretendeuano a douer volere fare la merchadantia: che per loro amore me piacesse affadigarme vno puocho: de dargli in scritto qualche fundamento cerca larte de arismetrica: chiamata vulgarmente labbacho. Unde io constretto per amor di loro: et etiandio ad vtilita di tuti chi pretendano a quella: segondo la picola intelligentia del inzegno mio: ho deliberato se non in tuto: in parte tanto satisfare a loro. acio che loro virtuosi desiderii vtile frutto receuere posseano. In nome di dio adoncha: toglio per principio mio el ditto de algorismo cosi dicendo.

1· Ute quelle cose: che da la prima origine hano habuto producimento: per raxone de numero sono sta formade. E cosi come sono: hano da fir cognoscude. Pero ne la cognitione de tute le cose: questa practica e necessaria. E per intrar nel proposito mio: primo sapi lectore: che quanto fa al proposito nostro: Numero e vna moltitudine congregata ouero insembrada da molte vnitade. et al meno da do vnitade. come e.2. el quale e lo primo e menore numero: che se truoua. La vnitade e quella cosa: da la quale ogni cosa si ditta vna. Segondario sapi: che se truoua numeri de tre maniere. El primo se chiama numero simplice. laltro numero articulo. El terzo se chiama numero

Libro de Marco Beltrami

FIG. 1. FIRST PAGE OF THE TREVISO ARITHMETIC

Boethius (p. 25), who wrote in the beginning of the sixth century, following the Greek models of Nicomachus (p. 186) and Euclid (p. 11). They are devoted to such matters as the theory of figurate numbers, of which the square and cube are all that now remain in elementary textbooks, and the cumbersome Greek ratio-system which testified to the ancient difficulty with fractions. Books of this class were written by such men as Bradwardin (p. 61), Albert of Saxony (p. 9), and Jordanus (p. 62).

2. *The algorisms* (*algorismi*). These were practical arithmetics,

Uoglio pero che tu intendi che ſono altri modi de moltiplicare per ſcachiero: li quali laſſaro al ſtudi o tuo: mettendo li exempli ſoi ſolamente in forma. come porai vedere qui ſotto
Or togli de fare lo predicto ſcachiero. zoe. 3 i 4. fia. 9 3 4. e nota de farlo per li quatro modi come qui da ſotto.

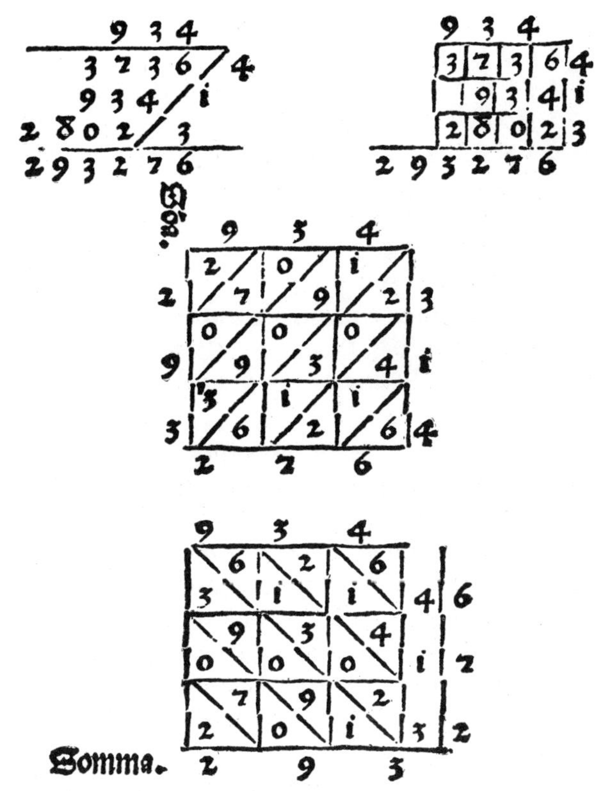

FIG. 2. FROM THE TREVISO ARITHMETIC, SHOWING VARIOUS FORMS OF MULTIPLICATION

```
         18949465889280
                 360072
    -----------------------
          3709893177 8,60
      132646261224960
  113696795335680000
  568483976678 40
  -----------------------
  6023172081684828160
```

100032 | 0
3126 | 6
duc.130

Et e compita Unae e venuto lo impromisso. zoe il prexio menzonado li. Che se lire 14616 onze 9 sazi 5. e $\frac{1}{3}$ valisse duc. 1903 g̃ 11 p $\frac{1}{4}$ $\frac{334531 2}{420864}$ che lire. 1000. e $\frac{1}{5}$ valeranno ducati. 130. g̃. 6. li quali sono vn quarto de vno ducato. Sich q̃lle razone e queste stano seguramente bene.

Auisando te. che quando hauerai da fare qualche raxone da importantia: e che tu dubiti: non porai pruouare piu seguramẽte: che voltare la toa raxone. al modo che hai visto ne le tre raxõe preditte. Unde per queste e per le altre raxone preditte: le quale sono in tuto numero quindexe: tu puo inten

FIG. 3. FROM THE TREVISO ARITHMETIC, SHOWING MULTIPICATION AND THE GALLEY FORM OF DIVISION

written to supply the mathematical knowledge necessary for business computations, and using the Hindu-Arabic numerals. These numerals were known in India, without the zero, as early as the third century B. C. They were gradually perfected, and by the time they reached Bagdad from India, in the eighth century A. D., they included the zero. An arithmetic employing these numerals was written about 800 A. D. by an Arab scholar, Mohammed ibn Musa, known by the name of al-Khowarazmi (from his birthplace, Khwarazm), and from the Latinized form of his name came the word *algorism*.

3. *The abacus arithmetics*. These were also commercial books; but since they used the Roman numerals, which were not suited to computation, the actual calculations were carried on by means of *calculi* (Latin, pebbles), *getons* (French for things thrown or cast, from the Latin *jacere*, to throw), or *counters*, from which English form we have expressions like 'cast an account.' The table on which the calculi were cast is still called a *counter* in our shops, but, like the sand tables used for computing, it was in early times called an *abacus*. Just at the opening of the Renaissance the contest was still waging between the algorists and the abacists. These arithmetics are not found, however, in Italy, because the merchants of that country abandoned the use of the counters long before this was done in other countries. In Germany the arithmetics frequently have in their titles the expression 'auff der Linien und Federn' (Riese, p. 138, 1522), or 'mit der ziffer unnd mit den zal pfenningen' (Rudolff, p. 151, 1526), 'auff der Linien' referring to the lines on which the counters ('zal pfenningen') were cast, and the 'Feder' referring to the pen with which a figure ('ziffer') was written.

Questo e lo registro de li quaderni de la presente opera.

Incomminciatia.de la
E per inten
numero:
che.2.e di
segondo
lo.3.poi
cauarcima
iongere a
f1.8. siche
Se tu vo
8 2 9
pruoua de
apresso.4 i.
.i.che auanza
9 3 7 5

Se.5.e
Segondo
se onza.i.
precedente
La terza
5
1 8 9 4
I ho bz
2 4)
Nicolo messe
1 5 2 0 9 1
E quante
Jonti a
Uno maran
e doue tu

Che zuoua la virtu a chi non se affadica:
Niente
A Treuiso :: Adi.10.Decembre :: .1478.

FIG. 4. LAST PAGE OF THE TREVISO ARITHMETIC

4. *The computi*. The computus or compotus was a treatise upon the Church calendar, containing such simple directions as were necessary for computing the dates of Easter and the other movable feasts. The chapter on the calendar, of which there is still some trace in our arithmetics, originated in the computus. (See Anianus, p. 31, 1488.)

The Treviso arithmetic is a good example of an algorism. Fig. 3 shows that multiplication was performed as it is to-day (but see Fig. 2), and that division was performed by the 'galley' method, so called because the work resembled in form an ancient galley with its sails set.

ISIDORUS OF SEVILLE. Ed. pr. 1472. Venice, 1483.

Born, probably at or near Cartagena, c. 560 or 570; died at Seville, April 4, 636. One of the most learned men of his time, Bishop of Seville, and writer on theology, philosophy, and the general learning of the Middle Ages.

Title. '¶Incipit epiſtola Iſidori iunioris hiſpalenſis epi-//ſcopi ad Braulionem cęſar auguſtanū epiſcopū.' (F. 1, r.)

With this is: '¶In chriſti nomine incipit liber primus ſancti // Iſidori hiſpalenſis epiſcopi de ſūmo bono.// Qd deus ſūmus ꝛ incōmutabilis ſit ¶Cap. I.' (F. 1, r., following the first f. 105.)

Colophon. '¶Finit liber etymologiarum // ¶Iſidori hiſpalenſis epiſcopi.' (F. 101, v.)

Colophon of the second part. '¶Finit liber tertius ꝛ vltimꝰ de sūmo bono // ſancti Iſidori hyſpalenſis epī: Impreſſus // Venetijs per Petrū loſlein de Langencen̄.// ¶.M.cccc.lxxxiij.¶' (F. 28, r.)

Description. Fol., 19.8 × 27.7 cm., printed in double columns, each being 6.7 × 23 cm. 129 ff. numb. + 4 unnumb. = 133 ff., 58 ll. Venice, 1483.

Editions. Augsburg, 1472; Venice, 1483 (here described); ib., 1485 (?); ib., 1493, fol.; Basel, 1577 (mentioned below). Also two editions s. l. a. (Strasburg, 1470 ?), and one s. l. a. (Cologne, 1476–78 ?). 497

This book of etymologies written by Isidorus, Bishop of Seville in the seventh century, is the standard authority upon the state of learning in Spain in that period. The subject of arithmetic is treated in book 3, beginning (f. 15): '¶Incipit liber tertius // ¶De vocabulo arithmeticę // diſciplinę ¶Cap. I.' The work consists entirely of the mediæval theory. The treatment is very brief (5 ff.) and is followed by a few pages on the calendar. Although appearing in 1472, this cannot be called the first printed arithmetic, since it touches so briefly upon the subject. It has therefore been placed after the Treviso arithmetic.

ISIDORUS OF SEVILLE. Ed. pr. 1472. Basel, 1577.

Title. 'Isidori // Hispalensis // Episcopi // Originum libri viginti/ex antiquitate eruti.// . . . Basileæ,//per Petrum Pernam.' (F. 1, r.)

Description. Fol., 20.5 × 31 cm., printed in double columns, each being 8.2 × 24.4 cm., 60 ll. With an edition of Capella. Basel, 1577.

ALBERT OF SAXONY. Ed. pr. c. 1478. Venice (?), c. 1478.

Born c. 1330. He lectured in the University of Paris, was Rector at Vienna in 1365, and was Bishop of Halberstadt from 1366 to 1390. He wrote several scientific works.

Title. 'Eccellētiſſimi magistri alberti de // ſaxonia tractatus pportionum incipit feliciter.' (F. 1, r.)

Colophon. 'Explitiunt pportiones magiſtri // alberti de ſaxonia.' (F. 9, r.)

Description. Fol., 19.7 × 28.1 cm., in double columns, each being 5.6 × 16.7 cm. 9 ff. unnumb., 39 ll. S. l. a. (Venice ?, c. 1478).

Editions. S. l. a. (Venice ?, c. 1478, here described); Padua, 1482 (12 ff.); ib., 1484 (12 ff.); ib., s. a., but before 1487; Venice, M.cccc.xxxlvii for 1487 (10 ff.); ib., 1494 (10 ff.); ib. (with another work), 1496; ib., 1496 (but no copies extant ?); Bologna (with another work), 1502; ib., 1506 (commentary by Vittori); Paris, s. a. An epitome by Padre Isidoro Isolani Milanese was published in Pavia in 1513, and again in 1522. Prince Boncompagni, in an elaborate discussion of the various editions (*Bulletino*, IV, 498), mentions this rare first edition, of which he knew but one other copy, that in the Biblioteca Ambrosiana at Milan. He was of the opinion that it was printed in Venice, in a type used before 1480. 501

Although the date is uncertain, this Tractatus may contest with the Treviso arithmetic (p. 3) and the Ars Numerandi (p. 23) the honor of being the first printed work devoted wholly to some phase of arithmetic. It treats of ratios according to the cumbersome method of Boethius as followed during the Middle Ages. It is purely theoretical and represents the university treatment of scientific arithmetic in that period. 501

Other works of 1472–1480. The mathematical activity in Italy during this period was very considerable. From 1472 to 1480 there were 38 mathematical works printed in the country. In the next decade there were 62 and in the next 100, with 13 of uncertain date between 1472 and 1500, making a total of 213 appearing in a period of less than thirty years.

Before the Treviso book there were printed at least three works which touched briefly upon arithmetic. These were (1) 'De re militari libri

XII,' by Robertus Valturius of Rimini, printed at Verona, 1472, fol., in the second book of which the author treats 'de arithmetica & militari geometria'; subsequent editions ib., 1483 (two editions this year); Paris, 1483 (with title changed); ib., 1534, fol.; ib., 1555 (French translation); (2) 'Speculum Majus,' by Vincent de Beauvais (Vincentius Bellovacensis), the greatest mediæval encyclopædia, printed at Strasburg by Mentelin, 1469(?)–1473, 10 vols., fol. The second part of this work contains one book (no. 16) on mathematics, in which is given a brief treatment of algorism (see p. 5), probably the first written in French (c. 1250), although M. Henry asserted that a MS. of c. 1275, which he edited, was entitled to that distinction. Incomplete editions were also published at Venice in 1484, 1493–4, 1591. The early editions of Priscian, 'De figuris et nominibus numerorum,' are mentioned under the publications of 1565. (3) The Etymologies of Isidorus, Augsburg, 1472 (p. 8). 501

In 1480 an anonymous work was issued from the Caxton press in London, entitled 'The Mirrour of the World or Thymage of the same.' Chap. 10 of this work began: 'And after of Arsmetrike and whereof it proceedeth,' and this was probably the first English printed matter upon the subject. There was a second edition, London, 1506, and a third s. l. a. (London, 1527 ?), fol. (See also Boethius, p. 25, and Faber, p. 62.) About 1480 there was published at Padua a folio work by Richard Suiseth (Suicetus, Swincetus, Swinshead, Suineshevedus, the first name possibly Roger or Raymund), entitled 'Opus aureum calculationum per Johanem de Cipro emendatum et explicit.', with subsequent editions at Pavia in 1488(?), 1497, 1498, fol., at Venice in 1505, 1520, and at Salamanca in 1520. I have seen an edition of this work, s. a., assigned to c. 1477. For Nicolaus Cusa, see p. 42.

GIORGIO CHIARINI. Ed. pr. 1481. Florence, 1481.

A Florentine arithmetician of the fifteenth century.

Title. 'Qvesto e ellibro che // tracta di mercatantie // et vsanze de paesi.' (F. 1, r. See Fig. 5.)

Colophon. 'Finito ellibro di tvcti // ichostvmi : cambi : mone //te : pesi : misvre : & vsanze // di lectere di cambi : & ter// mini di decte lectere che // nepaesi sicostvmaet in // diverse terre. Per me France//ſco di Dino di Iacopo Kartolaio Fiorē//tino Adi X di Dicembre MCCCCLXXXI. In Firenze Apreſſo // almusiſtero di Fuligno.' (F. 102, r.)

Description. 8°, 13.2 × 21.2 cm., the text being 7.1 × 14.1 cm. 3 ff. blank + 6 unnumb. + 96 numb. (Roman numerals) = 105 ff., 24 ll. Florence, 1481.

Editions. Florence, 1481, 8° (here described); s. a. (1498), 8°. The undated edition was not the first; Coppinger has shown that it was printed in 1498.

While this is not, strictly speaking, an arithmetic, it is the first printed book to give the customs relating to exchange in use among the Florentine merchants at the close of the fifteenth century. It is the source from which several later writers drew their material, and is particularly valu-

QVESTO E EL LIBRO CHE
TRACTA DI MERCATANTIE
ET VSANZE DE PAESI.

FIG. 5. TITLE OF CHIARINI'S WORK

able in showing the nature of the practical problems of the time. Copies of this first edition are extremely rare, but the work was well enough known for Paciuolo to appropriate some of the contents. There is a question as to its authorship.

EUCLID. Ed. pr. 1482. Basel, 1562.

Flourished at Alexandria c. 300 B.C. He was the author of the 'Elements,' the basis of most of the textbooks on geometry.

Title. 'Die Sechs Erſte Bůcher // Euclidis/ // Vom anfang oder grund // der Geometrj.// In welchen der rechte grund/ nitt allain der Geometrj // (verſteh alles kunſtlichen/ gwiſen/ vnd vortailigen ge-//brauchs des Zirckels/ Linials oder Richtſcheittes vnd // andrer werckzeüge/ ſo zu allerlaj abmeſſen dienſtlich) // ſonder auch der fürnemſten ſtuck vnd vortail // der Rechenkhunſt/ furgeſchriben vnd // dargethon iſt.// Auſs Griechiſcher ſprach in die Teütſch gebracht/ aigene-//tlich erklårt/ Auch mit verſtentlichen Exempeln/ gründ-//lichen Figuren/ vnd allerlaj den nutz fürangen ſtellen-//den Anhången geziert/ Dermaſſen vormals // in Teütſcher ſprach nie geſehen // worden.// Alles zu lieb vnd gebrauch den Kunſtliebenden Teütſchen/ ſo ſych der

Geo-//metrj vnd Rechenkunſt anmaſſen/mit vilfåltiger mühe vnd arbait //zum trewlichſten erarnet/vnd in Truckh ge-//geben/ Durch // Wilhelm Holtzman/genant Xylander/ //von Augſpurg. //Getruckt zů Baſel.' (P. 1.)

Colophon. 'Vollendet durch Jacob Kündig/zu Baſel/in // Joanns Sporini koſten/im jar 1562.//auff den dreyſzigſten tag des // Winmonats.' (P. 199.)

Description. Fol., 20 × 31.3 cm., the text being 12.5 × 25.7 cm. 14 pp. unnumb. + 185 numb. + 1 blank = 200 pp., 39–52 ll. Basel, 1562.

Editions. The editions of Euclid have not been considered in this work except in so far as they relate particularly to arithmetic. This is the first German edition. The Plimpton library contains the first edition of Euclid (Venice, 1482), but since this has no arithmetical work except Book V it has not been included in the list. Several manuscripts of Euclid are, however, included in the second part of this bibliography because of their value in tracing the changes in the forms of the numerals.

This edition of Euclid is mentioned because the editor has thought it necessary to add to Book II some arithmetical work. In particular he gives three forms of multiplication, first from left to right, then in the usual way, and finally for special cases in which the short processes are involved. He also considers the division of numbers in given ratios, the extraction of roots, and a few other semi-algebraic calculations.

Euclid's 'Elements' contain much work upon the Greek theory of numbers, besides what appears in Book V, and several books were published in the sixteenth century, embodying this material. These are mentioned later.

Other works of 1482. In 1482 appeared the first German arithmetic, if we except the 'Ars Numerandi' (p. 23). It was written by Ulrich Wagner, a Nürnberg Rechenmeister, and was printed by Heinrich Petzensteiner at Bamberg. Only nine small pieces of parchment proof sheets remain. They contain the following colophon: 'Anno dñi . . . 1482 kl'16. Iunij p. Henr. peczensteiner Babenberge: finit Ulrich wagner Rechēmeister zu Nürnberg.' 'Ludus Arithmomachiæ,' by John Sherwood (Shirewode), was published at Rome in the same year. (See also p. 63, and on Albert of Saxony see p. 9, c. 1478.)

PROSDOCIMO DE BELDAMANDI, AND LIVERIUS.

Ed. pr. 1483. Padua, 1483.

PROSDOCIMO DE BELDAMANDI was born at Padua c. 1370–1380, and died in 1428. He was educated at the University of Padua, and also taught there. He wrote on arithmetic, music, and astronomy.

JOHANNES DE LIVERIUS (LIVERIIS, LINERIIS) was a Sicilian writer on astronomy who flourished c. 1300–1350.

Title. 'Profdocimi de beldamandis algo-//rifmi tractatus perutilis ꝛ neceffarius // foeliciter incipit. qui de generibus cal-// culationum fpeciē preteri.t nullaꝫ.q̄ faltē // neceffaria ad h⁹ art⁹ ꝯgnitōꝫ fuerat.' (F. 1, r. See Fig. 6.)

Colophon. 'Algorifmus. Profdocimi de beldamādis // vna cum minuciis. Johānis de liueriis. hic // felicitē'finit Impreff⁹ padue. Anno .1.4.8.3 // die zz. februarii.' (F. 21, v.)

Description. 4°, 15 × 20.8 cm., the text being 9 × 14.7 cm. 27 ff. numb., 32 ll. Padua, 1483.

Editions. Padua, 1483, fol. (here described); Venice, 1540, 8° (see p. 15). Boncompagni could learn of only three copies of the first edition, and seven of the second.

This rare work was written for the Latin schools, and is a good example, the first to appear in print, of the non-commercial algorisms of the fifteenth century. It follows 'Bohectius' (Boethius) in defining number and in considering unity as not itself a number, as is seen in the facsimile of the first page. Prosdocimo then treats of the fundamental operations with integers, including mediation (division by 2, which the author places before duplation or multiplication by 2), progressions, and the roots. The treatment of fractions is left to Liverius: 'Incipit Algorismvs de mi-//nutijs tam vulgaribus quàm phyficis magi-//ftri Ioannis de Liuerijs Siculi.' The 'vulgar minutes' were the common fractions, and the 'physical' were the sexagesimal fractions. Towards the end of the 'Algorismus de integris' in the 1540 edition, the date of composition appears: '... per Profdocimum de Beldamandis,de Padua anno domini .1410. die .10. Iunij compilata fufficiant.' (F. D 5, v.) The work of Prosdocimo contains the first reference that I have seen to a slate. 'Indigebat etiam calculator semper aliquo *lapide*, vel sibi conformi super quo scribere atque faciliter delere posset figuras cum quibus operabatur in calculo suo.' (See Fig. 6.) It is probable, from this statement, that computers of his time actually erased the figures in the galley form of division (see the Treviso arithmetic, p. 3),

Prosdocimi de beldamandis algorismi tractatus perutilis et necessarius foeliciter incipit. qui de generibus calculationum speciē preteri. et nullaȝ. q̄ saltē necessaria ad hꝰ artꝰ cognitōȝ fuerat

Inueni inq̄ȝ pluribus libris algorismi nūcupatis. mōs circa numeros operandi satis uarios: atq̄ȝ diuersos. q licet boni existerēt. atq̄ȝ veri erāt. tñ fastidiosi: tū ppt̄ ipaȝ regulaȝ multitudinē: tū ppt̄ eaȝ deleatōes. tū etiā ppter ipaȝ operationū pbatōes: vtȝ. s. bone fuerint uel ne. Erāt et etiam isti modi intm̄ fastidiosi: q̄ si in aliqꝰ calculo astroloico error ɔtigissȝ: calculatorē opatōȝ suam a capite incipere oportebat: dato q̄ error suus adhuc satis ppiquus existeret. et hoc ppt̄ figuras in sua opatōe deletas. Indigebat etiā calculator semp aliqꝰ lapide uel sibi ɔformi. sup quo scribere atq̄ȝ faciliter delere possȝ figuras cū qbus opabat̄ in calculo suo. Et qa hec oia sane fastidiosa atq̄ȝ laboriosa mihi uisa sunt: disposui libellū edere in quo oia ista abicerent̄: qui etiā algorismus siue liber de numeris denoiari poterit. Scias tñ q̄ in hoc libello ponē nō intēdo nisi ea q̄ ad calculū necessaria sunt. alia q̄ in aliis libris practice arismetrice tāgunt̄. ad calculū nō necessaria ppt̄ breuitateȝ dimitendo. ¶ Quia ergo libellus iste de numeris tractare hȝ: a diffinitōne nūeri ipm icboare nolo. ¶ Numerus ergo sᵐ Euclidē. 7ꝰ. sue Geometrie. et sᵐ Boectiū pꝰ sue arismetice sic diffinit̄. Nūerus. est multitudo siue quātitas discreta ex unitatibus pfusa: siue ex unitatibus aggregata. Et dī unitas illud: qꝰ unaqueq̄ȝ res dicit̄ una. Per hāc gꝰ diffinitōȝ nūeri hic potes. quoꝰ unitas nō ē nūerus. licȝ sit pncipiū nūeri. dato q̄ nūerꝰ etiā vocari possit lāgoꝰ sumenꝰ nūeȝ. s. p oī eo quo rē aliquā nūerate possumus. et isto largoꝰ accipit̄ nūerꝰ in pcessu huius libelli. in qꝰ etiā vnitas

Nūs quid. Euclides. Boetius. Vnitas q̄d

FIG. 6. FIRST PAGE OF PROSDOCIMO DE BELDAMANDI

as the Hindus had done on their sand or dust abacus, instead of canceling them in the manner explained in the early printed arithmetics.

The best discussion of the lives and works of Beldamandi and Liverius is in the Boncompagni *Bulletino*, vol. XII.

Other works of 1483. Valturius (p. 10). In this year the second German arithmetic was printed at Bamberg, only one (incomplete) copy being known. It was possibly written by Ulrich Wagner (p. 12), and, like the 1482 work, it was printed by Petzensteiner, as appears from the following colophon: 'In zale Xp̄i .1483. kl'.17. des Meyen Rechnung // in mancherley weys in Babenberg durch henr9 // petzenſteiner begriffen: volendet.'

PROSDOCIMO DE BELDAMANDI, AND LIVERIUS.

Ed. pr. 1483. Venice, 1540.

See p. 13.

Title. 'Algorismvs de In//tegris Magistri Prosdoci-//mi Debeldamandis Pataui ſimul cū algoriſmo de de-//minutijs ſeu fractionibus magiſtri Ioānis de Liuerij // ſiculi. Reintegratus ab erroribus cōmiſſis a ſcri-//ptoribus, a me Federico Delphino artium, & // medicine doctore, mathematicarum diſci//plinarū in celeberrimo gymnaſio Pata//uino publico profeſſore, additis ali// quibus verbis, in aliquibus locis,// pro maiori claritate. Et da-// tus impreſſioni ad inſta-//tiam meoꝝ ſcholariū // nunc algoriſmū // maxime de-//ſideran-//tium.// Venetijs. M. D. XXXX.' (F. 1, r.)

Colophon. 'Venetijs per Ioannem Antonium de Vulpinis de Ca-//ſtrogiufredo. Anno domini .M. D. XXXX. //die octauo menſis Aprilis.' (F. 52, v.)

Description. 8°, 9.6 × 14.5 cm., the printed part being 7.1 × 12.5 cm. 44 ff. unnumb., 30 ll. Venice, 1540. (This particular copy has 8 ff. more, the sheet F appearing in duplicate.)

See p. 13.

RAPHAEL FRANCISCUS.

Ed. pr. 1484. S. l. a. (Florence, c. 1516).

RAFFAELE FRANCESCO. A Florentine philosopher of the latter part of the fifteenth century.

Title. 'Verificatio Vniversalis in // regulas Ariſtotelis de motu non rece-//dens a cōmuni Mathema-//ticoꝝ doctrina.' (F. 1, r. See Fig. 7.)

Description. 4°, 13 × 19.9 cm., the text being 9 × 16.7 cm. 8 ff. unnumb., 41–42 ll. S. l. a. (Florence, B. Zucchetta, c. 1516).

Editions. Pisa, 1484, 8°; this edition, s. l. a. (Florence, c. 1516).

This is a brief treatment of proportion, hardly worthy of ranking as an arithmetic. The applications relate to problems of Aristotle.

VERIFICATIO VNIVERSALIS IN regulas Ariſtotelis de motu non recedens a cõmuni Mathematicoꝝ doctrina.

FIG. 7. TITLE OF THE VERIFICATIO OF FRANCISCUS

PIETRO BORGHI. Ed. pr. 1484. Venice, 1484.

PIERO BORGI. A Venetian arithmetician; died after 1494.

Title. 'Qui comenza la nobel opera de // arithmethica ne la qual ſe tracta // tute coſſe amercantia pertinente // facta ⁊ compilata ꝑ Piero borgi // da venieſia.' (F. 2, numbered 1, r. See Fig. 8 for the first folio.)

Colophon. 'Nela inclita cita de venetia a çorni .2.// auguſto .1484. fu impoſto fine ala pre-//ſente opera.' (F. 118, v.)

Description. 4°, 14.3 × 19.3 cm., the text being 8.4 × 13.4 cm. 2 ff. unnumb. + 116 numb. = 118 ff., 37–38 ll. Venice, 1484.

Editions. Venice, 1484, 4° (here described); ib., s. a., which Riccardi thought might be earlier than 1484, since Ratdolt, the printer, published books in Venice as early as 1476; ib., 1488, 4° (see p. 19); ib., 1491, 4° (see p. 20); ib., 1501, 4°; ib., 1505; ib., 1509, 4°; ib., 1517, 4° (see p. 20); ib., 1528, 4° (see p. 21); ib., 1534, 4° (see p. 21); ib., 1540, 4° (see p. 21); ib., 1550, 4° (see p. 22); ib., 1551; ib., 1560; ib., 1561, 4°; ib., 1567; ib., 1577. It was at one time thought on the testimony of Maittaire that there was an edition of 1482, but it has been shown by several bibliographers that the first edition is that of 1484. 495; 501

This is the very rare first edition of Borghi's treatise, the second commercial arithmetic printed in Italy and long thought to be the first. This particular copy belonged to Count Paolo Vimercati-Sozzi.

The text of the first edition is closely followed in that of 1488 (see p. 19), except for the index, 'Tauola de li capitoli ꝯtegnudi i q̄ſta opa' (f. 118, r.). This does not appear in the second edition, at least in Mr. Plimpton's copy. The letters S H S U which appear twice are thought to stand for J H S U, Jesus, possibly changed on account of some conjectured pronunciation. They appear the second time on f. 118, v., in connection with a set of verses beginning as follows:

'S H S U

Quanto latua memoria et alto ingegno
vaglia ne larithmetica hai moſtrato
nel preſente volume compilato
petro borgo date veneto degno.'

In the verses appears the name of the printer:

'Ma limpreſſor de auguſta Errardo experto
di lopera preſente ſtampatore
degno e non di mediocre laude certo.'

This folio, as already stated, does not appear in the Plimpton copy of the 1488 edition, and since the later printer was not the same it probably never appeared after 1484.

This work is more elaborate than the Treviso arithmetic, and had far greater influence on education. More than any other book it set a standard for the arithmetics of the succeeding century, and none of the early textbooks deserves more careful study. Borghi first treats of notation (see Fig. 9), carrying his numbers as high as 'numero de million de million de million,' and making no mention whatever of the Roman numerals. In the same spirit he eliminates all of the mediæval theory of numbers, asserting that he does this because he is preparing a practical book for the use of merchants. 'Et nota che ſono // nūeri de piui maniere ſi cho-//me dichiara Boetio in el ſuo // de arithmetiche Ma volen-//do hora tratar de quelle choſ//ſe che ſolo amerchadāti apertien: pero tratādo ſolo de quelli // che ale choſſe merchādateſche ſono neceſſarij io laſſero ogni // altra maniera de numeri.' (F. numb. 1, v.)

The sequence is now peculiar, for multiplication is the first operation treated. ('¶Che coſſa ſia moltiplichar,' f. numb. 7, r.) First comes the table, arranged in the column form, unlike the Boethian type of arithmetic, which preferred the square array. In addition to the products through '9 uia 10 fa 90,' the products of 12, 16, 20, 24, 32, 36, by 2 . . . 10, are given, these having been necessary on account of the monetary tables of the time. The author then gives the method of checking by casting out 7's and 9's. ('℄Dela pruoua del .7.', 'Dela pruoua del .9.', f. numb. 8.) Then follows multiplication 'per colonna' (i. e., by reference to the columns of the table, '℄Del multiplicar p cholona,' f. numb. 9, r.), with its checks by 7 and 9, and 'per crocetta' (our 'cross multiplication,' '℄Del multiplicar per chroxeta,' f. numb. 13, v.), showing that these were the common methods in Venice. Division is then explained by the galley form ('℄Como ſe die partir p batelo chō leſuo pruoue,' f. numb. 20, r.), our present method, then known as the method of giving, 'a danda,' and described by Paciuolo and Calandri (pp. 54 and 47), not being mentioned. Then follow addition (although this was used in multiplication), subtraction, denominate numbers, common fractions (also beginning with multiplication), rule of three ('De la riegola del .3.'), partnership, barter, alligation, and false position. The rule of three had been developed many centuries earlier by Oriental arithmeticians. It was one of the inheritances from the Arabs, and was not improbably learned by the Venetian traders through their contact with the East. Partnership was to the fifteenth what the corporation is to the twentieth century, and it is only very recently that "partnership involving time" was thought to be a necessary subject of study. Barter, a subject until fifty years ago common in American textbooks, was necessary at a time when currency was not so plentiful as now. Alligation was a practical topic in connection with the coinage of money in the days when minting was not the monopoly of great centralized governments. The rule of false position was an Oriental device which we have now replaced by the equation; it was found in schoolbooks in various countries until the second half of the nineteenth century. The problems are generally practical for the time, and they reveal some interesting facts concerning business customs at the close of the fifteenth century.

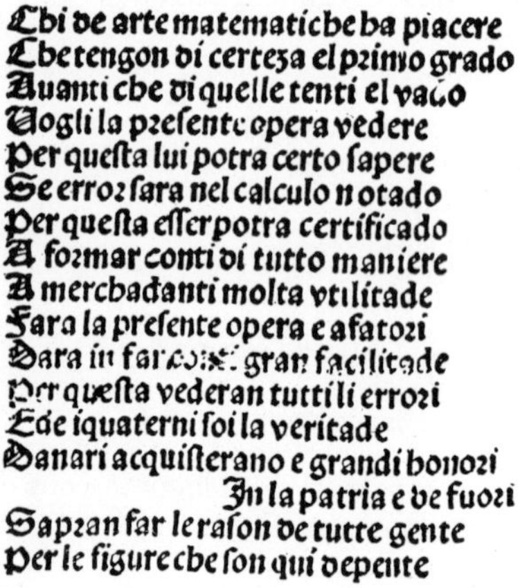

Chi de arte matematiche ha piacere
Che tengon di certeza el primo grado
Auanti che di quelle tenti el vaco
Vogli la presente opera vedere
Per questa lui potra certo sapere
Se error sara nel calculo notado
Per questa esser potra certificado
A formar conti di tutto maniere
A merchadanti molta vtilitade
Fara la presente opera e afatori
Dara in far conti gran facilitade
Per questa vederan tutti li errori
Ede iquaterni soi la veritade
Danari acquisterano e grandi honori
In la patria e de fuori
Sapran far le rason de tutte gente
Per le figure che son qui depente

FIG. 8. FIRST PAGE OF THE FIRST EDITION OF BORGHI'S ARITHMETIC

Other works of 1484. Albert of Saxony, p. 9, c. 1478; Vincent de Beauvais, p. 10.

PIETRO BORGHI. Ed. pr. 1484. Venice, 1488.

See p. 16.

Title. 'Chi de arte matematiche ha piacere // Che tengon di certeza el primo grado // Auanti che di quelle tenti el vado // Vogli la preſente opera vedere // Per queſta lui potra certo

Como ſi formano milion

L million adoncha ſe die formar per ſette figure in queſto modo . 1000000 . perche laſeptima figura tien elluogo demiara demiara;cper che millemiara ,fano vno million :et eſſendo in quel luogo lafigura che riprexenta vno pero bene edito vno milion. Ma.iqueſtp modo . 1 100000 diria vno milion e cento milia : perche oltra el million:in luogo de centenara de miara:ſono lafigura che riprexenta vno ſi che bene edito vno million e cento milia . Ma in queſto modo. 1 1 10000.diria vno milion e cento e diexe milia perche oltra elmilion e cento milia : in luogo dele dexene demiar:ſono lafigura che riprexenta vno:ſi che bene edito vno milion e cento ediexe milia. Ma in queſto modo. 1 1 1 1000.diria vnmilion cento e vndexe milia per che oltra elmilion cento ediexe milia:in luogo de numeri demiar ſono lafigura che riprexenta vno:ſi che bene edito vnmilion cento e vndexe milia. Ma in queſto modo. 1 1 1 1 100.diria vnmilion cento e vndexe milia e cento:perche oltra elmilion cento e vndexe milia: in luogo de elſimplice centenar:ſono lafigura che riprexenta vno:ſi che bene editovnmilion cento e vndexe milia e cento. Ma in queſto modo. 1 1 1 1 1 10.diria vnmilion. cento e vndexe milla cento ediexe:perche oltra elmilion cento e vndexe milia e cento : i luogo dele ſimplice dexene:ſono lafigura che riprexenta vno. Ma in queſto modo. 1 1 1 1 1 1 1.diria vnmilion cento e vndexe milia cento e vndexe.per che anche in luogo dele ſimplice vnita .ſono lafigura che riprexenta vno .ſi che bene edito vnmilion cento e vndexe milia cento e vndexe.et choſi procedendo perfina .9999999. ponendo ſempre aiſuo luogi quele figure reprexentante queli numeri ouero dexene ocentenara.che ſinomina et cetera.equeſto baſta cercha lo amaiſtramento del numerar.ben che in infinitum ſiporia proceder .ma chom vna zeneral figura miſorcero dichiarir quanto poteſſe achader . et ſarano queſto ſotto poſta

1000000
1 100000
1 1 10000
1 1 1 1000
1 1 1 1 100
1 1 1 1 1 1 0
1 1 1 1 1 1 1
9999999

FIG. 9. FROM BORGHI'S ARITHMETIC, 1488 EDITION

ſapere // Se error ſara nel calculo notado // Per queſta eſſer potra certificado // A formar conti di tutto maniere // A merchadanti molta vtilitade // Fara la preſente opera e afatori // Dara in far conti gran facilitade // Per queſta vederan tutti li errori // Ede iquaterni ſoi la veritade // Danari acquiſterano e grandi honori // In la patria e de fuori // Sapran far le rafon de tutte gente // Per le figure che fon qui depente.' (F. 1, r.)

'Qui comēza la nobel opera de // arithmeticha ne laqual fe tracta // tute coffe amercantia pertinen-//te facta ꝛ compilata per Piero // borgi da Veniefia.' (F. 2, r.)

Colophon. 'Stampito in Veniexia per zouāne de Hall' 1488.' (F. 95, v.)

Description. 4°, 14.8 × 20.7 cm., the text being 12.4 × 14.9 cm. 95 ff., 44 ll. Venice, 1488.

Editions. See p. 16. This is the third edition, and is nearly as rare as the first. Proctor mentions three books from the press of the printer, John Leoviller, of Hall (Halle ?).

The text is practically verbatim with that of the first edition (p. 16).

PIETRO BORGHI. Ed. pr. 1484. Venice, 1491.

See p. 16.

Title. The title page is the same as that of 1488, except for the letters SHSU which precede the former.

Colophon. 'Nela inclita citade venetia a zorni .22.// ottubero 1491. u impofto fine ala pre//fente opera.// Libro dabacho.' (F. 100, v.)

Description. 4°, 15.4 × 20.8 cm., the text being 11.7 × 16.8 cm. 100 ff. unnumb., 40–43 ll. Venice, 1491.

See p. 16. This is the fourth edition.

PIETRO BORGHI. Ed. pr. 1484. Venice, 1517.

See p. 16.

Title. This is substantially the same as that of the 1484 edition already described.

Colophon. '¶Stampata in Venetia per Iacomo pentio da Lecho ad inftā//tia de Marchio Seffa & Piero di Rauani compagni // anno dn̄i .1517. adi .25. de zugno.' (F. 100, v.)

Description. 8°, 15.5 × 21.6 cm., the text being 13.3 × 17 cm. 2 ff. blank + 100 numb. = 102 ff., 41 ll. Venice, 1517.

See p. 16.

PIETRO BORGHI. Ed. pr. 1484. Venice, 1528.

See p. 16.

Title. This is substantially the same as that of the 1484 edition already described.

Colophon. '¶Stampato in Venetia per Frãcefco Bindoni, & Mapheo // Pafyni compagni. Nel anno .M.D.XXVIII.// Adi .XVIII. Del mefe di Zenaro.' (F. 100, v.)

Description. 8°, 15.1 × 20.5 cm., the text being 13.4 × 17.1 cm. 100 ff. numb., 41 ll. Venice, 1528.

See p. 16.

PIETRO BORGHI. Ed. pr. 1484. Venice, 1534.

See p. 16.

Title. This is substantially the same as that of the 1484 edition already described.

Colophon. '¶Stampato in Vinegia per Frãcefco Bindoni, & Mapheo // Pafini compagni. Nel anno .M.D.XXXIIII.// Adi .25. Del mefe di Settembre.' (F. 100, v.)

Description. 8°, 15.4 × 20.9 cm., the text being 13.5 × 17.1 cm. 100 ff. numb., 41 ll. Venice, 1534.

See p. 16. The various editions changed but little.

PIETRO BORGHI. Ed. pr. 1484. Venice, 1540.

See p. 16.

Title. 'Pietro Borgo // Libro de Abacho.// Chi d'arte Mathematice ha piacere // Che tengon di certezza il primo grado // Auanti che di quelle tenti il vado // Vogli la prefente opera vedere.// Per quefta lui potra certo fapere // Se error fara nel calculo notado // Per quefta effer potra certificado // A formar conti di tutte maniere.// A merchadanti molta utilitade // Fara la prefente opera e à fattori // Dara in far conti gran felicitade // Per quefta uederan tutti gli errori // E delli quaterni fuoi la ueritade // Danari acquiftaranno, e grandi honori.// In la patria e di fuori // Sapran far le raggion de tutte gente // Per le figure che fon qui depente.// Auenga che alquanto per me fu promeffo affai ſufficientemente // alla promeſſa ſatisfaceſſe, niente dimancho per ſatisfar alle pre-//giere di qualch'uno, e maſſime di alcuni Impreſſori, iquali era-//no per ſtampar la preſente Opera, ho uoluto alquanto ampliar la di qualche gentilezza oltra quello che prima ꝑmiſſe, benche // di quello che ſe potria dir, queſto ſia vna minima parte, pero cli // chi uoleſſe metter pur la centiſinia parte di quello che ſi potria // poner, el ſaria molto piu la gionta di quello che ſia tutta l'opera // inſi. Et pero pro nunc mi paſſo con alcune coſette aggionte nel // ligar de metalli, lequal principiano a carte .77. & anchora in fin // de l'opera con dieci caſi aſſai piaceuoli & leggiadri comincian//do a carte .98. da quello che dice. Le vno che compra tre pezze // de panno per ducati .70. &c. Et ſe le mente di quelli, iquali me // hanno pregato non fuſſino à ſuo modo ſatisfatte, prego quelli // me habbino per iſcuſato.' (F. 1, r.)

Colophon. 'Stampato in Venetia per Bernardino de Bindoni.// Ne l'anno .M.D.XL. Del meſe di Ottober.' (F. 100, v.)

Description. 4°, 15.5 × 20.9 cm., the text being 13.2 × 17 cm. 2 ff. unnumb. + 98 numb. = 100 ff., 39–40 ll. Venice, 1540.

See p. 16. This is the eleventh edition. The title is considerably extended and some changes are made in the text, chiefly in the way of added matter.

PIETRO BORGHI. Ed. pr. 1484. Venice, 1550.

See p. 16.

Title. The title page is substantially identical with that of the 1540 edition already described. It bears the date, 'Anno Domini M. D. L.'

Colophon. '¶Stampato in Vinegia per Franceſco Bindoni, & // Mapheo Paſini. Nell Anno .M D L.// Adi .21. Del meſe di Nouembrio.' (F. 100, v.)

Description. 4°, 15.6 × 20.9 cm., the text being 12.9 × 17.9 cm. 100 ff. numb., 41 ll. Venice, 1550.

See p. 16. This is the twelfth edition, and at least five subsequent editions appeared in the sixteenth century.

ANONYMOUS. Ed. pr. c. 1485. S. l. a. (Cologne ?, c. 1485).

Title. 'Ars numerandi. Incipit cōpendiofus tractatul9 quin //tupliciū dcīonū numeāliū in quo docet' // luculēt' qūo ordīant' variātur ɔponūt' // et abinuicem deriuātur dictōnes nume// rales.' (F. 2, r.; see Fig. 10.) Without abbreviations this would appear as follows: 'Incipit compendiosus tractatulus quintuplicium dictionum numeralium in quo docetur luculenter quomodo ordinantur variantur componuntur et ab invicem derivantur dictiones numerales.'

Description. 4°, 14 × 20.6 cm., the text being 10.8 × 15.2 cm. 6 ff. unnumb. (1 blank), 35 ll. S. l. a. (Cologne ?, c. 1470–1485).

Editions. There was no other edition. The date of this rare book is uncertain. The style of type has led to the assertion that it was printed in Mainz by Fust and Schoeffer, about 1470. The book is not mentioned by Hain, nor is any copy known in the French libraries. Coppinger believes that it was printed in Cologne by Ulrich Zell in 1485, and in this he is followed by Zell's biographer, Merlo. The British Museum catalogue gives this date and printer, but questions each. A comparison of the water-marks in this copy with those of the fifteenth century which are described in standard treatises (e. g., Sotheby, E. L., *Principia Typographica*, London, 1858, vol. III) fails to throw any light upon the date. Riccardi attributes it to Zell, c. 1471, who had been an apprentice of Guttenberg, but had left Mainz at the sacking of the city in 1462. 502

The book is not strictly speaking an arithmetic, but a treatise on grammatical usage as applied to numbers. A considerable portion of the text is occupied with the distinction between ordinals and cardinals, and the methods of using them.

Other works of 1485–1487. Albert of Saxony, 1487, p. 9; c. 1478. In 1485 there was published at Bologna, edited by Pietro Almadiano of Viterbo, a quarto work by Nicolò de Orbelli (Nicolaus Orbellis) entitled 'Compendium considerationis matematice quo ad aritmeticam et geometriam sunt necessaria.' Orbilli's 'Cursus librorum philosophie naturalis,' which appeared in 1494, 4°, and at Basel in 1503, 4°, contained 2 pp. on arithmetic. See Isidorus, p. 8, 1483. 473; 495

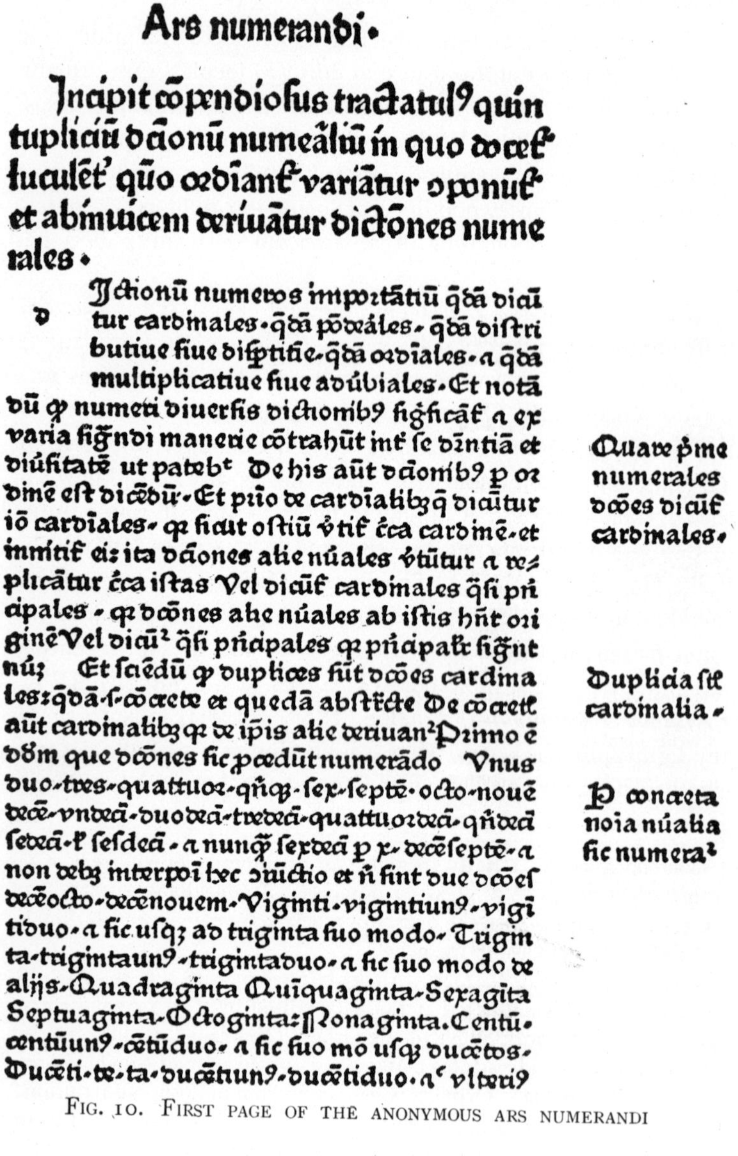

Ars numerandi.

Incipit cōpendiosus tractatul9 quin tupliciū dictionū numealiū in quo docet luculēt' quō ordiant' variātur cponūt' et abinuicem deriuātur dictōnes numerales.

Dictionū numeros importātiū q̄dā dicūtur cardinales. q̄dā pōderāles. q̄dā distributiue siue dispetitie. q̄dā ordiales. et q̄dā multiplicatiue siue adúbiales. Et notādū q numeri diuersis dictionib9 sig̃ficāt et ex varia sig̃ndi manerie cōtrahūt int' se dr̄ntiā et diu̇sitatē ut patebt De his aūt dcōib9 p ordinē est dicēdū.

Quare p̄me numerales dcōes dicūt' cardinales.

Et p'io de cardialib3 q̄ dicūtur iō cardiales. q sicut ostiū v̇tit' c̄ca cardinē. et innitit' ei: ita dcōnes alie nūales v̇tūtur et replicātur c̄ca istas Vel dicūt' cardinales q̄si pricipales. q dcōnes alie núales ab istis hn̄t originē Vel dicū² q̄si pn̄cipales q pn̄cipalr sig̃nt nū;

Duplicia sūt cardinalia.

Et sciēdū q duplices sūt dcōes cardinales: q̄dā .s. cōcrete et quedā abstr̄cte De cōcret aūt cardinalib3 q de ipis alie deriuan²

P cōcreta noia núalia sic numera²

Primo ē dsm que dcōnes sic pcedūt numerādo Vnus duo. tres. quattuor. q̄nq3. sex. septē. octo. nouē decē. vndecī. duodecī. tredecī. quattuordecī. q̄ndecī sedecī. l' sesdecī. et nunq̄ sexdecī p x. decēseptē. et non deb3 interpoi hec diūctio et ñ sint due dcōes decēocto. decēnouem. Viginti. vigintiun9. vigitiduo. et sic usq3 ad triginta suo modo. Triginta. trigintaun9. trigintaduo. et sic suo modo de alijs. Quadraginta Quīquaginta. Sexagita Septuaginta. Octoginta: Nonaginta. Centū. centūun9. cētūduo. et sic suo mō usq3 ducētos. Ducēti. te. ta. ducētiun9. ducētiduo. et' vlteri9

Fig. 10. First page of the anonymous Ars numerandi

ANICIUS MANLIUS SEVERINUS BOETHIUS.

Ed. pr. 1488. Augsburg, 1488.

BOETIUS. Born at Rome c. 480; died there October 25, 524. He was a Roman senator, a philosopher, and the last of the great Latin writers.

Title. 'Arithmetica boetij.' (F. 1, r.)

Colophon. 'Finit arithmetica Boetij bene re//uiſa ac fideli ſtudio emendata Im//preſſa per Erhardū ratdolt viri ſo-//lertiſſimi eximia īduſtria ⁊ mira im-//primēdi arte: qua nuꝑ

Incipiunt duo libri de Arithmetica anitij manilij seuerini Boetij viri clarissimi et illustrissimi ex cõsulis: ordinarij: patricij: ad patricium simmachum.

N dandis accipiendisque muneribus ita recte officia precipue inter eos q sese magni faciunt estimant si liquido ɔstabit nec ab hoc aliud qd liberalius afferret inuentũ: nec ab illo vnq̃ qd iucundius beniuolẽtia cõplecteret acceptũ. Hec ipse cõsiderans: attuli non ignaua opum põdera quibus ad facinus nihil instructius est: cũ habendi sitis incanduit: ad meritũ nihil vilius cũ ea sibi victor anim' calcata subiecit: sed ea que ex grecarũ opulentia litterarũ in romane orationis thesaurum sũpta couexim'. Ita enĩ mei quoq3 operis mihi ratio cõstabit: si que ex sapientie doctrinis elicui: sapientissimi iudicio cõprobent. Uides igit vt tam magni laboris effectus tuu3 tantũ expectet examẽ: nec in aures pdire publicas nisi docte sentẽtie astipulatione nitat. In quo nihil mirũ videri debet: cũ id opus qd sapiẽtie inuenta persequit: non auctoris s3 alieno incũbit arbitrio. Suis qppe instrumentis res rationis expenditur: cũ iudicium cogit subire prudentis. Sed huic munusculo: nõ eadem que ceteris imminent artibus munimenta cõstituo. Neque enim fere vlla sic cũctis absoluta partib' nullius indiga suis tantũ est sciẽtia nixa psidijs: vt nõ ceterarũ quoq3 artiũ adiumenta desideret. Nã in effigiandis marmore statuis: alius excidende molis labor est: alia formã de imaginis ratio: nec eiusdẽ artificis manus politi operis nitor expectat. At picture manibus tabule cõmisse fabrorum. cere rustica obseruatione decerpte: colorũ fuci mercatorũ solertia perquisiti: lintea operosis elaborata textrinis: multiplicem materiã prestant. Nõne idem quoque bellorũ visitur instrumẽtis: Hic spicula sagittis exacuit: illi validus thorax nigra gemit incude. Ast alius: crudi vmbonis tegmina pprij labor: orbi infigenda mercatur. tam multi artibus ars vna perficit. Ast nostri laboris absolutio lõge ad faciliorẽ currit euentu3. Tu enĩ solus manu3 supremo operi impones: in quo nihil de decernentiũ necesse est laborare cõsensu. Qualibet enĩ hoc iudiciũ multis artibus probet excultũ vno tamẽ cumulat examine. Experiare igitur licet quantũ nobis in hoc studio longis tractus ocijs labor adiecerit. An rerũ subtilium fugas exercitate mentis velocitas cõphendat. vtrũ ieiune macies orationis ad ea que sunt caligantibus impedita sententijs expedienda sufficiat. Qua in re mihi alieni quoque iudicij lucra queruntur. Cum tu vtrarũque peritissimus litterarũ: possis grate orationis expertibus quantu3 de nobis iudicare audeant: sola tantũ pnunciatione prescribere. At nõ al-

a 2

nulla scientia est, que aliqd ab altera non desiderat adiumenti

sola arithmetica nullius indiget adiumẽ

FIG. 11. FIRST PAGE OF THE 1488 BOETHIUS

Tetragona. Longitudo. Secūda vnitas.

Prima vnitas. — Latitudo. — Latitudo. — Tertia vnitas.

1	2	3	4	5	6	7	8	9	10
2	4	6	8	10	12	14	16	18	20
3	6	9	12	15	18	21	24	27	30
4	8	12	16	20	24	28	32	36	40
5	10	15	20	25	30	35	40	45	50
6	12	18	24	30	36	42	48	54	60
7	14	21	28	35	42	49	56	63	70
8	16	24	32	40	48	56	64	72	80
9	18	27	36	45	54	63	72	81	90
10	20	30	40	50	60	70	80	90	100

Secūda vnitas. Longitudo. Tetragona.

Ratio atq3 expositio digestę formu
lę. Cap.27.

SI igit duo pma late
ra ppositę formulę q̄
faciūt āgulū:ab vno
ad. 10.et. 10. pceden
tia respiciāt:⁊ his sub
teriores ordies cōpa
rent:qui scilicet a.4.angulum incipi
entes:in vigenos terminū ponunt:
duplex id est prima species multipli
citatis ostenditur:ita vt primus primū sola superet vnitate: vt duo vnū
secūd⁹ scdm binario supuadat:vt q̄
ternari⁹ binariū. terti⁹ tertiū tribus:
vt senari⁹ ternariū.qrtus qrtū q̄ter-
narij numerositate transcendat: vt
8 q̄ternariū :⁊ p eādē cūcti sequētiā
sese minoris pluralitate pretereant.
Si vero terti⁹ angulus aspiciat:q ab
9.inchoās lōgitudinē latitudinēq3
tricenis altrinsecus numer⁹ extēdit:
et hic cū pma latitudine et lōgitudi-
ne cōparetur: triplex species multi-

Fig. 12. Multiplication table, 1488 Boethius

venetijs nūc // auguſtȩ excellet nominatiſſimus.// Anno dn̄i .M.cccc. lxxxviij. Men-//ſis maij die vigeſima.' (F. 48, r. Fig. 14.)

Description. 4°, 15.1 × 20.8 cm., printed in double columns, each being 5.1 × 14.7 cm. 47 ff. unnumb. + 1 blank = 48 ff., 40 ll. Augsburg, 1488.

Editions. Augsburg, 1488, 4° (here described); Cologne, 1489; Leipzig, 1490; Venice, 'Opera,' 1491–92 (see p. 28); Paris, 1496, 4°; Venice, 1497; ib., 1499 (bearing also the date 1497), fol.; s. l. a. (the Compendium of Muris, c. 1500); Paris, 1501, fol.; ib., 1503 (see p. 29); ib., 1507, 4°; ib., 1510 (see p. 30), fol.; ib., 1511, 4°; ib., 1514, fol.; Vienna, 1515 (the Muris 'Compendium,' in Tannstetter's works); Paris, 1521 (see p. 31), fol.; ib., 1522, fol.; 1528, fol.; Paris, 1530, 8°, 'De differentiis topicis libri quatuor'; Basel, 1536, 8°; Basel, 1546, 'Opera,' fol.; ib. and Paris, 1549; Basel, 1553, 'ajectis explic. per J. Scheubelium,' 8°; Paris, 1553; Basel, 1570; Venice, 1570, 'Opera,' fol. There are undoubtedly various other editions of the Arithmetic, or the Epitome by Faber Stapulensis, these works often being bound with such treatises as the Arithmetic of Jordanus Nemorarius. Murhard (I, 160) and Rogg (p. 137) mention an edition of Faber's 'Compendium,' s. l., 1480, but it is not given by other bibliographers. (See Hain, I, 468; Brunet, *Man.*, I, 1059; Graesse, *Trésor*, I, 464; Riccardi, I, 1, 159; Boncompagni, *Bulletino*, XII, 148.) 502

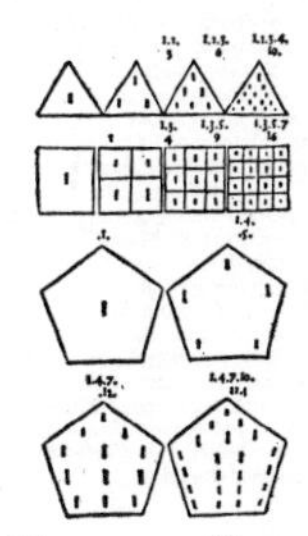

FIG. 13. FIGURATE NUMBERS, 1488 BOETHIUS

The text is practically that followed by Friedlein in his standard edition of the 'Opera' of Boethius (Leipzig, 1867), except as to numerals. Here, as in the later manuscripts, the Arabic characters have replaced the Roman of the original text. (See Fig. 12.)

The arithmetic of Boethius was based upon the Greek work of Nicomachus (fl. c. 100 A.D.), and related only to the theory of numbers, the Ἀριθμητική, as distinguished from the practical calculations, the

Λογιστική, and from the later algorismus (p. 5). Boethius gave an elaborate theory of ratios and devoted much attention to figurate numbers, such as the triangular, square, pentagonal, and cubic. (See Fig. 13.) The work was the standard in the Church schools throughout the Middle Ages.

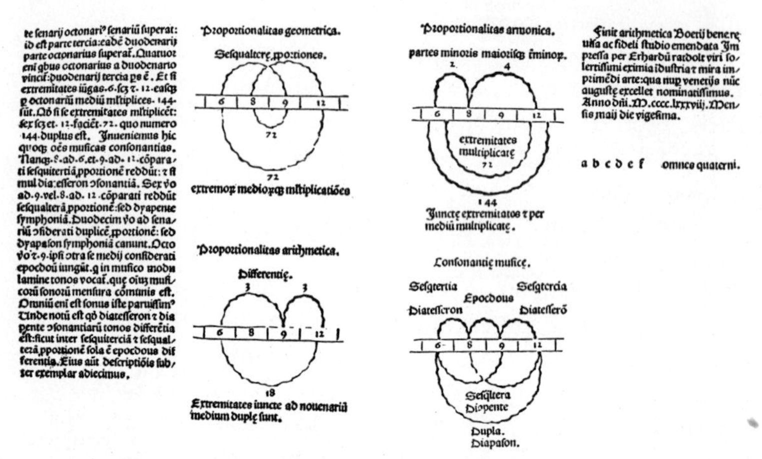

te ſenarij octonari⁹ ſenariū ſuperat: id eſt parte tercia: eadē duodenarij parte octonarius ſuperat. Quatuor eni qbus octonarius a duodenario vincit: duodenarij tercia ps ē. Et ſi extremitates iūgas. 6. ſcʒ ꝛ. 12. eaſq; p octonariū mediū mltiplices. 144. ſūt. Qð ſi ſe extremitates mltiplicēt: ſex ſcʒ et. 12. faciēt. 72. quo numero 144. duplus eſt. Inueniemus hic quoq; oēs muſicas conſonantias. Nanq; 8. ad. 6. et. 9. ad. 12. cōparati ſeſquitertiā pportionē reddūt: ꝛ ſimul diateſſeron ꝯſonantiā. Sex vo ad. 9. vel. 8. ad. 12. cōparati reddūt ſeſqualterā pportionē: ſed dyapente ſymphoniā. Duodecim vo ad ſenariū ꝯſiderati duplicē pportionē: ſed dyapaſon ſymphoniā canunt. Octo vo ꝛ. 9. ipſi ꝯtra ſe medij conſiderati epocdoū iungūt. q in muſico modulamine tonos vocat. quę oīuʒ muſicorū ſonorū menſura cōmunis eſt. Omniū eni eſt ſonus iſte paruiſſim⁹. Unde notū eſt qð diateſſeron ꝛ diapente ꝯſonantiarū tonos differētia eſt: ſicut inter ſeſquiterciā ꝛ ſeſqualterā pportionē ſola ē epocdous differentia. Eius aūt deſcriptiōis ſubter exemplar adiecimus.

Proportionalitas geometrica.

Seſqualterę pportiones.

6 | 8 | 9 | 12

72

72

extremoꝝ medioꝝq; mltiplicatiões

Proportionalitas arithmetica.

Differentię.

3 3

6 | 8 | 9 | 12

18

Extremitates iuncte ad nouenariū medium duplę ſunt.

Proportionalitas armonica.

partes minoris maioriſq; ī minoꝝ.

2 4

6 | 8 | 9 | 12

extremitates multiplicatę

72

144

Junctę extremitates ꝛ per mediū multiplicatę.

Conſonantię muſicę.

Seſqtertia Seſqtercia

Epocdous

Diateſſeron Diateſſerō

6 | 8 | 9 | 12

Seſqltera Diapente

Dupla. Diapaſon.

Finit arithmetica Boetij bene reuiſa ac fideli ſtudio emendata Impreſſa per Erhardū ratdolt viri ſolertiſſimi eximia īduſtria ꝛ mira imprimēdi arte: qua nup venetijs nūc auguſtę exceller nominatiſſimus. Anno dñi. M. cccc. lxxxviij. Menſis maij die vigeſima.

a b c d e f omnes quaterni.

FIG. 14. LAST TWO PAGES OF THE 1488 BOETHIUS

BOETHIUS.

Ed. pr. of the Arithmetic, 1488. Venice, 1491–92.

See p. 25.

Title. See Fig. 15.

Colophon. 'Impreſſis venetijs per Joannē de Forli-//uio et Gregorium fratres. Anno ſalutis .M//cccc.lxxxxj. die xxvj, menſis Martij.' (F. 352, r.)

On f. 256 (220 as numbered in the book), at the end of the Geometry is the following: 'Venetijs Impreſſum Boetij opus ꝑ Joānez ꝛ Gre//goriū de gregorijs fratres fęlici exitu ad finē vſqʒ ꝑductu // accuratiſſimeqʒ emēdatū Anno humane reſtaurationis. // 1492. die .18. Auguſti. Auguſtino Barbadico Sereniſſi//mo Venetiarum principe Rem pu. tenēte.'

Description. Fol., 21.7 × 32 cm., printed in double columns, each being 6.6 × 24.1 cm. 3 ff. unnumb. + 345 numb. + 1 blank = 349 ff., 66–70 ll. Venice, 1491–92 (see the colophons).

Editions. This is the *editio princeps* of the works of Boethius. For other editions see p. 27.

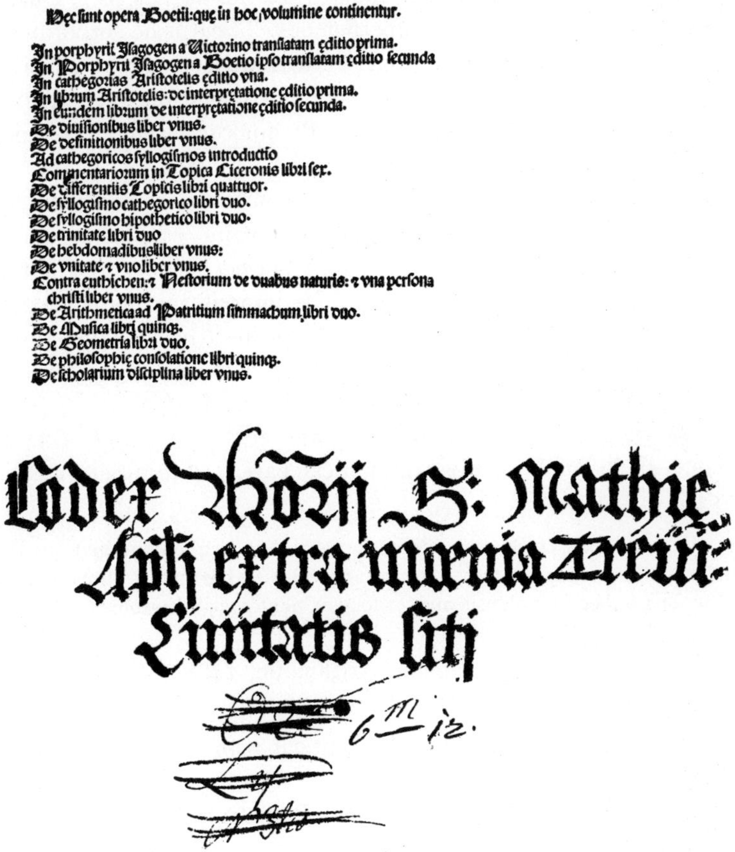

Hęc sunt opera Boetii: quę in hoc volumine continentur.

In porphyrii Isagogen a Victorino translatam ędītio prima.
In Porphyrii Isagogen a Boetio ipso translatam ędītio secunda
In cathegorias Aristotelis ędītio vna.
In librum Aristotelis: de interpretatione ędītio prima.
In eundem librum de interpretatione ędītio secunda.
De diuisionibus liber vnus.
De definitionibus liber vnus.
Ad cathegoricos syllogismos introductio
Commentariorum in Topica Ciceronis libri sex.
De differentiis Topicis libri quattuor.
De syllogismo cathegorico libri duo.
De syllogismo hipothetico libri duo.
De trinitate libri duo
De hebdomadibus liber vnus:
De vnitate et vno liber vnus.
Contra eutichen: et Nestorium de duabus naturis: et vna persona christi liber vnus.
De Arithmetica ad Patritium simmachum libri duo.
De Musica libri quinque.
De Geometria libri duo.
De philosophię consolatione libri quinque.
De scholarium disciplina liber vnus.

Codex monrij S: Mathie
Apli extra mœnia Treuir:
Ciuitatis siti

Fig. 15. Title page, 1491–92 Boethius

BOETHIUS. Ed. pr. of the Arithmetic, 1488. Paris, 1503.

See p. 25.

Title. 'In hoc libro contenta.// Epitome/ compendiofaqȝ introductio in libros // Arithmeticos diui Seuerini Boetij: adiecto fa-//miliari commentario dilucidata.// Praxis numerandi certis quibufdam regulis // conftricta.// Introductio in Geometriam breuiuſculis an-//notationibus explanata. ſex libris diſtincta.// Primus de magnitudinibus et earū circūſtan-//tiis.// Sc'dus de cōſequētibus/ cōtiguis/ & cōtinuis.// Tertius de punctis.// Quartus de lineis.// Quintus de ſuperficiebus.// Sextus de corporibus. // Liber de quadratura circuli.// Liber de cubicatione ſphere. // Perſpectiua introductio.// Inſuper Aſtronomicon.' (F. 1, r.)

Colophon. 'Abſolutum in almo Parhiſiorum ſtudio/ //Anno dn̄i qui numero definuit omnia // 1503.' (F. 48, r.)

On f. 84, v. is the following: 'Geometrici introductorij:// ſexti / etvltimi libri finis.// Editi anno domini // milleſimo quin-//gēteſimo pri//mo : viceſi-//ma quin//ta no-//uem//bris.' And on f. 111, v., the following: 'Id opus impreſſerūt Volphgangus // hopilius et Henricus ſtephanus // ea in arte ſocii in Almo pari-//ſiorum ſtudio Anno Chri//ſti Celorum totiuſqȝ // nature cōditoris.// 1503. Die vice//ſimaſepti-//ma Iu-//nij.'

Description. 8°, 20 × 26.8 cm., the text being 15.6 × 22.3 cm. 48 ff. numb. in the arithmetical part, 113 in all (1 blank), 54 ll. Paris, 1503.

The editions of Boethius differ more or less in the combinations of works which they contain. See p. 27. This is the first edition of the Jacobus Faber Stapulensis and Jodocus Clichtoveus 'Epitome.'

BOETHIUS. Ed. pr. of the Arithmetic, 1488. Paris, 1510.

See p. 25.

Title. This is substantially identical with that of the 1503 edition already described.

Colophon. '¶Abſolutum in almo Pariſiorum ſtudio/ // Anno domini qui numero definiuit // omnia 1503. Et emiſſum ex offi-//cina Henrici ſtephani Anno // Christi ſaluatoris // omnium 1510 de-//cima quinta // die Mar-//tij.' (F. xlviii, r.)

Description. Fol., 19.5 × 27.9 cm., the text being 15.4 × 25.6 cm. 48 ff. numb., 46–56 ll. Paris, 1510.

This is one of the editions containing the commentary of Jodocus Clichtoveus on the 'Epitome' of Boethius by Jacobus Faber Stapulensis. (See p. 27.)

BOETHIUS. Ed. pr. of the Arithmetic, 1488. Paris, 1521.

See p. 25.

Title. 'Divi Severi-//ni Boetii Arithmetica,// dvobvs discreta libris; adie-//cto commentario, mysticam nvme-//rorum applicationem perſtringente, declarata.// (Woodcut with initials of the printer: S.D.C.). Vænundatur apud Simonem Coli-//nævm, e regione ſcholæ Decretorum.' (F. 1, r.)

Colophon. 'Excudebat Simon Colinæus, Pariſijs, Anno MDXXI/1521 Quinto//Idus Iulias.' (F. 139, v.)

Description. Fol., 19.9×28.2 cm., the text being 13.9×26 cm. 4 ff. unnumb. + 136 numb. = 140 ff., 40–52 ll. Paris, 1521.

Fig. 16. Printer's device, 1521 edition of Boethius

This is the first edition of Boethius with the commentary of Girardus Ruffus: '¶Girardi Rvffi, in duos arithmeticæ Boetii libros, commentarivs.' (F. 5, r.) This commentary greatly exceeds the text in extent, and as to ponderosity it leaves little to be desired. As a piece of typography, however, this is one of the best editions of Boethius.

ANIANUS and JOHANNES SACROBOSCO.
Ed. pr. 1488. Strasburg, 1488.

Anianus was a fifteenth-century astronomer and poet, of Strasburg.

Johannes de Sacrobosco (Sacrobusto, Sacrobuschus, Holywood, Holybush, Holywalde, Holyfax, Halifax) was born at Halifax (Holywood), Yorkshire; died at Paris in 1244 or 1256. He studied at Oxford and lectured at Paris. He wrote on astronomy and algorism.

Title. 'Cōpotus manua//lis magr̄i aniani. // metricus cū ɔmēto // Et algoriſmus.' (F. 1, r. Fig. 17.) F. 45 begins, 'Incipit textus algoriſmi.'

Colophon. 'Impreſſum Argn̄. per Johāmem pryß.// Anno domini .i488 .i8. kall.' decembris.' (F. 44, r.)

Description. 4°, 13.9 × 19.5 cm., the text being 8.8 × 14.4 cm. 53 ff. numb. + 2 unnumb. = 55 ff., 31–34 ll. Strasburg, 1488.

Editions. The editions of Sacrobosco's *Algorismus* were as follows: Strasburg, 1488, 4° (here described); s. l. a. (Venice?, 1490?), 4°; Venice, 1501 (see p. 35); Paris (edited by Clichtoveus), 1498; ib., 1503; ib., 1510; Vienna, 1517; Cracow,

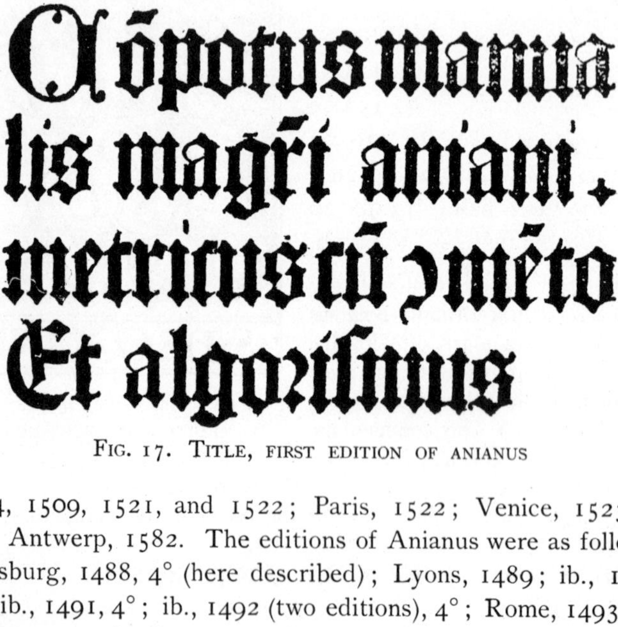

Cōpotus manua
lis magrī aniani.
metricus cū ꝯmēto
Et algoꝛiſmus

FIG. 17. TITLE, FIRST EDITION OF ANIANUS

1504, 1509, 1521, and 1522; Paris, 1522; Venice, 1523 (p. 35); Antwerp, 1582. The editions of Anianus were as follows: Strasburg, 1488, 4° (here described); Lyons, 1489; ib., 1490, 4°; ib., 1491, 4°; ib., 1492 (two editions), 4°; Rome, 1493, 4°; Paris, 1494, 4°; ib., 1498, 4°; s. l. a. (Paris?, c. 1495, 4°, see p. 33); s. l. a. (Basel, c. 1500), 4°; Rouen, s. a. (1502), 8°; Paris, 1501, 4°; ib., 1502, 8°; Lyons, 1504 (see p. 35); ib., 1509, 4°; Paris, 1508; **502** ib., 1511; ib., 1515; ib., 1519, 4°; ib., 1529; ib., 1530, 4°; Lyons, 1540, 4°; Frankfort, 1549; Wittenberg, 1550; ib., 1568; Antwerp, s. a. (c. 1558); ib., 1559. There are probably others s. l. a., and some appear under the name of Sacrobosco. Of this first edition Boncompagni knew only the Munich copy. (*Bulletino*, XII, 126 n.)

The first part of this rare book is the *Compotus Manualis* of Anianus, and the second is the *Algorismus* of Sacrobosco, described later. It is probably the first book on mathematics printed in Strasburg (but see p. 10), and it is the first edition of each of the two treatises mentioned, and the first printed work on the computus, the arithmetic of the Church calendar. In the work of Anianus appears for the first time in print the original of the rhyme beginning

'Thirty days hath September,'

the English version of which is said to have been first published in the 1590 edition of Grafton's *Chronicles*. It also appeared in an arithmetic published anonymously in 1596. The Latin form of Anianus is as follows:

'Junius aprils feptember et ipfe nouember
Dant triginta dies reliquis fuꝑadditur vnus,
De quorum numero februarius excipiatur.' (F. B 8.)

It is found in various forms in other computi, manuscript and printed. Anianus also gives for the first time in print the astronomical formula: 'Sunt Aries, Taurus, Gemini, Cancer, Leo, Virgo, Libraque, Scorpio, Arcitenens, Caper, Amphora, Pices,' which appeared in the works of Bede under the title 'Verfus Prifciani, de Astronomia,' as 'Hinc Aries, Taurus, Gemini, Cancer, Leo, Virgo, Libra, Scorpius, Arcitenens, Capricornus, & Vrna, Qui tenet & Pifcis.' (1563 edn., vol. I, 517.)

ANIANUS. Ed. pr. 1488. S. l. a. (Paris ?, c. 1495).

See pp. 31, 502.

Title. 'Cōpotus cum // commento.' 'Liber qui Compotus infcribitur: vna cum figuris et ma/ //nibus neceffariis tam in fuis locis q̃3 in fine libri pofitis.// Incipit feliciter.' (F. 1, r.)

Description. 4°, 13.1 × 19.4 cm., the text being 9.3 × 13.1 cm. 39 ff. unnumb. + 1 blank = 40 ff., 35 ll. S. l. a. (Paris ?, c. 1495).

Editions. See p. 32. This rare and interesting edition was published about 1495, possibly by Mich. le Noir at Paris, although there is no date or place of publication given. It is one of the best examples of the mediæval computus that appeared in print. Unlike the first edition (p. 31, 1488) it contains a number of illustrations showing the use of the hand and fingers in assisting in calendar reckoning, the title of 'Com-

Liber qui Compotus inscribitur: vna cum figuris et ma/
nibus necessariis tam in suis locis q̄ȝ in fine libri positis.
Incipit feliciter.

Ux orta est iusto Psalmista Ista ver
ba possunt dupliciter considerari. Primo pōt
dici de deo q̄ est lux vera. ideo dicebat dauid.
lux orta est iusto. Et de ista luce dicit Johā.i.
Erat lux vera q̄ illuminat omnē hominē veniē/
tē in hunc mundū. Secūdo ponit de scientia.
Et dicit lux quasi scientē reddētē lucidū. quia facit hominē sciē/
tem esse lucidū. In quibus verbis ad cōmendationē scīe duo bre
uiter tangūtur. Primo eñ tangit sciētie altitudo preciosa p hoc
quod dicit lux. Secundo largitudo gloriosa per hoc q dicit or/
ta est. Primū probatur auctoritate et ratiōe. auctoritate Isidori
sic dicentis. Scientia est fons indeficiens. bonitatis via. sui sal/
uatoris cognitio. Ratiōe sic. illud est validū et preciosum quod
de inualido ꝛ imperfecto facit validum ꝛ pfectum. sciētia ē huius
modi. ergo ꝛc. maior est manifesta. minor declaratur p phm ter/
cio de aīa sic dicentē: Aīa in pncipio sue creationis ē tanq̄ȝ tabu
la rasa in qua nihil depictū est. depingibilis tm̄ sciētiis ꝛ virtuti/
bus. Primum probat auctoritate boetii ꝛ ratōe in prologo aris
metrice. Scientia ē coꝝ que vera sunt ꝛ imutabilis essentie no
straqȝ comphensione veritatis. Ratiōe sic: illud est tanq̄ȝ sūmū
bonū quod habet largitionē gloriosam scīa ē hmōi. ergo ꝛc. ma
ior est vera. minor pbatur p diffinitionē sciētie q̄ talis ē. Scīa est
qdā habitus aīe rōnalis nō innatus sed acq̄situs oīm hūanaꝝ re
rum indagatrix ꝛ totius humane vite gubernatrix. q scīa sit ha/
bitus ptȝ. qr scīa est aliquod existens in aīa. sed omne illud quod
est in aīa aut ē hītus aut potētia aut passio. Et hoc testat Aristo.
in secūdo ethicoꝝ. q scīa nō sit passio ptȝ qr passiones sunt in vo
luntate scīa nō ē hmōi. ergo ꝛc. q nō sit potētia ptȝ. qr q̄libȝ po/
tentia fit a natura. sicut irascibilis. ꝛ cōcupiscibilis. ꝛ sic relinqui
tur q nō sit potentia. q sit hītus aīe rōnalis ptȝ p p̄dicta. q aūt
a ij

FIG. 18. BEGINNING OF ANIANUS, EDITION OF C. 1495

putus manualis' being thus justified. This differs from the 1488 edition in the notes of the commentator. The original text is, however, substantially unchanged. 502

ANIANUS. Ed. pr. 1488. Lyons, 1504.

See pp. 31, 502.

Title. 'Compotus cū // commento.' (F. 1, r.)

Colophon. '❡Liber compoti cum cōmento finit feliciter // Impreſſus Lugduni per Claudiū nour//ri. Anno domini .M.ccccc.iiij. // die .iiij. Octobris.' (F. 32, r.)

Description. 8°, 14.5 × 20 cm., the text being 8.9 × 14.1 cm. 32 ff. unnumb., 34–40 ll. Lyons, 1504.

See p. 32. In this edition there is a curious misprint in the calendar verse given on p. 25 (f. 13, r.). It here begins 'Julius (instead of Junius) aprilis ſeptembre & ipſe nouember.' Like the edition of c. 1495 (p. 33), this is well illustrated. It differs from the 1488 edition in the notes of the commentator, but is practically identical with that of c. 1495.

JOHANNES SACROBOSCO. Ed. pr. 1488. Venice, 1501.

See pp. 31, 502.

Title. 'Algoriſmus Domini Ioā-//nis De Sacro Buſco // Nouiter Impreſſuȝ.// Cum Gratia Et Priuilegio.' (F. 1, r.)

Colophon. 'Impreſſum Venetijs per Bernardinum Venetum // De Vitalibus: Anno Dn̄i .M.CCCCC.I.// Die Tertio Men̄. Februarij.' (F. 8, r.)

Description. 4°, 14.1 × 19.6 cm., the text being 10.5 × 16.8 cm. 8 ff. unnumb., 39 ll. Venice, 1501.

See p. 32.

JOHANNES SACROBOSCO.
Ed. pr. 1488. S. l. a. (Venice, 1523).

See pp. 31, 502.

Title. 'Algoriſmus Domini Joannis // de Sacro Buſco noui- //ter impreſſum.' (F. 1, r.)

Colophon. 'Impreſſum Venetiis per Melchiorem Seſſam & Petrum // de Rauanis Socios. Anno domini .M.D.XXIII.// die .XXIIII. Octobris.' (F. 8, r.)

Description. 4°, 14.8 × 18.6 cm., the text being 10.3 × 16.8 cm. 8 ff. unnumb., 40 ll. S. l. a. (Venice, 1523).

JOHANN WIDMAN (?). Ed. pr. c. 1488. Leipzig, c. 1488.

Born at Eger, Bohemia, c. 1460. He was a student at Leipzig in 1480, A.B. in 1482, bachelor of medicine in 1485, A.M. in 1486. He evidently received the doctor's degree about the same time, for he wrote a medical work in 1497 with the title, 'Tractatus clarissimi medicina ꝑ doctoris Johānis widman . . . de pustulis . . .' That he gave lectures on algebra, possibly the first at Leipzig, is proved by a passage found by Wappler in an old Dresden manuscript: 'Quare hodie hora secunda post sermonem atque Baccelaureorum celebrata disputatione Magister Jo. W. De. Eg. Aporismata et Regulas Algobre resumpturus pro hora atque loco conuenienti cum audeturis concordabit . . .'

Title. 'Algorithmus Linealis.' (F. 1, r.) 'Ad euitādum multipli//ces Mercatorum erro-//res et alteri⁹ . . .' (F. 2, r.)

Description. 4°, 14 × 19.5 cm., the text being 7.8 × 14.5 cm. 14 ff. unnumb., 31–34 ll. Initials in red, by hand. Leipzig, c. 1488.

Editions. Leipzig, s. a. (c. 1488, here described); ib., possibly 1490, 1493; 1516; 1517, and several others.

This rare treatise, the first printed work on calculation by the aid of counters ('apud noſtras appellata eſt calculatio'), is of unknown authorship, but was probably written by Widman. (*Abhandlungen*, V, 152.) At the end of the book is the device of Martin of Würzburg (Martinus Herbipolis), and the book was printed by him, probably c. 1488. After a brief introduction on the use of counters ('projectiles' as they are here called), the author treats of the following topics: De Additione, De Subtractione, De Duplatione, De Mediatione, De Multiplicatione, De Diuiſione, De Progreſſione, De radicum extractione, De radicum extractione in Cubicis. The book closes with the words: 'Et tantū de Radicum extractione et vltima huius Algorithmi ſpecie Et ꝑ conſequens de toto Algorithmo.' There are no applied problems, and the only computations with abstract numbers are performed 'on the line' (i. e. by the 'projectiles' on a line abacus), whence the name 'Algorithmus linealis.' The work is illustrated by woodcuts.

Other works of 1488. **495** Suiseth (p. 10, c. 1480); Borghi (p. 16, 1484).

JOHANN WIDMAN. Ed. pr. 1489. Pforzheim, 1500.

See above.

Title. 'Behennd vnd hüpſch // Rechnung vff allen // kauffmanſchafften.' (Woodcut of a schoolroom.) (F. 1, r.)

Colophon. 'Gedruckt zů Pfortzheim von Thoman // anſzhelm Im Jubel Jar als man zalt 1500 // Got ſey lob.' (F. 163, v.)

Description. 16°, 10 × 13.2 cm., the text being 6.5 × 10.4 cm. 11 ff. blank + 1 unnumb. + 162 numb. = 174 ff., 26 ll. Pforzheim, 1500.

Editions. Leipzig, 1489, 8°; Pforzheim, 1500, 8° (here described); ib., 1508, 8° (p. 39); Hagenau, 1519, 8° (p. 40);

11

gezogen vß hebreiſcher zungen oder iudſcher gleich als vil in ſich beſchlieſſen als die taffel im quadrat/welche dann die ander geſetzt iſt als dann hie hernach ietliche an ir ſelbſt form klerlichen beſchreiben iſt.

1	2							
2	4	3						
3	6	9	4					
4	8	12	16	5				
5	10	15	20	25	6			
6	12	18	24	30	36	7		
7	14	21	28	35	42	49	8	
8	16	24	32	40	48	56	64	9
9	18	27	36	45	54	63	72	81

Lern wol mit fleiß das ein mal ein So wirt dir alle rechnung gemein

1	2	3	4	5	6	7	8	9
2	4	6	8	10	12	14	16	18
3	6	9	12	15	18	21	24	27
4	8	12	16	20	24	28	32	36
5	10	15	20	25	30	35	40	45
6	12	18	24	30	36	42	48	54
7	14	21	28	35	42	49	56	63
8	16	24	32	40	48	56	64	72
9	18	27	36	45	54	63	72	81

FIG. 19. FROM THE 1500 WIDMAN

Augsburg, 1526, 8° (p. 40). The title of the first edition was as follows: 'Behēde vnd hubsche // Rechnung auff allen // kauffmanschafft,' and the colophon (f. 236, r.), 'Gedruckt In der Furstlichen Stath // Leipezick durch Conradū Kacheloffen // Im 1489 Iare.'

106

Nū mach yetlichs für sich selb mit seinem gelt nach der regel so kumt als hienach stat

Scherpes	424	13	4	
Profloret	303	13	$1\frac{1}{2}$	Sūma fl
Floret	fl 204	ß 3	hlr 9	1161 ß 1
Negregant	122	13	$1\frac{1}{2}$	hlr $5\frac{1}{2}$
Primera	65	10	$7\frac{1}{2}$	
Secanda	40	7	6	

Wechsel

FIG. 20. EXCHANGE. FROM THE 1500 WIDMAN

This is the second edition, and is even more rare than the first. It was unknown to Boncompagni when he printed his 'Intorno ad un Trattato d'Aritmetica di Giovanni Widmann di Eger' in the *Bulletino*, IX, 188, the best discussion of this arithmetic that has appeared.

Widman's arithmetic was the first great German textbook on the subject, although minor works had already appeared before 1489. It is in the main a practical treatise, with good problems, and it set the standard for Germany much as Borghi's book did for Italy. Among its noteworthy features is the use of the plus and minus signs for the first time in a printed work. (See Fig. 21.) These are not used, however, as signs of operation, but as symbols of excess or deficiency in warehouse measures. The book is illustrated (see Fig. 20) with pictures showing mercantile customs, and with geometric diagrams. Widman acknowledges his indebtedness to men like Sacrobosco, 'als da lert Joannes defacrobufto vñ ander mer,' although his work shows no dependence upon the 'Algorismus' named.

After the 'Inhalt difz buchs in einer gemein,' Widman devotes $2\frac{1}{2}$ pages to 'Numeratio,' $2\frac{1}{2}$ to 'Additio,' $2\frac{1}{2}$ to 'Subtrahiren' (including denominate numbers in these topics), 1 to 'Dupliren,' 2 to 'Medieren' (i.e. multiplying and dividing by 2), $11\frac{1}{2}$ to 'Multipliciren,' 5 to division, $2\frac{1}{2}$ to progressions, and 14 to roots (4 referring to cube root). He then takes up fractions in the same order, this work being followed by compound numbers and proportion. He then gives a large number of type problems, *regulae* as he calls them, although they are not stated in the form of rules as we now know them. These include the 'Regula detri' (rule of three, treated as distinct from proportion), and the regulae fusti, detri conversa, positionis, equalitatis, legis, augmenti, plurima, sentenciarum, suppositionis, residui, excessus, collectionis, quadrata, cubica, reciprocationis, lucri, pagamenti, and alligationis.

Other works of 1489. Boethius, p. 27, 1488; Anianus, p. 32, 1488.

JOHANN WIDMAN. Ed. pr. 1489. Pforzheim, 1508.

See p. 36.

Title. 'Behend vnd hüpfch // Rechnung vff allen // Kauffmanfchafften.' (F. 1, r.)

Colophon. '⸿Gedruck zů Pfhorzheim von Thoman // Anfzhelm Im iar als man zalt 1508.' (F. 161.)

Description. 8°, 9.7 × 13 cm., the text being 6.5 × 10 cm. 7 ff. blank + 161 numb. = 168 ff., 26 ll. Pforzheim, 1508.

Editions. See p. 37. This is the third edition of this famous arithmetic, and is by the same publisher as the second (1500), but is from different type. It is about as rare as the first edition.

JOHANN WIDMAN Ed. pr. 1489. Hagenau, 1519.

See pp. 36, 504.

Title. 'Behend vnd hüpſch // Rechnung vff allen // Kauffmanſchafften.' (F. 1, r.)

Colophon. 'Getruckt zů Hagenaw durch Thoman // Anshelm. Im iar als man zalt // 1519.' (F. 151, r.)

Description. 8°, 9.7 × 14.3 cm., the text being 6.8 × 11 cm. 1 f. unnumb. + 151 numb. = 152 ff., 20–29 ll. Hagenau, 1519.

Editions. See p. 37. This is the fourth edition of Widman's Arithmetic. As an inscription on the fly leaf says, this copy was presented to Prince Baldasarre Boncompagni by Ludwig Kunze as a 'liber rariſsimus.'

See p. 39.

JOHANN WIDMAN. Ed. pr. 1489. Augsburg, 1526.

See pp. 36, 504.

72
4 + 5 Wiltdu das wyſ-
4 — 17 ſen oder deßgley-
3 + 30 chen/So ſum̃ier
4 — 19 die zenttner vnd
3 + 44 lb vnnd was auß
3 + 22 —iſt/das iſt mi-
zentner 3 — 11 lb nus dz ſetz beſon-
3 + 50 der vnnd werden
4 — 16 4539 lb (So
3 + 44 du die zendtner
3 + 29 zů lb gemachett
3 — 12 haſt vnnd das /
3 + 9 + das iſt meer
darzů Addiereſt) vnd 75 minus. Nun
ſolt du für holtz abſchlahen allweeg für
ain legel 24 lb. Vnd das iſt 13 mal 24.
vnd macht 312 lb darzů addier das —
das iſt 75 lb vnd werden 387. Dye ſub-
trahier von 4539. Vnd bleyben 4152
lb. Nun ſprich 100 lb das iſt ein zentner
pro 4 fl 1/8 wie kum̃en 4152 lb vnd kumē
171 fl 5 ß 4 heller 4/5 Vñ iſt recht gmacht

Pfeffer
B

FIG. 21. FROM THE 1526 WIDMAN

Title. 'Behennde vnnd // hübſche Rechnūg auff allen // Kauffmanſchafften.' (Woodcut showing two men seated at a reckoning table.) 'M. D. XXVI.' (F. 1, r.)

Colophon. 'Getruckt zů Augſpurg durch // Haynrich Stayner // M, D, XXVI.' (F. 192, r.)

Description. 9.4 × 13.7 cm., the text being 6.7 × 11 cm. 2 ff. unnumb. + 190 numb. = 192 ff., 25 ll. Augsburg, 1526.

Editions. See p. 37. This copy has a note in the handwriting of Prince Boncompagni, together with his collation of the book. It is interesting not only for its rarity but for the contemporary coloring of the woodcuts.

In the other editions described the third line of Fig. 21 reads 3 + 36, as it should.

PETRUS DE ALLIACO. Ed. pr. 1490. Augsburg, 1490.

ALYACO, HELIACO, D'AILLY. Born in Compiègne in 1350; chancellor of the University of Paris, Bishop of Cambray, and Cardinal. He died in 1420.

Title. 'Cōcordātia aſtronomie cū theologia // Cōcordātia aſtronomie cū hyſtorica // narratione. Et elucidariū duorꝝ pre- //cedentium: dn̄i Petri de Aliaco car//dinalis Cameracenſis.' (F. 1, r.)

Colophon. 'Explicit tractatus de cōcordia aſtronomice veritatis ⁊ narrationis hiſtorice // a dn̄o Petro cardinali Cameracen. completus in ciuitate Baſilien̄. anno xp̄i // 1414: menſis. Maij die decima.' (F. 33, r.) 'Opus concordantie aſtronomie cum theologia necnon hyſtorice verita: nar//ratione explicit feliciter. Magiſtri Joannis angeli viri peritiſſimi diligēti cor//rectione. Erhardiqꝫ Ratdolt mira imprimendi arte: qua nuper Venetijs nūc // Auguſte vindelicorum excellit nominatiſſimus. 4. nonas Januarij. 1490.' (F. 55, v.)

Description. 4°, 15.6 × 20.4 cm., the text being 11.1 × 14.7 cm. 56 ff. unnumb., 39 ll. Augsburg, 1490.

Editions. There was no other edition.

This work has been included in the list because, while chiefly astronomical, it throws considerable light upon the early Computi. It was written to show the relation between theology and astronomy, and hence it has an important bearing upon the study of the mediæval calendar.

ALONSO DELATORE. Ed. pr. 1489. Seville, 1538.

A Spanish savant of the fifteenth century.

Title. 'Viſiō delectable de // la philoſophia ⁊ ar//tes liberales: meta//phiſica: y philoſo-//phia moral .:. // M.d.xxxviij.' (This is surrounded by an elaborate woodcut.) (F. 1, r.)

Colophon. '℄Feneſce el libro llamado viſion delectable dela Philoſophia ⁊ ar//tes liberales. Es impreſſo enla inſigne y muy leal ciudad // de Seuilla en caſa de Juan Crōberger.//Año de .M.d.xxxviij.' (F. lxxij, r.)

Description. Fol., 19 × 26.9 cm., the text being 15.3 × 24.3 cm. 72 ff. numb., 42 ll. Seville, 1538.

Editions. Seville, 1489, fol.; ib., 1538, fol. (here described).

This rare treatise is an encyclopædia, with chapters devoted to the various arts and sciences. The arithmetic is found in Chapter IV (see Fig. 22), and consists of only two pages of theoretical discussion.

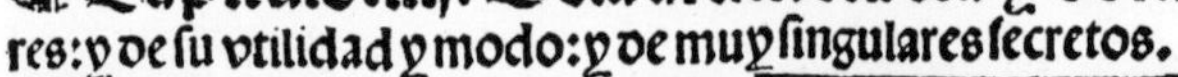

Parte primera.

¶ Capitulo.iiij. Dela arismethica y de sus inuẽtores: y de su vtilidad y modo: y de muy singulares secretos.

Passando ya τ atrauessando este sendero: vinieron encima del mõte: a do se começaua vn marauilloso camino: el q̃l los guio en vn lugar de casas τ palacios muy singulares: τ ala puerta de la villa hallarõ vna muy sagacissima τ muy ꝓfunda dõzella d̃ sciencia. La qual aunq̃ los miẽbros cubriesse cõ abito feminil: parescia debaxo d̃ aq̃l ascõder coraçõ de muy penetrãte τ muy ingenioso varõ. y en la diestra tenia vn grafio d̃ hierro: y enla siniestra vna tabla ẽblãqueada:

FIG. 22. FROM THE 1538 ALONSO DELATORE

NICOLAUS CUSA. Ed. pr. c. 1490. Strasburg ?, c. 1490.

NICOLAUS CUSANUS, NICOLAUS CHRYPFFS OR KREBS. Born at Kues on the Mosel in 1401; died at Todi, Umbria, August 11, 1464. He held positions of honor in the Church, including the bishopric of Brescia. He was made a cardinal in 1448. He wrote several other works on mathematics.

Title. Usually known as the 'Opuscula.' The work begins: 'Prohemium.// [I]n hoc volumine ꝯtinentur certi tractatus ꝛ libri altiſſime ꝯtemplatō//nis et doctrine: a preclare memorie preſtantiſſimo doctiſſimoq3 viro // Nicolao de Cuſa.' (F. 1, v.)

Description. Fol., 17.5 × 25.4 cm., the text being 11.7 × 18 cm. 163 ff. unnumb. + 1 blank = 164 ff., 45 ll. S. l. a. (Strasburg ?, c. 1490. Some bibliographers place it as early as 1480 and others assign it to Milan as late as 1505.)

Editions. Cusa's 'Opuscula varia' first appeared s. l. a. (Strasburg?, c. 1490 or earlier). His 'Opera' appeared at Paris in 1511 and 1514 (see below), and again at Basel in 1565.

This contains fifteen of Cusa's tractati, including 'Reparatio kalendarij,' 'De Apice theorie' (4 ff.), 'De mathematicis complementis,' 'De mathematica perfectione.' Of these mathematical chapters the first two are of some interest in the history of arithmetic, the others referring chiefly to mensuration.

NICOLAUS CUSA. Ed. pr. c. 1490. Paris, 1514.

See p. 42.

Title. 'Hęc in hoc fecũdo vo//lumine contenta.//

Dialogus de ignoto. 2.
Dialogus de annunciatione. 3.
Excitationũ libri X. 7.
Coniectura de nouiffimis diebus.// 1.
Septem epiftolæ. 3.
Reparatio Calẽdarii. 22.
Correctio Tabularum Alphonfi. 29.
De tranfmutationibus geometricis. 33.
De Arithmeticis complementis. 54.
De mathematicis complementis. 59.
Complementum theologicum. 92.
De prefectione mathematica. 111.'

(Woodcut of printing-press, with the words: 'Prelũ Afcẽfianũ.') 'Venũdantur cum cęte//ris eius operibus in Aedibus Afcenfianis.' (F. I, r.)

Colophon. 'Emissvm est hoc librorvm Cvsae opvs // egregivm Parisiis: ex officina Ascen//siana anno Christi pientissimi om// nivm Redemptoris MDXIIII, octa//va Assvmptionis semper San//ctae semperqve Virginis // Christi Deiqve Matris // Mariae. qva patroci//nante apvd Filivm // portvm salv-//tis spera-//mvs et // veniae.' (F. CXVI, r.)

Description. Fol., 20 × 29 cm., the text being 12.7 × 26.8 cm. 114 ff. numb. (Roman numerals) + 2 unnumb. = 116 ff., 46 ll. Paris, 1514.

Editions. See above.

This second volume of the Paris edition of Cusa's works, edited by Faber Stapulensis, contains the 'tractati' already mentioned on p. 43.

JOHANN WIDMAN (?) Ed. pr. c. 1490. S. l. a. (c. 1490).

See p. 36.

Title. 'Algorithmus Integroꝝ//Cum Probis annexis.' (F. 1,r. See Fig. 23.)

Description. 8°, 14.9 × 20.6 cm., the text being 7.8 × 14 cm. 12 ff. unnumb., 29–31 ll. S. l. a. (c. 1490).

Editions. The various bibliographies assign different dates, but I presume there was only this one edition. De Morgan (p. 99), whose judgment as to the dates of such early works was unreliable, estimated this as "hardly later than 1475," adding, "I think this is the oldest book in my list." Wappler, who has critically investigated the matter (*Abhandlungen*, V, 158) believes that Widman wrote this work, the 'Algorithmus linealis' (p. 36), and also the 'Algorithmus Minutiarum Phisicarum.'

The work opens with a quotation from Boethius, the same indeed as the opening sentence of Sacrobosco's Algorismus. After treating of numeration, addition, subtraction, duplation, mediation, multiplication, and division, the author takes up progressions, roots, and the proofs of the various processes. There are no applications in the book.

FIG. 23. TITLE OF THE ALGORITHMUS INTEGRORUM

Other works of 1490. Anianus, p. 32, 1488; Boethius, p. 27, 1488. There also appeared c. 1490, at Leipzig, an anonymous work edited by Norico, entitled 'Arithmeticae Textus communis.'

ANONYMOUS. Ed. pr. c. 1491. S. l. a. (c. 1491).

Title. See Fig. 24.

Colophon. '¶Finis trium Algoriſmoꝝ cum propor//tionum vel Mercatorum regula.' (F. 10, r.)

Description. 4°, 13.7 × 19.4 cm., the text being 8.6 × 15.6 cm. 10 ff. unnumb., 34–36 ll. S. l. a. (c. 1491).

Editions. This is not the edition described by De Morgan (p. 99), because the title has 'addita etiam regula,' instead of 'addita regula.' The word 'regula' instead of 'regla' at the

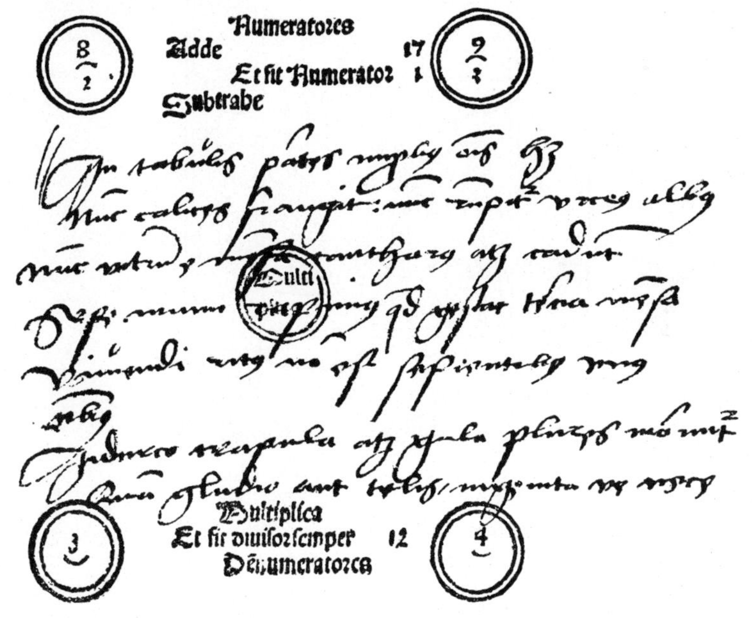

Algorismus nouus de integris cōpendiose sine figurarum (more Italorum) deletione compilatus. Artem numerandi omnemque viam calculandi enucliatim breuissime edocens. vna cū Algorismis de minutijs vulgaribus videlicet et phisicalibus Addita etiam regula proportionū tam de integris q̄ fractis que vulgo mercatorum regula dicit̄ Quibus habitis. quis modica adhibita diligentia omnem calculandi modum facillime adipisci potest

Fig. 24. Title page of the algorismus of c. 1491

end shows it to be different from the one described by Brunet (edition of 1860). It is probably no. 827 of Hain, and the internal evidence makes 1491 the probable year of the composition or publication. De Morgan says that three editions are known, and Günther on the authority of Chasles gives an edition at Cologne, c. 1510.[504]

This anonymous work is of the class of the arithmetics of Muris, Peurbach, Ciruelo, and other mediæval and early Renaissance writers.

FIG. 25. TITLE PAGE OF CALANDRI

It contains a brief treatment of the 'species' (the fundamental operations) with integers, omitting 'duplatio' and 'mediatio' but including progression and roots. This is followed by a discussion of common fractions ('Algoriſmus nouus de // minutijs vulgaribus'), a single page on sexagesimal fractions ('de minutijs Phiſicalibus'), and a page on proportion. It is one of the first books to identify proportion with the rule of three, or merchants' rule as it was often called. ('De regula proportionum // Sive aliter Regula Mercatorum dicta.') It is not as practical as the elaborate title would seem to indicate.

PHILIPPI CALANDRI. Ed. pr. 1491. Florence, 1491.

A Florentine arithmetician of the fifteenth century.

Title. 'Pictagoras arithmetrice introductor.' (Woodcut of

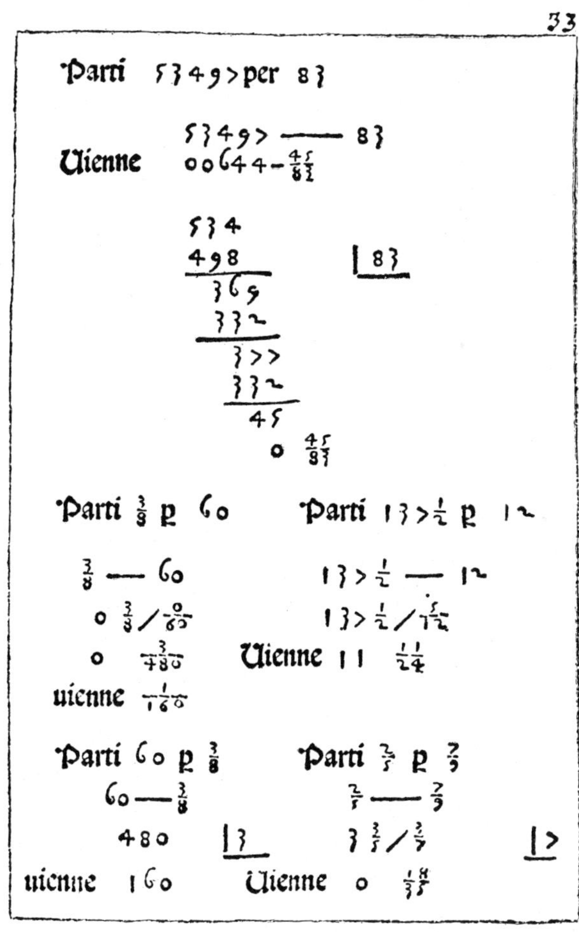

FIG. 26. FROM CALANDRI'S ARITHMETIC

Pythagoras.) (F. 1, v. Fig. 25.) 'Philippi Calandri ad nobilem et ſtudioſuȝ Julia//num Laurentii Medicē de arimethrica opuſculū.' (F. 2, r.)

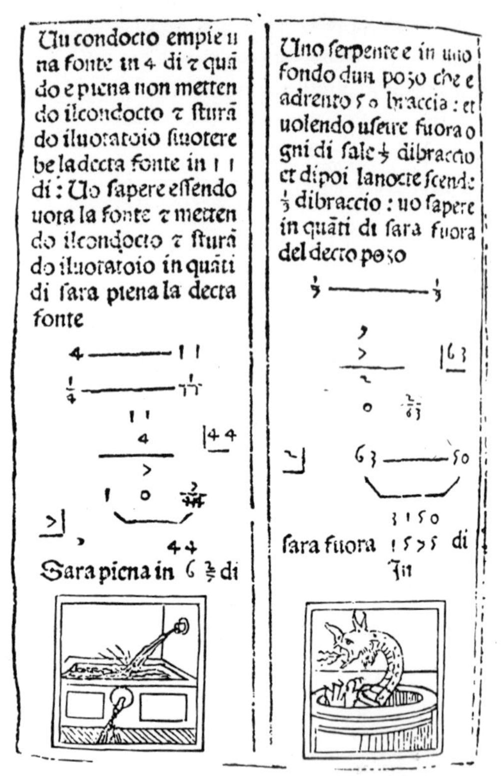

Vn condocto empie u
na fonte in 4 di ⁊ quã
do e piena non metten
do il condocto ⁊ sturã
do il uotatoio si uotere
be la decta fonte in 11
di : Uo ſapere eſſendo
uota la fonte ⁊ metten
do il condocto ⁊ sturã
do il uotatoio in quãti
di ſara piena la decta
fonte

4 —— 11

1/4 —— 1/11

Sara piena in 6 2/7 di

Vno ſerpente e in uno
fondo dun pozo che e
adrento 50 braccia : et
uolendo uſcire fuora o
gni di ſale 1/3 di braccio
et dipoi la nocte ſcende
1/7 di braccio : uo ſapere
in quãti di ſara fuora
del decto pozo

1/3 —— 1/7

63 —— 50

3150

ſara fuora 1575 di

FIG. 27. FROM CALANDRI'S ARITHMETIC

Colophon. 'Impreſſo nella excelſa cipta di Firenze per ß // Lorenzo de Morgiani et Giouanni // Thedeſco da Maganza fi// nito a di primo di // Gēnaio 1491.' (F. 104, v. Fig. 30.)

FIG. 28. FROM CALANDRI'S ARITHMETIC

Description. 8°, 9.8 × 13.2 cm., printed in double columns, each 3.3 × 10.7 cm. to 11.5 cm. 104 ff. unnumb., 9–26 ll. Florence, 1491.

Editions. Florence, 1491, 8° (here described); ib., 1518, printed by Bernardo Zuchetta, 4°.

Of this rare book, the first in De Morgan's list, Mr. Plimpton possesses two copies. It is beautifully printed, and is practical in its presentation of the operations, but traditional in its problems. It is

98

Eglie una torre che e alta 40 braccia τ dap pie uipassa uno fiume che e largho 30 brac cia. uo sapere quanto sara lungha una fune che sia appicata alla ri ua del fiume τ alla ci ma della torre

40 —— 30
40 30
1600
900
la radice di 2500 sara lunga 50 brac cia

Eglie unalbero insu la riua dun fiume elqua le e alto 50 braccia el fiume e largho 30 bra cia τ per fortuna di uē to siruppe intal luogo che lacima dellalbero toccaua lariua del fiu me. Uo sapere quante braccia sene ruppe τ quanto nerimase ritto

50 —— 30
2500 2500 30
100 900
1600
rimase ritto 16 brac cia τ 34 braccia sene ruppe

FIG. 29. FROM CALANDRI'S ARITHMETIC

the first printed Italian arithmetic with illustrations accompanying problems, and the first to give long division in the modern form (Fig. 26) known to the Italian writers by the name 'a danda.' Indeed Calandri gives only the 'a danda' method, omitting the galley form, and is therefore fully a century ahead of his time. De Morgan's statement that he uses a divisor diminished by 1 is incorrect, as will be seen from Fig. 26. Figs. 27 and 29 show that the problems of the cistern, the snail (serpent) in the well, the length of the hypotenuse, and the broken tree were familiar in Calandri's time.

Impresso nella excelsa cipta di Firenze per Lorenzo de Morgiani et Giouanni Thedesco da Maganza finito a di primo di Genaio 1491

FIG. 30. COLOPHON OF CALANDRI

Other works of 1491. Anianus, p. 32, 1488; Boethius, p. 27, 1488; Borghi, p. 16, 1484.

FRANCESCO PELLOS OR PELLIZZATI.

Ed. pr. 1492. Turin, 1492.

A native of Nice, living in the latter half of the fifteenth century.

Title. 'Sen ſegue de la art de arithme-//ticha. et ſemblãt-ment de ieume-//tria dich ho nominatȝ Cõpendiõ // de lo abaco.// 1 2 3 4 5 6 7 8 9 0.' (F. 1, r. Fig. 31.)

Colophon. 'Complida es la opera. ordinada. he condida // Per noble Frances pellos. Citadin es de Niſa. . . . Impreſſo in Thaurino lo preſent cõpendiõ de abaco per mei/ //ſtro Nicolo benedeti he meiſtro Jacobino ſuigo de ſancto ger//mano. Nel anno .1492. ad. Di .29. de ſeptembrio.' (F. 80, r.)

Description. Sm. 4°, 13.8 × 20.9 cm., the text being 9.2 × 15.2 cm. 80 ff. numb., 39 ll. Some of the initials have been inserted by hand, in red. Turin, 1492.

Editions. There was no other edition.

This is one of the rarest arithmetics known to exist. (Brunet, IV, 475; Graesse, *Trésor*, V, 100; Riccardi, I, 2, 256.) The only good description of the work is that given by Boncompagni in the *Atti dell' Accad. Pontif. de' nuovi Lincei*, XVI, 161, 332, evidently after examining this copy, since it bears a note in his handwriting.

Pellos first considers the fundamental operations with integers, following this by a treatment of proportion, square root, and cube root. He then discusses the subject of fractions in much the same order, the rule of three, certain rules relating to weights, time, money, and other measures, and such topics as partnership, barter, interest, alloys, and the rule of false position, single and double. He closes the work with a chapter on mensuration, or as he calls it, 'De la art de ieumentria' ('ieumetria' in the title), and gives a number of interesting woodcuts. The chief interest of the book attaches, however, to the fact that Pellos came very near the invention of decimal fractions, and that he actually used the decimal point as is shown in the illustration (Fig. 32). It cannot be said, however, that he had any conception of the real value of the decimal fraction as such, the first book devoted to this subject being 'La Disme' of Stevin (1585), hereafter described. Pellos simply uses the decimal point to indicate division by some power of ten, writing a common fraction in the quotient. Thus, to divide 425 by 70, Pellos would divide 42.5 by 7, writing the result $6\frac{5}{70}$.

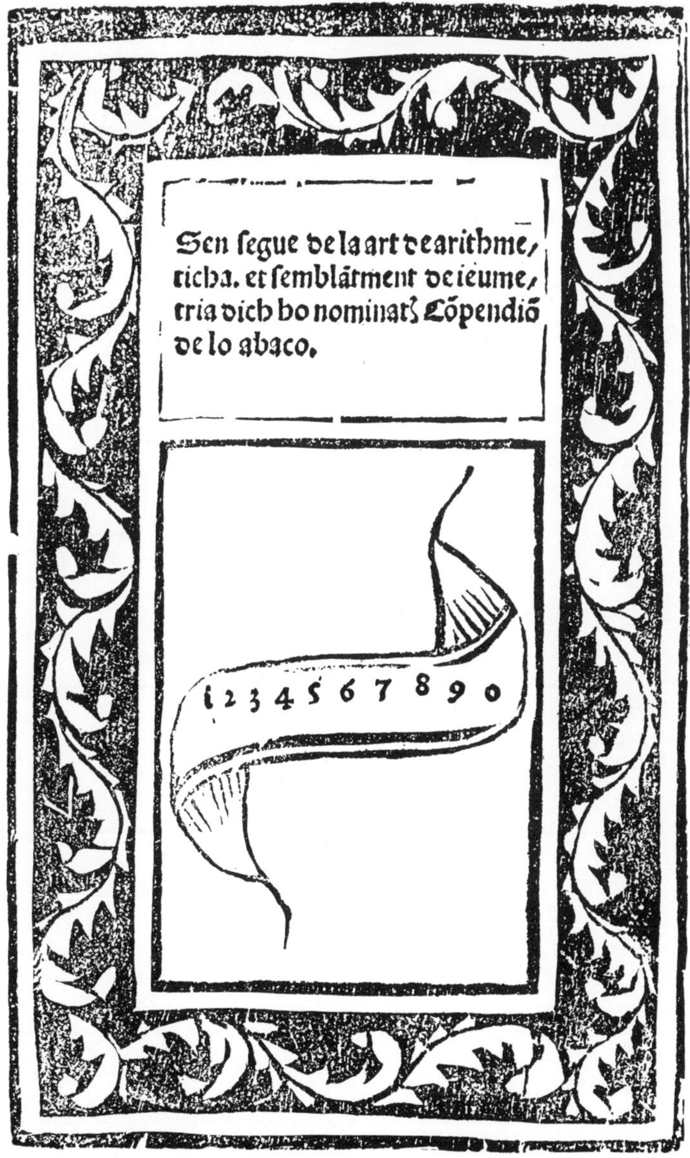
Sen ſegue de la art de arithme,
ticha. et ſemblãtment de ieume,
tria dich ho nominatz Cõpendiõ
de lo abaco.

1 2 3 4 5 6 7 8 9 0

FIG. 31. TITLE PAGE OF PELLOS

11

Partir per 2 0

7 9 6 5 4 8 3 9 . 7

quocient 3 9 8 2 7 4 1 9 $\frac{17}{20}$

Partir per 3 0

5 8 3 6 0 4 . 3

quocient 1 9 4 5 6 4 $\frac{23}{30}$

Partir per 7 0

9 5 3 7 9 1 . 9

quocient 1 3 6 2 5 5 $\frac{69}{70}$

Partir per 1 0 0

6 9 7 6 5 . 8 7

quocient

Partir per 4 0 0

7 8 9 6 5 . 7 3

quocient 1 9 7 4 1 $\frac{173}{400}$

Partir per 3 0 0 0

8 7 6 5 8 . 7 9 1

quocient 2 9 2 1 9 $\frac{1791}{3000}$

Et en aqsta maniera podes fayre abe tons semblas partimes

b iij

FIG. 32. FROM THE PELLOS ARITHMETIC

GEORG VON PEURBACH.

Ed. pr. 1492. Wittenberg, 1534.

PURBACH, PEUERBACH. Born at Peuerbach, Upper Austria, May 30, 1423; died at Vienna, April 8, 1461. He studied under Johann von Gmünden (see p. 117), Nicolaus Cusa (see p. 42), and other great teachers, and later he became professor of mathematics at Vienna, where Regiomontanus (Johannes Müller, of Königsberg) was his pupil. His interests were almost entirely in astronomy.

Title. 'Elemen//ta Arithmetices // Algorithmvs de // numeris integris auctore // Georgio Peurbachio.// De Nvmeris Fractis,// Regulis communibus & // Proporcionibus.// Cum præfatione Philippi // Melanchthonis.// M. D. XXXIIII.' (F. 1, r.)

Colophon. 'Impressvm Vitebergae // per Iosephvm Clvg.// Anno M. D. XXXIIII.' (F. 39, v.)

Description. 8°, 10.8 × 15.6 cm., the text being 6.4 × 11.2 cm. 39 ff. unnumb. + 1 blank = 40 ff., 22–25 ll. Wittenberg, 1534.

Editions. S. l., 'Explicitum est hoc opus anno Christi dom. 1492.' 4°; Vienna, s. a., c. 1500; Leipzig, 1503, 4°; ib., 1507, 4°, and probably ib., 1510 and ib., 1511; Vienna, s. a. but earlier than 1511, 4°; ib., 1511, 4°; ib., 1512; Nürnberg, 1513; Vienna, 1515, 4°; ib., 1520, 4°; Wittenberg, 1534, 8° (here described); ib., 1536, 8°; ib., 1538; Venice, 1539, 8° (see below); Frankfort, 1544. 504

The arithmetic of Peurbach went by various names, as 'Opus Algorithmi,' 'Institutiones in arithmeticam,' 'Elementa arithmetices,' and 'Introductorium in arithmeticam.' It is a brief treatise on the fundamental operations with integers and fractions, and contains a few simple applications. Peurbach was too profound a mathematician to have considered it a work of any importance, but it is probable that he wrote it for the benefit of students who were not yet prepared to take up his work in astronomy. 504

GEORG VON PEURBACH. Ed. pr. 1492. Venice, 1539.

See above.

Title. 'Elementa // Geometriæ ex Evclide // ſingulari prudentia collecta à Ioáne Vo-//gelin profeſſore Mathematico in

//ſchola Viennenſi.// Arithmeticæ practicæ per Georgium // Peurbuchium Mathematicum.// Cum præfacione Philippi // Melanchthonis.' (Woodcut with motto: 'Dissimilivm. Infida. Societas.') (F. 1, r.) The arithmetic of Peurbach begins on f. 32: 'Elementa // Arithmetices.// Algorithmvs de nv-//meris integris, fractis, Regulis // communibus, & de Pro-//portionibus. // Authore Georgio Peurbachio.// Omnia recens in lucem ædita fide & // diligentia ſingulari.// Cum præfatione Philip. Melanth.'

Colophon. 'Venetijs Ioan. Anto. de Nicolinis de Sabio.// Sumptu uero D. Melchioris Seſſæ. Anno // Domini M D XXXVIIII.// Menſe Ianuario.' (F. 67, v.)

Description. 8°, 10.3 × 15.8 cm., the text being 8 × 12.2 cm. 1 f. unnumb. + 68 numb. = 69 ff., 29 ll. Venice, 1539.

See p. 53.

Other works of 1492. Anianus, p. 32, 1488; Boethius, p. 27, 1488.

Works of 1493. Anianus, p. 32, 1488; Anonymous (see Widman), p. 36, c. 1490; Isidorus, p. 8, 1483; Vincent de Beauvais, p. 10.

LUCA PACIUOLO, DE BORGO SAN SEPOLCRO.

Ed. pr. 1494. Venice, 1494.

PACIOLUS, PATIULUS, PACIOLI. Born in Borgo San Sepolcro, Tuscany, c. 1445–1450; died soon after 1509. Not an original mathematician, but the compiler of several works.

Title. See Fig. 33.

Colophon to the part on arithmetic: 'Et ſi ſequenti ꝑti ꝑncipali Geoᵉ. finis decima nouembris īpoſitus fuerit: huic tamen ꝑti: die vigeſi//ma eiuſdem īpoſitus fuit. M°.cccc.lxliiij. Per eoſdem correctorem ⁊ impreſſorem vt i fine Geoᵉ. hr̄.' (F. 232, numb. 224, v.) There is also the following date on f. 1, v., 'M°.ccc.°lxliiij°.xxª. Nouembris.'

Description. Fol., 21.5 × 30.5 cm., the text being 19 × 24.2 cm. 8 ff. unnumb. + 224 numb. = 232 ff. in the part on arithmetic; 76 ff. numb. in the part on geometry; making a total of 308 ff., 56–60 ll. Venice, 1494.

Editions. Venice, 1494, fol. (here described); Toscolano, 1523, fol. (p. 58).

Sũma de Arithmetica Geo- metria Pꝛopoꝛtioni & Pro- poꝛtionalita.

Continentia de tutta lopera.

De numeri e miſure in tutti modi occurrenti.
Pꝛopoꝛtioni e ꝓpoꝛtiõalita anotitia del.5º de Eucli
de.e de tutti li altri ſoi libꝛi.
Chiaui ouero euidentie numero.13.ꝑ le q̃tita continue ꝓpoꝛtiõali del.6º.e.7º de Euclide extratte.
Tutte le ꝑti delalgoꝛiſmo:cioe releuare.ꝑtir.multi-
plicar.ſũmare.e ſotrare cõ tutte ſue ꝓue ĩ ſani e rot-
ti.e radici e pꝛogreſſioni.
De la regola mercanteſca ditta del.3.e ſoi fõdamen-
ti con caſi exemplari per cº mº 8.Ꝯ guadagni:perdi
te:tranſpoꝛtationi:e inueſtite.
Partir.multiplicar.ſummar.e ſotrar de le pꝛopoꝛtio
ni e de tutte ſoꝛti radici.
De le.3.regole del catayn ditta poſitiõe e ſua oꝛigie.
Euidentie generali ouer concluſioni nº66.abſoluere
ogni caſo che per regole.oꝛdinarie nõ ſi podeſſe.
Tutte ſoꝛte binomii e reciſi e altre linee irratiõali del
decimo de Euclide.
Tutte regole de algebꝛa ditte de la coſa e loꝛ fabꝛi-
che e fondamenti.
Compagnie ĩ tutti modi.e loꝛ partire.
Socide de beſtiami.e loꝛ partire
Fitti:peſciõi:cottimi:liuelli:logagioni:egodimenti.
Baratti ĩ tutti modi ſemplici:compoſti:e col tempo.
Cambi reali.ſecchi.fittitii.e di minuti ouer comuni.
Meriti ſemplici e a capo danno e altri termini.
Reſti.ſaldi.ſconti.de tempo e denari ela recare a un
di piu partite
Oꝛi.argẽti.eloꝛo affinare.ecarattare.
Molti caſi e ragioni ſtraoꝛdinarie varie e diuerſe a
tutte occurentie commo nella ſequente tauola ap-
pare oꝛdinatamente de tutte.
Oꝛdine a ſaper tener ogni cõto e ſcripture e del qua
derno in vinegia.
Tariffa de tutte vſançe e coſtumi mercanteſchi in tut
to el mondo.
Pꝛatica e theoꝛica de geometria e de li.5.coꝛpi regu-
lari e altri dependenti.
E molte altre coſe d grandiſſimi piaceri e frutto cõmo diffuſamente per la ſequente tauola appare.

Fig. 33. Title page of the 1494 Paciuolo

This volume, the first great general work on mathematics printed, includes treatises on arithmetic, algebra, and geometry, each being considered from a somewhat scientific rather than practical standpoint. The arithmetic, for example, gives the various methods in multiplication (see Fig. 34) and division, instead of emphasizing the one or two most prominent in business circles. In the same way Paciuolo's treatment of

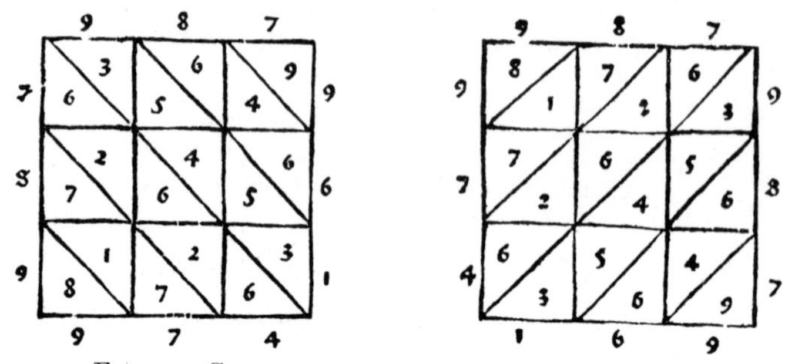

FIG. 34. GELOSIA MULTIPLICATION, 1494 PACIUOLO

the rule of three, the rule of false ('El cataym'), partnership, pasturage, barter, exchange, and interest, while nominally practical, was too elaborate for the mercantile schools. His was the first printed work to illustrate the finger symbolism of number (Fig. 35). Paciuolo copies without hesitation, practically verbatim, from the work of Chiarini (p. 10), and doubtless laid under contribution, after the manner of his time, various other works of his predecessors. In algebra he used the common symbolism of the time for the unknown quantities and for roots, but he made use of no symbols of operation. This part of the treatise relates chiefly to surd numbers. In geometry he follows Euclid's Book I very closely. but departs quite radically from the subsequent books. The work had a great influence on subsequent writers, including the English Tonstall (p. 132). Paciuolo's training had fitted him to write a treatise of this nature. He had been a tutor in the family of a Venetian merchant, had traveled extensively, had come in contact with practical mathematicians, and had studied the ancient mathematics in the cloisters; and traces of all these influences are seen in his work. In 1497 Paciuolo wrote at Milan a work entitled 'Divina proportione,' which was published at Venice in 1509 (p. 87). He also published an edition of Euclid at Venice in 1509. 504

Other works of 1494. Albert of Saxony, p. 9, 1478; Anianus, p. 32, 1488.

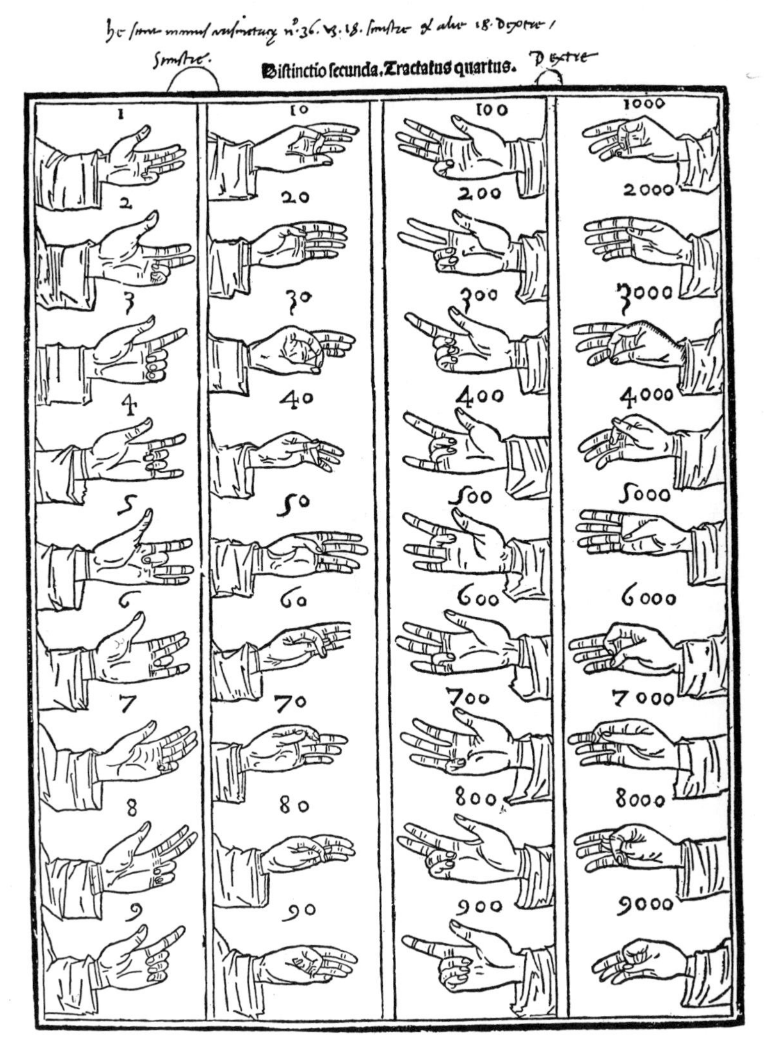

FIG. 35. FINGER SYMBOLISM, 1494 PACIUOLO

LUCA PACIUOLO, DE BORGO SAN SEPOLCRO.

Ed. pr. 1494. Toscolano, 1523.

Title. 'Summa de // Arithmetica geo//metria. Proportioni: et proportionalita: // Nouamente impressa In Toscolano su la riua dil Benacense et // vnico carpionista Laco: Amenissimo Sito: de li antique ꝛ // euidenti ruine di la nobil cita Benaco ditta illustra-//to: Cum numerosita de Imꝑatorij epithaphij // di antique ꝛ perfette littere sculpiti do-//tato: ꝛ cuꝫ finissimi ꝛ mirabil co-// lone marmorei: inumeri // fragmenti di alaba-//stro porphidi ꝛ serpentini. Cose certo // lettor mio diletto oculata fi-//de miratu digne sot-//terra se ritro//uano.// Continentia de tutta lopera:' // (The rest of the title page is substantially identical with that of the 1494 edition.) (F. 1, r.)

Colophon to the part on arithmetic: '¶Et si sequēti ꝑti ꝑnci-pali Geoᵉ. finis decima nouembris īpositus fuerit: huic tamen ꝑti die vigesima // eiusdē impositus fuit .M°.ccccc.xxiij. Per eosdem correctorē impressorem vt in fine Geoᵉ. hētur.' (F. 232, v.).

Description. Fol., 21.1 × 30.4 cm., the text being 18.2 × 23.5 cm. 9 ff. unnumb. + 223 numb. = 232 ff. Toscolano, 1523.

Bound with this is the 'Tractatus Geometria. Pars secunda principalis huius operis ꝛ, primo eius diuisio.' This part of the book contains 75 ff., besides the index.

Editions. See p. 54. De Morgan has shown (p. 2), that there are slight differences between the copies printed in 1523, proving that a second impression was necessary in that year.

See p. 56, and Fig. 36.

PEDRO SÁNCHEZ CIRUELO.

Ed. pr. 1495. Paris, 1505.

Born at Daroca in Aragon, c. 1470; died at Salamanca in 1560. One of the most learned men of his time. He was professor of philosophy at Alcalá.

Title. 'Tractatus Arithmethice // Practice qui dicitur // Algo-rismus.' (A large woodcut with the initials D. R. and the inscription 'Alaventvre. Tout Vient. Aponit. Qvi. Pevt. Atendre. Denis. Roce.') (F. 1, r.)

Ad illustrissimũ Principem Gui. Ubaldũ Urbini Ducem Montis feretri: ac Durantis Comitem. Grecis latinisq; litteris Ornatissimum: & Mathematice discipline cultorem ferventissimum: Fratris Luce de Burgo sancti Sepulchri: Ordinis minorum: & sacre Theologie Magistri. In artem Arithmetice & Geometrie Prefatio

A quantita Magnanimo Duca: e si nobile & excellente cosa: che molti philosophi per questo giudicata lanno alla substantia para: e concessa coeterna. Peroche hanno cognosciuto per vero modo alchuna cosa in rerum natura sensa lei nõ potere existere. Per laqual cosa de lei intendo (con ladiuto de colui che li nostri sensi regge) tractarne: non che per altri priscihi e antichi philosophi nonne sia copiosamente tractato: e in Theorica e practica. Ma perche lor dicti gia alli tempi nostri sonno molto obscuri: e da molti male apresi: e alle practiche vulgari male applicati: dilche in loro operationi molto variano: e con grãdi e laboriosi affanni mettano in opera: si de numeri cõmo de misure: onde di lei parlando non intendo se non quanto che alla practica e operare sia mestiero: mescolandoci secondo i luoghi oportuni anchora la theorica: e causa de tale operare: si de numeri cõmo de geometria. Ma prima accio meglio quello che sequita se habia apprendere: essa quantita diuidiremo secondo el nostro proposito: e diuidendola a dlaschuno suo membro assegnaremo sua propria e vera diffinitione e descriptione. E allora poi seguira quello che Aristotile dice in secondo posteriorum. Tunc enim maxime scitur aliquid cum habetur suum quid est &c.

Diffinitiones/et diuisio discrete/et continue quantitatis.

Articulus primus. Prime distinctionis.

Dico adoncha. La quantita essere immediate bimembre: cioe continua e discreta. La continua e quella le cui parti sonno copulate e gionte a certo termine commune: como sonno legni: ferro: e sassi: &c. La discreta ouerámente numero e quella le cui parti non sonno gionte ad alchuno termine cõmune: cõmo e. 1. 2. 3. &c. Di che prima de la discreta: cioe del numero: e poi de la continua: cioe geometria: quãto allo intento aspecta chiaramente tractaremo.

Diffinitio numeri proprijssima. Articulus secundus.

Numero: è (secondo ciaschuno philosophante) vna moltitudine de vnita cõposta: & essa vnità non è numero: ma ben principio de ciaschuno numero: e de quella mediantè laqle ogni cosa e dicta essere vna. E secondo seuerino Boetio in sua musica: è la vnita ciaschun numero in potentia: & passiz in la sua arithmetica regina e fondamento dogni numero lapella. Laqual piu magnificandola in le cose naturale disse in quello che fa de vnitate & vno. Omne qd est: ideo est: quia vnum numero est. E sie anchora el numero in infiniti membri diuiso: per quello che esso Aristotile dice: cioe. Siquid infinitum est: numerus est. E per la tersa petitione del septimo de Euclide: la sua serie in infinito potere procedere: & quocũq; numero dato: dari potest maior: vnitatem addendo. Ma noi pigliaremo quelle parte a noi piu note e accõmodate. E pero dico con glialtri alchuno essere primo: et e quello che solo da la vnita e numerato: e non ha altro numero: che integralmente aponto lo parta. Altro è ditto cõposto: et è quello che da altro numero è mesurato: ouero numerato. Exemplum primi. Cõmo. 3. 7. 11. 13. e. 17. &c. Exẽplũ secũdi. Cõmo. 4. chel doi lo mesura e numera: e. 8. chel. 2. e. 4. El. 12. 14. 18. e simili: tutti sonno ditti numeri composti: nõ solo che cõstino

A

FIG. 36. FIRST PAGE OF TEXT, 1523 PACIUOLO

Colophon. 'Arithmetice practice ſeu Algoriſmi tractatus a Petro ſanchez // Ciruelo nouiter compilatus Explicit Impreſſus Pariſius In // Bellouiſu. Anno dn̄i, 1505. Die .29. Aprilis.' (F. 14, r.)

Description. 4°, 13.2 × 18.4 cm., the text being 9.7 × 14 cm. 14 ff. unnumb., 35–39 ll. Paris, 1505.

Editions. Paris, 1495, 4°; ib., 1505, 4° (here described); ib., 1509, 4°; ib., 1513, 4° (see below). Ciruelo also wrote a 'Cursus quattuor mathematicarvm artiū liberaliū,' Paris, 1516; ib., 1523; ib., 1526; ib., 1528; Alcalá, 1516, fol.; ib., 1518; ib., 1523, fol.; ib., 1526; ib., 1528. He also edited Bradwardin's arithmetic, Paris, 1495; ib., 1502. 505

Ciruelo treats very briefly of the fundamental operations with integers, common fractions, and denominate numbers. Following the Spanish custom he uses *cuento* for million. There is little that is noteworthy in the book, and, like Peurbach, Ciruelo could not have taken his contribution to algorism very seriously.

Other works of 1495. Anianus, c. 1495, p. 32, 1488. There also appeared in 1495 the treatise of Herodianus, 'De notis Græcorum Arithmeticis Græce,' Venice. Of this work there was an edition in 1525, and a Latin edition published at Basel in 1600, but it hardly deserves to be classed as an arithmetic.

PEDRO SÁNCHEZ CIRUELO. Ed. pr. 1495. Paris, 1513.

See p. 58.

Title. 'Tractatus arithmetice // Practice qui dicitur algoriſmus.// (A large woodcut with the name Jehanlambert.) Venundantur Parrhiſijs a Johāue Lā//berto eiuſdem ciuitatis bibliopola in ſtē-//mate diui claudij manente iuxta gymna-//ſium coquereti.' (F. 1, r.)

Colophon. '¶Arithmetice practice ſeu Algoriſmi tractatus a Petro ſan//chē Ciruelo nouiter compilatus Explicit Impreſſus Pariſius // per Anthonium Auſſourt. Pro Johāne Lamberto eiuſdem ci-//uitatis bibliopola in ſtemmata diui claudij manente iuxta gym-//naſium coquereti. Anno dn̄i .1513. Dievero .21. menſis Martij.' (F. 12, r.)

Description. 4°, 13.1 × 19 cm., the text being 9.6 × 14.8 cm. 12 ff. unnumb., 43–46 ll. Paris, 1513.

Editions. This is the fourth edition (see p. 60), and is an exact reprint of the 1505 copy, save as to spelling and pagination. As a piece of typography it is much better than its predecessors.

THOMAS BRADWARDIN. Ed. pr. 1495. Paris, c. 1510.

BRAGWARDINE, BRANDNARDINUS, BREDWARDYN, BRADWARDYN, DE BRADWARDINA, DE BREDWARDINA. Born at Hertfield (Hartfield) in the diocese of Chichester, c. 1290; died at Lambeth, August 26, 1349. On account of his learning he was called 'Doctor profundus.' He was professor of theology at Oxford, and died as Archbishop of Canterbury. He wrote four works on mathematics.

Title. 'Arithmetica thome brauardini.//Olivier Senant// Venum exponuntur ab Oliuiario ſenant in vico diui Jacobi ſub si-//gno beate Barbare ſedente.' (F. 1, r.)

Colophon. 'Explicit arithmetica ſpeculatiua thōe brauardini bn̄ re-//uiſa et correcta a Petro ſanchez Ciruelo aragonenſi ma//thematicas legēte Pariſius, īpressa ꝑ Thomā anguelart.' (F. 6, v.)

Description. 4°, 20.3 × 27.8 cm., printed in double columns, each being 6.8 × 19.8 cm. 6 ff. unnumb., 61 ll. Paris, c. 1510.

Editions. Paris, 1495, 4°; ib., 1496, fol.; ib., 1498, 4°; ib., s. a. (c. 1500); ib., 1502, 4°; Valencia, 1503, fol.; Paris, 1504; ib., 1505, 4°; ib., s. a. (c. 1510, here described); ib., 1512; ib., 1530; Wittenberg, 1534, 8°; ib., 1536, 8°. His 'Tractatus de proportionibus' appeared in several editions, as follows: Paris, 1495; Venice, 1505, fol.; Vienna, 1515 (p. 117); and a commentary by Vittori appeared at Bologna in 1506. Some of these editions contained two or three of his works in one volume, and it is probable that his arithmetic and his treatise on proportion appeared in other editions than those mentioned. [505]

Bradwardin was one of the earliest English mathematicians after Bæda and Alcuin. His arithmetic is of the Boethian type, relating to the theory of numbers. He gives much attention to the ancient theory of ratios and to figurate numbers.

BOETHIUS, JORDANUS NEMORARIUS, AND FABER STAPULENSIS. Ed. pr. 1496. Paris, 1496.

For the biography of BOETHIUS see p. 25.

JORDANUS NEMORARIUS (JORDANUS DE SAXONIA) was born at Borgentreich, in the diocese of Paderborn, and died in 1236. He studied at Paris and was the greatest mathematician of his time save Leonardo Fibonacci of Pisa.

JACOBUS FABER STAPULENSIS (JACQUES LE FÈVRE D'ESTAPLES) was born at Estaples, near Amiens, in 1455, and died at Nérac in 1536. He was a priest, vicar of the bishop of Meaux, lecturer on philosophy at the Collège Lemoine in Paris, and tutor to Charles, son of François I. He wrote on philosophy, theology, and mathematics.

Title. See Fig. 37.

Description. Fol., 20 × 29 cm., the text being 13.9 × 27.2 cm. 72 ff. unnumb., 60–63 ll. Paris, 1496.

Editions. The arithmetic of Jordanus went through various editions as follows: Paris, 1496, fol. (here described); ib., 1503, fol.; ib., 1507, fol. (p. 65); ib., 1510, fol.; ib., 1514 (p. 65). Rogg speaks of an edition of 1480, but I do not know of it. Jordanus also wrote an 'Algorithmus demonstratus,' published anonymously at Nürnberg in 1534, 4°. De Morgan, following Schönerus, attributed it to Regiomontanus, but the evidence shows that he only revised it and it may be due to Jordanus. His 'De Ponderibus,' edited by Apianus, was published at Nürnberg in 1533, and at Venice in 1565. An interest attaches to the 1496 edition in that it is the first printed work with which a Scotchman's name is connected, the printer being David Lauxius of Edinburgh, then working in Paris. 495;502

The greater part of this volume is devoted to the ten books on arithmetic by Jordanus Nemorarius, with the commentary of Jacobus Faber Stapulensis. The work of Jordanus is similar to that of Boethius, and is concerned only with theory of numbers. In particular, the Greek theory of ratios, as elaborated during the Middle Ages, is extensively treated.

The second part consists of the work of Jacobus Faber Stapulensis on music, in four books.

The third part is the Epitome of the Arithmetic of Boethius: '¶Jacobi Fabri Stapulenſis Epitome in duos libros Arithmeticos // diui Seuerini Boetij ad Magnificum dn̄um Joannem Stephanum // Ferrerium Epiſcopum Verſellenſem.'

The fourth part, consisting of four and a half pages, is a description of the arithmetical game of Rithmimachia, possibly by Shirewode (John Shirwood, Bishop of Durham, who died in 1494), but usually ascribed to Faber Stapulensis. An edition appeared at Erfurt in 1577, 4°.

In hoc opere contenta.
Arithmetica decem libris demonstratã
Musica libris demonstrata quattuor
Epitome i libros arithmeticos diui Seuerini Boetij
Rithmimachie ludus q̃ & pugna nũeroꝝ appellatꝰ

G. Gonterius Cabilonensis: in
laudẽ Arithmetices & Musices.

Tempore iam multo docte latuere sorores:
Quas retinet comites flaua minerua suas.
Nũc placide terras post tempora multa reuisunt.
Grata quoq; ante alias Gallica terra placet.
His olim celebris fuit omnis Acaica tellus.
Pythagora patriam diffugiente samon.
Hellada nunc linquunt: et doctas palladis vrbes.
Sequaniosq; petunt/parhisiosq; lares.
Hec venit omnimoda numerorum cincta cãterua.
Atq; docet numeris quidquid in orbe situm est.
Altera dulcisono cantu/fidibusq; canoris
Edomuisse viros traditur atq; feras:
Que sua Pierijs tenet vnica nomina musis
Q̃ nichil hac musis gratius esse solet.
Attamen artificem stapule misere marine:
Qui leta hoc studijs fronte dicaret opus.
Hoc solum studium atq; hec illi cura: iuuare/
Irritus et ne sit/dispereatq; labor.

Fig. 37. Title page of the 1496 Boethius

Other works of 1496. Albert of Saxony, p. 9, c. 1478; Bradwardin, p. 61, 1495; Z. Lillius, 'De origine et laudibus scientiarum,' Florence, 4° (one page on arithmetic). There was also published at Paris, s. a.

(c. 1496), an anonymous treatise entitled 'De arte numerādi cōpediū putile īcipit feliciter. (Q)uoniam rogatus a pluribʒ compēdium artis numerandi ac breuē tractalutu . . .'

Works of 1497. Boethius, p. 27, 1488 (the colophon of the 1497

FIG. 38. RITHMIMACHIA, 1496 BOETHIUS

edition has the incorrect date Mccccl xxxvii); Suiseth, 1497 and 1498, p. 10, c. 1480.

Works of 1498. Anianus, p. 32, 1488; Bradwardin, p. 61, 1495; Chiarini, p. 11, 1481; Anonymous, 'Enchiridion sive tractatus de numeris integris, fractis,' etc., 4° (doubtless the work published at Deventer in 1499, p. 67).

BOETHIUS, JORDANUS NEMORARIUS, AND FABER STAPULENSIS. Ed. pr. 1496. Paris, 1507.

See p. 62.

Title. 'In hoc opere contenta.// Arithmetica decem libris demonſtrata // Muſica libris demonſtrata quattuor // Epitome ī libros arithmeticos diui Seuerini Boetij // Rithmimachie ludus q̄ ꝛ pugna nūeroꝝ appellat.' (F. 1, r.)

Colophon. '℄Impreſſum Pariſij in officina Henrici ſtephani e regione Schole decretorum ſita.// Anno Chriſti ſiderum conditoris 1507. Decimo die Nouembris.' (F. 78, r.)

Description. 8°, 20 × 26.6 cm., the text being 18.1 × 21.5 cm. 78 ff. unnumb. + 1 blank = 79 ff., 60 ll. Paris, 1507.

Editions. See pp. 27, 62, for the editions of Boethius and Jordanus.

Like most of the works from the press of Stephanus, this book is beautifully printed. It contains the 'Elementa Arithmetica' of Jordanus, with the demonstrations of Faber Stapulensis, the Epitome of Boethius by Faber Stapulensis, the 'Rithmimachia,' the commentary on Sacrobosco's astronomy by Faber Stapulensis, and the first four books of Euclid 'a Boetio in latinum tranſlate.'

BOETHIUS, JORDANUS NEMORARIUS, AND FABER STAPULENSIS. Ed. pr. 1496. Paris, 1514.

See p. 62.

Title. 'In hoc opere contenta // Arithmetica decem libris // demonſtrata.// Muſica libris demōſtrata // quatuor.// Epitome in libros Arith-//meticos diui Seuerini // Boetij.// Rithmimachie ludus qui // et pugna numerorū ap-//pellatur.'// (Surrounded by an elaborate woodcut with the following wording: 'Haecsecvndariaes//tetcastigat//issimaexofficina//aemissio' //) '℄Hęc ſecundaria ſuperiorū operum æditio // venalis habetur Pariſijs: // in officina Henrici Stephani e regione ſchole Decretorum.' (F. 1, r.)

Colophon. '℄Has duas Quadriuij partes et artium liberalium precipuas atq; duces cū quibuſdam ammini-//cularijs adiectis: curauit ex ſecunda recognitione vna formulis emēdatiſſime mandari ad ſtudiorum // vtilitatem Henricus Stephanus ſuo grauiſſimo labore et ſumptu Parhiſijs Anno ſalutis domini: // qui omnia in numero atqȝ harmonia formauit 1514. abſolutumqȝ reddidit eodē anno: die ſeptima // Septembris / ſuum laborem vbicunqȝ valet ſemper ſtudioſis deuouens.' (F. 71, v.)

Description. Fol., 19.8 × 28.4 cm., the text being 17.7 × 26.8 cm. 71 ff. unnumb., 62 ll.

Editions. See p. 62. This edition is practically identical with that of 1496. It is the second edition of this combination of works and the fourth of Faber's Epitome.

MARTIANUS MINEUS FELIX CAPELLA.

Ed. pr. 1499. Vincenza, 1499.

Flourished c. 475. He was probably born at Carthage, and he lived at Rome.

Title. See Fig. 39.

Colophon. 'Martini Capellæ Liber finit: Impreſſus Vincentiæ Anno Salutis // M.ccccxcix. xvii. Kalendas Ianuarias per Henricum de Sancto // Vrſo Cum gratia & priuilegio decem annorum: ne imprima-//tur neqȝ cum Commentatiis: neqȝ ſine: & cætera: quæ in ipſo pri//uilegio continentur. Laus Deo & beatæ Virgini.' (F. 123, v.)

Description. Fol., 20.5 × 30.3 cm., the text being 12.1 × 22.3 cm. 124 ff. unnumb., 37 ll. Vincenza, 1499.

Editions. Vincenza, 1499, fol. (here described); Modena, 1500, fol. (p. 67); Vienna, 1516, fol.; Basel, 1532, fol.; Leyden, 1539; Basel, 1577, fol. (p. 68); Leyden, 1592, 8° (p. 68); ib., 1599, 8°; and later. An Italian translation was published at Mantua in 1578. For bibliography, see Boncompagni's *Bulletino*, XV, 506.

This work is a medley of prose and verse, and forms a kind of encyclopedia of the arts and sciences as known for about a thousand years. It was highly esteemed in the Middle Ages as a textbook. The seventh book is on the Greek theory of arithmetic. It treats of the various classes of numbers, such as plane and solid, and mentions the supposed mysteries of the smaller numbers, the monad suggesting one God, the dyad good and evil, the triad the Trinity, and so on.

Other works of 1499. Boethius, p. 27, 1488; Suiseth, p. 10, c. 1480; Anonymous, 'Enchiridion Algorismi sive tractatus de numeris integris,' Deventer, 4° (p. 64, 1498); George of Hungary, 'Arithmetica summa tripartita,' s. l., reprinted at Budapesth in 1894.

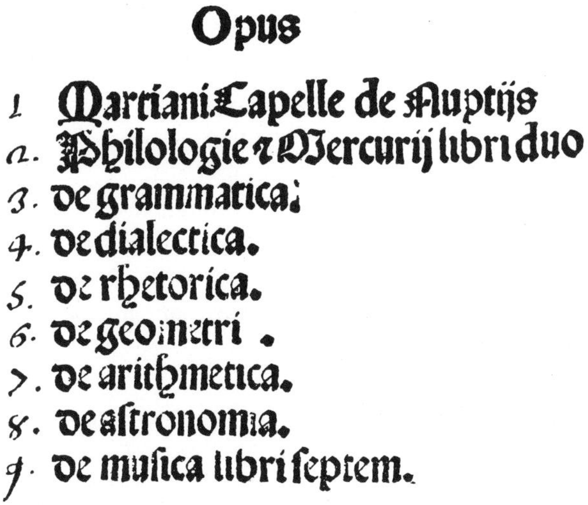

Opus

1 Martiani Capelle de Nuptijs
2. Philologie & Mercurij libri duo
3. De grammatica;
4. De dialectica.
5. De rhetorica.
6. De geometri .
7. De arithmetica.
8. De aſtronomia.
9. De muſica libri ſeptem.

FIG. 39. TITLE PAGE OF THE 1499 CAPELLA

MARTIANUS MINEUS FELIX CAPELLA.

Ed. pr. 1499. Modena, 1500.

See p. 66.

Title. 'Opvs.// Martiani Capellæ // de nvptiis phi//lologiæ et //mercvrii//liberi//dvo.//De gramatica. Liber. Tertius.//De dialectica. Liber. Quartus.//De Rhetorica. Liber. Quintus.//De geometria. Liber. Sextus.//De Arithmetica. Liber. Septimus.//De aſtronomia. Liber. Octauus.//De muſica. Liber. Nonus.' (F. 1, r.)

Colophon. 'Martiani Capellæ Liber finit. Impreſſus Mutinæ. Anno Salutis. M.//CCCCC. Die .XV. Menſis Maii. Per Dionyſiu. Berthocum.' (F. 100, r.)

Description. Fol., 20 × 29 cm., the text being 13.7 × 24 cm. On f. 69, v., the part on arithmetic begins, and occupies 10 ff. 100 ff. in the entire book, unnumb., 42 ll. Modena, 1500.

MARTIANUS MINEUS FELIX CAPELLA.

Ed. pr. 1499. Basel, 1577.

See p. 66.

Title. 'Isidori // Hispalensis // Episcopi // Originum libri viginti // ex antiquitate eruti.// Et // Martiani Capellæ // De nuptijs Philologiæ & Mercurij // Libri nouem.// Vterque, præter Fulgentium & Veteres Grammaticos, va-//rijs lectionibus & ſcholijs illuſtratus // Opera atq̃ Industria // Bonaventvrae Vvlcanii Brvgensis.// Cum gratia & priuilegio Caeſareae Maieſtatis. // Basileæ,// per Petrvm Pernam.' (F. 1, r.)

Description. Fol., 20.5 × 31 cm., printed in double columns, each being 8.2 × 24.4 cm., 7 ff. + 240 columns + index + 550 columns. The part on arithmetic in the work of Capella begins in column 155 and covers 12 pp., or as here numbered 24 columns, 60 ll. Basel, 1577.

Editions. See p. 66.

This edition includes the works of both Isidorus (p. 8) and Capella, and is an excellent specimen of printing. One interesting feature of Capella's work is the evidence that it gives of the use of the abacus in the fifth century. We are still quite uncertain as to the history of this method of calculating in the centuries following Capella.

MARTIANUS MINEUS FELIX CAPELLA.

Ed. pr. 1499. Leyden, 1592.

See p. 66.

Title. 'M. Capella.// Martiani // Minei Capellæ // Carthaginensis // de Nvptiis Philolo-//giæ, & ſeptem artibus // Liberalibus // Libri Novem // optime castigati.// Lvgdvni,// Apud Bartholomæum Vincentium.// 1592.' (P. 1.)

Colophon. 'Lugduni,// Excvdebat // Stephanus Seruain.// 1592.' (P. 416.)

Description. 8°, 10 × 15.7 cm., the text being 7.1 × 12.8 cm. 4 pp. blank + 18 unnumb. + 396 numb. = 418 pp., 30 ll. Leyden, 1592.

Editions. See p. 66.

Algorithmus linealis cũ pulchris cõditõibus Regule detri: septẽ fractionũ: reglis socialibus. et semp exẽplis idoneis Recte sicut in scolis Nurnbergeñ. arithmetricorũ docet In florentissimo studio Lipczensi nup editus Nõ minus litteris eruditis q̃ Mercatoribus utilis et maxime incipientibus.

Lectori

Aurea succincte pateat tibi regula detri
Frangere quo valeas queq3 minuta vafer
A socijs dictas quo possis prẽdere normas
Huius vilescant non tibi dona libri
Hijs nurnberga nitet numerandi insignis ab arte
Huic arti multum contulit illa boni

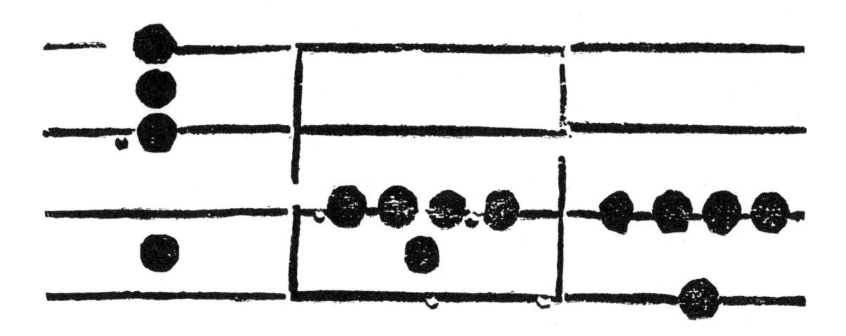

FIG. 40. TITLE PAGE OF LICHT

BALTHASAR LICHT.

Ed. pr. 1500. Leipzig, s. a. (1500).

A German Rechenmeister of c. 1500.

Title. See Fig. 40.

Colophon. 'Impreſſum Lipczk per Melchiar Lotter.' (F. 15, v.)

Description. 4°, 14 × 19.9 cm., the text being 9.3 × 16.2 cm. 15 ff. unnumb., 22–37 ll. Leipzig, s. a. (1500).

Editions. Leipzig, s. a. (1500), here described; ib., 1509; 1513; Leipzig, 1515, 4°, which may be the 'Algorithmus linealis,' s. a. and 1505, by Lotter, referred to by De Morgan (p. 101). There was also an 'Algorithmus linealis, Impressum Lipzik per melchiorem Lotter Anno xc̄,' probably printed in 1490, Lotter having printed in Leipzig from 1490 to 1512. (See contra, *Abhandlungen*, V, 154, n., 152, n., and cf. Widman, c. 1490, p. 44.) On f. 1, v., the dedicatory epistle closes with the words 'Vale ex noſtra academia Lyptzen̄ Anno 1500,' which throws much doubt on the conjecture that Licht's work appeared earlier.

This is a brief treatise on the line abacus, one of the earliest of the type represented also by Huswirt (see p. 73).

LEONARDUS PORTIUS.

Ed. pr. c. 1500. S. l. a. (Venice ?, c. 1500).

A Venetian jurist of the fifteenth century.

Title. 'Leonardi // de Portis ivrisconsvlti Vi//centini de sestertio pe//cvniis ponderibvs et // mensvris antiqvis // libri dvo.' (F. 1, r. See Fig. 41.)

Description. 4°, 14.4 × 19.8 cm., the text being 10 × 14.8 cm. 37 ff. unnumb., 30 ll. Venice (?), c. 1500.

Editions. S. l. a. (Venice ?, c. 1500, here described); Florence, 1514 (?); Basel, 1520, 4°; ib., 1530, 8°. 498; 505

A work on ancient measures, using the Roman numerals throughout, except in the index. Such treatises are of value in studying the history of arithmetic, but are not, in general, included in the bibliographical lists of this work.

Other works of c. 1500. Anianus, c. 1500, p. 32, 1488; Boethius, c. 1500, p. 27, 1488; Bradwardin, c. 1500, p. 61, 1495; Capella, 1500, p. 66, 1499; Peurbach, c. 1500, p. 53, 1492. Widman, 1500, p. 37, 1489. There also appeared about this time, s. l. a., an anonymous 'Algorithmus minutiarum vulgarium,' blackletter, 4° (Libri, 1861 cat., 483), and an anonymous 'Ars numerandi,' 5 ff., 4°, a title given to several books of this period (see p. 23), including 'De arte numerandi siue arismetice (perfectionis) summa quadripartita' (*Abhandlungen*, I, 24; Brunet, *Man.*, 6 (1), 458).

LEONARDI
DE PORTIS IVRISCONSVLTI VI
CENTINI DE SESTERTIO PE
CVNIIS PONDERIBVS ET
MENSVRIS ANTIQVIS
LIBRI DVO.

FIG. 41. TITLE PAGE OF LEONARDUS PORTIUS

GEORGIUS VALLA. Ed. pr. 1501. Venice, 1501.

Born at Piacenza in 1430; died at Venice in 1499. He was a physician and philologian.

Title. 'Georgii Vallae Placentini viri cla-//riss. de expetendis, et fvgiendis // rebvs opvs, in qvo haec // continentvr.// De arithmetica libri .iii. ubi quædam a Boetio prætermiſſa tractantur.// De Muſica libri .v. ſed primo de inuentione, & commodiatate eius.// De Geometria libri .vi. in quibus elementorum Euclidis difficultates omnes fere // exponuntur, ubi etiā de Mechanicis ſpiritalibus, Catoptricis, ac Opticis, deq; // quadrato circuli habetur tractatus.// De tota Aſtrologia libri .iiii. in qua fabrica,uſusq́; aſtrolabi exaratur, & quæ ſi-//gnorum in exhibendis medicaminibus ſit habenda obſeruatio.// De Phyſiologia libri .iiii. ubi &

Metaphyſices q̄dā lectu q̃ digniſs. utiliſſimaq́;.// De Medicina libri .vii. ubi de ſimplicium natura per ordinem litterarum.// Problematum liber unus.// De Grammatica libri .iii.// De Dialectica libri .iii.// De Poetica liber unus.// De Rhethorica libri .ii.// De Morali Philoſophia liber unus.// De Oeconomia, ſiue adminiſtratione domus libri .iii. in quibus de Architectu-//ra, req́; ruſtica eſt locus.// Politicon unicum uolumen, ubi de iure ciuili, ac pontificio primum, Mox de le//gibus in uniuerſum, Inde de re militari agitur.// De Corporis commodis, & incommodis libri .iii. quorum primus totus de ani-//ma, Secūdus de corpore, Tertius uero de urinis ex Hippocrate, ac Paulo ægi-//neta, deq́; Galeni quæſtionibus in Hippocratem.// De Rebus externis liber unus, ac ultimus, ubi de Gloria, Amplitudine, & cæte-//ris huiuſmodi.// Hæc ſummatim, ſed inſunt, & alia plurima,quæ legēdo licet cognoſcere.' (F. 1, r.)

Colophon. 'Venetiis, in aedibus Aldi Roma-//ni impensa, ac studio Ioan-//nis Petri Vallæ filii pi-//entiss. mense Decem-// bri. M.D.I.' (Not in this copy.)

Description. Fol., 28.8 × 43.7 cm., the text being 18.8 × 32.7 cm. 3 ff. blank + 308 unnumb. = 311 ff., 55 ll. Venice, 1501.

Editions. There was no other edition.

The first book consists of 23 brief chapters on the general value and nature of mathematics (18 pp.); the second book, of 18 chapters on the Greek classification of numbers (17 pp.); the third, of 20 chapters on figurate numbers, proportions, and the fancied properties of each number of the first decade (27 pp.); and the fourth, of 13 chapters on the operations (13 pp.). There is nothing that is noteworthy in the treatment. Works on the value of mathematics were quite common at this time, while all university treatises on arithmetic were devoted chiefly to the Greek theory. The thirteen pages devoted to the operations were a rather generous allowance for the time, especially as each page has as much matter as six or eight pages of an ordinary octavo arithmetic of that period.

Valla also wrote a treatise on the astrolabe, 'Insignis philosophi Nicephori Astrolabii expositio' (Paris, 1554), and published an edition of Euclid (Venice, 1492). His collected commentaries, but without the arithmetic and other original works, appeared in Venice in 1498.

ARNALDO DE VILLA NOVA.

Ed. pr. 1501. Venice, 1501.

ARNAULD DE VILLENEUVE, ARNALD BACHUONE. Born in 1248, at Villa Nova (Catalonia), or possibly Villeneuve, near Montpellier; died in 1314, shipwrecked on the Mediterranean. He is known principally for twenty works on alchemy. He lectured on philosophy and medicine at Barcelona and Paris, and was later a celebrated physician.

**Computus Eccleſiaſticus ⁊ Aſtrono-
micus Editus a Magiſtro Arnaldo de villa Noua No-
uiter Impreſſum.**

Cum Gratia Et Priuilegio.

FIG. 42. TITLE PAGE OF ARNALDO DE VILLA NOVA

Title. 'Computus Ecclesiasticus ꝛ Astrono-//micus Editus a Magistro Ar-//naldo de villa Noua No-//uiter Impressum.// Cum Gratia Et Priuilegio.' (F. 1, r. See Fig. 42.)

Colophon. '¶Impressum Venetijs per Bernardinū Venetū de Vitalibus.//Anno Dñi.M.CCCCC.J. Die .xvij. Meñ. Februarij.' (F. 11, v.)

Description. 4°, 14.4 × 20.9 cm., the text being 10.5 × 16.6 cm. 11 ff. unnumb., 37–39 ff. Venice, 1501.

Editions. There was no other edition.

This is a good example of the works on the ecclesiastical calendar in use in the Middle Ages. It employs only the Roman numerals and gives no treatment of computation. In spite of the words 'nouiter impressum,' I know of no earlier edition, and indeed these words were not infrequently used when a book was first printed.

JOHANN HUSWIRT, Sanensis.

Ed. pr. 1501. Cologne, 1501.

A German arithmetician of c. 1500. The name Sanensis suggests his birthplace as Sayn in the Westerwald, and the problems relate to places in that vicinity. Nothing is known of his life.

Title. See Fig. 43.

Colophon. 'Enchiridion algorismi sagaci cura studioqʒ ꝑ Johānē huswirt sanēsez // elaboratus. caracteri p° ꝯmissus Colonie In officina felicis memorie ho//nesti viri Henrici Quentell. Anno reꝑatoris humane seruitutʒ Mcccccі.' (F. 20, r.)

Description. 4°, 14.3 × 20.6 cm., the text being 9 × 15.4 cm. 20 ff. unnumb., 25–47 ll. Cologne, 1501.

Editions. Cologne, 1501, 4° (here described); ib., 1503; ib., 1504, 4° (p. 77); 1507; 1554; and a French edition (Chasles). It was published with historical notes by Professor Wildermuth, at Tübingen, in 1865. Mr. Plimpton's copy has the bookplates of Dr. Kloss and Chasles.

This is the earliest treatise on algorism printed at Cologne. It is divided into four 'tractati,' and includes the fundamental operations through evolution ('Tractatus Primus'), a brief treatment of abacus or line reckoning ('¶Tractatus Secundus de proiectilibus'), common fractions ('Tractatus Tertius'), rule of three, partnership, and over twenty miscellaneous rules ('Tractatus quartus de regulis mercatoꝝ' etc.). In the algoristic treatment of integers Huswirt places 'duplatio (doubling) after multiplication, and 'mediatio' (halving) after division;

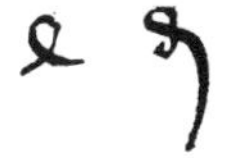

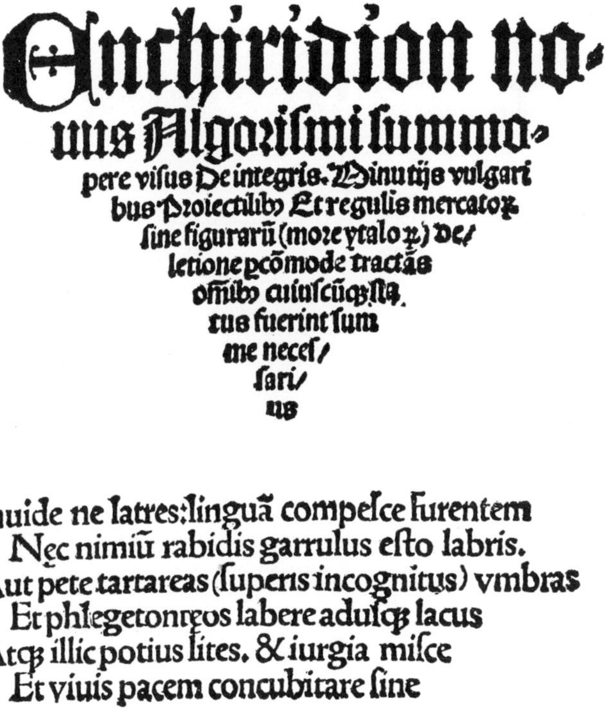

Enchiridion no-
uus Algorismi summo-
pere visus De integris. Minutijs vulgari
bus Proiectilibꝫ Et regulis mercatoꝝ
sine figurarũ (more ytaloꝝ) deletione pcõmode tractãs
omnibꝫ cuiuscũqꝫ sta-
tus fuerint sum
me neces-
sari-
us

Inuide ne latres: linguã compesce furentem
Nec nimiũ rabidis garrulus esto labris.
Aut pete tartareas (superis incognitus) vmbras
Et phlegetonteos labere adusqꝫ lacus
Atqꝫ illic potius lites. & iurgia misce
Et viuis pacem concubitare sine

FIG. 43. TITLE PAGE OF THE 1501 HUSWIRT

but when he is dealing with counters and with fractions he places them before multiplication, because they are needed there in abacus calculating. It is interesting to see how these chapters on doubling and halving, of which we have traces in ancient Egypt, persisted throughout the Middle Ages and well into the sixteenth century.

As in several other works of this period, there is evidence of the difficulty of finding a generally acceptable name for the character 0, a difficulty not yet removed in the English language. Huswirt gives four names to this tenth character: 'Decimo ꝟo theca. circulꝰ cifra. ſiue figura nihili appellat'.'

There is also noticeable in this work a tendency which is seen in other arithmetics of the time, to name a group of problems after some well-known type. For example, Huswirt's sixth rule is that of the fleeing hare ('Regula Sexta de lepore fugiente,' f. 16, r.), although it has nothing to do with the hound and hare, but relates to a traveler going from Cologne to Rome, and followed five days later by another traveler who in due time overtakes him. ('¶Ambulat q̄dam de Colonia verſus romā et ambulat q̊tidie 9 miliaria. alius aūt ꝑſequit' ipſum poſt 5 dies,' etc.)

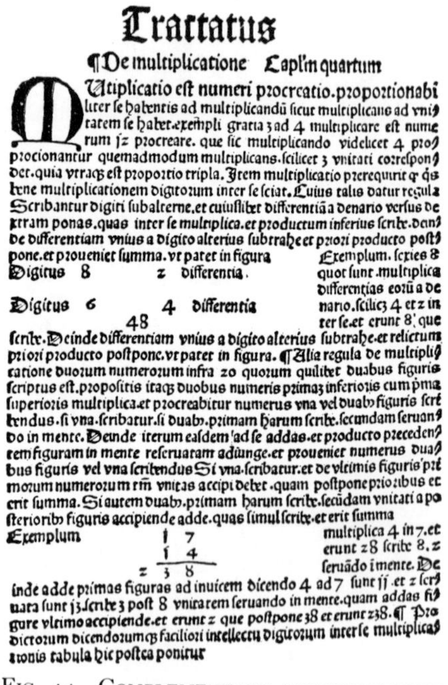

Fig. 44. Complementary multiplication, Huswirt (1501)

Other works of 1501. Anianus, p. 32, 1488; Sacrobosco, p. 32, 1488; Boethius, p. 27, 1488; Borghi, p. 16, 1484; H. Torrentini, 'Elucidarius carminum et hystoriarum,' Deventer, 4° (with a chapter on arithmetic), with subsequent editions as follows: ib., 1503, 4°; Strasburg, 1505, 4°; Hagenau, 1507, 4°; ib., 1510, 4°; ib., 1512, 4°; Strasburg, 1514, 4°; 1515, 4°; Strasburg, 1518, 4°; Paris, 1530, 8°; ib., 1535, 8°; Cologne, 1536, 8°; Paris, 1550, 8°.

Works of 1502. Anianus, p. 32, 1488; Albert of Saxony, p. 9, c. 1478; Bradwardin, p. 6, 1495; Nicolò Calvino, a work on arithmetic and geometry, Milan, of which no extant copy is known (see Riccardi, part I, col. 213).

JOHANN HUSWIRT, Sanensis.

Ed. pr. 1501. Cologne, 1504?

See p. 74.

Title. The same as in the edition of 1501 (see p. 74).

Colophon. The last folio, with the colophon, is missing from this copy.

Description. There are a few changes in type, but otherwise this edition, which is probably that of 1504, is line for line identical with that of 1501.

ANTON BARTHOLOMEO DI PAXI.

Ed. pr. 1503. Venice, 1503.

PASI. A Venetian writer of the fifteenth and sixteenth centuries.

Title. 'Tariffa de pexi e mesvre.//con gratia et privilegio.' (F. 1, r.) 'Prohemio del prestantissimo miser Bartho-//lomeo di Paxi da Venetia.' (F. 1, v.) 'Qvi comincia la vtilissima opera chiama-//ta taripha laqval tracta de ogni sorte // de pexi e misvre conrispondenti per tvto // il mondo fata e composta per lo excelen//te et eximio miser Bartholomeo di Paxi da // Venetia.' (F. 2, r. Fig. 45.)

Colophon. 'Stampado in uenefia per Albertin // da lifona uercellefe regnante lin-//clyto principe mifer Leonardo lo//reduno. Anno domini. 1503. A di//26. del mefe de Iuio. Finis.' (F. k 5.)

Description. 4°, 15.5 × 21.1 cm., printed in double columns, each being 5.3 × 15.9 cm. 156 ff. unnumb., 33–38 ll. Venice, 1503.

Editions. Venice, 1503, 4° (here described); ib., c. 1510, 4° (p. 79); ib., 1521, 8°; ib., 1540, 8° (p. 79); ib., 1557, 8° (p. 80).

The book is not a textbook on arithmetic, but a collection of information useful to merchants, relating to the measures of weight, value, length, etc., of the various cities and countries with which Venice had trade relations. It is valuable as leading to an understanding of the contemporary arithmetics of Italy, and historians could find much useful information as to the prices and the material of trade by examining this and similar works. An inspection of Fig. 45 will give some idea of the scope of Paxi's Tariffa.

QVI COMINCIA LA VTILISSIMA OPERA CHIAMATA TARIPHA LA QVAL TRACTA DE OGNI SORTE DE PEXI E MISVRE CONRISPONDENTI PER TVTO IL MONDO FATA E COMPOSTA PER LO EXCELENTE ET EXIMIO MISER BARTHOLOMEO DI PAXI DA VENETIA.

HAVERemo adūq; prima a dechiarare a uoſtre excellētie tutte lerobe che ſe uédano ĩ Venetia a pexo groſſo & quelle che ſe uédano a pexo ſotile e de lordine di pexi de li arzenti:& de le cōditione di panni de lane franceſche fatte in Venetia:& etiam dele conditione di panni de ſeda: e panni doro:& del ordine del uédere de le ſpecie:& de le ſue tare:& del ordine di pexi dele farine e biſchoti: & del ordine de le miſure di uini:& del ordine del uender del oio:& in che modo e pexo ſe uendeno iguadi:& del ordine di frutti che ſe uendeno a nome di ſter: e che pexo hano cadaun ſter:e come reſponde el pexo groſſo con tuta la Italia:& tuto el leuante & ponente:& etiam come ipexi ſubtili e pexi groſſi reſpondeno con molte terre de Italia : de dalmatia e de leuante:& come reſpondeno le meſure di panni de lana con tuta Italia:e con tuto el leuante:& etiam come reſpondeno le meſure di panni de ſeda:e panni doro e darzento con tuta Italia:& cō tuto el leuante & ponente:& come reſpondeno le meſure dingilterra zoe la uirga da londra:e de la taripha dalixādria:& etiam quella de damaſcho & la taripha da leppo.e come torna la ſporta dalixādria con molte terre de leuante e de ponēte e de la Italia:& come el canter forfori dalixandria reſponde con alchune altre terre de leuante de ponente e de Italia:& etiam come reſponde el canter zeroui con alchune terre del leuante del ponente e de Italia:& come reſpōde el cento de le mene con alchune terre del leuante e molte de Italia & del ponente:& come reſpondeno tuti icantera de leuante e del ponente con el pexo ſubtile da Venetia: & come reſpōde el canter da napoli de reame con molte terre del leuante del ponēte e con molte de Italia:& come reſponde el canter de Conſtātinopoli con molte terre de leuante & etiam de Italia:& come reſponde el canter de

FIG. 45. THE BEGINNING OF THE 1503 PAXI a ii

ANTON BARTHOLOMEO DI PAXI.

Ed. pr. 1503. S. l. a. (Venice ?, c. 1510).

See p. 77.

Title. 'Tariffa De Peſi e meſure cor-//reſpondenti dal Leuante al Ponēte : da // vna terra a laltra : e a tutte le parte del // mondo : con la noticia delle robe // che ſe trazeno da vno Paeſe // per laltro. Nouamente // cō diligentia Ri-//ſtampata .·. ✠' (F. 1, r.)

Colophon. '¶Finiſſe il prohemio de // Miſſer Bartholomio di // Paſi da Veneſia.// Finis.' (F. 218, v.)

Description. 8°, 10.2 × 15.1 cm., printed in double columns, each being 4.1 × 12.5 cm. 218 ff. numb., 30 ll. S. l. a. (Venice ?, c. 1510.)

See p. 77. This is the second edition of this popular 'Tariffa.' Since it was one of the first books of its kind to appear in Venice, its five editions are easily explained.

ANTON BARTHOLOMEO DI PAXI.

Ed. pr. 1503. Venice, 1540.

See p. 77.

Title. 'Tariffa // de i pesi, e misvre // corriſpondenti dal Leuante al Ponente : // e da una terra, e luogo allaltro, qua ſi ꝑ // tutte le parti dil Mondo : con la dichia-//ratione, e notificatione di tutte le robbe : // che ſi tragono di uno paeſe per laltro.// Composta per M. Bartholomeo di // Paſi da Vinetia. Con la ſua // Tauola copioſiſſima, e faci-//liſſima a trouare ogni coſa // per ordine, nuouamēte // fatta : e con ſomma // diligēza reuista,// e stāpata.// In Vinetia. M.D. XL.' (F. 1, r.)

Colophon. 'In Vinegia. Nelli caſe di Pietro di Nicolini da Sabbio.// Ne glianni dilla ſalutifera Circonciſione dil no-//ſtro Signore. M. D. XL.// Dil meſe di Genaio.' (F. 212, r.)

Description. 8°, 10.3 × 14.7 cm., printed in double columns, each being 3.5 × 12.3 cm. 11 ff. unnumb. + 1 blank + 200 numb. = 212 ff., 30 ll. Venice, 1540.

See p. 77.

ANTON BARTHOLOMEO DI PAXI.

Ed. pr. 1503. Venice, 1557.

See p. 77.

Title. The title page is practically identical with that of the 1540 edition, except for the date: 'In Vinegia per Paolo Gherardo. // M. D. LVII.' (F. 1, r.)

Colophon. 'In Vinegia per // Comin da Trino.// M. D. LVII.' (F. numbered 200, r.)

Description. 8°, 10.6 × 15 cm., printed in double columns, each being 3.7 × 12.3 cm. 11 ff. unnumb. + 200 numb. = 211 ff., 30 ll. Venice, 1557

See p. 77.

BOETHIUS, JODOCUS CLICHTOVEUS, AND FABER STAPULENSIS. Ed. pr. 1503. Paris, 1503.

For the biographies of BOETHIUS and FABER STAPULENSIS see pp. 25, 62. JODOCUS CLICHTOVEUS was born at Nieuport, Flanders; died at Chartres, September 22, 1543. He was educated at the Sorbonne, and was canon of Saint-Jean, at Chartres. Like Faber Stapulensis, he was known chiefly as a commentator.

Title. 'In hoc libro contenta // Epitome/ cōpendiofaqʒ // introductio in libros // Arithmeticos diui Seuerini Boetij : adie// cto familiari comētario dilucidata.// Praxis numerandi certis quibufdam re-//gulis cōftricta.// Introductio īgeometriā : ex libris diftīcta // Prim⁹ de magnitudinib⁹ & earū // circūftantiis.// Secūdus de cofequentibus/ conti-//guis/ & cōtinuis.// (Surrounded by an elaborate woodcut.) ¶Tertius de pūctis. ¶Quartus de lineis. ¶Quītus de fuperficieb⁹.// ¶Sextus de corporibus. ¶Liber de quadratura circuli. ¶Liber de cubica//tione fphere. ¶Perfpectiua introductio. ¶Infuper aftronomicon.' (F. 1, r.)

Colophon. 'Id opus impreffērūt Volphgangus // hopilius et Henricus ftephanus // ea in arte focii in Almo pari-//fiorum ftudio Anno Chri//fti Celorum totiufqʒ // nature cōditoris.// 1503. Die vice//fimafepti-//ma Iu-//nij.' (F. cxi, v.)

Description. Fol., 19 × 26.6 cm., the text being 16 × 22.6 cm. 112 ff. numb., 47–54 ll. Paris, 1503.

Editions. This is the first edition of this combination of works, the second (somewhat changed) appearing at Paris c. 1507 (see below), and the third ib., 1510. The epitome appeared as a 'Compendium arithmetices Boethii,' s. l., in 1480, and, with the arithmetic of Jordanus, at Paris, in 1496. There was an edition of Faber Stapulensis, Clichtoveus, and others, at Cologne, c. 1515, 4°. Scheubel published an edition at Basel in 1553, and a work entitled 'Arithmetica Boethi epitome acced. Christiani Morisani Arithmetica' also appeared at Basel in 1553 (pp. 27, 182, 260). The 'Praxis numerandi' of Clichtoveus, of which this is the first edition, appeared separately at Paris in 1510, fol. 498

The copy here described is bound with the 1509–10 edition of Bovillus (see p. 89). It consists of a brief introduction by Faber Stapulensis, and the arithmetic of Boethius with the commentary of Clichtoveus on Faber's epitome. This is followed by the geometry and perspective of Faber Stapulensis.

BOETHIUS, JODOCUS CLICHTOVEUS, AND FABER STAPULENSIS. Ed. pr. 1503. Paris, c. 1507.

See p. 80.

Title. 'Introductio//Jacobi fabri Stapulēſis in Arithme//cam Diui Seuerini Boetij pariter ⁊ Jordani // Ars ſupputādi tam per calcu-//los q̄ȝ notas arithmeticas ſuis quidem regulis elegāter expreſſa // Judoci Clichtouei Neoportuenſis.// Queſtio haud indigna de numerorū //et ꝑ digitos ⁊ ꝑ articulos finita ꝑgreſſione ex Aurelio Auguſtino // ℂEpitome rerum geometricaruȝ ex Geometrico introductorio // Caroli Bouilli.// ℂDe quadratura Circuli Demonſtratio ex Campano.' (F. 1, r.)

Description. 4°, 12.6 × 17.4 cm., the text being 10.6 × 14.5 cm. 32 ff. unnumb., 45–46 ll. Paris, c. 1507. The dedicatory epistle is dated 'Data āno ſalutis // noſtre Milleſimo q̇ngēteſimo ſeptimo tercio calēdas Iunij,' that is, 1507. It was evidently printed at Paris, but it is without date or place.

Editions. See above.

The introduction by Faber Stapulensis to the arithmetic of Boethius and Jordanus was very popular in the university of Paris at the opening of the sixteenth century. It is, like Boethius, purely theoretical. The author begins with a dissertation 'de vtilitate arithmetice difcipline,' and then gives an epitome of the two works. This is followed by the 'compendium' of Clichtoveus, merely a set of rules for the operations. Books of this character, evidently intended as the bases of lectures to university students, show in what a hopeless state the Boethian arithmetic found itself at the end of the Middle Ages.

GREGORIUS REISCH. Ed. pr. 1503. 495 Strasburg, 1504.

Born at Balingen, Württemberg; died at Freiburg, 1523. He was a student at Freiburg in 1487, and took his bachelor's and master's degrees there. He then entered the Carthusian order and became prior of the cloister at Freiburg, and confessor of Maximilian I.

Title. 'Aepitoma omnis phylosophiae. ali-//as Margarita phylosophica tractans // de omni genere fcibili: Cum additionibus: Quę in alijs non habentur.' (Large woodcut representing the liberal arts. F. 2, r. Plate II.)

Colophon. 'Explicit phylofophica Margarita. Caftigatione acri // In nobili Heluecioru͞ ciuitate Argentina Chalchogra-//phatu͞: Per Ioanne͞ Gruninger Ciue͞ Arge͞tinu͞: ı͞ vigilia // Mathię Anno incarnationis Saluatoris M.ccccc.iiij.// Valete & Plaudite.' (F. 289, v.)

Description. 4°, 15.1 × 20 cm., the text being 11.5 × 15.6 cm. 2 ff. blank + 289 unnumb. = 291 ff., 45 ll. The illustrations are hand-colored. Strasburg, 1504. 505

Editions. Freiburg, 1503, 4°; 495 Strasburg, 1504, 4° (here described); Freiburg, 1504, 4°; another edition, s. l., by Schott (Freiburg), 1504; Strasburg, 1508; Basel, 1508, 4° (p. 83); ib., 1512, 4°; Strasburg, 1512, 4°; ib., 1515, 4°; Basel, 1517, 4°; Paris, 1523 (first Finaeus edition); Basel, 1535, 4° (p. 84); ib., 1583, 4°; Venice, 1594; ib., 1599; ib., 1600. The three Venetian editions (1594, 1599, 1600) are Italian translations by Giovanni Paolo Gullucci, and contain the additions by Orontius Finaeus, and also the introduction by Faber Stapulensis to the arithmetics of Boethius and Jordanus (see p. 62), Clichtoveus

A. Geometry

B. Arithmetic

Plate II. From the margarita philosophica

on arithmetic (see p. 80), and 'Questione di S. Agostina della progressione dei numeri per li digiti, et per li articoli.' Hartfelder (*Zeitsch. f. Gesch. des Oberrheins*, II, 170) has shown that the assertion of Hain, Poggendorff, and others, that it appeared in 1496, is incorrect.

This was the first modern encyclopedia to appear in print. It contains a compendium of the trivium, the quadrivium, and the natural and moral sciences. It is made up of twelve books, of which the fourth, consisting of fifteen folios in the present edition, is on arithmetic. The author first considers the definition of arithmetic, and then gives the mediæval classification of number, including the system of ratios as set forth by Boethius and his followers. The second part of the work contains a short treatment of algorism, including the fundamental operations and roots. The third tractatus relates to common fractions and the fourth to physical or sexagesimal fractions. The arithmetic closes with a treatment of line reckoning, giving the four fundamental operations and the rule of three. The illustrations are particularly interesting. (See Plate II.)

Other works of 1503. Boethius, p. 27, 1488; Bradwardin, p. 61, 1495; Faber Stapulensis, p. 62, 1496; Huswirt, p. 74, 1501; Jordanus, p. 62, 1496; Orbellis, p. 23, 1485; Peurbach, p. 53, 1492; Sacrobosco, p. 32, 1488; Torrentini, p. 76, 1501; Anonymous, 'Textus arithmeticæ comunis, cum Conradi Norici commentatione,' Leipzig, fol. **505**

Works of 1504. Anianus, p. 35, 1488; Bradwardin, p. 61, 1495; Johannes Carolus (see Landshut, below); Huswirt, p. 74, 1501; Reisch, p. 82, 1503; Sacrobosco, p. 32, 1488; Johann Karl von Landshut (Lanzut), 'Algorithmus integrorum,' Leipzig (see also p. 97, 1513, 1515); Henricus Stromer, 'Algorithmus linealis cum Regula de Tri,' Leipzig, 4°, with other editions in 1510; 1512, 4°; 1514; Leipzig, 1516, 4°; ib., 1517, 4°; 1520; Cracow, 1536. **506; 508**

GREGORIUS REISCH. Ed. pr. 1503. Basel, 1508.

See p. 82.

Title. 'Margarita philoſophica // cū additionibus nouis : ab auctore ſuo // ſtudioſiſſima reuiſiōe tertio ſup additis. // Jo. Schottus Argen̄. lectori. S. // Hanc emo/ non preſſam mendaci ſtigmate/ Lector :// Pluribus aſt auctam perlege : doctus eris. // Baſileę. 1508' (F. 1, r.)

Colophon. '☾Tertio induſtria complicū Micha//elis Furterij/ et Joānis Scoti//ſtudioſiiſſime preſſa. Ba-//ſileę.i4.Kal'.Mar// tias. Anno Chriſti.// 1508.' (F. 308, r.)

Description. 4°, 15.2 × 21.9 cm., the text being 12.7 × 16.7 cm. 309 ff. unnumb., 42 ll. In this edition the leading initials are inserted by hand and the illustrations are colored. Basel, 1508.

See p. 83.

GREGORIUS REISCH. Ed. pr. 1503. Basel, 1535.

See p. 82.

Title. 'Marga-//rita philosophica, rati-//onalis, Moralis philoſophiæ princi-//pia, doudecim libris dialogice cōple-//ctens, olim ab ipſo autore recognita: // nuper aūt ab Orontio Fineo Delphi//nate caſtigata & aucta, unà cum ap-//pendicibus itidem emēdatis, & quā // plurimis additionibus & figuris, ab // eodem inſignitis. Quorū omni-//um copioſus index, uerſa // continetur pagella.//Vireſcit uulnere uirtus.// Basileae 1535.' (Surrounded by an elaborate woodcut.) (P. 1.)

Colophon. 'Basileae excvdebat Henricvs // Petrus, ac Conradi Reſchij impenſis. An• // M. D. XXXV.' (P. 1577.)

Description. 4°, 15 × 20.8 cm., the text being 9.6 × 13.8 cm. 78 pp. unnumb. + 1498 numb. + 1 blank = 1577 pp., 26–30 ll. Basel, 1535.

Editions. See p. 82. Finaeus dates the dedicatory epistle 'Pariſijs ex regali collegio Nauarræ. 1523,' and his first edition appeared in that year. This edition gives only part of the elaborate engravings found in the earlier ones. It is, however, much better printed, being set in clear Roman type and having a more open page.

See p. 83.

THEODOR TZWIVEL. Ed. pr. 1505. Cologne, 1507.

A German arithmetician of c. 1500, from Monte Gaudio (Mongavensis), Westphalia.

Title. See Fig. 46.

Colophon. '¶Algorithmi. qui ars dicitur numerandi. de integris // per figurarum (more Alemānorum) deletioneȝ. Nec//nō de ꝑportionibꝰ ingenioſi Pythagoriſte Theodo//rici Tzwyuel. poſt plurimā praxin iam tandē in hoc // ꝯpendiuȝ reducti finis adeſt. quod et puplicā ob vti//litatem in magiſtrali artis impreſſori e taberna inge-//nuorum liberorum Quentell iterato diſſeminari pro

Arithmetice
opuſcula duo Theodorici tzwi
uel de numeroꝝ praxi(que algorithmi dicunt)vnū de integris per
figurarū(more alemanoꝝ)deletionē. Alterū de proportionibus cu
ius vſus frequens in muſicā harmonicā Seuerini Boetij

De vtilitate huius libelli Tetraſtichon
ioannis Murmellij Ruremundenſis

Si quis arithmeticēs optat cognoſcere praxin
Pythagore numeros diſcere ſi quis amat
Scire mathematicen ſi vult, ſi deniqꝫ quicqꝫ
De ſophia, hunc modicum comparet ere librū

FIG. 46. TITLE PAGE OF TZWIVEL

//curauit. Anno a natali dominico Milleſimo quin//geneteſimo-ſeptimo.' (F. 9, v. See Fig. 47.)

Description. 4°, 13.9 × 20 cm., the text being 8.8 × 14.5 cm. 10 ff. (1 blank), 46 ll. Cologne, 1507.

Editions. Günther mentions a Münster edition of 1505, but I have not seen it; there was a Cologne edition in the same year; Cologne, 1507, 4° (here described).

The work is divided into two parts, the first beginning as follows: '¶Algorithmus de integris ꝑ figurarū (more alemanoꝝ) deleti-//onē artē numerādi enucleatim ꝯpendioſeqȝ ꝑdocens.' (F. 2, v.) This part contains a brief explanation of the writing of numbers and the fundamental operations. The second part begins as follows: '¶Algorithmꝰ de ꝓportiōibꝫ cuius vſus frequēs in muſicam har//monicam Seuerini Boetij' (f. 8, v.), and two pages treat of the operations with the mediæval 'proportiones' or ratios. A comparison of the title pages represented on pp. 45, 75, 85, and of the works to which they belong, at least two of which were printed in Cologne, leads to the belief that the expression 'per figurarū (more alemanorꝝ) deletionē' (by the deletion of figures in the German way) refers to a contemporary North German custom of not actually canceling the figures in the galley division, as the Italians did.

Other works of 1505. Anonymous (see Licht), p. 70, 1500; Borghi, p. 16, 1484; Bradwardin, p. 61, 1495; Ciruelo, p. 60, 1495; Licht, p. 70, 1500; Suiseth, p. 10, c. 1480; Torrentini, p. 76, 1501; Anonymous, 'Tractatus perutilis in arithmetica speculativa,' Paris; Hermannus Buschius, 'Enchiridion novus Algorismi,' Cologne, s. a. (1504), 4° (see also p. 106, 1514); Nicolaus Horem, 'Tractatus proportionū,' Venice, in a volume with Bradwardin and Suiseth (see also p. 116, 1515); 'Algorithmus linealis Baccalariū Wolfgangum Monacenseni,' Leipzig, 4°; Georg Leunbach, an arithmetic.

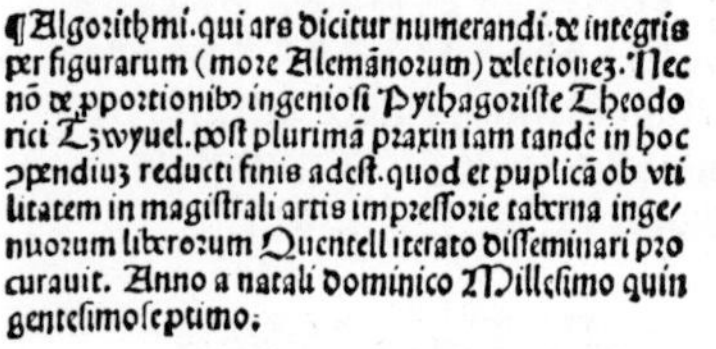

¶Algorithmi. qui ars dicitur numerandi. de integris per figurarum (more Alemānorum) deletionez. Nec nō de ꝓportionibꝫ ingenioſi Pythagoriſte Theodorici Tzwyuel. poſt plurimā praxin iam tandē in hoc ꝯpendiuȝ reducti finis adeſt. quod et puplicā ob vtilitatem in magiſtrali artis impreſſorie taberna ingenuorum librorum Quentell iterato diſſeminari procurauit. Anno a natali dominico Milleſimo quingenteſimoſeptimo.

FIG. 47. COLOPHON OF TZWIVEL

Works of 1506. Albert of Saxony, p. 9, c. 1478; Anonymous, p. 10, 1480; Vittori, p. 9, Albert of Saxony, c. 1478, and p. 61, Bradwardin, 1495; Pietro Borriglione, 'Arismetices praxis,' Turin, 22 ff., with a second edition, ib., 1523; Raphael Maffei, 'Commentarii Urbani,' Rome, fol., an encyclopedia containing a book (no. 35) 'De scientiis mathematicis,' which includes a little arithmetic; Maffei's work was also printed in 1527, and at Paris in 1511, 1515, 1526, and 1530, and at Basel in 1559.

ANONYMOUS. Ed. pr. 1507. Leipzig, 1507.

Title. See Fig. 48.

Colophon. '¶Impreſſum Liptzck per Baccalariū Vuolfgangū //Monacenſem Anno noſtre redemptionis .1507.' (F. 27, r.)

Description. 4°, 14.5 × 19.5 cm., the text being 9.5 × 15.3 cm. 28 ff. unnumb. + 1 blank = 29 ff., 28–34 ll. Leipzig, 1507.

Editions. Leipzig, 1507 (here described); ib., 1509, 4°; s. l. a. (Nürnberg ?, c. 1510).

This resembles several of the works on algorism appearing about this time, such as Widman's (?), Licht's, and Huswirt's. It contains a very brief treatment of the fundamental operations, including duplation and mediation. In division, only a single example is given, that of 1456 ÷ 12. After a similarly brief treatment of fractions, the Boethian propor-

Algorithmus de Integris. Mi
nucijs vulgaribus ac ppor
tionibus Cum annexis
detri falsi alijsq;
Regulis

FIG. 48. TITLE OF THE 1507 *Algorithmus*

tions (ratios) are taken up: 'Sequitur Algorithmus proportionū.' Then follow the Rule of Three and several other rules now entirely forgotten, such as 'Regula legis,' 'Regula augmenti,' 'Regula plurima,' 'Regula pulchra,' and 'Regula falſi.' Such 'regulae' were not stated like modern rules but consisted of groups of similar problems.

Other works of 1507. Boethius, p. 27, 1488; Faber Stapulensis, c. 1507, p. 81, 1503; Huswirt, p. 74, 1501; Peurbach, p. 53, 1492; Torrentini, p. 76, 1501.

Works of 1508. Anianus, p. 32, 1488; Reisch, p. 82, 1503; Widman, p. 39, 1489; Hieronimus de Hangest, 'Liber proportionum,' Paris, 4°. [507]

LUCA PACIUOLO. [54] Ed. pr. 1509. Venice, 1509.

Title. After a vocabulary and index the work begins on f. 1, r., as follows: 'Excellentiſſimo principi Ludouico mariæ Sfor. Anglo Mediolanen//ſium duci: pacis et belli ornamento fratris Lucæ pacioli ex Burgo ſancti // Sepulchri ordinis Minorum: Sacræ theologiæ pfeſſoris. De diuina pro//portione epiſtola.'

FIG. 49. FROM THE 1507 *Algorithmus*

Colophon. '¶Venetiis Impreſſum per probum virum Paganinum de paganinis de // Brixia. Decreto tamen publico vt nullus ibidem totiqꝫ dominio annorum // XV. curiculo imprimat vel īprimere faciat. Et alibi impreſſum ſub quouis // colore in publicum ducat ſub penis in dicto priuilegio contentis. Anno Re//demptionis noſtre.M.D.VIIII. Klen. Iunii. Leonardo Lauretano Ve.//Rem.Pu. Gubernante. Pontificatus Iulii.II. Anno.VI.' (F. 27, r., of Part 3. A similar colophon appears at the foot of f. 35, v., of Part 1.)

Description. 8°, 20.5 × 28.3 cm., the text being 9.1 × 21.5 (with marginal drawings). 1 f. blank + 2 unnumb. + 90 numb. 93 ff., 52–57 ll. Venice, 1509.

Editions. There was no other edition.

With some hesitancy this book has been included, the only justification being the fact that there are several pages devoted to the discussion of proportion in general, including the arithmetical, geometric, and astronomic. Paciuolo excludes the other forms of ancient proportion given by 'Platone e Ariſto. e yſidoro ī le ſue ethimologie. El ſeuerin Boetio in ſua arithmetica.' (F. 5, r.) Most of the treatise is, however, devoted to geometry.

CAROLUS BOVILLUS. Ed. pr. 1509–10. Paris, 1509–10.

Bouvelles, Boüelles, Bouilles, Bouvel. Born at Saucourt, Picardy, c. 1470; died at Noyon, c. 1553. Canon and professor of theology at Noyon.

Title. See Fig. 50.

Colophon. 'Libelli De Mathematicis Svpple-//mentis Finis Anno Salutis Humane // 1509 Ianuarij Die Deci//maoctaua // ¶Editum est vniversvm hoc volvmen Ambianis in edibvs Re// uerendi in Christo Patris Franciſci De Hallevvin Eiuſdem Loci Pontificis ∴ // Et emiſſum ex officina Henrici ſtephani. Impenſis eiuſdem et Ioannis parui in chalcotypa // arte ſociorum Anno Christi Saluatoris omnium 1510. Primo Cal. Februarij.// Parisiis.' (F. 198, v.)

Description. Fol., 19.7 × 27.4 cm., the text being 15.1 × 25.8 cm. 198 ff. numb., 53–54 ll. Paris, 1509–10.

Editions. Fontès has described (Toulouse *Mem.* (9) VI, 155–167, for 1894) a quarto of 1510, published by Stephanus of

H S

¶ Que hoc volumine continẽtur.
Liber de intellectu.
Liber de ſenſu.
Liber de nichilo.
Ars oppoſitorum.
Liber de generatione.
Liber de ſapiente.
Liber de duodecim numeris
Epiſtole complures.

¶ Inſup mathematicũ opus quadripartitũ ¶ De Numeris Perfectis ¶ De Mathematicis Roſis ¶ De Geometricis Corporibus ¶ De Geometricis Supplementis

FIG. 50. TITLE PAGE OF BOVILLUS

Paris, entitled, ‘Caroli Bovilli liber de numeris perfectis.’ This book is, however, without title page, and is of course simply ff. 172–180 of this work. I know of no other editions.

The first part on numbers, the ‘Liber de duodecim numeris,’ begins on f. 148, v., and ends on f. 171, r., with the colophon: ‘Libri dvodecim nvmerorvm finis: editi // in domo- R. P. Franciſci de Hallevvin/ Pontificis Ambi//anenſis. Anno ab autore numerorum Incarnato.// 1510: Maij decimaſexta.’ It relates solely to the mystery of numbers and to the Greek theory. The part on perfect numbers, numbers which equal one plus the sum of their factors, begins on f. 172, r., and ends on f. 180, r., with the following colophon: ‘¶Liber perfectorvm nvmerorvm finis.// Perfecto/ trinoqȝ deo laudes in gentes. Anno domi-//ni/ 1509: Ianuarij 4.’ The dates of the several parts vary from October 25, 1509, to January 18, 1509 (1510 N. S.).

Other works of 1509. Anianus, p. 32, 1488; Anonymous, p. 87, 1507; Borghi, p. 16, 1484; Ciruelo, p. 60, 1495; Licht, p. 70, 1500; Sacrobosco, p. 32, 1488.

Other works of 1510. Anonymous, p. 87, 1507; Anonymous, c. 1510, p. 46, 1491; Boethius, p. 30, 1488; Bovillus, p. 89, 1509; Bradwardin, c. 1510, p. 61, 1495; Clichtoveus, p. 81, 1503; Faber Stapulensis, p. 62, 1496; Jordanus, p. 62, 1496; Peurbach, p. 53, 1492; Sacrobosco, p. 32, 1488; Stromer, p. 83, 1504; Paxi, c. 1510, p. 79, 1503; Torrentini, p. 76, 1501. 507

Works of 1511. Anianus, p. 32, 1488; Boethius, p. 27, 1488; Maffei (Maphæus), p. 86, 1506; Peurbach, p. 53, 1492; Simon Eisenmann, ‘Enchiridion arithmetica,’ Leipzig, fol. 507

JUAN DE ORTEGA. Ed. pr. 1512. Rome, 1515.

JOHN DE LORTZE. A Spanish priest of the Dominican order, from Aragon. He was still living in 1567.

Title. ‘Svma // de Arithmetica: Geometria // Pratica vtiliſſima: ordina//ta per Johane de Or//tega Spagnolo // Palentino.// Cum Priuilegio.’ (Surrounded by an elaborate woodcut. F. 1, r.)

Colophon. ‘Impreſſo in Roma per Maſtro Stephano Guilleri de Lorena // anno del noſtro Signore 1515 adi 10 de Nouēbre Regnante Leo//ne Papa decimo in ſuo Anno tertio.’ (F. 116, r.)

Description. Fol., 20.6 × 30.2 cm., the text being 13.1 × 22.7 cm. 2 ff. unnumb. + 114 numb. = 116 ff., 32–38 ll. Rome, 1515.

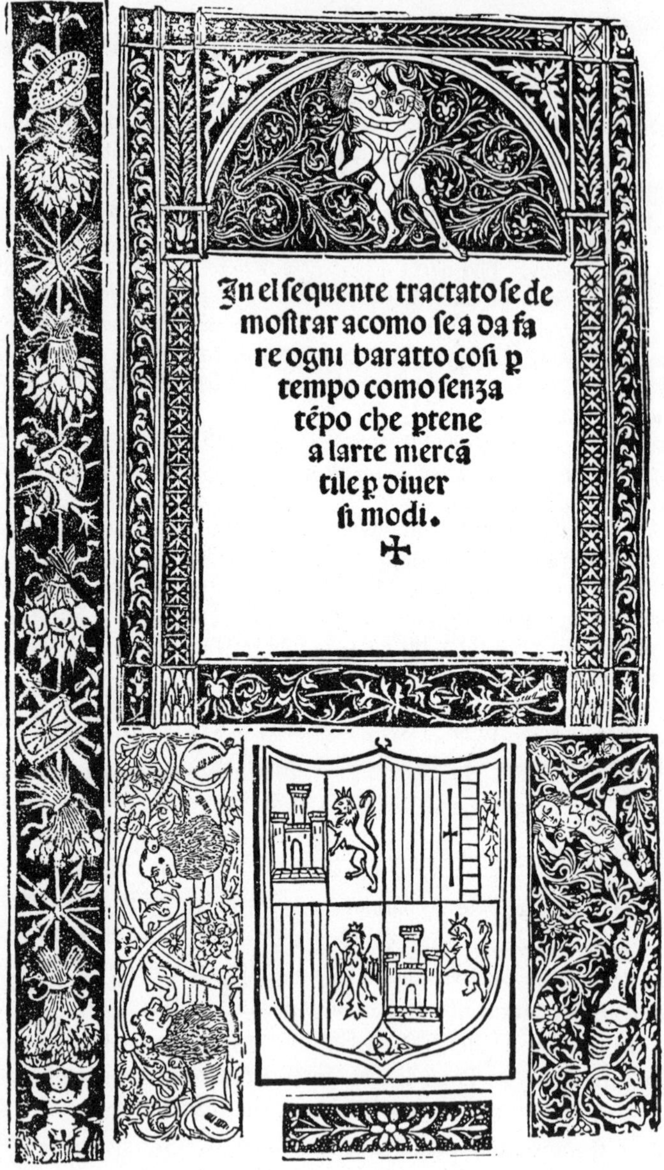
In el sequente tractato se de
mostrar acomo se a da fa
re ogni baratto cosi p̄
tempo como senza
tēpo che p̄tene
a larte mercā
tile p̄ diuer
si modi.

Fig. 51. From the 1515 Ortega, introduction to barter

Editions. Barcelona, 1512; Lyons, 1512, 4°; ib., 1515, 4°; Rome, 1515, fol. (here described); Messina, 1522; 1534; Seville, 1536; ib., 1537; Paris, 1540 (?); Seville, 1542, 4° (see below); ib., 1552 (p. 94); s. l. (?), 1552; Granada, 1563, 4°. The Lyons edition of 1512 was the first book on commercial arithmetic printed in France. It differs somewhat from the Rome edition of 1515, but the latter differs only a little from the first (Barcelona) edition. Mr. Plimpton's copy of 1515 belonged to Prince Boncompagni and has his collation on the cover. It is beautifully printed and is one of the best examples of the early Italian mathematical typography.

This is one of the most celebrated arithmetics written in Spain in the sixteenth century. It is a purely commercial book, beginning with notation, taking up the four processes with integers, the progressions, the roots, and the checks on operations, and the same operations in the same order with fractions, and then discussing the business rules. These last include exchange, rule of three, profit and loss, partnership, testament problems, barter, alloys, false position, and a little mensuration.

Other works of 1512. Bradwardin, p. 61, 1495; Peurbach, p. 53, 1492; Reisch, p. 82, 1503; Stromer, p. 83, 1504; Torrentini, p. 76, 1501; I. Furst, 'Novus . . . algorithmus.'

JUAN DE ORTEGA. Ed. pr. 1512. Seville, 1542.

See p. 91.

Title. 'Tratado // ſubtiliſſimo de Ariſmeti-//ca y de Geometria : cō-//pueſto y ordenado // por el reuerendo // padre frāy Juā // de Ortega d' // la orden d' // los pre//dicadores.// 1542 // 1234567890.' (The whole is surrounded by an elaborate woodcut border.) (F. 1, r.)

Colophon. 'Sue impreſſo el preſente libro // re Ariſmetica y Geometria (agora nueuamēte // corregido y emendado) en caſa d'Jacom // crōberger: enla muy noble y muy leal // ciudad de Seuilla : a cinco dias // de deziembre de. M.d. y // quarēta y dos añoz.' (F. 232, v.)

Description. 4°, 14.5 × 20.5 cm., the text being 11.3 × 17.2 cm. 232 ff. numb., 34 ll. Seville, 1542. 507

See above.

JUAN DE ORTEGA. Ed. pr. 1512. Seville, 1552.

See p. 91.

Title. 'Tractado // Subtiliſſimo d'Ariſmetica y de Geome// tria. Compueſto por el reuerēdo padre // fray Juan de Hortega, d'la orden // de los predicadores.// Elhora de nueuo emendado con mucha // diligētia por Gonçalo Buſto d'muchos // errores que auia en algunas im//preſſiones paſſadas.// ¶Van añadidas en eſta impreſſion las // prueuas deſde reduzir haſta partir que- //brados. Y en las mas de las figuras de // geometria ſus prueuas, con ciert os aui-//ſos ſubjetos al Algebra. Y al fin deſte tra //ctado: 13. exemplos de arte mayor.// 1552.' (Title page is printed in red, and is surrounded by an elaborate woodcut in black.) (F. 1, r.)

Colophon. 'Hizo fin el tractado de Ariſmetica Y // Geometria, que compuſo y ordeno el reuerendo padre // fray Juan de Hortega, de la orden de los predica-//dores. Fue impreſſo ēla muy noble ꝛ muy leal // ciudad de Seuilla por Juā canalla, enla // collacion de ſant Juā. Acaboſe a diez // y feys dias del mes de Abril del // anō de nueſtro criador y redē//ptor Jeſu Christo de mill // ꝛ quinientos ꝛ cin-//quenta y dos // años.' (F. 223, r.)

Description. 8°, 14.6 × 20 cm., the text being 11.9 × 17.4 cm. 232 ff. numb. + 7 unnumb. = 239 ff., 33 ll. Seville, 1552.

See p. 93. The 'arte mayor' mentioned in the title is algebra.

JODOCUS CLICHTOVEUS. Ed. pr. 1513. Paris, 1513.

See p. 80.

Title. See Fig. 52.

Colophon. '¶Expletum eſt hoc opuſculum & ex officina emiſ- //ſum/ in alma Pariſiorum academia : āno domi-//ni (qui omnia numero definiuit) decimoter//tio ſupra milleſimū & quingenteſimū/ //decimaſexta die Decembris. Per // Henricū ſtephanū/ artis excu//ſoriȩ librorū ſedulū & indu//ſtriū opificē/ e regione // ſchole Decretorū // habitan-//tem.' (F. 43, v.)

Description. 4°, 14 × 19.5 cm., the text being 12 × 15.7 cm. 41 ff. numb. + 3 unnumb. + 2 blank = 46 ff., 42 ll. Paris, 1513.

Editions. There was no other edition.

This is, I believe, the first separate treatise on the mystery of numbers to appear in print. Paciuolo had included a good deal of such material in his Summa of 1494, and about a century later Bungus published a monumental treatise upon the subject, but Clichtoveus was a pioneer in the publication of a separate work. The result of his labors is properly

¶De myſtica numerorum ſignificatione opuſculum: eorum pręſertim qui in ſacris litteris vſitati habentur/ſpiritualē ipſorum deſignationem ſuccincte elucidans.

FIG. 52. TITLE PAGE OF THE 1513 CLICHTOVEUS

included in a list of arithmetics, for, while there is nothing of computation in the work, it is not unrelated to the number theories of the mediæval writers and even of the Pythagoreans.

Clichtoveus discusses, as is usual among such writers, the religious significance of one ('Quid vnitas/ numerorum fons et origio deſignat. Cap. I.') and the numbers of the first decade. He also mentions several larger numbers which were supposed to have some scriptural significance, not forgetting, of course, 666, 'the number of the beast.'

There is also in this work a chapter, generally unrecognized by writers on the history of the subject, on finger-reckoning: 'Quomodo antiqui: numeros omnes per certas digitorū & manuum figurationes/ ſignificare ſunt ſoliti.' Cap. XXVIII.

JOANNES MARTINUS BLASIUS, Villagarciensis. 508

Ed. pr. 1513. Paris, 1513.

A Spanish astrologer and arithmetician of c. 1500. In this edition the author's name appears as 'Ioannes Martinus Blaſius dioceſis Pacēſis,' and in the 1519 edition as 'Ioannes Martinus Silecevs (and Sciliceus) Diocesis Pacēſis.'

Title. See Fig. 53.

Colophon. 'Explicit liber Arithme//tices practice magri Joannis Martini Blaſij Vil-//lagarciēſis: Pariſijs edit9 in honeſtiſſima Belua-//corū paleſtra: impreſſus vero a calcographorum ex-//pertiſſimo Thoma Kees: Vveſalienſe expenſis pro//biſſimorum

FIG. 53. TITLE PAGE OF THE 1513 BLASIUS

virorum: Joannis Parui et Joannis // Lambert. Anno domini. 1513. in vigilia diui Jo-//annis baptiſte.' (F. 26, r.)

Description. Fol., 19.5 × 28.1 cm., printed in double columns, each being 6.5 × 21.3 cm. 26 ff. unnumb., 64–66 ll. Paris, 1513.

Editions. Paris, 1513, fol. (here described); ib., 1514, large 8°; ib., 1519 (see below); ib., 1526, fol.

Although an algorism, the work is mediæval in character. The author first discusses the fundamental operations with integers, including series and roots as was the custom, but not considering duplation and mediation as distinct topics. He is one of the earliest writers to adopt the spelling *substractio,* for subtraction, a custom more or less followed by the Dutch and English arithmeticians for several generations. Blasius closes his 'primus tractatus' with a discussion of compound numbers. 'Tractatus secundus' considers computations with counters, or 'nummi supputatorii.' The 'tertius tractatus' is devoted to common fractions, 'fractiones vulgares'; the 'quartus tractatus' to sexagesimal fractions, 'fractiones phisicae'; and the 'quintus tractatus' to the rule of three ('Prima regula ꝛ fūdamētalis quā detri dicunt.') There are no practical applications of any value.

This first edition differs greatly from the third (1519) described below. It includes only the algorism, while the latter consists of two parts, the first being on Boethian arithmetic, and the second being substantially identical with the 1513 edition.

Other works of 1513. Albert of Saxony, p. 9, c. 1478; Ciruelo, p. 60, 1495; Gyraldus, p. 254, 1553; Licht, p. 70, 1500; Peurbach, p. 53, 1492; Johann Karl von Landshut (Lanzut), 'Algorithmus linealis,' Cracow, with editions, ib., 1515, 1519, 4° (see p. 83, 1504). 498; 508; 509

JOANNES MARTINUS BLASIUS, Villagarciensis.
Ed. pr. 1513. Paris, 1519.

See p. 95.
ORONTIUS FINAEUS, editor. See p. 160.

Title. 'Arithmetica // Ioannis // Martini, Scili-//cei, in theoricen, et praxim // ſciſſa, nuper ab Orontio Fine, Del//phinate, ſumma diligentia caſtigata, lon-//geqʒ caſtigatius q̄ prius, ipſo curā-//te impreſſa: omni hominū // conditioni perq̄ // vtilis,// & neceſſaria.// Vireſcit vulnere virtus.' Surrounded by an elaborate woodcut, with the following on four sides: 'Emissa ex officina Henrici Stephani, e regione // scholae Decretorvm Commorantis, // vbi et vaenalis reperitvr. // Parisiis anno Christi. 1519.') (F. 1, r.)

Description. Large 8°, 20.3 × 28.1 cm., the text being 13.3 × 20 cm. 64 ff. numb., 52 ll. Paris, 1519.

Editions. See p. 97.

The first half of this rare work, not found in the 1513 edition, is one of the best exponents of the Boethian arithmetic of the time. Finaeus, the editor, refers to the author's work in these words: 'Hanc Ioãnes Martinus, Sciliceus, Hiſpanus, vir Mathematicarũ peritus, noſtra tẽpeſtate Pariſijs edidit.' The author shows a good knowledge of the ancient writers, mentioning particularly Pythagoras, Nicomachus, Euclid, Apuleius, and Boethius, together with Jordanus, Faber Stapulensis, and Clichtoveus. The distinctive superiority of this part of the work lies in the clearness and arrangement of the illustrations of the various classes of numbers defined. The theory of numbers ends on f. 24, v., the practical part beginning on f. 25, v. Among the most noteworthy features of this half of the work is a 'Tabvla mvltiplicationis et divisionis' with all products to 50 × 50.

RAGGIUS FLORENTINUS. Ed. pr. 1514. Florence, 1520.

A Florentine mathematician of the fifteenth century.

¶ In hoc opuſculo hec continentur.

¶ Quid ſit proportio & quot eius ſpecies

¶ Quo intellectu compoſitio & diuiſio proportionum accipiatur & male opinantium confutationes

¶ Que maior minorue proportio dicenda ſit

¶ Quid propinquitas & remotio

¶ Confutationes argumentorum calculatoris

Fig. 54. Title page of Raggius

Title. See Fig. 54.

Colophon. '¶Impreſſum Florẽtiæ Bernardum Zucchettã // Anno. M.D.XX. Ianuarij. XV.' (F. 11, v.)

Description. 4°, 13.4 × 20.1 cm., the text being 9.2 × 15 cm. 11 ff. unnumb., 34–37 ll. Florence, 1520.

Editions. Florence, 1514, 4°; ib., 1520, 4° (here described).

This work consists of a theoretical treatment of proportion. While partly arithmetical, this treatment relates to the fundamental theory, and is equally applicable to geometry. The book is dedicated to the illustrious Giovanni Salviati, uncle of Cosimo I, Grand Duke of Tuscany.

GUILLIELMUS BUDAEUS. Ed. pr. 1514. Florence, 1562.

GUILLAUME BUDÉ. Born at Paris in 1467; died at Paris, August 23, 1540. Son of Jean Budé, grand audiencer of France. He became secretary to Louis XII, master of requests to François I, royal librarian, and ambassador to Leo X. He was a man of great erudition, and was instrumental in founding the Collège de France.

Title. 'Trattato // delle Monete // e Valvta loro,// Ridotte dal coſtume antico, all'vſo mo-//derno, Di M. Guglielmo // Bvdeo. // Tradotto per M. Giouan Bernardo // Gualandi Fiorentino.// In Fiorenza; // Apresso I Givnti // MDLXII.// Con licenza, & Priuilegio.' (P. 1.)

Colophon. 'In Fiorenza apreſſo gli heredi di // Bernardo Giunti // 1562.' (P. 318.)

Description. 8°, 10.5 × 17.2 cm., the text being 6.8 × 13.1 cm. 8 pp. unnumb. + 3 blank + 309 numb. = 320 pp., uncut, 28 ll. Florence, 1562.

Editions. The dedicatory epistle is dated 'Da. Viterbo il xxx. d'Agoſto. MDLXI,' so that this is the first (as it is the only) edition of Gualandi's translation of Budaeus. I have not compared it with the 'Libri de asse et partibus ejus,' Paris, 1514; second edition, Venice, Aldus, 1522. Although on the title page it is called a translation, the various books, six in number, begin 'Trattato delle // Monete // di M. Gio. Bernardo // Gvalandi Cittadino // Fiorentino,' leading to the belief that it may have been rewritten by Gualandi. 508

This treatise is purely historical, describing in a prolix manner the ancient measures, a subject of interest to arithmeticians in the sixteenth century on account of the great number of tables of denominate numbers in use in Italy, France, and Germany.

There was also a work by Budaeus entitled 'Minervæ Aragoniæ Assis Budeani supputatio compendiaia ad monetam ponderaque et mensuras Hispanię nostrę,' etc., published at Saragossa in 1536, 8°.

JOHANN BÖSCHENSTEYN.

Ed. pr. 1514. Augsburg, 1514.

BESCHENSTEIN, BOESCHENSTAIN, BOSSENSTEIN, BOECHSENTEIN, BUCHSENSTEIN, POSCHENSTEIN, BESENTINUS, etc. Born at Esslingen, Swabia, in 1472; died in 1532. He taught Hebrew at the universities of Ingolstadt and Heidelberg, and also at Antwerp and Nürnberg. Luther and Melanchthon were among his pupils.

Title. See Fig. 55.

Colophon. 'Getruckt in der Kayſerlichen ſtat Augſpurg durch // Erhart ŏglin Anno 1514 Jar.' (F. 24, r.)

Description. 4°, 14.4 × 19.2 cm., the text being 8.9 × 14.8 cm. 24 ff. unnumb., 30 ll. Augsburg, 1514.

Editions. Augsburg, 1514, 4° (here described); ib., 1516; ib., 1518. Böschensteyn is also said to have published at Augsburg in 1514 'Ein New geordnet Rechenbüchlein auf den linien mit Rechenpfennigen,' 4°, but this is doubtless Köbel's work (p. 102).

This is one of the more interesting of the early German arithmetics. It is mercantile in character, and presents in condensed form the essentials of business arithmetic. Among the peculiarities of the book is the use of 'figures' for 'species.' Böschensteyn gives seven of these fundamental operations: 'Das scind nun die Siben figuren,' 'Die Erſt figur Numeratio,' 'Die Ander figur Additio,' etc. He includes Duplatio and Mediatio, and he checks all of his work by casting out nines. His applications are chiefly in the 'Regula de Try,' partnership, and 'Regula Fuſti' (where he gives his rule in verse).

JAKOB KÖBEL. Ed. pr. 1514. Augsburg, 1514.

KOBEL, KOBELIUS, KOBILINUS. Born at Heidelberg in 1470; died at Oppenheim, January 31, 1533. He studied at Cracow, where Copernicus was his fellow-student. He was a man of varied attainments, meeting with success as a Rechenmeister, printer, engraver, woodcarver, poet, and public official.

Title. See Fig. 56.

Colophon. 'Getruckt tzů Augſpurg durch Erhart ŏglin. // Anno M.D.XIIII.' (F. XXIIII, r.)

Description. 4°, 13.7 × 18.9 cm., the text being 9 × 15 cm. 6 ff. unnumb. + 24 numb. (in Roman) = 30 ff., 30–35 ll. Augsburg, 1514.

Ain New geordnet Rech
en biechlin mit den zyffern
den angenden schülern zů nutz In
haltēt die Siben species Algorith=
mi mit sampt der Regel de Try/vnd sechs regeln d̄
prtich/vñ der regel Fusti mit vil andern gůten fra=
gen den kündern zum anfang nützbarlich durch
Joann Böschensteyn von Esslingen priester
neülych auß gangen vnd geordnet.

FIG. 55. TITLE PAGE OF THE 1514 BÖSCHENSTEYN

Editions. Köbel's Rechenbuch appeared under such varied titles and in such different combinations with his other books that it is difficult to say whether a given edition is a new work or merely a revision. It will aid the student if he recognizes in the first place that Köbel wrote three distinct books, (1) the 'Rechenbüchlein,' (2) 'Mit der Kryden,' (3) the 'Vysierbuch.' The 'Rechenbüchlein' first appeared at Augsburg in 1514, 4°; the 'Vysierbuch,' a treatise on gauging, at Oppenheim, s. a. (1515); and 'Mit der Kryden' at Oppenheim in 1520.

When the 1518 edition of the Rechenbüchlein (p. 108) appeared the title was changed, and a few pages were slightly altered. The 1531 edition (p. 108), however, shows many changes, certain chapters being entirely rewritten, and others considerably expanded. Although bearing a similar title, this might with some justice be called a different treatise; and yet it is so manifestly a revision of the 1514 work that it may more properly be classed as a new edition.

The three books were sometimes published as one and sometimes separately. The following list of editions is, therefore, probably incomplete, and it should be understood that any book mentioned may have been published with some other one.

Editions of the Rechenbüchlein: Augsburg, 1514, 4° (p. 100); Oppenheim, 1514, 4° (p. 106); ib., s. a. (c. 1515); Augsburg, 1516, 4°; 1517, 4°; Oppenheim, 1518, 4° (p. 108; 'zům Dritte male gebeſſert,' and hence the third revision, although at least the sixth edition); two other editions before 1520 (Unger); Oppenheim, 1522, 8°; ib., 1525, 12°; Frankfort, 1527, 8°; ib., 1531, 8° (p. 108); Oppenheim, 1531; ib., 1532; ib., 1535; Frankfort, 1537, 8° (p. 110); ib., 1544, 8°; ib., 1549; ib., 1564 (p. 111); ib., 1573; 1575; Frankfort, 1584, 8°.

Editions of 'Mit der Krydē od' Schreibfedern/ durch die zeiferzal zu rechē // Ein neüw Rechēpüchlein/ den angenden Schulern d' rechnūg zu erē getruckt': Oppenheim, 1520; Frankfort, 1537, 8° (p. 110); probably included in various other editions of the Rechenbüchlein.

Ain New geordnet Rech
en biechlin auf den linien
mit Rechen pfeningen: den
Jungen angenden zu heis
lichem gebrauch vnd hend
eln leychtlich zu lernen
mit figuren vnd exempeln
Volgt hernach klär-
lichen angezaigt.

Fig. 56. Title page of köbel's *Rechenbiechlin* (1514)

55	LV	90	XC	4000	IIII^{M}
56	LVI	91	XCI	5000	V^{M}
57	LVII	92	XCII	6000	VI^{M}
58	LVIII	93	XCIII	7000	VII^{M}
59	LIX	94	XCIIII	8000	VIII^{M}
60	LX	95	XCV	9000	IX^{M}
61	LXI	96	XCVI	10000	X^{M}
62	LXII	97	XCVII	20000	XX^{M}
63	LXIII	98	XCVIII	30000	XXX^{M}
64	LXIIII	99	XCIX	1400	M CCCC
65	LXV	100	C		M IIII^{C}
66	LXVI	101	CI	1500	M V^{C}
67	LXVII	102	CII		M D
68	LXVIII	103	CIII	1514	M V^{C} XIIII
69	LXIX	104	CIIII	1600	M D C
70	LXX	105	CV		M VI^{C}
71	LXXI	106	CVI	1612	M VI^{C} X II
72	LXXII	107	CVII	1700	M DCC
73	LXXIII	108	CVIII		M VII^{C}
74	LXXIIII	109	CIX	1715	M VII^{C} XV
75	LXXV	110	CX	1800	M DCCC
76	LXXVI	111	CXI		M VIII^{C}
77	LXXVII	112	CXII	1820	M VIII^{C} XX
78	LXXVIII	112	CXIII ꝛc.	1900	M VIIII^{C}
79	LXXIX	200	CC		
80	LXXX	300	CCC		
81	LXXXI	400	CCCC		
82	LXXXII	500	V^{C}		
83	LXXXIII	600	VI^{C}		
84	LXXXIIII	700	VII^{C}		
85	LXXXV	800	VIII^{C}		
86	LXXXVI	900	IX^{C}		
87	LXXXVII	1000	M		
88	LXXXVIII	2000	II^{M}		
89	LXXXXI	3000	III^{M}		

Dem nach mag ſtu die bayde zale durch ainn ander lernen erckennen vnd rechen.

FIG. 57. A PAGE FROM KÖBEL'S *Rechenbiechlin* (1514)

bedeüt diß figur der selben tayl ains.

$\frac{I}{IIII}$ Diesse figur ist vñ bedeüt ain fiertel von ainez gantzen/also mag man auch ain fünfftail/ayn sechstail/ain sybentail oder zwai sechstail 2c. vnd alle ander brüch beschreiben/Als $\frac{I}{V}$ | $\frac{I}{VI}$ | $\frac{I}{VII}$ | $\frac{II}{VI}$ 2c.

$\frac{VI}{VIII}$ Diß sein Sechs achtail/das sein sechstail der acht ain gantz machen.

$\frac{IX}{XI}$ Diß Figur betzaigt ann newn ayilfftail das seyn IX tail/der XI ain gantz machen.

$\frac{XX}{XXXI}$ Diß Figur betzaichet/zwentzigk ainundreyssigk tail/das sein tzwentzigk tail der ainsundreissigk ain gantz machen.

$\frac{II^{C}}{IIII^{C}.LX}$ Diß sein tzwaihundert tail/der Fierhundert vnd sechtzigk ain gantz machen.

Auß den obgeschriben Figuren/vñ Exempeln magstu leichtlich lernen all brüch/die dir in deinen rechen fürkommen/wie du die. versteen/anschreiben vnnd außsprechen solt/Vnd wil dir nün hyernach orden/vnnd setzen etlich Regeln/Fragen/vnnd Exempel in gebrochen tzalen/wie man die durch die Regel Try rechenn vud aufflösen solle.

Die Erst Regel So dir ain frag fürkombt/in deren die erst zale geprochen ist/vnd die mittel vnd letste tzale gantz pleiben/So mustu die erst tzale auch brechen/in den brüch der bey ir geschriben stet Des geleichen müstu die letst tzal auch brechen durch den selben brüch vnd geschicht das auß d̃ vrsach dye weyl alweg dye erst vnnd letst tzale in bedeütniß gelcich sein sollen/vnnd so das also geschicht/So solt

FIG. 58. A PAGE FROM KÖBEL'S *Rechenbiechlin* (1514)

Editions of the 'Vysierbuch': Oppenheim, s. a. (1515), 4° (p. 113); Frankfort, 1527; ib., 1531, 8° (p. 108); Oppenheim, 1531; ib., 1532; s. l., 1584; probably included in various other editions of the Rechenbüchlein (p. 111, 1564). 511

This is the first edition of this well-known arithmetic. As already stated, the title was occasionally changed, but the work was essentially but little altered. It is a purely commercial book, with all of the operations performed by counters as was still the custom of the time in most parts of Germany. Köbel treats of the rule of three ('die Gulden Regel, die von dem Walen de Try genant wirt'), partnership, reduction, inheritances, and exchange. The fundamental operations include progressions, and Roman numerals are used except in the section on notation (Fig. 57). Köbel makes a curious use of the Arabic method of writing fractions, the terms being written in Roman, as in the case of

$$\frac{\text{II}^{\text{C}}}{\text{IIII}^{\text{C}}.\text{LX}} \text{ for } \frac{200}{460} \text{ (see Fig. 58).}$$

Altogether, Köbel was a vigorous writer, and his Gothic style shows him to have been no more a follower of the Italian arithmeticians than Dürer and Holbein were of the Italian artists.

Other works of 1514. Boethius, p. 27, 1488; Grammateus, p. 123; Jordanus, p. 65, 1496; Stromer, p. 83, 1504; Torrentini, p. 76, 1501; Hermannus Buschius, 'Algorithmus linealis Proiectiliū: de ītegris perpulchris Arithmetice artis regulis: earundemque probationibus claris exornatus,' Vienna, 4°, 4 ff. (also catalogued as anonymous, and as written by Johannes Cusanus; see p. 43, c. 1490, for Nicolaus Cusa; see also p. 86, 1505). There was also published c. 1514 an anonymous work entitled 'Arithmeticæ practicæ Tractatus qui dicitur Algorismus, cum additionibus utiliter adjunctis,' Paris, 4°. 511

JAKOB KÖBEL. Ed. pr. 1514. Oppenheim, 1514.

See p. 100.

Title. The title page is missing. Fol. A ij begins: 'Dem Ernueſten Dietherichen // Remerer von Wormbs:genant von Dalburgk: // meinen beſundern günſtigen lieben Junck-//herzen/ Eubeütich Jacob Kŏbel/ //dieſſer zeit Statſchreiber zů Op-// penheim/ mein willig dinſt // allerzeyt bereydt // zů vor.'

Colophon. 'Getrückt zů Oppenheym.// Anno. ꝛc M.CCCCC. XIIII.' (F. no. XXIIII. See Fig. 59.)

dreien zelen/vnd wz ym über drey überig bleybt/ das
heyß dir sagen/vnd schreib es eygentlich auff/ Noch
dem heiß yn das selb sein gelt noch ein mal gewißlich
mit fünfen zelen/vñ was ym als dañ aber mals über
fünff über bleibt/heiß dir auch sagen/dz schreib aber
auff/ Züm Lesten/heiß dir dasselbig gelt widerumb
mit syben zelen/vnd was ym als dañ abermals über-
bleibt schreib auch vff/ So dz also gewyß volbracht
wirt/so lüg wie viel züm ersten über drey überblyben
ist/ Finstu dañ .i. überigk/ so lege vff die linien LXX.
Findestu aber .ii. so lege dar für vff die linien CXL.
Findestu gerad iij. überig/ so lege vff die linien CCX.
Dar nach lüg was noch dez andern zelẽ mit fünffen
über blieben ist. Vnd als manich mal als du eins fin-
dest/als offt leg XXI. vff die linien. Züm dritten hab
acht/vie viel einiger zale noch syben überblyben ist/
so du das weist soltu alweg vor eins XV. vf die linien
legen/So das also volnbracht so thu die drey zale zu
sammen/iñ ein Sum̃/vnd was darauß wirt/ douon
soltu als dick du kanst hundert vnd funfzehen douon
zyhen/was als dañ am lestẽ überbleibt/ so viel ist der
Sum̃ des gelts gewessen dz der iñ seynẽ seckel gehabt
hat/vnd ist recht gemacht.

Also wil ich diessem Rechen-
biechlein/ meinem verheiß noch/ genüg gethon/vnd
ein Ende geben/vñ so ich empfind mich do mit icht
nütz geschafft haben/ wil ich kunftigklich mit gottes
vnd der Hochgelertẽ hilff dz meren vnd bessern/auch
von Vysiren/Veld zu messen/ꝛc/vnd andere notturf-
tige ding dẽ leyen zü bewslichem gebrauch Tolmet-
schen /vnd auß der Arismetrick vnd Geometrei offen
baren.

Getrückt zü Oppenheym.
Anno. ꝛc M. CCCCC. XIIII.

FIG. 59. LAST PAGE OF KÖBEL'S *Rechenbiechlin* (OPPENHEIM, 1514)

Description. 4°, 14.1 × 19.6 cm., the text being 9.2 × 15.8 cm. 5 ff. unnumb. + 24 numb. = 29 ff., 28–34 ll. Oppenheim, 1514.

See p. 106. It is a curious fact that the first and second editions should have appeared in the same year at two different places.

JAKOB KÖBEL. Ed. pr. 1514. Oppenheim, 1518.

See p. 100.

Title. See Fig. 60.

Colophon. 'Alſo Endet ſich Seliglich d' Drit // Trůck diſʒ Rechenbüchleins/ Zů Oppenheym zůſam-//men geſegt/ vnd volendt vff Dinſtag des Hey-//ligen Crüg Erhôhungs tag.// Anno dñi .1518.' (F. XLVI, r.)

Description. 4°, 13.7 × 18.4 cm., the text being 9.2 × 14.8 cm. 4 ff. unnumb. + 46 numb. (I to XLVI) = 50 ff., 32–33 ll. Oppenheim, 1518.

See p. 106.

JAKOB KÖBEL. Ed. pr. 1514. Frankfort, 1531.

See p. 100.

Title. 'Ein new geor-//denet Künſtlich Rechenbůchlin/ Ja-//cob Kôbels/ Stattſchreiber zů Oppen-//heim/ Auff den Linien vnd Spacien/ mit Rech-//enpfeñingen. Den angehnden Schůlern Rech//nens gantz leichtlich zů lernen. Vnd zů Kellerei//en Ampten/ Kauffmanſchafften/ vnd Krâme-//reien dienlich vnd brâuchlich. Mit vilen erklâ-//rungen/ Leren/ Regeln vnd Exempln.// ☞ Mehr dañ vorm̃als ie getruckt/ //gebeſſert/ vnd zůgeſetzt.// Im Jare M. D. XXXI.// Viſir Bůchlin/ //Den Jungen/ angehnden/ // Leijſchen Viſirern gantz leichtlich zů lernen/ //verſtehn vnd Rechnen. Wie mañ ein // Viſirrůt machen/ vnnd damit ein // iedes vaſz viſirenn/ ſolle Zů // ende diſes Rechenbůch//lins angehenckt.' (F. 1, r.)

Colophon. '¶Getruckt zů Franckenfurt/ am Mein. In Ver// lag vnnd Gemeynſchafft des Ernhaffetnn // vnd Fürnemen Herrn Jacoben Kô-//bels/ Stattſchreiber zů Oppen-//heim. Bei Chriſtian Ege-//nolffen. Im Mertzen/ // Nach der geburt // Chriſti. // M.D.XXXI. Jar.' (F. CXII, r.)

Das new Rechē
püchlein Wie mann vff den
Linien vnnd Spacien / mit
Rechēpfenningē / Kauffmanschafft
vnd Tegliche handelungē / leichtlich
rechē lernē mage / zům Drittē male /
gebessert vñ zů Oppenheym getruckt.

1518

Zů dem Kauffer.

¶ Wer do begert leichtlich zu lern
Wie mañ Rechen sol / vñ wüst gern /
Jñ kauffmāschafft / gewyn / verlůst /
Das er nit arbeit thet vmb sůst
Sein ampt / vnd Kellery / vßricht /
Das er nit würd / veracht vernicht /
Der Kauff diß Buch vñ leß mit fleiß
Vnd merck / was ich yn vnderweiß /
Dem kom̃ er noch mit freiem můt
Behůt vns got / Seel / Leib / vñ gůt.
AMEN.

FIG. 60. TITLE PAGE OF KÖBEL'S *Rechēpüchlein* (1518).

Description. 8°, 9.7 × 14.9 cm., the text being 6.9 × 11.5 cm. 112 ff. numb. in Roman, 28–29 ll. Frankfort, 1531.

Editions. See p. 102.

It is often stated that the 'Jacob's staff' used by surveyors was first described by Köbel in this year (1531) and that it received its name in his honor. The name was old before this time, however, as applied to some form of surveying instrument, for in the Margarita Philosophica (1503, Bk. VI, tract. II) is a conversation between a master and his disciple, in which it is mentioned: 'MAG. baculo quē Iacob dicunt. DIS. Qualis is eſt baculus?' The master thereupon describes the instrument, and a picture of it is given.

JAKOB KÖBEL. Ed. pr. 1514. Frankfort, 1537.

See p. 100.

Title. 'Zwey rech-//enbůchlin: vff der // Linien vnd Zipher/ Mit eym angehenck-//ten Viſirbuch/ ſo verſtendtlich für//geben/ das iedem hieraufz on // eiñ lerer wol zulernen.// ℂDurch den Achtbarn vnd wol erfarnen // H. Jacoben Kőbel Statſchreiber // zu Oppenheym.// Franc. Chriſt. Egen.' (F. 1, r.)

Colophon. 'Ende/ Im Iar M.D.XXXVII.' (F. numb. 144, r.)

Description. 8°, 9.5 × 15 cm., the text being 7.1 × 11.9 cm. 8 ff. unnumb. + 9–144 numb. = 144 ff., 30 ll. Frankfort, 1537.

Editions. See p. 102.

This is the earliest of Mr. Plimpton's copies containing the three books written by Köbel, (1) 'Rechenbüchlein,' (2) 'Mit der Krydē,' (3) the 'Vysierbuch.' The combination of the three in a single volume formed one of the best books of the time, giving the operations both with counters and according to algorism. The latter is given in the part entitled 'Mit der kreiden // odder ſchreibfederen/ durch // die ziffer-zal zu rechen/ Ein new Rechen-//bůchlin/ den angehnden ſchůlern der // rechnung zu eren getruckt.' (F. 106.) In this work Köbel also includes the usual business problems of the period and the chapter on the calendar required by the Church schools. An unusually complete treatment of gauging is given in the part entitled 'Eyn new Viſir // Bůchlin/ den Leyen/ zu // leichtem vnd begreiflichem verſtandt // verordnet/ Durch H. Jacob // Kőbel Stattſchreiber zu // Oppenheym.' (F. 95, r.) It is much more complete than the 1514 edition of the 'Rechenbüchlein' (see p. 106), and is substantially the same as the

1531 edition (see p. 108) except that it contains the third part, 'Mit der Kreiden,' which the latter does not.

The Hindu-Arabic numerals were still considered difficult ('den die Zifer zal am erſten zulernen ſchwere,' f. 9, v.), and teachers still felt it better to begin with the common Roman characters ('wil ich zům erſten die ſelb Teutſche zal . . . hie anzeygen vn̄ erkleren').

JAKOB KÖBEL. Ed. pr. 1514. Frankfort, 1564.

See p. 100.

Title. 'Rechenbůch/ //Auff Linien vnd Ziffern. //Mit einem Viſir bůchlin/ Klar // vnd verſtendtlich fůrgeben. // Gerechnet Bůchlin/ auff alle // Wahr vnd Kauffmanſchafft / Müntz/ // Gewicht/ Elen/ vnd Maſz/ viler Land // vnd Stett verglichen.// Durch H. Jacob Köbel. // (Woodcut.) Cum Gratia & Priuilegio.// Franckfurt/ Bei Chr. Egen. Erben // M. D. LXIIII.' (F. 1, r.)

Colophon. 'Getruckt zu Franckfurt // am Mayn bey Chriſtian E-//genolffs Erben.// M. D. LXIIII.' (F. 194, v.)

Description. 8°, 9.1 × 15.2 cm., the text being 6.3 × 12.2 cm. 12 ff. unnumb. + 194 numb. = 206 ff., 28–29 ll. Frankfort, 1564. (There were at least eight Frankfort editions, of which this is the sixth.)

Von verkerten fragen. 57
acht haſt/ vnnd jhm alſo nachkompſt/ wir-
ſtu leichtlich lernen rechnen/ alles das dir in
gmeinem kauffen vnd verkauffen fürkom-
met in gantzen zalen.
¶ Ein ander Exempel.

Ein Fraw oder Haußmutter gehet auff
den marckt/ kaufft vberhaupt ein Körblin
mit Rebnerbyrn/ darumb gibt ſie achtzehen
pfenning / ſo ſie heim kompt / findet ſie im
körblin hundert vnd achtzig byrn / Iſt die
frag/ wie vil byren ſie vmb ein pfenning ha-
be? Thu/ als obgelert/ ſo kompt dir zehen/
Alſo vil byren hat ſie vmb einen pfenning / Vnd iſt wolfeyl
drumb.
¶ Ein anders/ von Heringen.
Ein Kuchenmeyſter geht an einem Frei-
J v

FIG. 61. FROM KÖBEL'S *Rechenbuch* (1564)

Editions. See p. 102.

The first 165 ff. are the same as in the 1537 edition (see p. 110). The rest of the book is not found in the 1537 edition, or in any other edition in the Plimpton library. It consists of a description of foreign and domestic money, with numerous illustrations of coins. This begins

von Wandern.

¶ Von wandern vber Landt.

ZWen Bürger auß Oppenheym/ Einer Son Heynrich/der ander Contz võ Treber genant/ wolten mit einander gen Rom gehen/Vnd Heynrich was alt / vnd mocht einen tag nicht mehr denn zehen meilen gehen/Aber Contz von Treber was jung vnd starck/der mocht einen tag 13. meilen gehen/ Deßhalben gieng Son Heynrich neun tag ehe auß Oppenheym denn Contz von Treber / Also war Son Heynrich Contzen 90. meilen fürgangen/ ehe Contz angehabẽ hat außzugehen.

Nun ist die frag / in wie vil tagen Contz von Treber Son Heynrichen vbergangen/ vnd die zwen zusamen kommen seind?

Die vnd dergleichen frag wirstu also ent scheiden. Zum ersten leg die meilen/die Son Heynrich Contzen fürgangen ist/ das seind 90. meilen/vff dein Rechenbanck/ Vnd nim dann die meilen/die Conrat einen tag mehr hat

FIG. 62. FROM KÖBEL'S *Rechenbuch* (1564)

(f. 165, r.): 'Von Frembden vnd Hie//lăndiſchen Mŭntzen/ So diſer zeit in // Teutſch vnnd Welſchen landen/ inn aller // Kauffmanſchafft vnd Gewerb/ Hăndeln/ //viler Land art im brauch/ geng/ gibig oder // verruffen Mŭntzen/ . . .' The book closes (f. 184, v.) with a set of tables and (f. 192, v.) a 'Regiſter.'

JAKOB KÖBEL. Ed. pr. 1515. Oppenheim, s. a. (1515).

See p. 100.

Title. 'Eyn New geordēt // Vyſirbůch. Helt yñ.// Wie mañ vff eins yden Lands Eych // vñ Maſz/ ein gerecht Vyſirůt machē

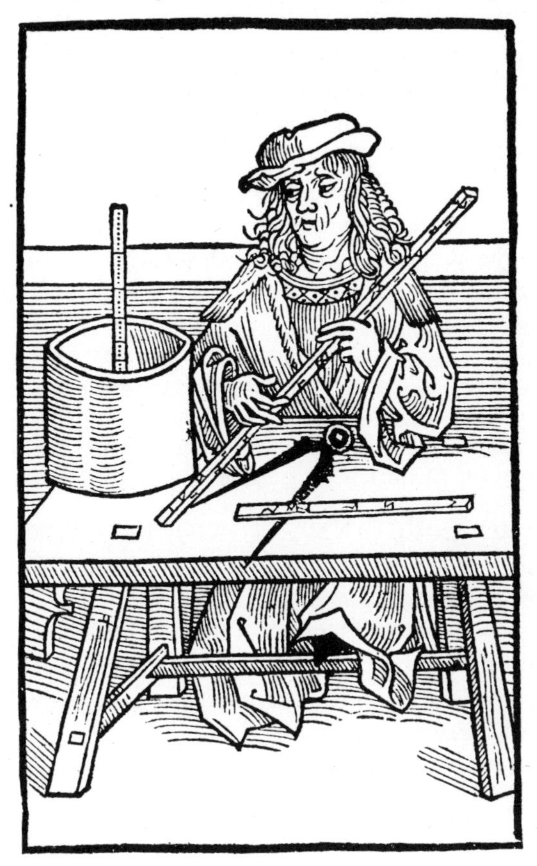

Getruckt zu Oppenheym

FIG. 63. LAST PAGE OF KÖBEL'S *Vysirbůch* (1515)

// vñ do mit ein ygklich onbekant Vaſz // vyſieren/ auch ſeynen inhalt erlernen // ſolle. Den anhebenden Schůlern Vi//ſirens

Leichtlich/ mit Figuren vnnd // Exempeln/ zů lernen/ angezeigt. // Angehengt Tafeln.// Die Erſten Fyer halten yn̄ gerechet/ // ſo eyn Fůder weins kaufft wirt/ vmb // Guldē zu XXVI. Od' XXIIII. A ℔. ꝛc.// Was die Ome/ das Fyrtel/ vnnd die Maſz gelten.// Die Andern Tafeln/ Zeygē an/ Ver//anderūg vn̄ wechſelūg einer Müntz // durch die ander/ als ꝯ. in̄ ℔. ꝛc. // Gedruckt zu Oppenheym.' (F. 1, r.)

Colophon. 'Gedruckt zu Oppenheym.' (F. 32, v. See Fig. 63.)

Description. 4°, 14.4 × 19.7 cm., the text being 9.4 × 14.5 cm. 4 ff. unnumb. + 28 numb. = 32 ff., 30–32 ll. Oppenheim, s. a. (1515). There is no date on the title page nor in the colophon, but the prefatory statement closes with the words, 'Volnendet vff dornſtag noch Letare. Anno & c. 1515.'

Editions. See p. 106.

The work is semiarithmetical, quite as much so as the chapters on mensuration in our textbooks; chapters, indeed, which owe their origin in no small degree to these treatises on gauging so often appended to the old arithmetics. This work is illustrated with quaint woodcuts showing the use of the 'Vyſirſtab' or gauging measure. (See Fig. 63.)

GIROLAMO AND GIANNANTONIO TAGLIENTE.

Ed. pr. 1515. s. l. (Venice), 1525.

Venetian arithmeticians of c. 1500.

Title. 'Opera che // insegna // A fare ogni Ragione // de Mercātia // et a pertegare le Terre // Con arte giometrical // Intitolata Componimēto // di arithmetica // Con gratiª & preuilegio // M. D. XXV.' (F. 1, r.)

Description. 8°, 10.3 × 15.1 cm., the text being 7.5 × 12.5 cm. 91 ff., 28–33 ll. S. l. (Venice), 1525. 495; 511; 512

Editions. Venice, 1515, 8° (De Morgan having erred in saying 'apparently before 1500'); ib., 1520; ib., 1523, 8°; ib. (s. l.), 1525, 8° (here described); ib., 1526; ib., 1527; ib., 1528, 4°; s. a. (1530?), 8°; Venice, 1541; Milan, 1541, 8° (p. 115); s. l., 1547, 8°; Milan, 1548; Venice, 1548 (with probably a second Venetian edition, 1548, under the title 'Thesoro universale de abacho,' by 'lucha ātonio de Uberti,' 8°); ib., 1550; ib., 1554; ib.,

1557; ib., 1561; ib., 1564; ib., 1567; ib., 1570; Milan, 1570; ib., 1576; ib., 1579; ib., 1586. Riccardi also mentions eleven other editions, s.a., and four such appeared in the Boncompagni sale, and four in the Fisher sale of 1906. The work also appeared, and is frequently catalogued, without the authors' names.

These various editions have been the object of critical study by E. A. Cicogna and Prince Boncompagni. The former set forth his results in his *Saggio di bibliografia Veneziana*, Venice, 1847, p. 218, ascribing the work to Girolamo Tagliente 'con l'ajuto del suo consanguineo Giannantonio Tagliente.' In the edition of 1525, here described, only the former name appears, the text beginning, '¶Al benigno lettore // Hieronymo Tagliente.' Boncompagni's investigations, setting forth the differences in the various editions, appeared in the *Atti dell' Accademia Pontif. de' Nuovi Lincei*, XVI, 139, 147, 155, 304. See also Riccardi, I, 2, 484, and Boncompagni's *Bulletino*, XIII, 247. 495; 511; 512

There was also a treatise published by the Taglientes entitled 'Regole di mercatura intitolato componimento di arithmetica,' Venice, 1524, 8°, probably another edition of this work. See also the treatise on bookkeeping mentioned on p. 141, 1525.

The book opens with a brief treatment of notation and finger symbols. Then follow in order the multiplication table, the proof of sevens, various methods of multiplication, division by the galley method, addition chiefly of denominate numbers, subtraction, the operations with fractions in the same order, exchange, rule of three, and applied problems. There are numerous interesting woodcuts, and such familiar problems as those of the couriers, the testament, and the sale of eggs are given with illustrations. In spite of the arrangement of topics, there were few textbooks so influential as this in shaping the subsequent teaching of arithmetic.

GIROLAMO AND GIANNANTONIO TAGLIENTE.

Ed. pr. 1515. Milan, 1541.

See p. 114.

Title. 'Libro // dabaco che in//segna a fare // ogni ragione mercadantile, & // pertegare le terre cō l'arte di // la Geometria, e altre no//bilisime raginoe ſtra-//ordinarie cō la Ta-//riffa come reſpon//deno li peſi & // Monede de molte terre del mon-//do con la inclita citta di Vene-//gia. Elquel Libro ſe chiama //Theſauro vniuerſale.' (F. 1, r.)

Colophon. 'Stampato in Milano per Io. Antonio da Borgho.// Nell'anno del. M. D. XLI.' (F. 80, v.)

Description. 8°, 10 × 14.6 cm., the text being 7.7 × 12.3 cm. 80 ff. unnumb. + 7 blank = 87 ff., 23–30 ll. Milan, 1541.

Editions. See p. 114.

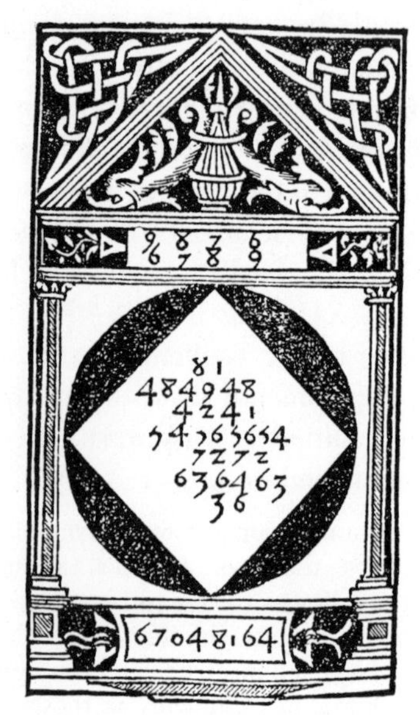

Fig. 64. From the 1541 Tagliente

This differs but little from the 1525 edition, the 'Opera che insegna' (see p. 114), except in having a set of tariff tables at the end: '¶Qui comenza el terzo Libro di la fruttifera opera // chiamata la Tariffa' (f. K iiii, v.). There is some slight change in the phraseology, particularly at the beginning of the various sections. For two curious forms of multiplication see Fig. 64. Such arrangements of the work in multiplication were quite common, particularly in the early Spanish and Italian arithmetics of the first half of the sixteenth century. That they should have found place in a popular mercantile treatise is, however, rather surprising.

VARIOUS AUTHORS. Ed. pr. 1515. Vienna, 1515.

JOANNES DE MURIS (JEAN DE MEURS, MURS, MURIA) was born in Normandy, c. 1310; died after 1360. He wrote on arithmetic, astronomy, and music.

THOMAS BRADWARDIN. See p. 61.

NICOLAUS HOREM (NICOLAS ORESME) was born at Caen (?) c. 1323; died at Lisieux, July 11, 1382. He taught in the Collège de Navarre at Paris, and in 1377 became Bishop of Lisieux. He wrote also an 'Algorismus Proportionum,' in which the idea of fractional exponents first appears.

GEORG VON PEURBACH. See p. 53.

JOANNES DE GMUNDEN (JOHANN VON GMUNDEN, JOHANN WISSBIER? NYDEN? SCHINDEL? JOHANNES DE GAMUNDIA) was born c. 1380, at Gmunden on the Traunsee, or Gemünd in Lower Austria, or Gemünd in Swabia; died at Vienna, February 23, 1442. He was educated at Vienna, and taught there, being the first professor of mathematics alone in Austria.

Title. See Fig. 65.

Colophon. 'Impreſſum Vienne per Joannem Singrenium // Expenſis vero Leonardi ⁊ Luce Alantſe // fratrum Anno domini M.ccccc.xv.// Decimono die Maij.' (F. 54, r.)

Description. 4°, 14.3 × 18.4 cm., the text being 9.7 × 15.5 cm. (without the marginal references). 54 ff. unnumb., 26 ll. Vienna, 1515.

Editions. There is no other edition of this combination of works. See pp. 53, 61, 118, for the individual treatises.

This interesting work consists of five parts. The first is the arithmetic of Joannes de Muris. While it is called an extract from the arithmetic of Boethius, it is merely suggested by that treatise and is really the work of Muris. This part of the work begins on f. 2 with the following title: 'Incipit Arithmetica cōmunis ex // diui Seuerini Boetij Arithmetica per M. Joannem // de muris compendioſe excerpta. // Prohemium.' De Morgan (p. 3) mentions a possibly earlier edition, s. l. a., 4°. (See also p. 86, 1505.)

The second part of the volume begins on f. 17, v., and is the work on proportion by Thomas Bradwardin. It is a theoretical treatment of the subject, and has the following title: 'Tractatus breuis proportionū: ab-//breuiatus ex libro de Proportionibus. D. Thome // Braguardini Anglici.'

The third part begins on f. 27, v., and is a treatise by Nicolaus Horem, with the following title: 'Tractatus de Latitudinibus forma-// rum ſcdm doctrinā magiſtri Nicolai Horem.' This subject attracted considerable attention in the latter part of the Middle Ages.

The fourth part is the algorism of Peurbach, and has the following title on f. 37, v.: 'Opuſculū Magiſtri Georgij // Peurbachij doctiſs.' As already stated (p. 53), this work takes up the four fundamental operations and progressions, giving merely a theoretical discussion of the subject.

The fifth part (f. 44, v.) begins: 'Incipit tractatus de Minucijs phi-//ſicis: compoſitus Vienne Auſtrie per magiſtrum // Joannem de Gmunden.' This is the treatise of Gmunden on sexagesimal fractions, or, as they were called in the Middle Ages, physical fractions. These fractions, still used by us in our degrees (or hours), minutes, and seconds, served the purposes of the later decimal fractions. They were carried much farther than is now the case, a number like $3^{\circ}\ 15'\ 40''\ 15'''\ 45^{\text{IV}}$ meaning merely $3+\frac{15}{60}+\frac{40}{60^2}+\frac{15}{60^3}+\frac{45}{60^4}$, or $3\ \frac{75221}{288000}$. The present symbolism (°, ′, ″) is relatively modern.

The book is particularly interesting because it combines in one volume five well-known books by mediæval writers. In no other

Contenta in hoc libello.

Arithmetica communis.

Proportiones breues.

De latitudinibus formarum.

Algorithmus. M. Georgij Peurbachij in integris.

Algorithmus Magiſtri Joannis de Gmunden de minucijs phiſicis.

Fig. 65. Title page of the Vienna work of 1515

single volume could the inadequacy of the mediæval treatment of mathematics be better seen. Indeed, a manuscript of 1515, found by Gerhardt in the Wolfenbüttler Bibliothek, expressly states that the lectures on arithmetic given in the universities of that period were based on the above works of Muris, Bradwardin, Peurbach, and Joannes de Gmunden. (*Monatsberichte der K. P. Akad. d. Wissensch. zu Berlin*, 1867, p. 43.)

JOANNES DE MURIS. Ed. pr. 1515. Mainz, 1538.

See p. 117.

Title. 'Arithme//ticae specvlativae // Libri duo Ioannis de Muris ab in-//numeris erroribus quibus hacte-//nus corrupti, & uetuſtate fer//mè perierant diligen-//ter emendati,// Pvlcherrimis qvoqve exemplis, Formisq; nouis declarati & in // uſum ſtudioſæ iuuentutis Mogun-//tinæ iam reccens ex-//cuſi.// Mogvntiae excvdebat // Ivo Scoffer anno.// M. D. XXXVIII.' (F. 1, r.)

Colophon. 'Mogvntiae excvdebat // Ivo Scheffer anno.// M. D. XXXVIII.' (F. 90, v.)

Description. 8°, 9.6 × 14.5 cm., the text being 6.6 × 11.9 cm. 4 pp. unnumb. + 3–88 numb. + 5 blank + 1 with woodcut = 96 pp. Mainz, 1538.

Editions. This is the second dated edition of the arithmetic of this popular mediæval teacher. (See p. 117.) It is more complete than the one of 1515, but it does not, like the latter, give the marginal references to Boethius, upon which it is so largely based. (See also Boethius, p. 27.)

See p. 117.

JOANNES FŒNISECA. Ed. pr. 1515. Augsburg, 1515.

An Augsburg teacher of c. 1500.

Title. See Fig. 66.

Colophon. 'Impreſſa Auguſtę Vindelicorum/ communibus impenſis Io/ //annis Miller atqȝ Ioannis fœniſecę. Anno a natiuita //te domini. M.D. XV. ad. IIII. Cal.' Maias.' (F. 20, r.)

Description. 4°, 15.1 × 21 cm., the text being 11 × 13.8 cm. 20 ff. unnumb., 7–39 ll. Augsburg, 1515.

Editions. There was no other edition.

This is an extract from a larger volume, for the folios have been numbered by hand 40–59, and the register begins with 'aa i.' Only two pages (aa ii, v., and aa iii, r.) are devoted to 'Arithmetica,' and these relate only to the Boethian system. The rest of the book is devoted chiefly to geometric figures, the mediæval astronomy, and music. Such a book shows the superficiality and general emptiness of the work of the schools that were supposed to stand for culture in the period of the early Renaissance.

40

Opera Ioannis Fœniſecæ Augū.
hec in ſe habent.

Quadratum ſapientię: continens in ſe ſeptem
artes liberales veterum.
Circulos biblię.iiii.in quibus metaphyſica
moſaica.
Commentaria horum.
Ad hęc/libri rubrica inferius
ſignati: neceſſarii ſunt.

ſermo	Grammatica	don.alex.gua.laſca.Poetę
	Logica	
	Rhetorica	
εθοϛ	Monaſtica	
.i.mos	Oeconomica	
	Politica	
	Alcarithmus ſubalternus	nouus
	Arithmetica	boetius
	Geometria	boetius
mathēatica	Perſpectiua ſubalterna	petrus iacobi
	Muſica	boetius
	Aſtronomia	boetius ephemerides
	Geographia	ptolemęus Hiſtorici
	Phyſica	
philoſophia		
	Medicina ſubalterna	dioſcorides
theologia	Metaphyſica	biblium triplex

aa i.

Fig. 66. Title page of Fœniseca's *Opera* (1515)

ANONYMOUS. Ed. pr. c. 1515. Leipzig, s. a. (c. 1515).

Title. 'Melchiar Lotthervs Ivnior candido lectori salutem.' (Line 1.) 'Articularis Bedae preſbyteri numerorū computatio.' (Line 24.)

Description. One sheet, 28 × 36.4 cm., the text being 21.6 × 26.7 cm. 42 ll. Printed on one side of a single sheet.

Editions. There was no other edition, and only this copy is known of this one.

This broadside was published by Lotter c. 1515, and is a brief statement of the numerical finger-symbolism of the ancients, particularly as described by the Venerable Bede. The symbolism is practically the same as that described by such writers as Paciuolo and Aventinus (see pp. 57 and 136.)

GASPAR LAX. Ed. pr. 1515. Paris, 1515.

Born at Sariñena, Spain, c. 1487; died at Saragossa, February 23, 1560. He taught at Paris and Saragossa. His only works are the two here described.

Title. See Plate III. 'Proportiones magiſtri Gaſparis // lax aragonenſis de ſarinyena . . .' (F. 101, r. Separately catalogued, see below.)

Colophon. 'Explicit Arithmetica ſpeculatiua Magiſtri Gaſparis Lax Aragonenſis de Sarinyena duode-//cim libris demonſtrata. Impreſſa Pariſius opera ac characteribus Magiſtri Nicolai de la barre.//Expenſis honeſti viri Hemundi le feure Bibliopole Pariſius in vico diui Jacobi ſub ſigno Creſcē//tis albi vitam degentis. Anno Domini .1515. Die vero .13. Menſis Decembris.' (F. 100, v.)

Description. Fol., 19.8 × 26.5 cm., the text being 14.1 × 21.8 cm. 100 ff. unnumb., 59–61 ll. Paris, 1515.

Editions. There was no other edition.

A very prolix treatment of theoretical arithmetic, based on Boethius and his mediæval successors. As the title shows, Lax was a Spanish teacher, one of several from the southern peninsula who taught in the University of Paris in the fifteenth century. Among the others were Rollandus (originally from Lisbon, mentioned later in connection with the manuscripts) and Ciruelo (p. 58). All of the contributions of these scholars were of this general theoretical character. De Morgan facetiously remarks, 'For anything that appears the author (Lax) could not count as far as 100.'

GASPAR LAX. Ed. pr. 1515. Paris, 1515.

See above.

Title. 'Proportiones magiſtri Gaſparis //lax aragonenſis de ſarinyena.// Venundātur Pariſius In vico diui //Jacobi ab Emundo le feure ſub ſi-//gno creſcentis albi vitam degente.' (F. 101, r.)

Colophon. 'Expliciunt proportiones Ma//giſtri Gaſparis Lax Aragonen//ſis de Sarinyena impreſſe Pa-//riſius opera Magiſtri Nicolai // de la barre pro Emundo le feure // Anno dñi M. d. xv. die vo vi. mē//ſis Octobris.' (F. 26, 126 of the whole book, r.)

Description. Fol., 19.5 × 26.6 cm., printed in double columns, each being 7 × 21.5 cm. 26 ff. unnumb., 66 ll. Paris, 1515.

Editions. There was no other edition.

Bound with the 'Arithmetica Speculativa' (p. 121). This is a prolix treatment of mediæval ratios after the Boethian manner, and as such it ranks with works like those of Bradwardin (p. 61), Jordanus (p. 62), and Faber Stapulensis (p. 82).

Other works of 1515. Boethius, p. 27, 1488; Bradwardin, p. 61, 1495; Köbel, p. 102, 1514; Lanzut, pp. 83, 97, 1504, 1513; Licht, p. 70, 1500; Ortega, p. 93, 1512; Peurbach, p. 53, 1492; Torrentini, p. 76, 1501; Juan Andrés, 'Sumario breve de la practica de la arithmetica,' Valencia (from the book it appears that it was written in Saragossa in 1514; it was reprinted at Seville in 1537); V. Rodulphus Spoletanus, 'De proportione proportionvm dispvtatio,' Rome, 4°.

Works of 1516. Johann Böschensteyn, p. 100, 1514; Capella, p. 66, 1499; Ciruelo, p. 60, 1495; Köbel, p. 102, 1514; Stromer, p. 83, 1504; Widman, p. 36, c. 1488. 511

PIETRO MARIA BONINI. Ed. pr. 1517. Florence, 1517.

A Florentine writer of the first half of the sixteenth century.

Title. 'Lvcidario darithmetica.' (Large woodcut. F. 3, r. Fig. 67.)

Colophon. '¶Impreſſo nella excelſa cipta di Firenze per // Gianſtephano di Carlo da Pauia // adi 7 di Gennaio.' (F. 18, r.)

Description. 8°, 10.9 × 15.9 cm., the text being 8 × 12.7 cm. 19 ff. unnumb., 24 ll. Florence, 1517.

Editions. There was no other edition. This interesting volume was known to De Morgan only by hearsay when he wrote his *Arithmetical Books*. It came into his possession, however, after that work was published, as is shown by his autograph on the title page (see Fig. 67). It is not often mentioned by bibliographers, and is one of the rare books of the century. In the Boncompagni sale (no. 1441) there is mentioned

Arithmetica speculatiua magistri Gasparis Lax Aragonensis de sarinyena duodecim libris demonstrata.

hemon le feure

Uenundatur Parisius in vico diui Jacobi ab Emundo le feure sub signo crescentis albi vitam degente.

PLATE III. TITLE PAGE OF LAX

an edition of 1547; but this is a misprint for 1517, as appears from Riccardi, vol. I, col. 153–4. Riccardi mentions only three copies known to him or to Boncompagni, but there was one in the Fisher sale (London, 1906).

¶LVCIDARIO DARITHMETICA.

FIG. 67. TITLE PAGE OF BONINI'S *Lvcidario*

The book is a small octavo, the first two-thirds being given to mercantile problems on exchange and the reduction of money. The last part treats exclusively of mensuration: 'Speculationi geometriche di piu sorte: & prima laquadratura del triangolo.'

Other works of 1517. Anonymous, 'Algorithmus linealis' (see Widman), p. 36, c. 1488; Borghi, p. 16, 1484; Feliciano, p. 146, 1526; Köbel, p. 102, 1514; Reisch, p. 82, 1503; Sacrobosco, p. 32, 1488; Widman, p. 36, c. 1488.

HENRICUS GRAMMATEUS.

Ed. pr. 1518. Frankfort, 1535.

HEINRICH SCHREIBER; HENRICUS SCRIPTOR; Latinized Greek, GRAMMATEUS. Born at Erfurt, at least as early as 1496. He describes himself as 'Henrich Grammateus // von Erffurt/ der ſiben freien künſten Meyſter.' He was a student at Cracow and at the University of Vienna (1507). The dates of his birth and death are unknown, but a record at Vienna reads: 'Anno domini millesimo quingentesimo septimo in festo sanctorum Tibureii et Valeriani martirum . . . Henricus Scriptoris de Erfordia.' He also taught at Vienna.

Title. See Fig. 68.

Description. 8°, 9.2 × 15.2 cm., the text being 6.7 × 11.3 cm. 96 ff. unnumb., 31 ll. Frankfort, 1535.

Editions. Vienna, 1518, 8°; Frankfort, 1535 (here described); s. l. a. (Frankfort, 1544); Frankfort, 1572. [495; 496; 498; 513]

Grammateus also published an 'Algorithmus proportionum una cum monochordi generis Dyatonici compositione . . .', Cracow, 1514, 4°; 'Libellus de compositione regularum pro vasorum mensuratione,' Vienna, 1518; 'Behend unnd khunstlich Rechnung nach der Regel und welhisch practic,' Nürnberg,

1521, 8°, an extract from the work here described; 'Algorismus de integris Regula de tri cum exemplis,' Erfurt, 1523; 'Eyn

Eyn new künstlich behend vnd gewiß Rechenbüchlin vff alle Kauffmanschafft.
Nach
Gemeynen Regeln de tre.
Welschen practic. Regeln falsi.
Etlichen Regeln Cosse.
Proportion des gesangs / in Diatonio außzutheylen monochördū / Orgelpfeiffen / vnd andre Instrumēt / durch erfindung Pithagore.
Büchhalten durch das Zornal / Kaps vnd Schuldtbüch.
Visier rüten zu machen durch den Quadrat / vnd Triangel / mit andern lustigen stucken der Geometrei.
M. Henricus Grammateus.

Fig. 68. Title page of the 1535 Grammateus

kurtz newe Rechenn unnd Visyrbuechleynn gemacht durch Heinricum Schreyber,' Erfurt, 1523.

That the book was written at Vienna in 1518 appears by the dedication to 'Dem Edlen fürfichtigen weifen Johan//fen Tfchertte einer des Senats zu Wien,' which ends: 'Gebē // zu Wi//en in O-//sterreich im jar // nach der geburt vn-//sers Seligmachers. M. D. XViij.' In the chapter on bookkeeping is the date 1535, so that probably the work was revised for this edition.

Additio.

Alhie sein zu addiren die quantitet eines na= mens/als N. mit N: prima mit prima/secunda mit secūda/tertia mit tertia ꝛc. Vnd mañ brauchet solche zeichen als + ist mehr/vnd — /minder/in welcher sein zu mercken drei Regel.

Die Erst Regel.

Wann ein quantitet hat an beyden orten + oder — so sol mann solche quantitet addirn hin zū gesatzt das zeychen + oder —

als 9 pri. + 7 N. 6 pri. — 4 N.
6 pri. + 5 N. 8 pri. — 10 N.
Facit 15 pri. + 12 N. 14 pri. — 14 N.

Die ander Regel.

Ist in der obern quantitet + vnd in der vndern —/vnd + übertrifft —/so sol die vnder quantitet von der obern subtrahirt werden/vñ zu dem übrigen setz + So aber die vnder quā titet ist grösser/so subtrahir die kleinern vō der grössern/vñ zu dem das do bleibend ist/setze —

als 6 pri. + 6 N: 4 pri. + 2 N.
12 pri. — 4 N. 6 pri: — 6 N.
18 pri. + 2 N. 10 pri. — 4 N.

Die dritt Regel.

So in der obgesatzten quantitet würt fundē — vnd in der vndern +/vnd — übertrifft +/ so subtrahir eins von dem andern/vnd zum ü= brigen schreib — Ist es aber/das die vnder quā titet übertrifft die obern/so ziehe eins von dem andern/vnd zu dem ersten setze + als

FIG. 69. FROM THE 1535 GRAMMATEUS

The work is for the most part a mercantile arithmetic, the operations being given according to both the abacists (with counters) and the algorists (by the Hindu-Arabic numerals), and a chapter on bookkeeping being appended. Grammateus gives, however, some consideration to the theory of numbers, the rules of the Coss (algebra), music, bookkeeping, and gauging. In the treatment of the 'Regula falsi,' or rule of false position, the signs + and – are first found in this connection (f. E iij). (See Widman, p. 40, 1489, who uses them for another purpose, and Vander Hoecke, p. 183, 1537.) Grammateus also uses these signs in writing algebraic binomials, as shown in Fig. 69. It is interesting to know that Rudolff (p. 150) learned algebra from Grammateus, as he states in the following words: 'Ich hab von meister Heinrichen so Grammateus genennt / der Cofs anfengklichen bericht emphangen. Sag im darumb danck.'

Other works of 1518. Böschensteyn, p. 100, 1514; Calandri, p. 48, 1491; Feliciano, p. 146, 1526; Köbel, p. 102, 1514; Riese, p. 139, 1522; Torrentini, p. 76, 1501; Pérez de Oliva, 'Dialogus in laudem Arithmeticæ,' Paris.

ANONYMOUS. Ed. pr. 1519. Venice, 1519.

Title. See Fig. 70.

Colophon. 'Venetijs in Edibus Petri Liechtenftein // Anno virginei partus .1519.' (F. 12, v.)

Description. 4°, 15.6 × 20.4 cm., the text being 10.1 × 16.3 cm. 12 ff. (1 blank), 33–38 ll. Venice, 1519.

Cōputus nouus ⁊ ecclesiasticus: totius fere Astronomie fundamentum pulcherrimum con/tinens. Clerico non minus vtilis ꝙ neces/sarius: cum additionibus quibus/dam nouiter apressis.

* * *

Anno. 1519. Venetijs in Edibus Petri Liechtenstein.

FIG. 70. TITLE PAGE OF THE *Cōputus nouus*

Editions. There was no other edition, so far as I know, although there are several anonymous computi, and some may be the same as this.

This is one of the rare works setting forth the computus as it was taught in the Church schools of the Middle Ages. (See p. 7.)

The verses on the calendar mentioned on p. 33 (Anianus, 1488) here appear as follows:

> 'Aprilis: Junius: September atqꝫ Nouember
> Hij trigenta habent vnum reliqui ſuperaddunt
> Februarius vigint inſuper octoqꝫ dies.' (F. 9.)

Other works of 1519. Feliciano, p. 146, 1526; Blasius, p. 97, 1514; Widman, p. 37, 1489.

JOHANNES FRANCISCUS PICUS MIRANDULA.

Ed. pr. 1520. S. l., 1520.

A nephew of Pico de Mirandola, and biographer of his uncle. He was murdered in 1533. Like his uncle he was a savant of reputation.

Title. 'Ioannis Francisci Pici Mirandvlae domini, et // Concordiae comitis, examen vanitatis do//ctrinae gentivm, et veritatis Chri-//stianae disciplinae,//distinctvm in libros sex, qvorvm tres // omnem philosophorvm sectam vni-//versim, reliqvi Aristoteleam // et Aristoteleis armis // particvlatim im-//pvgnant.// vbicvnqve avtem Christiana et // asseritvr et celebratvr // disciplina.' (F. 1, r.)

Description. Fol., 20.6 × 30.6 cm., the text being 16.9 × 24 cm. 6 ff. unnumb. + 208 numb. + 1 blank = 215 ff., 44 ll. S. l., 1520.

Editions. There was no other edition. The dedication (f. 2, v.) bears date M. D. XX, and the privilege M. D. XIX.

The book hardly deserves place in a list of this kind. It has, however, been included because of the following brief chapters on the nature of arithmetic: 'Quod ſuper mathematicis artibus arithmetica & geometria, ſuperqꝫ mediis Aſtrologia & muſica, gentium philoſophi non conueniunt. Cap. vii'; 'De opinione pythagoricorū, & de ratione et philolai & poſtidonii . . . Cap. ix'; 'Quid aduerſus arithmeticā facultatē pyrrhonii diſputauerint. Cap. vii' (of liber III).

ANONYMOUS.

S. l. et a. (c. 1520).

Title. 'Von dem Råchnen auff den Linien.' (Running headline.)

Thirteen fragments of proof sheets of an unknown German arithmetic, three duplicates. The date is purely conjectural. The work was of at least 46 pages, since the folios were numbered and part of f. 23

is among the fragments. The title of the book probably appears in the running headline above given, although this may be the title of only part of the work. There was at least one illustration of line reckoning in the book. The lines are 6.4 cm. in length, but there is no complete page among the fragments. Several anonymous works have already been mentioned, with some such title as 'Algorismus linealis,' and possibly this is one of them. It would probably be possible to identify it if one should examine the types and compare the fragments with possible originals.

ESTIENNE DE LA ROCHE, Villefranche.

Ed. pr. 1520. Lyons, 1520.

Born at Lyons, c. 1480.

Title. See Fig. 71.

Colophon. '❡Cy finiſt lariſmetique de maiſtre Eſtienne de la roche dict ville franche natif de Lyon // ſus le roſne. Imprimee par Maiſtre guillaume huyon. Pour Conſtantin fradin mar-//chant ⁊ libraire du dict Lyon. Et fut acheuee lan .1520. le 2e. de Juing.' (F. 234, r.).

Description. Fol., 17.3 × 25.5 cm., the text being 13 × 21 cm. 1 f. blank + 4 ff. unnumb. + 230 numb. = 235 ff., 49 ll. Lyons, 1520.

Editions. Lyons, 1520, fol. (here described); ib., 1538, fol. (see p. 130).

This is the best of the early French arithmetics. Since it is semi-mercantile in character, it was naturally printed at Lyons, then the commercial center of France, the theoretical books being usually printed at Paris under the influence of the Sorbonne. De la Roche gives a very complete treatment of the operations with integers, fractions, and compound numbers, and a large number of business applications. Perhaps no arithmetic published in France in the sixteenth century gives a more comprehensive view of the science and art of arithmetic and of the applications of the subject. Unfortunately, however, de la Roche took much of his work bodily from a manuscript of his master, Chuquet, which he had in his possession, and which has since been published.

Other works of 1520. Köbel, p. 102, 1514; Peurbach, p. 53, 1492; Raggius, p. 98, 1514; Stromer, p. 83, 1504; Suiseth, p. 10,tc. 1480; Tagliente, p. 114, 1515; Anonymous (Tagliente ?), 'Libro de Abaco,'

L'arismethiquē nouuellement composee par
maistre Estienne de la roche dict Villefrāche natif de Lyō
sus le Rosne diuisee en deux parties dont la p̄miere tracte
des p̄prietes p̄fectiōs et regles de la dicte sciēce: cōme le nōbre entier: Le nōbre rout: La regle de troys: La regle dune
faulse position: De deux faulses positiōs: dapposition et re
motiō: de la regle de mediatiō entre le plus et le moīs: de la
regle de la chose: et de la quātite des p̄gressiōs et p̄portiōs.
La secōde tracte de la practique dicelle applicquee en fait
de mōnoyes: en toutes marchādises cōme drapperie: espi-
cerie: mercerie et en toutes aultres marchādises qui se ven-
dent a mesure au poiz ou au nōbre: en cōpaignies et en tro-
ques: es chāges et merites: en fin dor et dargent et en lava-
leur diceux. En argēt le roy et en fin dargēt dore. Es dene-
raulx allyages et essaiz. tant de lor que de largēt. Et en geometrie applicquee aux ars mechāiques cōme aux massons
charpētiers et a tous aultres besongnās en art de mesure.

Cum Priuilegio.

FIG. 71. TITLE PAGE OF THE FIRST EDITION OF DE LA ROCHE

Venice, 8°; Anonymous, 'Libretto de Abaco,' s. l., 8°; Anonymous, 'Algorismus novus de integris, de minutiis vulgaribus, de minutiis physicis,' Augsburg. To an edition of Sacrobosco's Sphere the commentator, Johannes Guyion, prefixed a treatise on arithmetic, 'De quantitate discreta,' Avignon, s. a., c. 1520. There was published, s. l. a., possibly at Lyons in this year, 'Le liure des gectz grandement profitable pour messeigneurs les marchans et aultres,' 4°, a work on counter reckoning (see p. 7). 513

ESTIENNE DE LA ROCHE, Villefranche.

Ed. pr. 1520. Lyons, 1538.

See p. 128.

Title. 'Larifmetique & Geometrie de maiftre // Eftienne de la Roche dict Ville Fran//che, Nouuellement Imprimee & des faultes corrigee,// a la qvelle font adiouftees les Tables de diuers comptes, auec leurs Ca-//nons, calculees par Gilles Huguetan natif de Lyon, Par lefquelles on pourra facil-//lement trouuer les comptes tous faictz, tant des achatz que uentes de toutes mar-// chandifes. Et principalement des marchandifes que fe uendent, ou achetent a la // mefure, cōme a Laulne, a la Canne, a la Toyfe, a la Palme, au Pied, & aultres fem-//blables. Au poix, cōme a la Liure, au Quintal, au Millier, a la Charge, au Marc,// & a Lonce, a la Piece, au Nōbre, a la Douzaine, a la Groffe, au cent, & au Millier.// Auec deux Tables feruantz aulx Librayres uendeurs & acheteurs de papier. En-//femble une Table de defpence, a fcauoir a tant pour iour, combien on defpēd Lan // & le Moys, & a tant le moys, combien reuient lan & le iour, & a tant pour an, cō-//bien on defpend tous les moys, & a combien reuient pour chafcun iour.// Davantaige, les Tables du fin dor & dargent, pour fcauoir (fcelon que le Marc de billon tiendre//daloy, ou de fin) combien il uauldra de poix de fin or, ou dargent fin.// On les uend . . . lenfeigne de la Sphære,// cheulx Gilles & Jacques Huguetan freres.// 1538.' (F. 1, r.)

Colophon. '¶Cy finift Larifmetique ⁊ Geometrie de maiftre Eftienne de la Roche dict Villefranche // Imprime a Lyon par maiftre Jacques myt Lan. 1538.' (F. 160, r.)

Description. Fol., 21.1 × 33 cm., the text being 16.3 × 30.6 cm. 2 ff. unnumb. + 158 numb. = 160 ff., 59–60 ll. Lyons, 1538.

Editions. See p. 128.

The Huguetan referred to is the one mentioned on p. 188.

BÆDA. Ed. pr. 1521. Basel, 1563.

'The Venerable Bede' was born in England, probably near Wearmouth, Durham, c. 673, and died in 735. He was the most distinguished scholar of his time, and his works cover all the branches of learning then known.

Title. 'Opera // Bedae // Venerabi-//lis Presbyte-//ri, Anglo-saxonis: vi-//ri in divinis atqve hv-//manis literis exercitatiſsimi: omnia in octo to-//mos diſtincta, prout ſtatim poſt Præfa-//tionem ſuo Elencho enu-//merantur.// Addito Rerum & Verborum Indice // copioſiſsimo.// Cum Cæſareæ Maieſtatis gratia & priuile-//gio, Regisque Galliarum ad // decennium.// Basileæ, per Ioannem // Heruagium, Anno M. D. LXIII.' (Surrounded by an elaborate woodcut with inscriptions.) (F. 1, r.)

Description. Fol., 24.2 × 38 cm., the text being 17.7 × 28.6 cm., printed in double columns, each 8.5 cm. wide, 61 ll. 152 pp. unnumb. + 271 numbered by columns (i.e., 2 numbers to a page) = 423 pp. in vol. 1. 8 vols. bound in 4. Basel, 1563. Only the first volume, containing the arithmetic, is described here.

Editions. Some of his arithmetical work is said to have been published in 1521, fol.; 1525 (in part, see p. 140); 1529 (in part, see p. 159); Paris, 1544–45 (first edition of the Opera); ib., 1554; Basel, 1563, fol. (here described). See also c. 1515, anonymous. His 'Historia Ecclesiastica' appeared as early as 1473.

The first volume contains the 'De Arithmeticis nvmeris liber' (cols. 98–116), with little save an elaborate multiplication table and a dialogue on number, names, and symbols; 'De Arithmeticis proportionibvs' (cols. 133–146), with the 'Propoſitiones ad acuendos iuuenes' often attributed to Alcuin, but certainly not Bæda's; 'De ratione calcvli' (cols. 147–158), chiefly multiplication tables of Roman money; 'De nvmerorvm divisione' (cols. 159–163); 'De loqvela per gestvm digitorvm, et temporvm ratione' (cols. 164–181), or, as the headline states it, 'De indigitatione,' giving us almost our only knowledge of the finger reckoning or symbolism of the Middle Ages in western Europe, and possibly spurious;

'De ratione vnciarvm' (cols. 182–184), a treatise on Roman fractions; an extensive treatment of the calendar and the computus, with a description of the astrolabe.

FRANCESCO GHALIGAI. Ed. pr. 1521. Florence, 1552.

A Florentine arithmetician of the first part of the sixteenth century. He died February 10, 1536.

Title. 'Practica // d'Arithmetica. // di // Francesco Ghaligai // Fiorentino. // Nuouamente Riuiſta, & con ſomma // Diligenza Riſtampata. // In Firenze // Appreſſo i Givnti // M. D. LII.' (F. 1, r.)

Colophon. 'In Firenze // Appreſſo i Givnti // M. D. LII.' (F. 114, r.)

Description. 4°, 15.2 × 20.3 cm., the text being 12 × 17.3 cm. 2 ff. unnumb. + 112 numb. = 114 ff., 37–38 ll. Florence, 1552.

Editions. Florence, 1521, 4° (see Boncompagni's *Bulletino*, VII, 486; XIII, 249); ib., 1548, 4°; ib., 1552, 4° (here described). The 1521 edition is entitled 'Summa De Arithmetica,' but it is the same as the 1552 edition here described. Some bibliographers mention other editions, as of 1540, 1551, 1562, 1572, 1582, 1591, but Boncompagni's careful investigation, supported by Riccardi's, throws doubt upon all these. [514; 515]

The book is written in the general style of the Italian works of the sixteenth century, more or less resembling Borghi (p. 16). It was intended for the use of merchants, and contains a large number of practical problems showing the conditions of trade at the time of its publication. Books 10 to 13 relate to algebra, and their chief interest attaches to the symbolism employed.

Other works of 1521. Boethius, p. 31, 1488; Grammateus, p. 124, 1518; Paxi, p. 77, 1503; Sacrobosco, p. 32, 1488.

CUTHBERT TONSTALL. Ed. pr. 1522. London, 1522.

Tunstall. Born at Hackforth, Yorkshire, in 1474; died November 18, 1559. He was educated at Oxford, Cambridge, and Padua, was a man of great learning and energy, and held important positions in the Church and State. He was bishop of London, and later of Durham.

Title. 'De arte svppvtandi // libri qvattvor // Cvtheberti // Tonstalli.' (Surrounded by a woodcut. F. 1, r. See Fig. 72.)

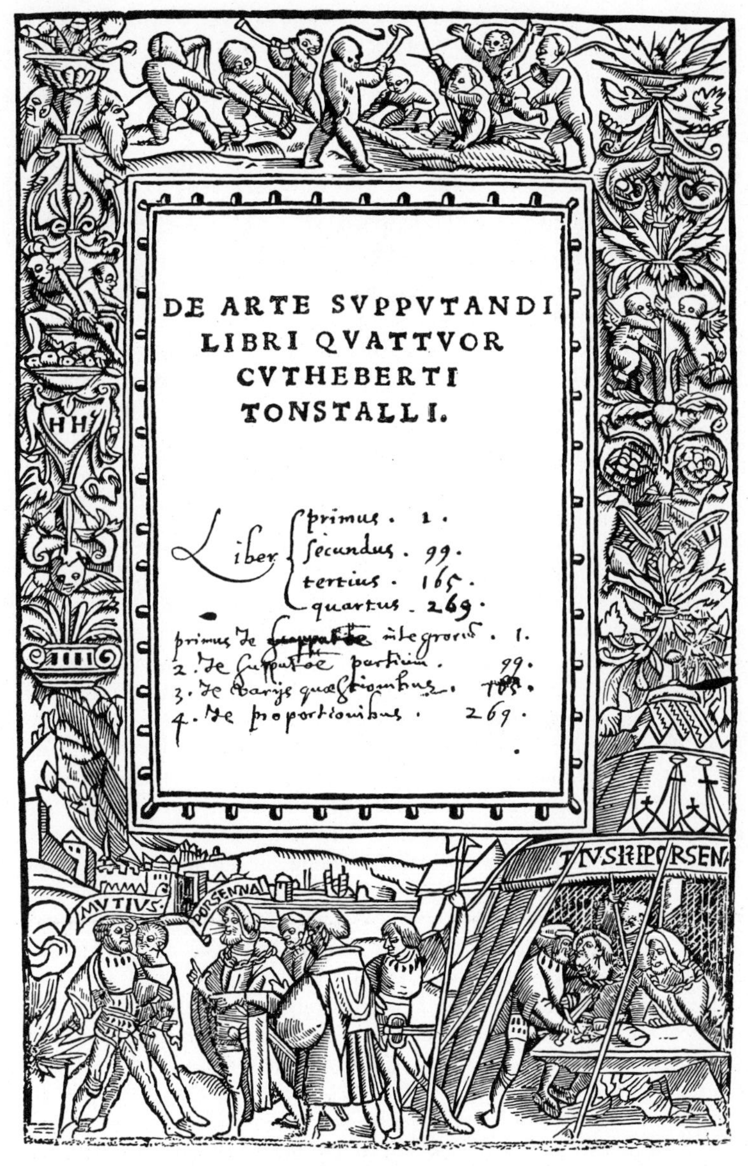

DE ARTE SVPPVTANDI LIBRI QVATTVOR CVTHEBERTI TONSTALLI.

Liber primus. 1. secundus. 99. tertius. 165. quartus. 269.

primus de supputatione integrorum. 1.
2. de supputatione partium. 99.
3. de varijs quaestionibus. 165.
4. de proportionibus. 269.

FIG. 72. TITLE PAGE OF THE FIRST EDITION OF TONSTALL

Colophon. 'Impress. Londini in aedibvs Ri-//chardi Pynsoni. Anno ver-//bi incarnati .M.D.XXII.// Pridie idvs octo-//bris. Cvm pri-//vilegio a // rege in-//dvl-//to.' (F. 202, r.)

Description. 4°, 15.7 × 20.9 cm., the text being 10.5 × 16.4 cm. 202 ff. unnumb., 29 ll. London, 1522.

Editions. London, 1522, 4° (here described); Paris, 1529, 4° (p. 135); ib., 1535; ib., 1538, 4° (p. 135); Strasburg, 1543; ib., 1544, 8° (p. 136); ib., 1548; ib., 1551.

This is the first edition of the first book wholly on arithmetic that was printed in England. (But see p. 10, 1480.) In the dedicatory epistle Tonstall states that in his dealing with certain goldsmiths he suspected that their accounts were incorrect, and he therefore renewed his study of arithmetic so as to check their figures. On his appointment to the See of London he bade farewell to the sciences by printing this book in order that others might have the benefit of a work which he had prepared for his own use. The treatise is in Latin, and, although it was written for the purpose of supplying a practical handbook, is very prolix and was not suited to the needs of the mercantile class. It is confessedly based upon Italian models, and it is apparent that Tonstall must have known, from his residence in Padua and his various visits to Italy, the works of the leading Italian writers. The book includes many business applications of the day, such as partnership, profit and loss, and exchange. It also includes the rule of false, the rule of three, and numerous applications of these and other rules. It is, however, the work of a scholar and a classicist rather than a business man.

The word 'supputandi,' in the title, was not uncommon at that time. Indeed there was some tendency to use the name 'supputation' for arithmetic and to speak of calculations as 'supputations.'

Tonstall dedicates the work to his friend Sir Thomas More, whose talented daughter Erasmus addressed as 'Margareta Ropera Britanniæ tuæ decus,'—ornament of thine England. More speaks of Tonstall in the opening lines of his *Utopia*: 'I was colleague and companion to that incomparable man Cuthbert Tonstal, whom the king with such universal applause lately made Master of the Rolls; but of whom I will say nothing; not because I fear that the testimony of a friend will be suspected, but rather because his learning and virtues are too great for me to do them justice, and so well known, that they need not my commendation unless I would, according to the proverb, "Show the sun with a lanthorn."' The *Utopia* was first printed in 1516, so this sonorous praise was written some years before Tonstall's arithmetic appeared.

Some idea of the prolixity of the treatise may be obtained from the number of closely-printed quarto pages assigned to certain topics. The chapter 'De Numeratione' fills 11 pages, 'De Additione' 14 pages, 'De Svbdvctione' 15 pages, 'De mvltiplicatione' 14 pages, 'De partitione' 27 pages (the old galley method being used exclusively), and so on for the other subjects. Some 66 pages, for example, are given to the theory of ratio and proportion.

The title page was engraved by Holbein, and was evidently printed after the book was completed, because in this copy the errata appear on the reverse of the first folio. The work was printed by Richard Pynson, the successor to Caxton.

Following the arithmetic is an appendix: 'Appendix ex Bvdaei libro de as-//ſe excerpta: in qua priſca Latinorum et Gręcorū //ſupputatio, ad æſtimationem pecunię, tum Gallicæ,// tum Angli-//cæ reuocatur.'

Tonstall also published a work in 1518: 'In Lavdem matrimonii oratio,' second edition in 1519, now very rare.

CUTHBERT TONSTALL. Ed. pr. 1522. Paris, 1529.

See p. 132.

Title. 'De arte svppvtandi libri qvatvor//Cvthberti Tonstalli. (Picture of a tree from which is falling a broken branch, and the words: Noli altum ſa//pere, ſed time.) Parisiis ex officina Roberti Stephani // M.D.XXIX.' (P. 1.)

Colophon. 'Parisiis//excvdebat Robertvs Stepha-//nvs. Ann. M.D.XXIX. Prid. id. ivn.' (P. 279.)

Description. 4°, 13 × 18.8 cm., the text being 9.4 × 15.9 cm. 271 pp. numb. + 8 unnumb. + 2 blank = 281 pp., 36–38 ll. Paris, 1529.

Editions. See p. 134.

The text is the same as in the first edition of 1522. There has been added, however, a second appendix with the following title: 'Gvlielmi Bvdaei Parisiensis,// secretarii regii, breviari-//vm de asse.'

CUTHBERT TONSTALL. Ed. pr. 1522. Paris, 1538.

See p. 132.

Title. 'De arte svp-//putandi libri qua-//tuor, Cutheberti // Tonſtalli.// (Large woodcut.) Parisiis.// Ex officina Roberti Stephani.// M.D.XXXVIII.' (P. 1.)

Colophon. 'Excvdebat Robertvs Stephanvs Parisiis, // ann. M. D. XXXVIII.// xvi. cal. novemb.' (P. 259.)

Description. 4°, 14.4 × 20.1 cm., the text being 9.5 × 16 cm. 259 pp. numb., 39 ll. Paris, 1538.

See p. 134.

CUTHBERT TONSTALL. Ed. pr. 1522. Strasburg, 1544.

See p. 132.

Title. 'De arte // svppvtan//di, libri qvatvor // Cvthberti Tonstalli,// hactenus in Germania nus-//quam ita impreſſi.// Ioan. Stvrmivs.// Arithmeticam Cvthbertvs // Tonſtallus præ cæteris dilucide & pure tradidit: atqȝ // ita tradidit, ut ars ipſa dum hic author extat, con-//tenta ſcriptore, doctorem non maximopere aliquem // requirat. Non nego, poſſe ex alijs quoque diſci: // ſed hic docet erudite, perſpicue latine, id quod non fa//ciunt cæteri: nec abest longe à perfectione, qui eius // præcepta intelligit.// Argentorati, ex offi.// Knobloch. per Georg. Machærop.' (P. 1.)

Colophon. 'Argentorati, ex officina // Knoblochiana, per Ge-// orgivm Machaero-//poevm, mense // febrvario // anno,// M. D. XLIIII.' (P. 478.)

Description. 8°, 9.5 × 15 cm., the text being 6.8 × 11.4 cm. 25 pp. blank + 453 numb. = 478 pp., 26 ll. Strasburg, 1544. This copy is bound with the arithmetic of Victorius Strigelius.

Editions. See p. 134. This is the same as the edition of 1529, having the second appendix there mentioned. It is the second Strasburg edition. It is interesting to see that the classical influence on the Continent was such that seven out of the eight editions appeared in Paris or Strasburg.

JOHANNES AVENTINUS.

Ed. pr. 1522. Regensburg, 1532.

THURNMAYER. Born at Abensberg, Bavaria, July 4, 1477; died at Regensburg, January 9, 1534. He wrote on history.

Title. See Fig. 73.

Colophon. 'Ratiſponę apud Ioannem Khol // Anno. MD-XXXII.' (Large woodcut and date, 'Io. Kol 1532.') (F. 12, r.)

Description. 4°, 14.2 × 19.5 cm., the text being 9.9 × 14 cm. 12 ff. unnumb., 26 ll. Regensburg, 1532.

Editions. Nürnberg, 1522; Regensburg, 1532, 4° (here described).

The book is primarily a treatise on numerical finger symbolism, and contains the most complete explanation of that subject extant. It gives

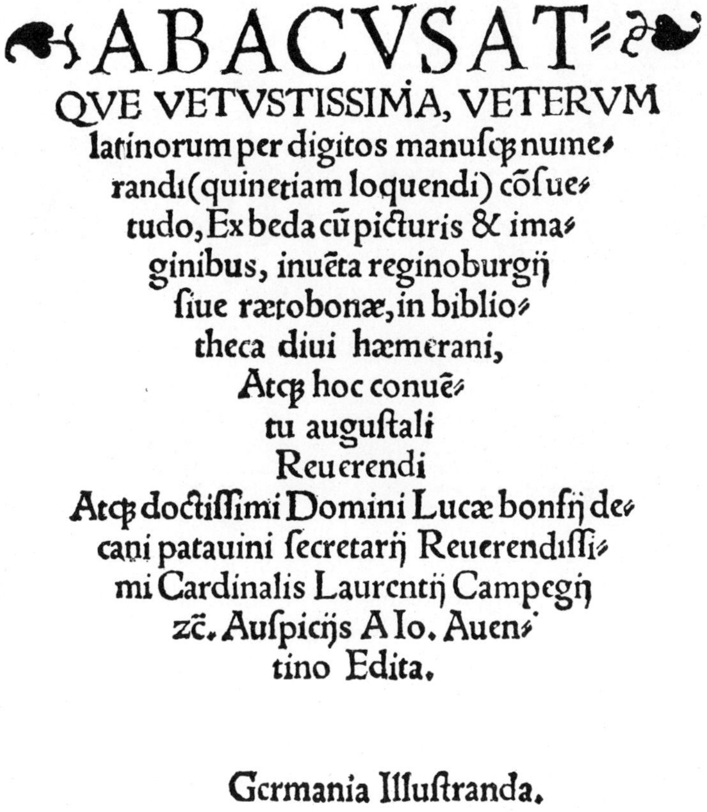

ABACVSAT-
QVE VETVSTISSIMA, VETERVM
latinorum per digitos manuſcꝫ numerandi (quinetiam loquendi) cõſue-
tudo, Ex beda cũ picturis & ima-
ginibus, inuẽta reginoburgij
ſiue rætobonæ, in biblio-
theca diui hæmerani,
Atcꝫ hoc conuẽ-
tu auguſtali
Reuerendi
Atcꝫ doctiſſimi Domini Lucæ bonfij de-
cani patauini ſecretarij Reuerendiſſi-
mi Cardinalis Laurentij Campegij
zč. Auſpicijs A Io. Auen-
tino Edita.

Germania Illuſtranda.

FIG. 73. TITLE PAGE OF THE 1532 AVENTINUS

illustrations showing the representation of the numbers up to one million by means of the fingers and arms (see Fig. 74). This finger symbolism is found in the works of Bæda, it was practical in both the East and

West during the Middle Ages, and it is mentioned by several sixteenth-century arithmeticians. (Compare Fig. 74 with Fig. 35, p. 57.)

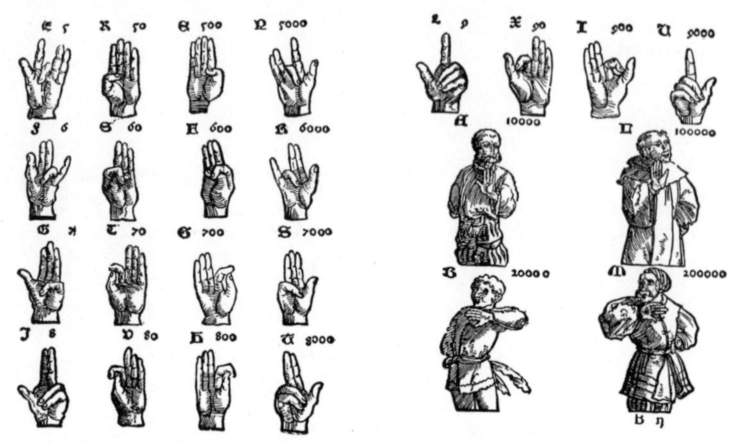

FIG. 74. FINGER SYMBOLISM FROM AVENTINUS

ADAM RIESE. Ed. pr. 1522. Leipzig, 1538.

RYSE, RIS, RIES. Born at Staffelstein, near Bamberg, c. 1489; died at Annaberg, March 30, 1559. One of the most celebrated Rechenmeisters of the sixteenth century, and the most influential of all the Germans in replacing the counter reckoning ('auff der Linien') by the written computations ('auff Federn'). 514

Title. 'Rechnung auff // der Linien vnd Federn // Auff allerley handtirung ge-//macht/ durch Adam Riſen.// (Woodcut of counting house, with reckoning on the line abacus.) Item auffs new vberſehen vnd // an viel ŏrten gebeſſert.// M.DXXXVIII.' (F. 1, r.)

Colophon. 'Gedruckt zu Leiptzigk // durch Valentinum // Schumann. 1538.' (F. 63, v.)

Description. 8°, 9.7 × 15.1 cm., the text being 7.2 × 11.5 cm. 63 ff. unnumb. + 2 blank (with manuscript notes) = 65 ff., 28–33 ll. Leipzig, 1538.

Editions. In order to make clear the different editions of Riese's work, it is necessary to distinguish between the four arithmetics which he published. These were as follows:

1. 'Rechnung auff der linihen gemacht durch Adam Riesen vonn Staffelsteyn // in massen man es pflegt tzu lern in allen rechenschulen gruntlich begriffen anno 1518.' (Graesse, followed by Unger, p. 50, who knew of no extant copy.) A second edition appeared in 1525 ('Getruckt tzu Erffordt durch Mathes Maler M. CCCCCxxv Jar,' 8°, 43 ff.), and a third in 1527. This was embodied in his second arithmetic which is here described. The work is rare. [495]

2. 'Rechnung auff // der Linien vnd Federn,' Riese's best-known work. The title of the first (1522) edition was as follows: 'Rechenung auff der linihen vnd federn in zal/mafs/vnd gewicht auff allerley handierung/ gemacht vnd zusamen gelesen durch Adam Riesen von Staffelstein Rechenmeister zu Erffurdt im 1522 Jar. Itzt vff sant Annabergk durch in fleyssig vbersehen/ vnd alle gebrechen eygentlich gerechtfertigt/ vnd zum letzten eine hübsche vnderrichtung angehengt.' The following editions of this work are known to me: Erfurt, 1522; ib., 1525, 8°; Nürnberg, 1527; 1528, 8°; Erfurt, 1529, 8°; ib., 1530; Leipzig, 1533, 8° (first edition containing Helm's Visirbuch; see p. 142); Frankfort, 1535, 12°; Annaberg, 1535, 8° (p. 141); 1536, 12°; Leipzig, 1538, 8° (here described); 1541; Frankfort, 1544, 8°; Leipzig, 1544; s. l., 1548, 8°; Leipzig, 1548, 8°; ib., 1550; Breslau, 1550; Frankfort, 1552, 8°; Leipzig, 1554; 1556, 12°; Frankfort, 1558, 8° (p. 141); Leipzig, 1562, 12°; Frankfort, 1564, 8°; Frankfort, 1565, 8° (p. 142); Frankfort (a. Oder ?), 1568, 8°; Stettin (Frankfort ?), 1570, 8°; Leipzig, 1571, 8° (p. 142); Frankfort, 1574; Magdeburg, 1579, 8°; Frankfort, 1581, 12°; ib., 1585, 8°; Leipzig, 1586; Frankfort, 1586; Wittenberg, 1587; Nürnberg, 1592, 8°; Frankfort, 1592, 8° (p. 143); Leipzig, 1598, 12°. There were several editions after 1600. It is possible that some of the editions here mentioned may be of Riese's fourth book, the titles being much alike and printers varying them from time to time. [496;514]

3. 'Ein Gerechent Büchlein,' first published at Leipzig in 1533; second edition in 1536. See p. 171.

4. 'Rechnung nach der lenge / auff den Linihen vnd Feder,' first published in 1550, 4°. See p. 250.

Kuckuck's remark, that over twenty-six editions of Riese's arithmetics appeared, greatly underestimates the number. More than forty appeared in the sixteenth century alone, and several were published in the seventeenth century.

This was probably the most popular commercial arithmetic of the sixteenth century. So firmly did it impress itself upon the schools that 'nach Adam Riese' is a common expression in Germany to-day, nearly four hundred years after the first of his books appeared. It was to Germany what Borghi's book was to Italy and Recorde's to England. It differed from Riese's first book in that it emphasized computation by the aid of the Hindu-Arabic numerals instead of the counters. There is no other book that gives as good a picture of the sixteenth-century mercantile problems of Germany, and of the methods of solving them.

Other works of 1522. Albert of Saxony, p. 9, c. 1478; Boethius, p. 27, 1488; Budaeus, p. 99, 1514; Köbel, p. 102, 1514; Ortega, p. 93, 1512; Sacrobosco, p. 32, 1488; Francisco Pelacani, 'Arithmetica prattica,' Florence; Ludovico Vincento (Vincentino) degl' Arrighi, 'La operina . . . da . . . bellissima Ragione di Abbacho,' Rome, 4°, with editions at Venice in 1532, 1533, chiefly on chirography.

Works of 1523. Borriglione, p. 86, 1506; Ciruelo, p. 60, 1495; Grammateus, p. 124, 1518; Paciuolo, p. 54, 1494; Reisch, p. 82, 1503; Sacrobosco, p. 35, 1488; Tagliente, p. 114, 1515. Rodrigo Fernández de Santaella (or Valencia, see p. 269, 1555), 'Ars cōputandi,' Saragossa, fol. There was also written in this year, but published s. l. a., a work by Vincenzo Barziza entitled 'Operetta nouamente composta,' 8°, 39 ff., containing a few mercantile rules and tables. 516

Works of 1524. Feliciano, p. 145, 1526; Tagliente, p. 115, 1515.

Works of 1525. Herodianus, p. 60, 1495; Köbel, p. 102, 1514; Riese, p. 139, 1522; Rudolff, p. 151, 1526; Tagliente, p. 114, 1515; Bede et al., 'Valetius Probus et Petrus Diaconus de notis Romanorum, Demettius Alabaldus de minutiis, ponderibus et mensuris, Ven. Beda de computo per gestum digitorum,' etc., Venice, 4°; Angelus Mutinens (i.e. of Modena), 'Thesavro de Scrittori opera artificiosa le quale con grandissima arte, si per pratica come per geometria insegna . . .,' s. l., M.D.XXxv, has four folios at the end relating to arithmetic, with the note 'Angelus Mutinens composuit,' and there seems to have been an edition in 1525, and another s. a. published at Rome; Giovanni

Tagliente published two editions of a work chiefly on bookkeeping, 4°, 24 ff., Venice, beginning, 'Considerando io Ioanni Taiente quanto e necessaria cosa ali nostri magnifici gẽtilhomeni & ad altri mercatanti.'

ADAM RIESE. Ed. pr. 1522. Annaberg, 1535.

See p. 138.

Title. The title page is missing.

Colophon. '¶Nach diſer vnderrichtung kanſtu auffs be//hendeſt alle Exempel in der Falſi machen: Wo̊l//leſt ſolch Bůchlin vnnd kurtze erklerung ietzt/ // welches ich zum andern mal laſſe auſzge-//hen/ zu danck an nemen/ wil ich ver-//dienen/ vnd dir auffs eheſt ich // mag die Practica nach al-//lem fleiſz herauſzſtrei//chen. Datum//auff ſanct // Annaberg/ // Dinſtag nach // Martini. Im Iar // M. D. XXV.' (F. 55, r.) On f. 69, v., is the following colophon: 'Alſo iſt kürtzlich // beſchriben vnd // begriffen // die Confection // der Viſier rů̊ten mit // Irer übung vnd gebrauch.// ¶End:// An. M. D. XXXV.// Im Christmonat.' This latter is evidently the date of printing of the entire book.

Description. 8°, 8.8 × 13.2 cm., the text being 6.8 × 11.6 cm. 69 ff. unnumb., 31 ll. Annaberg, 1535.

Editions. See p. 139.

This contains the 'Viſirbůchlin' of Erhart Helm, as in the 1533 edition, but it does not give his name. See the 1565 edition (p. 142).

ADAM RIESE. Ed. pr. 1522. Frankfort, 1558.

See p. 138.

Title. The title page is missing. Page numbered 2 begins as follows: 'Vorrede in diß Rechen-//bůch/ Adam Riſen.'

Colophon. '¶End.// Zu Franckfurt bei Chr. Egeb. erben/ // Anno 1558.' (F. numb. 87, r.)

Description. 8°, 9.4 × 15 cm., the text being 6.5 × 12.2 cm. 87 ff. numb. + 1 blank = 88 ff., 31 ll. Frankfort, 1558.

Editions. See p. 139.

This contains the 'Viſirbůchlin' of Erhart Helm, as in the 1533 edition.

ADAM RIESE. Ed. pr. 1522. Frankfort, 1565.

See p. 138.

Title. 'Rechenbůch/ Vff Lini//en vnnd Ziphren/ In allerley // Handtierung/ Geſchefften vnd Kauff-//mannſchafft. Durch Adam // Riſen.// Mit new en kůnſtlichen Regeln vnd Ex-// emplen gemehrt/ Innhalt fuͤrge-//ſtelten Regiſters.//Viſier vnd Wechſelrůten kunſtlich vnd // gerecht zumachen/ auß dem Quadrat/ // Durch die Arithmetic vnd Geometri.// Von Erhart Helm/ Mathema//tico zu Franckfurt/ be // ſchriben.// Alles von newem jetzund widerumb erſe-//hen vnd corrigirt.// (Woodcut of counting house) Franck, Bei Chr. Egen. Erben. 1565.' (F. 1, r.)

Colophon. 'M. D. LXV.' (F. 113, r.)

Description. 8°, 9.3 × 15.5 cm., the text being 6.5 × 11.9 cm. 105 ff. numb. + 8 unnumb. = 113 ff., 28 ll. Frankfort, 1565.

Editions. See p. 139. [517]

See p. 140. The Visirbuch of Helm begins, with no separate title page, on f. 77, v.: 'Viſirbůchlin // Hernach folget der ware // Proceſz/ vnnd kurtzeſt weg/ wie mann Viſir růlhen machen ſol/ auſz dem // Quadraten/ Auff alle Eich.' The name of the author, 'Erhart Helmen,' appears in the headlines of each folio recto. This part of the book is strictly speaking not an arithmetic, but it includes a few explanations of those processes that are necessary in gauging. It also includes a table of square roots to the equivalent of three decimal places, and a brief explanation of roots. It was published in separate form in 1529.

ADAM RIESE. Ed. pr. 1522. Leipzig, 1571.

See p. 138.

Title. 'Rechnung auff // der Linien vnd Federn/ // auff allerley Handtierung/ //Gemacht durch // Adam Riſen.// (Woodcut of Adam Riese, with motto: 'Anno 1550 Adam Ries Seins Alters Im LVIII.') Auffs newe durcheleſen/ vnd // zu recht bracht.// 1571.' (F. 1, r.)

Colophon. 'Zu Leipzig druckts // Hans Rhambaw/ //Im Jar// 1571.' (F. 94, v.)

Description. 8°, 9.7 × 15.6 cm., the text being 6.9 × 11.8 cm. 94 ff. unnumb., 24 ll. Leipzig, 1571.

Editions. See p. 139.

See p. 140. This is substantially identical with the 1538 edition except as to the title page.

ADAM RIESE. Ed. pr. 1522. Frankfort, 1592.

See p. 138.

Title. See Fig. 75.

Description. 8°, 9.5 × 15.5 cm., the text being 6.7 × 11.8 cm. 1 f. unnumb. + 79 numb. = 80 ff., 26–27 ll. Frankfort, 1592.

Editions. See p. 139. Bound with this is Helm's work of 1592 (described later).

FRANCESCO DAL SOLE. Ed. pr. 1526. Ferrara, 1546.

A French arithmetician, born c. 1490, and living in Ferrara at the time of writing his books.

Title. See Fig. 76.

Colophon. 'In Ferrara Nella Stampa di .M. Giouanni de bugl-hat &. M. Antonio // Hucher Compagni, Ad Inſtantia de .M. Rinaldo, cuoco dello Illuſtriſſi-//mo ſignor Duca, nel meſe di zenaro 1546.' (F. 42, v.)

Description. 4°, 14 × 19 cm., the text being 12.1 × 16.5 cm. 2 ff. unnumb. + 40 numb. = 42 ff., 31 ll. Ferrara, 1546.

Editions. Sole published a 'Libretto di Abaco' in Venice in 1526, 8°, and this is merely a revision of that work. A third edition appeared in 1564 (see p. 146).

This is little more than a primer of arithmetic. It contains the fundamental operations, a few of the more important applications, eight pages of products and roots, and several pages on astrology. The part on astrology includes some theory of the calendar, as may be seen by the title: 'Incominciano le regoline daſtrologia, p ritrouare ha quāti di et minute fa la luna, la lrā dñicale, et infinite gētileſſe, Delli circuli, elementi, et natura, del monde.' The most distinctive feature of the arithmetic is the combination of number and space concepts. For example, in addition the author considers not only abstract but compound numbers as well as geometric magnitudes. ('Regola dellæ additioni in generalita, tanto geometrica, quanto arithmetica. Ca. 6.') The same idea is carried out in the other fundamental operations.

Adam Risen
Rechenbüchlin/
auff Linien vnd Ziphren/ für die junge angehende Schüler/ in allerley Handthierung/ Geschäfften/ vnnd Kauffmanschafft/ mit neuwen künstlichen Regeln vnd Exempeln gemehrt.

Auff

Mancherley Landt vnnd Stätt Müntz/ Gewicht/ Elen vnd Maß gerichtet/ Inhalt angehengten Registers.

Jetzt alles von newem widerumb fleissig ersehen vnd corrigiert.

Franckf. bey Christ. Egen. Erben. 1592

FIG. 75. TITLE PAGE OF THE 1592 RIESE

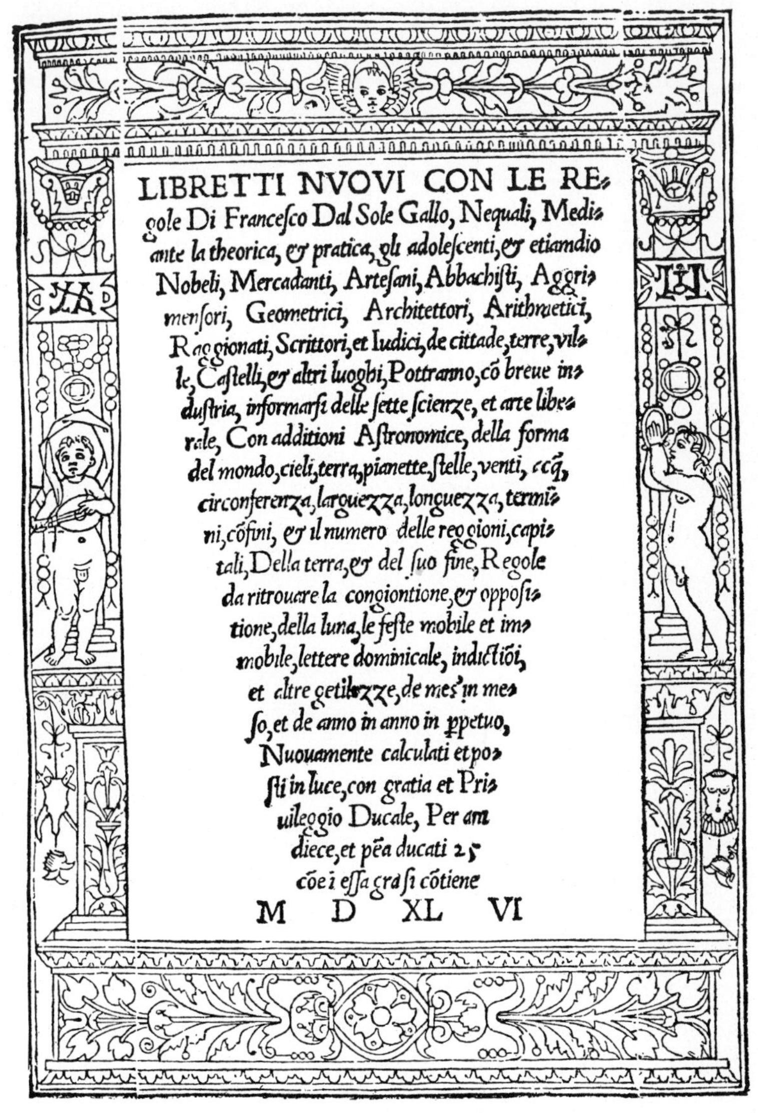

LIBRETTI NVOVI CON LE RE-
gole Di Francesco Dal Sole Gallo, Nequali, Medi-
ante la theorica, & pratica, gli adolescenti, & etiamdio
Nobeli, Mercadanti, Artesani, Abbachisti, Aggri-
mensori, Geometrici, Architettori, Arithmetici,
Raggionati, Scrittori, et Iudici, de cittade, terre, ville, Castelli, & altri luoghi, Pottranno, cõ breue industria, informarsi delle sette scienze, et arte libe-
rale, Con additioni Astronomice, della forma
del mondo, cieli, terra, pianette, stelle, venti, ecq̃,
circonferenza, larguezza, longuezza, termini, cõfini, & il numero delle reggioni, capi-
tali, Della terra, & del suo fine, Regole
da ritrouare la congiontione, & oppositione, della luna, le feste mobile et im-
mobile, lettere dominicale, indictiõi,
et altre getilezze, de mes in me-
so, et de anno in anno in ꝑpetuo,
Nuouamente calculati et po-
sti in luce, con gratia et Pri-
uileggio Ducale, Per ani
diece, et pẽa ducati 25
cõe i essa gra si cõtiene
M D XL VI

Fig. 76. Title page of the 1546 Francesco dal Sole

FRANCESCO DAL SOLE. Ed. pr. 1526. Ferrara, 1564.

See p. 143.

Title. ' Instrvtioni // et Regvle // di Francesco // dal Sole,// Francese.//Cittadino di Ferrara, Sopra il fon-//damento delle alme ſcientie d'Abbac-//co, Arithmetica, Geometria, Coſ-//mografia, & Mathematica, No-//uamente riſtampate, & con // particolare addittioni di // eſſo Authore,// aggionte.// In Ferrara, Apreſso Franceſco di Rossi //da Valenza.//M.D.LXIII.' (F.1,r.)

Description. 4°, 14.8 × 20 cm., the text being 11.1 × 15.5 cm. 4 pp. unnumb. + 71 numb. = 75 pp., 38 ll. Ferrara, 1564.

Editions. See p. 143.

Although the title of this edition is quite different from that of the first, as shown in Fig. 76, the work is essentially the same. A set of verses entitled 'Il Sole' (The Sun), a play upon the author's name, which appeared in the first edition, gives place to some Latin lines in this one.

FRANCESCO FELICIANO da Lazesio.
Ed. pr. 1526. Venice, 1526.

Born at Lazisa, near Verona; he was living in 1563.

Title. See Fig. 77. [517]

Colophon. '¶Stampato nella inclita Citta di Vinegia, apreſſo // ſanto Moyſe nelle caſe nuoue Iuſtiniane: Per // Frãceſco di Aleſſandro Bindoni, & Ma-//pheo Paſini, compagni. Nelli anni // del ſignore, 1527. Del meſe // di Zenaro. Regnante il // Sereniſſimo Princi-//pe meſſer An-//drea Gritti.// A B C D E F G H I K L M N O P Q R S T V.// Tutti ſono duerni.' (F. 80, r.)

Description. 4°, 15.1 × 20.8 cm., the text being 13.4 × 17.1 cm. 80 ff. unnumb., 41 ll. Venice, 1526.

Editions. Feliciano published two works, of which the first was 'Libro de Abaco,' [517; 518] and appeared in the following editions: Venice, 1517 (the colophon date is 1518), 8°; ib., 1519, 8°; ib., 1524, 8°; ib., 1532, 8°. His second work was a revision of his first, and is the one here described, and this appeared in the following editions: Venice, 1526, 4° (here described); ib., 1527, 4°; ib., 1536, 4° (p. 148); ib., 1545, 4° (p. 149); ib., 1550, 4° (p. 149);

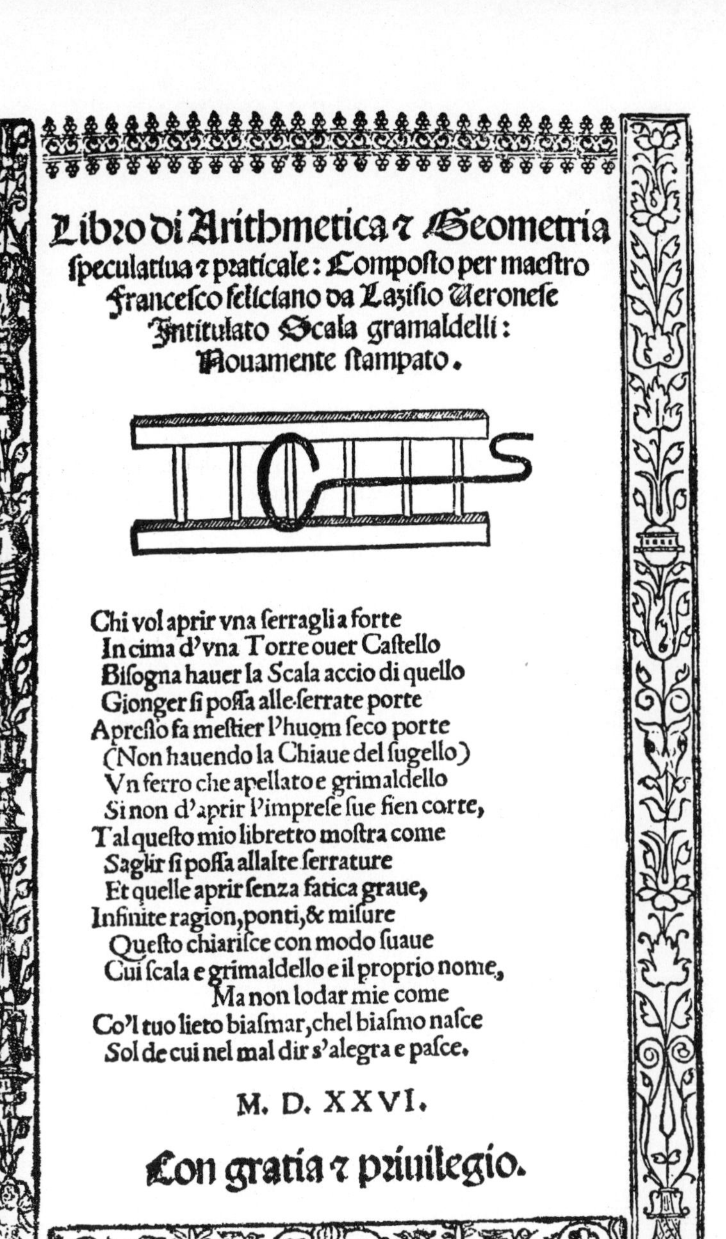

Libro di Arithmetica ⁊ Geometria ſpeculatiua ⁊ praticale: Compoſto per maeſtro Franceſco feliciano da Laziſio Veroneſe Intitulato Scala gramaldelli: Nouamente ſtampato.

Chi vol aprir vna ſeragli a forte
In cima d'vna Torre ouer Caſtello
Biſogna hauer la Scala accio di quello
Gionger ſi poſſa alle ſerrate porte
Apreſſo fa meſtier l'huom ſeco porte
(Non hauendo la Chiaue del ſugello)
Vn ferro che apellato e grimaldello
Si non d'aprir l'impreſe ſue fien corte,
Tal queſto mio libretto moſtra come
Saglir ſi poſſa allalte ſerrature
Et quelle aprir ſenza fatica graue,
Infinite ragion, ponti, & miſure
Queſto chiariſce con modo ſuaue
Cui ſcala e grimaldello e il proprio nome,
Ma non lodar mie come
Co'l tuo lieto biaſmar, chel biaſmo naſce
Sol de cui nel mal dir s'alegra e paſce.

M. D. XXVI.

Con gratia ⁊ priuilegio.

FIG. 77. TITLE PAGE OF FELICIANO

ib., 1560, 4° (p. 149); ib., 1560, 4° (another edition); ib., 1561 (the colophon of one 1560 edition); Verona, 1563, 8° (p. 150); s. l. (Venice ?), 1563, 4°; Venice, 1570. There were also seventeenth-century editions extending as late as 1692. For the 1602, 1629, and 1669 editions see pp. 150, 151. This first edition was evidently begun in 1526, but completed in January 1527, as the colophon shows. It is often assigned to the latter year. [496]

Feliciano's second work was highly esteemed as a textbook for schools. It follows the lines laid down by Borghi (p. 16), and the author acknowledges his indebtedness to him and to Paciuolo (p. 54). The first part of the book is commercial in character, and in the second part the author treats of roots, rule of false, and algebra, the third part being devoted to geometry from the practical side. More complete than the Treviso book, more modern than Borghi, more condensed and practical than Paciuolo, few books had greater influence on the subsequent teaching of elementary mathematics. The fanciful name, 'Scala grimaldelli,' is explained in the verses on the title page. Just as it is necessary in attacking a castle to have a ladder (scala) and a skeleton key (grimaldello) to open locks, so in attacking mathematics it is necessary to have a book that answers the same purposes.

FRANCESCO FELICIANO da Lazesio.

Ed. pr. 1526. Venice, 1536.

See p. 145.

Title. 'Libro di Arithmetica ꝛ Geometria // ſpeculatiua ꝛ praticale: Compoſto per maeſtro // Franceſco feliciano da Laziſio Veroneſe // Intitulato Scala Grimaldelli: // Nouamente ſtampato.' (F. r, 1. The rest is substantially as in the first edition, Fig. 77.)

Colophon. '¶Stampato nella inclita Citta di Vinegia, appreſſo // ſanto Moyſe nelle caſe nuoue Iuſtiniane: Per // Frãceſco di Aleſſandro Bindoni, & Ma-//pheo Paſini, compagni. Nelli anni // del ſignore. 1536. Del meſe // di Zenaro. Regnante il // Sereniſſimo Princi-//pe meſſer An-//drea Grittti.// ABCDEFGHIKLM NOPQRSTV.// Tutti ſono duerni.' (F. 80, r.)

Description. 4°, 15.6 × 20.8 cm., the text being 13.4 × 17.1 cm. 80 ff. unnumb., 41 ll. Venice, 1536.

See above.

FRANCESCO FELICIANO da Lazesio.

Ed. pr. 1526. Venice, 1545.

See p. 146.

Title. This is substantially the same as in the 1526 edition (p. 147).

Colophon. With the exception of the date (1545), this is substantially as in the 1536 edition.

Description. Substantially as in the 1536 edition. Venice, 1545.

Editions. See p. 146.

FRANCESCO FELICIANO da Lazesio.

Ed. pr. 1526. Venice, 1550.

See p. 146.

Title. This is substantially the same as in the 1526 edition.

Colophon. 'Stampato nella inclita Citta di Vinegia, per // Francefco Bindoni, & Mapheo Pafini,// Nelli anni del noftro Signore.// M. D. L.// Registro.// A B C D E F G H I K L M N O // P Q R S T V.// Tutti fonno duerni.' (F. 80, r.)

Description. 4°, 15.6 × 20.8 cm., the text being 13.4 × 17.2 cm. 80 ff. unnumb., 41 ll. Venice, 1550.

Editions. See p. 146.

FRANCESCO FELICIANO da Lazesio.

Ed. pr. 1526. Venice, 1560–61.

See p. 146.

Title. This is substantially the same as in the 1526 edition (p. 147), but bears the date M.D LX.

Colophon. 'Stampato nella Inclita Citta di Vinegia, Per Fran- //cefco de Leno. Nell' anno del N. Signore.// M. D. LXI.' (F. 79, r.)

Description. 4°, 15.9 × 21.1 cm., the text being 14.3 × 17.4 cm. 79 ff. unnumb., 41 ll. Venice, 1560 (colophon 1561).

Editions. See p. 146. Riccardi mentions two identical editions of this year, one of them, here described, with the colophon date 1561.

FRANCISCO FELICIANO da Lazesio.

Ed. pr. 1526. Verona, 1563.

See p. 146.

Title. This is substantially the same as in the 1561 edition.

Description. 4°, 14.9 × 19.4 cm., the text being 13.6 × 17.1 cm., 41 ll., 64 ff. unnumb. (part III missing). Verona, 1563.

Editions. See p. 146.

The third part, on geometry, is missing in this copy.

FRANCESCO FELICIANO da Lazesio.

Ed. pr. 1526. Verona, 1602.

See p. 146.

Title. This is substantially the same as in the 1526 edition except 'De nuouo riſtampato, & da molti errori corretto, & accreſciuto di molte coſe da M.// Fillipo Marcario Veroneſe Raſonato publico della Magnifica Città.// Con le gionta della Regola del Catain del medeſimo.// In Verona, Apreſſo Dionigi Filiberi. CIƆ IƆ C II.' (P. 1.)

Colophon. 'In Verona,// Nella Stamparia di Angelo Tamo. 1602.' (P. 284.)

Description. 4°, 14.7 × 19.2 cm., the text being 11.5 × 15.7 cm. 5 pp. blank + 7 unnumb. + 276 numb. = 288 pp., 32 ll. Verona, 1602.

FRANCESCO FELICIANO da Lazesio.

Ed. pr. 1526. Padua, 1629.

See p. 146.

Title. 'Scala // Grimaldelli // Libro di // Aritmetica, et Geometria // Speculatiua, & Practicale // Di M. Franceſco Feliciano Veroneſe.// Diviso in tre libri.// . . . In Padoua, Per Donato Paſquardi, & compagni. 1629.// Con licenza de' Superiori.' (F. 1, r.)

Description. 4°, 15 × 20.5 cm., the text being 11.4 × 16 cm. 4 pp. unnumb. + 276 numb. = 280 pp., 32–34 ll. Padua, 1629.

Editions. See p. 146.

FRANCESCO FELICIANO da Lazesio.
Ed. pr. 1526. Venice, 1669.

See p. 146.

Title. 'Scala // Grimaldelli // libro di // aritmetica, e geometria // Speculatiua, e Pratticale // Di M. Francesco Feliciano //Veronese.//Diviso in Tre Libri. . . . Di nuouo riſtampato, e da molti errori corretto, & accreſciuto di molte coſe da // M. Filippo Macario Veroneſe Raſonato publico della Magnifica // Città. Con l'aggionta della Regola del Catain del medeſimo.// Al Molt' Illuſtre Signor, e Padron Colendiſs. il Signor Gio: Battista Sorer.// Venetia, MDCLXIX.// Preſſo Gio: Giacomo Hertz.' (F. 1, r.)

Description. 4°, 16.1 × 22 cm., the text being 11.8 × 16 cm. 6 pp. blank + 6 unnumb. + 240 = 252 pp., 41 ll. Venice, 1669.

Editions. See p. 146.

See p. 148. It speaks well for this work of Feliciano's that this edition should have appeared one hundred and forty-three years after the book was first published.

CHRISTOFF RUDOLFF. Ed. pr. 1526. [517] Nürnberg, 1534.

Born at Jauer c. 1500, but the dates of his birth and death are unknown.

Title. 'Kunſtliche rech//nung mit der ziffer vnnd mit // den zal pfenningē // ſampt-// der Welliſchen Practica // vnd allerley vorteil // auff die Regel de Tri.// Item vergleichūg mancher-// ley Land vñ Stet // gewicht/ Elnmas //Muntz ec. Alles durch // Chriſtoffen Rudolff zu/ Wein verfertiger.// 1534.' (F. 1, r.)

Colophon. 'Getrukt zu Nůrmberg bey // Johan Petreio //im iar nach // der geburt Chriſti // M.D.xxxIIII.' (F, 120, r.)

Description. 8°, 10.2 × 14.9 cm., the text being 6.7 × 11.8 cm. 119 ff. unnumb. + 1 blank = 120 ff., 31 ll. Nürnberg, 1534.

Editions. Rudolff published three books as follows: (1) the Coss, an algebra, in 1525 (see p. 258 for the Stifel edition of 1553); (2) the Künstliche Rechnung, here described; (3) a collection of problems in 1530 (see p. 159). Of the Künstliche Rechnung the following editions appeared in the sixteenth century: Vienna (Nürnberg ?), 1526, 8°; Nürnberg, 1532, 8°; ib.,

1534 (here described); ib., 1537, 8°; ib., 1540 (below); ib., 1546; Nürnberg, 1553, 8° (below); ib., 1557 (p. 153); Vienna, 1561; Vienna (Augsburg ?), 1574; Augsburg, 1588, 8°. 517; 519

This work is an extension of the first part of the Coss, and is divided into three parts: (1) Grundbüchlein, the fundamental operations with abstract and concrete numbers, integers, and fractions, with and without the abacus; (2) Regelbüchlein, the rule of three ('Regel de Tri') and Welsch practice ('Wellisch rechnung'); (3) Exempelbüchlein, problems and results. It was one of the best-known of the practical arithmetics of that period. The rule of three is esteemed highly by Rudolff, for he says: 'sie beſchleuſzt in ſich die aller nützlichſte Regel, dadurch unzeliche rechnung in kauffen und verkauffen auſzgericht werde.' Of the Italian method of solving applied problems, the 'Welsch practice,' he says: 'Dieweil die Welliſch rechnung nichts anderes iſt, dañ ein geſchwinder auſzug in der Regel de Tri gegründet, wirt ſie auch derhalben practica geſprochē.'

Other works of 1526. Ciruelo, p. 60, 1495; Blasius, p. 97, 1513; Tagliente, p. 114, 1515; Widman, p. 37, 1489. Sterner mentions an anonymous Rechenbüchlein as printed this year at Nürnberg.

CHRISTOFF RUDOLFF.

Ed. pr. 1526. Nürnberg, 1540

See p. 151.

Title. This is substantially the same as in the 1534 edition, but bears the date 1540. (F. 1, r.)

Colophon. '℄Getruckt zu Nůrmberg bey Johañ // Petreo / Anno M. D. XL.' (F. 117, r.)

Description. 8°, 8.8 × 13.9 cm., the text being 6.8 × 12 cm. 117 ff. unnumb., 30 ll. Nürnberg, 1540.

Editions. See above.

CHRISTOFF RUDOLFF. Ed. pr. 1526. Nürnberg, 1553.

See p. 151.

Title. 'Kůnſtlich rech-//nung mit der ziffer vnd mit//den zal pfenningen/ ſampt der // Welliſchen Practica/ vnd allerley // fortheyl auff die Regel // De Tri.// Item vergleichung manch-// erley Gewicht/ Elnmas/ Můntz ꝛc. auff // etlich Landt vnd Stett. // Gemehrt mit 293 Exempeln/von man-//cherley Kauffhendeln/

mit erklerung/ wie // die ſelben zu machen vnd in die // Regel zu ſetzen ſein.// Auffs new widerumb fleiſſig vberſehen/ // vnd an vil orten gebeſſert.// Alles durch Chriſtoffen Rudolff zu // Wien verfertiget.// 1553.' (F. 1, r.)

Colophon. 'Gedruckt zu Nůrmberg/ durch // Gabriel Hayn.// 1553.' (F. 206, v.)

Description. 8°, 10 × 15.7 cm., the text being 6.8 × 11.8 cm. 206 ff. unnumb. + 2 blank = 208 ff., 24–26 ll. Nürnberg, 1553.

Editions. See p. 152.

As the title states, this is a revision of the 1526 book, with some added matter and a considerable number of new examples. The new matter begins on f. T 8. The book closes with a list of gauger's characters, 'die viſier ziffer,' not found in the 1534 edition, the integers being represented by what are practically the mediæval numerals, and the fractions being generally unit fractions.

CHRISTOFF RUDOLFF. Ed. pr. 1526. Nürnberg, 1557.

See p. 151.

Description. This edition of Rudolff's arithmetic is substantially verbatim with that of 1553 (p. 152). 8°, 9.5 × 15 cm., the text being 7 × 11.8 cm. 206 ff. unnumb., 26 ll. Nürnberg, 1557.

ANONYMOUS. Ed. pr. 1527. Cologne, 1527.

Title. 'Compendia-//ria artis nvmerandi ratio, et // expeditiſſima practicandi uia, figuris Arithme//ticis omnes numerorū formulas cōprehen//dens, additis quibuſdam, ut raris, ita // utilibus regulis.// Radicis Cvbicae Extractio.//

```
           1 3 8 8 6   1 5 1
Cubus    6 5 3 5 3 4   2 9 9 5 2
           .     .       .     .
-----------------------------------------
Radix      4     0       2     8
             1 2 1 2 0.    6
             9 6 4 8 0
             3 8 8 6 2 1 4 4.'        (F. 1, r.)
```

Colophon. 'Coloniae apvd Melchiorem // Noueſienſem Anno .M. .D. XXVII.//Menſe maio.' (F. 29, r.)

Description. 4°, 13.4 × 19.4 cm., the text being 8.9 × 14.3 cm. 29 ff. unnumb., 32 ll. Cologne, 1527.

Editions. There was no other edition. The dedicatory epistle is dated 'Coloniæ. Anno 1527. Calendis Maijs.'

This extremely rare little work, almost unknown to bibliographers, begins with a theoretical discussion of the nature of number and arithmetic. This is followed by a 'compendiaria . . . artis numerandi ratio in tres tractatus digesta.' Of these the first treats of the 'species' and the rule of three in integers; the second of fractions, and the third of business problems. The work is too theoretical to have had any influence on commercial arithmetic.

JOHANNES WOLPHIUS. Ed. pr. 1527. Frankfort, 1534.

A German mathematician, born c. 1500.

Title. 'Rvdi//menta Arithmetices // Authore Iohanne Vuolphio // Herſbrugienſe.// Elemen//tale Geometricvm, ex // Euclidis Geometria, à Ioanne Vœgelin,// Haylpronnenſi, ad omnium Mathe//matices studioſorum utili-//tatem decerptum. // Franc. Chriſtianus Ege-//nolphus excudebat.' (The title page is surrounded by an elaborate woodcut. F. 1, r.)

Colophon. At the end is the date 'M.D.XXXIIII.' (F. 56, r.)

Description. 8°, 9.7 × 15 cm., the text being 6.8 × 11.7 cm. 56 ff. unnumb., 24–29 ll. (The arithmetic occupies only 27 ff.) Frankfort, 1534.

Editions. Nürnberg, 1527; Frankfort, 1534, 8° (here described); ib., 1537; Strasburg, 1539, 8°; ib., 1540; Frankfort, 1548, 8° (below); ib., 1561.

This brief treatise on arithmetic covers the work required in some of the Latin schools of the sixteenth century. It contains little besides numeration and the fundamental operations, including duplation, mediation, the rule of three, and fractions. There are only a few applications, coinage and partnership being the most prominent.

JOHANNES WOLPHIUS. Ed. pr. 1527. Frankfort, 1548.

See above.

Title. 'Rvdi-//menta Arithmeti-//ces, Autore Ioanne Vuol-// phio Herſbru-//gienſe.// Elementa-//le Geometricvm,// Ex

Euclidis Geometria, a Ioanne // Vœgelin, Haylpronnenſi, ad o-// mnium Mathematices ſtudio-//ſorum utilitatem de-//cerptum.// Franc. Chri. Ege.' (F. 1, r.)

Colophon. At the end is the date 'M. D. XLVIII.' (F. 60, r.)

Description. 8°, 9.9 × 15.3 cm., the text being 8.2 × 11.8 cm. 60 ff. unnumb., 22–27 ll. (The arithmetic occupies only 28 ff.) Frankfort, 1548.

PETRUS APIANUS. Ed. pr. 1527. Ingolstadt, 1527.

PETER BIENEWITZ, or BENNEWITZ. Born at Leisnig, in 1495; died at Ingolstadt, April 21, 1552. He wrote chiefly on astronomy, and in his Cosmographia (1524) he first showed how to determine longitude by observing the distance of the moon from certain fixed stars. He was professor of astronomy at Ingolstadt, and was one of the few university professors of his time who gave instruction in arithmetic in the German language.

Title. See Fig. 78.

Colophon. 'Gedrückt vnd volendt zů Ingolſtadt // durch Georgium Apianum von Leyß-//nick/ jm Jar nach der geburt Chriſti // 1527. am 9. tag Auguſti.' (F. 299, v.)

Description. 8°, 10 × 14.8 cm., the text being 7 × 11.7 cm. 299 ff. unnumb. + 1 blank = 300 ff., 27 ll. Ingolstadt, 1527.

Editions. Ingoldstadt, 1527, 8° (here described); Frankfort, 1537, 8° (p. 157); ib., 1544, 8°; ib., 1564, 8°; ib., 1580. Graesse mentions an 'Arithmetica,' Leipzig, 1543, 8°, but questions it, and Romstöck does not give it in his article on Apianus in the *Astronomen, Mathematiker, und Physiker der Diöcese Eichstätt.*

Apianus follows Rudolff so closely as to give ground for comment. His arithmetic differs from the latter's chiefly in the arrangement of the matter. The work is largely commercial, and includes the fundamental operations and the ordinary rules and applications of the period. There is a chapter on counters at the end of the book. Indeed, Apianus advises their use, saying: 'die Sumering der Regiſter durch die rechenpfeñing auff der lini brauchſamer iſt dañ durch die federn oder kreide.' The work is also interesting on account of its quaint illustrations. The title page (p. 156) is noteworthy on account of the engraved 'Pascal triangle' a century before Pascal studied this numerical form, and some years before Stifel mentioned it, and because of the picture of line reckoning. I know of no example of the 'Pascal triangle' in print before this one, although the arrangement had doubtless long been more or less familiar to mathematicians.

Eyn Newe
Vnnd wolgegründte
vnderweysung aller Kauffmanß Rechnung in dreyen Büchern/mit schönen Re
geln vñ fragstucken begriffen. Sunderlich was fortl vnnd behendigkait in der
Welschē Practica vñ Tolletn gebraucht
wirdt/ des gleychen fürmalß wider in
Teützscher noch in Welscher sprach nie
gedrückt. durch Petrum Apianū
von Leyßnick/ð Astronomei
zū Ingolstat Ordina-
riū/ verfertiger.

FIG. 78. TITLE PAGE OF THE FIRST EDITION OF APIANUS

Other works of 1527. Feliciano, p. 146, 1526; Riese, p. 139, 1522, the first Nürnberg edition so far as I know; Tagliente, p. 114, 1515; W. Peer, 'Ain new guet Rechenbuchlein,' Nürnberg, 8°.

PETRUS APIANUS. Ed. pr. 1527. Frankfort, 1537.

See p. 155.

Title. 'Ein newe vnd wolge-//gründte vnderweiſung aller // Kauffmans Rechnung in dreien Bů//chern/ mit ſchönen Regeln vnd fragſtücken be-//griffen. Sunderlich was fortel vnnd behendig-//keit in der Welſchen Practica vnnd Tolle-//ten gebraucht wurt/ des gleichen vor//mals weder inn Teutſcher noch in // Weliſcher Spraach nie getruckt.// ℂDurch Petrum Apianum von Leyſznick der // Aſtronomei zů Ingolſtatt Ordinarium.// (Woodcut of merchants using counters.) Franc. Chri. Egen.' (F. 1, r.)

Colophon. 'Zu Franckfurt/ bei Chriſtian Egenolff/ // Anno Domini. M. D. xxxvij.// Im Herbſtmon.' (F. 183, v.)

Description. 8°, 9.8 × 15.3 cm., the text being 7 × 11.5 cm. 183 ff. unnumb. + 1 blank = 184 ff., 25–28 ll. Frankfort, 1537.

See p. 155.

JOHANNES FERNELIUS. Ed. pr. 1528. Paris, 1528.

JEAN FERNEL. Born at Clermont in 1497; died at Paris, April 26, 1558. He was a physician, with a taste for mathematics and astronomy. He wrote numerous works on medicine and mathematics.

Title. 'Ioannis Fer//nelii Ambianatis // de proportionibus Libri duo.// Prior, qui de ſimplici proportio-//ne eſt, & magnitudinum & nu-//merorum tum ſimplicium tum // fractorum rationes edocet.// Poſterior, ipſas proportiones cō-//parat: earúmq; rationes colligit.// Parisiis // Ex ædibus Simonis Colinæi // 1528.' (F. 1, r. Fig. 78.)

Colophon. 'ℂLibellorvm de proportionibvs, Ioanne // Fernelio Ambianate authore, finis.' (F. 28, v.)

Description. Fol., 22.2 × 31.5 cm., the text being 15.6 × 24.4 cm. 4 ff. unnumb. + 24 numb. = 28 ff., 44–45 ll. Paris, 1528.

Editions. There was no other edition.

This is one of the best of the sixteenth-century treatises on the mediæval proportion. It follows the Boethian treatment, as seen also in the work of Bradwardin.

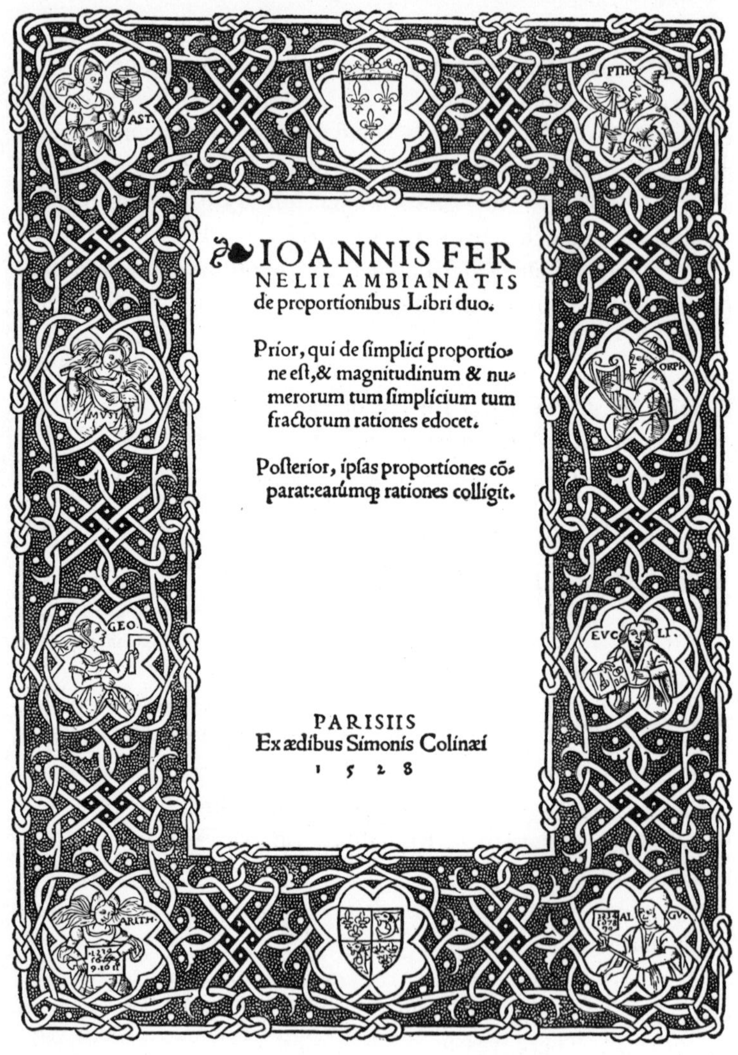

IOANNIS FER
NELII AMBIANATIS
de proportionibus Libri duo.

Prior, qui de ſimplici proportione eſt, & magnitudinum & numerorum tum ſimplicium tum
fractorum rationes edocet.

Poſterior, ipſas proportiones cõ-
parat: earúmq; rationes colligit.

PARISIIS
Ex ædibus Simonis Colinæi
1528

FIG. 79. TITLE PAGE OF FERNELIUS

Other works of 1528. Borghi, p. 16, 1484; Cassiodorus, p. 211, 1540; Ciruelo, p. 60, 1495; Riese, p. 139, 1522; Tagliente, p. 114, 1515; Christiernus Torchillus Morsianus, 'Arithmetica brevis et dilucida in quinque partes digesta,' Cologne, 8° (but see p. 182, 1536).

Works of 1529. Anianus, p. 32, 1488; Riese, p. 139, 1522; Tonstall, p. 135, 1522; Bæda, 'De natura rerum et temporum ratione libri duo,' Basel, cap. I being 'De computu vel loquela digitorum'; there was an edition by Noviomagus, Cologne, 1537 (see also pp. 131, 140, 263, 1521, 1525, 1554).

CHRISTOFF RUDOLFF. Ed. pr. 1530. Augsburg, 1530.

See p. 151.

Title. See Fig. 80.

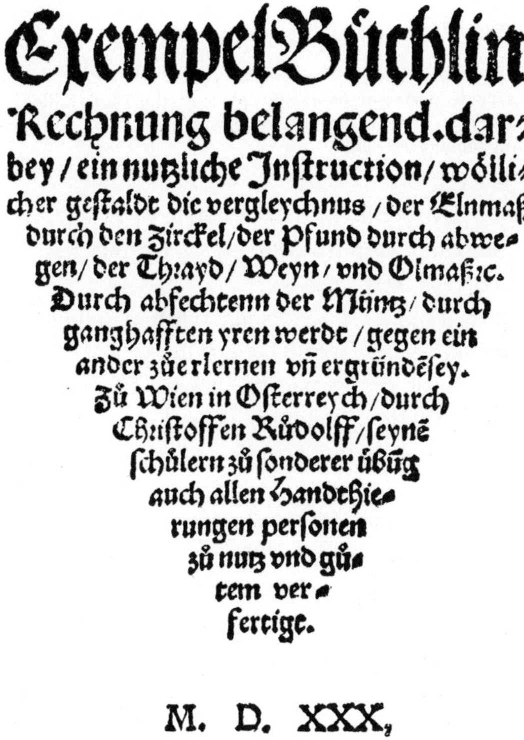

Exempel Büchlin
Rechnung belangend. dar=
bey/ ein nutzliche Instruction/ wölli=
cher gestaldt die vergleychnus / der Elnmaß
durch den Zirckel/der Pfund durch abwe=
gen/ der Thrayd/ Weyn/ vnd Olmaß ꝛc.
Durch abfechtenn der Müntz/ durch
ganghafften yren werdt / gegen ein
ander zu erlernen vñ ergründẽ sey.
Zu Wien in Osterreych /durch
Christoffen Rudolff/ seynẽ
schülern zu sonderer übũg
auch allen handtie=
rungen personen
zu nutz vnd gu=
tem ver=
fertige.

M. D. XXX,

FIG. 80. TITLE PAGE OF THE FIRST EDITION OF RUDOLFF'S *Exempel Büchlin*

Colophon. 'Getruckt in der lŏblichen Reychſtat Aug-//ſpurg/ durch Heynrichen Stayner/ //Volendet am 31 May im jar//M.D. XXX.' (F. 75, v.)

Description. 8°, 9.9 × 15.1 cm., the text being 7.4 × 12.2 cm. 75 ff. unnumb. + 2 blank = 77 ff., 28 ll. Augsburg, 1530.

Editions. Augsburg, 1530, 8° (here described); Nürnberg, 1538; ib., 1540. [517]

This is the third of Rudolff's works (see p. 151). As the name implies, it is merely a collection of problems, two hundred and ninety-two in number. Most of these problems are of a genuine business nature, and they furnish a good idea of the ordinary commercial needs of the first half of the sixteenth century in Germany.

ORONTIUS FINAEUS. Ed. pr. 1530–32. Paris, 1530–32.

ORONCE FINE. Born at Briançon in 1494; died at Paris, October 6, 1555. He was made professor of mathematics in the (later called) Collège de France in 1532. He wrote extensively on astronomy and geometry, but was not a genuine scholar.

Title. See Fig. 81.

Colophon. 'Excvsvm est avtem ipsvm opvs Pa//riſijs in uico Sorbonico, impenſis Gerardi Morrhij, & Ioannis Petri. Anno // M,D.XXXII.' (F. 216, r.)

Description. Fol., 24.1 × 37 cm., the text being 18.4 × 27.7 cm. 9 ff. unnumb. + 208 numb. = 217 ff., 48 ll. Paris, 1530–32.

The title page of the geometry appears on f. 49, r., with the date M. D. XXX; the cosmography on f. 101, r., with the date M. D. XXX; the horography on f. 157, r., with the date M. D. XXXI.

Editions. Paris, 1530–32, fol. (here described); ib., 1535; ib., 1542, fol. (p. 163); ib., 1544, 8° (p. 163); ib., 1554; ib., 1555, 4° (p. 163); Venice, 1587, 4° (p. 164). Leslie's statement that the work appeared in 1525 is unfounded. For the 'De rebus mathematicis,' 1556, see p. 279.

This is the first edition of the works of Finaeus, perhaps the most pretentious French mathematician of his time, and was published during the years 1530–32. The dedicatory epistle is dated 'Lutetiæ Pariſiorum Calendis Ianuarij 1531,' or 1532 new style. The part on arithmetic is

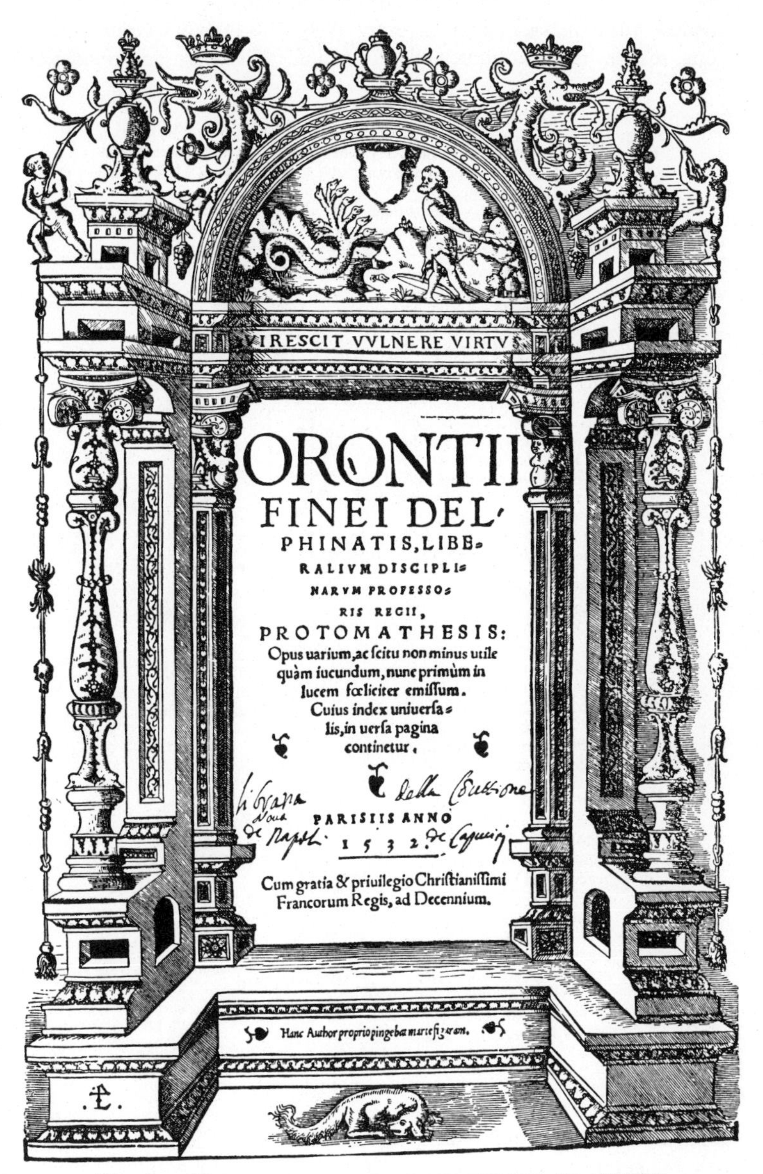

VIRESCIT VVLNERE VIRTVS

ORONTII
FINEI DEL-
PHINATIS, LIBE-
RALIVM DISCIPLI-
NARVM PROFESSO-
RIS REGII,
PROTOMATHESIS:
Opus uarium, ac ſcitu non minus utile
quàm iucundum, nunc primùm in
lucem fœliciter emiſſum.
Cuius index uniuerſa-
lis, in uerſa pagina
continetur.

PARISIIS ANNO
1532.

Cum gratia & priuilegio Chriſtianiſſimi
Francorum Regis, ad Decennium.

Hanc Author proprio pingebat marte figuram.

FIG. 81. TITLE PAGE OF THE FIRST EDITION OF FINAEUS

divided into four books dealing respectively with integers, common fractions, sexagesimal fractions, and proportion. There are no applications

ORONTII FINEI DELPH.

per 29 primi elementorum Euclidis facilè manifeſtatur. & angulus A B H, angulo A G F eſt æqualis (nam uterq; rectus) igitur per 4 ſexti eiuſdem Euclidis, fit ſicut H B ad B A, ita F G putei latitudo ad G A compoſitam ex G B & B A longitudinem, ſiue profunditatem.

Exemplum. Sit exempli gratia B H 20 partium, qualium latus quadrati eſt 60: B E autẽ metiatur, & ſit in exemplum 6 cubitorum, tot etiam cubitorum erit G F: ſunt enim latera perallelogrammi B E F G oppoſita, quæ per 34 eiuſdem primi ſunt inuicem æqualia. Duc igitur 6 in 60, fient 360: quæ diuide per 20, &. habebis pro quotiẽte 18. Tot igitur cubitorum erit A G: à qua ſi dempſeris A B trium uerbi gratia cubitorum, relinquetur B G deſyderata & in profundum depſſa putei lõgitudo 15 cubitorum.

Alia eiuſdem obſeruationis demonſtratio. IDEM QVOQVE SIC OBtinebis. Metire H E: ſitq; exempli cauſa 5 cubitorũ. Deinde multiplica 5 per 60, fient 300: hæc diui per 20, producentur 15, uelut antea. Bina nanq; triangula A B H et H E F ſunt rurſum æquiangula. quoniam angulus A H B angulo E H F ad uerticem poſito, per 15 primi Euclidis eſt æqualis. itẽ rectus qui ad B, recto qui ad E pariter æquat. reliquus igitur B A H reliquo H F E per 32 eiuſdem primi eſt æqualis. Vnde per ſuperius allegatã quartã propoſitionẽ ſexti, ſicut H B ad B A, ita H E ad E F, eidem B G per hypotheſim æqualem.

Notandum. Cum autem acciderit puteum rotundam habere figuram, habenda erit cõſyderatio diametri putealis orificij, & reliqua omnia ueluti prius abſoluenda.

Secundus modus metiendi profunda, per quadrantem. RELIQVVM ETS, VT eandem rerum in profundũ depreſſarum, per uulgatũ quadrãtem metiri doceamus altitudinẽ. Sit itaq; puteus circularis E F G H, cuius diameter ſit E F, aut illi æqualis G H. Adplica igitur quadrãtem ipſi putei orificio: in hũc modũ, ut finis lateris A D ad datũ punctum E conſtituatur. Leua poſtmodũ, aut deprime quadrãtem (libero ſemper demiſſo perpendiculo) donec radius uiſualis per ambo foramina pinnacidiorum ad inferiorem & è diametro ſignatũ terminũ H perducatur. Quo facto & immoto quadrãte, uide in quã

FIG. 82. FROM THE FIRST EDITION OF FINAEUS

worthy the name, and the work has little to commend it. Some interesting illustrations showing the use of the mediæval Quadrans are shown in Fig. 82.

ORONTIUS FINAEUS. Ed. pr. 1530–32. Paris, 1542.

See p. 160.

Title. 'Orontii // Finei Delphin. Re-//gii Mathematicarvm // Professoris: // arithmetica // practica, libris qva-//tuor abſoluta, omnibus qui Ma-//thematicas ipſas tractare volunt // perutilis, admodúmque neceſſa-//ria: Ex nouiſſima authoris reco-//gnitione, amplior, ac emenda-//tior facta.// Ædito tertia.// Parisiis.// Ex officina Simonis Colinæi.// 1542.// Cum gratia & priuilegio Chri-// ſtianiſſimi Francorum Regis.' (F. 1, r.)

Description. Fol., 20.9 × 30.1 cm., the text being 16 × 28.2 cm. 2 ff. unnumb. + 66 numb. = 68 ff., 40 ll. Paris, 1542.

See p. 160.

ORONTIUS FINAEUS. Ed. pr. 1530–32. Paris, 1544.

See p. 160.

Title. 'Orontii // Finæi Delphi-//natis, Regij Mathe-//maticarū Lutetiæ // Profeſſoris,// Arithmetica // Practica, in compendiū per Authorem // ipſum redacta, multiſq; acceſſionibus //locupletata: Ijs qui ad liberam quāuis,// nedū Mathematicā adſpirant philoſo-//phiā perutilis, admodúmq; neceſſaria.// Lvtetiae Parisiorvm // Apud Simonem Colinæum.// 1544.// Vireſcit vulnere virtus.' (F. 1, r.)

Description. 8°, 11 × 17 cm., the text being 8.6 × 12.8 cm. 95 ff. numb. + 1 unnumb. = 96 ff., 31 ll. Paris, 1544.

See p. 160.

ORONTIUS FINAEUS. Ed. pr. 1530–32. Paris, 1555.

See p. 160.

Title. 'Orontii Finaei // Delphinatis, Regii // Mathematicarum Lutetiæ // profeſſoris,// de arithmetica practi-//ca libri quator: Ab ipſo authore uigi-//lanter recogniti, multíſque //acceſsionibus recèns // locupletati.// Lvtetiae Parisiorvm,// apud Michaëlem Vaſcoſanum,// 1555.// Ex privilegio regis.// Vireſcit uulnere uirtus.' (F. 1, r.)

Description. 4°, 15.4 × 20.7 cm., the text being 11.6 × 17.2 cm. 4 ff. unnumb. + 72 numb. = 76 ff., 34–35 ll. Paris, 1555.

ORONTIUS FINAEUS. Ed. pr. 1530–32. Venice, 1587.

See p. 160.

Title. 'Opere//di//Orontio Fineo//del Delfinato://Diuiſe in cinque Parti;//Arimetica, Geometria, Coſmografia, & Oriuoli, //Tradotte//Da Coſimo Bartoli, Gentilhuome, & Academico Fiorentino://Et gli Specchi,//Tradotti dal Caualier Ercole Bottrigaro, Gentilhuomo Bologneſe.//Nuouamente poſte in luce: //con privilegio.//In Venetia, Preſſo Franceſco Franceſchi Seneſe, 1587.' (F. 1, r.)

Description. 4°, 14.9 × 20.9 cm., the text being 10.1 × 16.8 cm. 8 ff. unnumb. + 81 numb. = 89 ff. (in the part devoted to arithmetic), 35–39 ll. Venice, 1587.

Editions. See p. 160. That this is the first Italian edition appears in the printer's dedicatory epistle to Guidubaldo de' Marchesi del Monte, in which he mentions 'in queſta occaſione dell' hauere ſtampato l'opera d'Orontio nella noſtra Toſcana lingua,' with the date 'Di Venetia, il di 7. di Luglio, 1587.'

See p. 160.

ANDREAS ALCIATUS. Ed. pr. 1530. Hagenau, 1530.

Born at Alzano, near Milan, May 8, 1492; died there June 12, 1550. He was an Italian jurist.

Title. See Fig. 83.

Description. 8°, 10.3 × 15.4 cm., the text being 7.8 × 11.1 cm. 50 ff. and 2 plates unnumb., 28 ll. Hagenau, 1530.

Editions. There was no other separate edition, but the works of Alciatus appeared at Basel in 1571, 3 vol., fol.

The work of Alciatus extends only to f. D 1. The 'Oratio de legibus' of Melanchthon then extends to f. F 5. This is followed by 'Budæi qvædam de moneta Græca,' etc. The work of Alciatus is not an arithmetic, but a history of weights and measures. As such it is of value for the historical development of commercial mathematics. 517

Other works of 1530. Boethius, p. 27, 1488; Bradwardin, p. 61, 1495; Clatovenus, p. 292, 1558; Maffei, p. 86, 1506; Riese, p. 139, 1522; Tagliente, p. 114, 1515; Torrentini, p. 76, 1501; Johann Kolross, a primer entitled 'Enchiridion: das ist Handbüchlin tütscher

Orthographi . . . Auch wie mann die Cifer vnd tüdtsche zaal verston sol,' Basel, with another edition, ib., 1534, 8°. There was also an anonymous work entitled 'La vraye manière pour apprendre à chiffrer et compter,' published at Lyons, s. a., 12°, c. 1530.

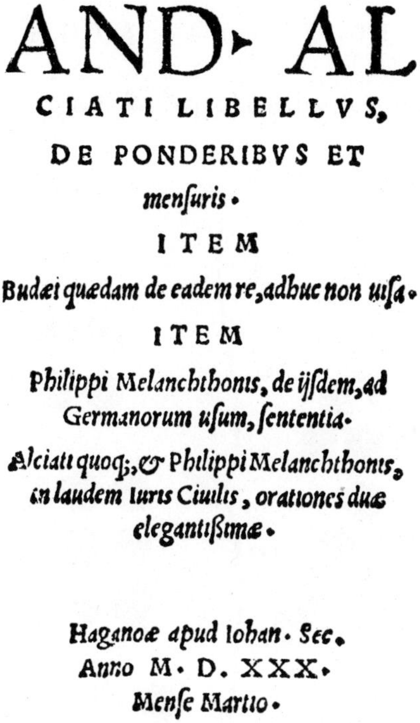

AND· AL
CIATI LIBELLVS,
DE PONDERIBVS ET
mensuris·
ITEM
Budæi quædam de eadem re, adhuc non uisa·
ITEM
Philippi Melanchthonis, de ijsdem, ad
Germanorum usum, sententia·
Alciati quoq;, & Philippi Melanchthonis,
in laudem Iuris Ciuilis, orationes duæ
elegantißimæ·

Haganoæ apud Iohan· Sec.
Anno M· D. XXX·
Mense Martio·

FIG. 83. TITLE PAGE OF ALCIATUS

JOACHIM FORTIUS RINGELBERGIUS.

Ed. pr. 1531. Leyden, 1531.

JOACHIM STERCK VAN RINGELBERGH. Born at Antwerp, c. 1499; died c. 1536. He taught philosophy and mathematics in various cities of Germany and France.

Title. 'Ioachimi // Fortii Ringel-//bergij Andouerpiani opera, // quæ proxima pagina //enumerantur.// Virtvte dvce // comite Fortvna.// Apvd Gryphivm // Lvgdvni, // anno // M. D. XXXI.' (P. 1.)

Colophon. ‘Lvgdvni apvd // Seb. Grypivm, // anno // M. D. XXXI.’ (P. 687.)

Description. 8°, 10.3 × 15.7 cm., the text being 8.4 × 12.2 cm. 687 pp. numb. + 30 = 717 pp., 22–29 ll. Leyden, 1531.

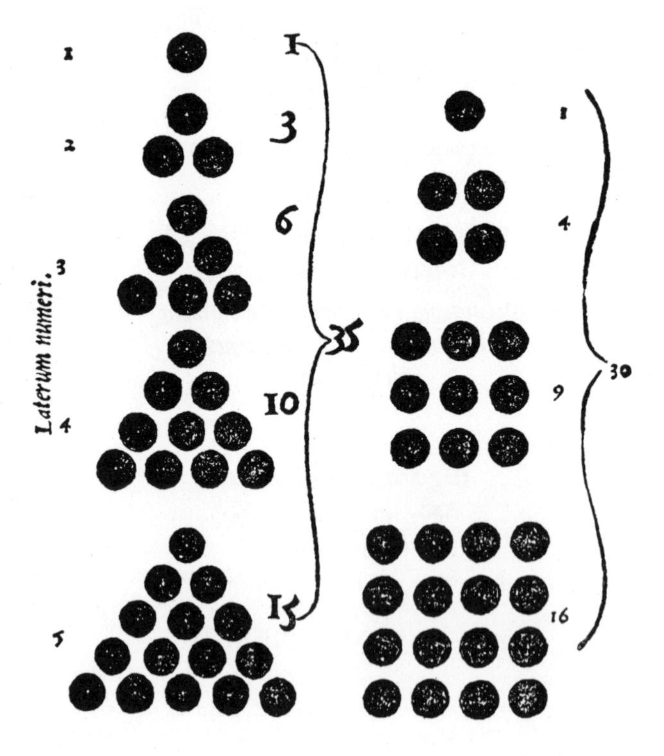

FIG. 84. FROM THE FIRST EDITION OF RINGELBERGIUS

Editions. Leyden, 1531, 8° (here described); ib. (at least the arithmetic part), 1539, 8°; Basel, 1541, 8° (see p. 167); Leyden,

1556, 8° (see p. 168). The 'Epistola ad Lectorem' is dated 'Louanij Idib. Augusti, Anno M. D. XXIX.'

This work of Ringelbergius is somewhat encyclopedic in character. The 'Liber de Ratione ſtudij,' with 'Annotationes' thereon and a 'Horoscopus libri ratione ſtudij' to show that it was written in an auspicious time, is followed by six other books. These relate to Grammar, Dialectics, Rhetoric, Mathematics, and Divination, closing with a book entitled 'Commvnis cvivsdam Naturæ ſunt.' The book on mathematics includes a chapter on arithmetic in which, in 17 pages (about 10 pages excluding the illustrations), the author treats of the Boethian ratios, figurate numbers (see Fig. 84), and the fundamental operations with figures and upon the line abacus. The part relating to astronomy had already been published at Basel in 1528, and the cosmography in Paris in 1529.

Other works of 1531. Köbel, p. 101, 1514; Juan Gutiérrez de Gualda, 'Arte breue y muy prouechos de cuenta castellana y arismetica,' Toledo, 4°; ib., 1539; Saragossa, 1557, 1564; Alcalá, 1570; H. C. Agrippa, 'De occulta philosophia libri tres,' s. l., with later editions, Cologne, 1533, fol.; 1541; Lugduni, 1550, 8°; s. l., 1565; Basel, 1567, 8°; s. l. a. (Paris, 1567 ?), included by De Morgan without much reason. 496;520

JOACHIM FORTIUS RINGELBERGIUS.

Ed. pr. 1531. Basel, 1541.

See p. 165.

Title. 'Ioachimi // Fortii Ringelber//gii Andoverpiani lvcvbra-//tiones, uel potius abſolutiſsima κυκλοπαίδιδæ: nem-//pe liber de Ratione ſtudij, utriusq; linguæ, Grāmatice, // Dialectice, Rhetorice, Mathematice, & ſublimioris // Philoſophiæ multa. Quorū ἔλεγχος ſub ſequenti pa//gina enumeratur. Atq; hæc omnia eo iudicio & // ordine ſunt tradita, ut uel ſola cuiq; // meliorum literarum ſtudioſo // ſatis ad ſummum inge-//nij cultum eſſe // poſſint.// Basileae.// Anno M. D. XLI.' (P. 1.)

Colophon. 'Basileae apvd Bartholo-//mevm Vvesthemervm // anno M. D. XLI.' (P. 797.)

Description. 8°, 10.3 × 15.7 cm., the text being 8.2 × 11.8 cm. 796 pp. numb. + 2 unnumb. = 798 pp., 25 ll. Basel, 1541.

See p. 166. Although the title differs from that of 1531, the work is the same.

JOACHIM FORTIUS RINGELBERGIUS.

Ed. pr. 1531. Leyden, 1556.

See p. 165.

Title. 'Ioachimi // Fortii // Rin-//gelbergii // Andoverpiani // Opera, // Quae proxima pagina enumerantur.// Lvgdvni,// Apud Ioannem Frellonium.' (P. 1. The rest of the page is torn off.)

Colophon. 'Lvgdvni,// ex officina typogra-//phica Michaelis // Sylvii,// M. D. LVI.' (P. 663.)

Description. 8°, 10.5 × 15.2 cm., the text being 9 × 13 cm. 4 pp. unnumb. + 4 blank + 660 numb. = 668 pp., 26–31 ll. Leyden, 1556.

See p. 166. Although the title differs from that of 1531, the work is the same.

MICHAEL PSELLUS. Ed. pr. 1532. Venice, 1532.

Called the Younger, to distinguish him from a philosopher of the same name who lived about 870 A.D. Born at Constantinople in 1020; died in a cloister in 1110. He studied at Athens and taught philosophy at Constantinople.

Title. See Fig. 85.

Description. 8°, 9.9 × 14.9 cm., the text being 7.4 × 11.4 cm. 1 f. blank + 104 unnumb. = 105 ff., 24–25 ll. Greek, Venice, 1532.

Editions. Venice, 1532, 8°, Greek (here described); Paris, 1538, Greek, 4°; Paris, 1545, Greek and Latin; Augsburg, 1554, 8°, Greek and Latin; Wittenberg, 1556, Latin; Basel, 1554 and 1556, 8°, Greek and Latin; Paris, 1557, 8°, Latin (p. 170); Wittenberg, 1560, Latin; Paris, 1585, Latin; Leipzig, 1590, 8°, Greek and Latin (p. 170); Heidelberg, 1591, Latin; Tours, 1592, Latin. [520]

Psellus was one of the last of the Greek writers on arithmetic. This part of his work is devoted solely to the theory of numbers, and it represents the arithmetical inheritance derived from the older Hellenic civilization. The treatise covers the mediæval Quadrivium — arithmetic, music, geometry, and astronomy — and is the only late Greek work on arithmetic that attracted attention in the Renaissance period. The arithmetic is merely a primer for the study of Nicomachus.

Other works of 1532. Aventinus, p. 136, 1522; Capella, p. 66, 1499; Feliciano, p. 146, 1526; Köbel, p. 102, 1514; Rudolff, p. 151, 1526; Stifel, p. 223, 1544; Vincento, p. 140, 1522; Johann Brandt, 'Kunst-

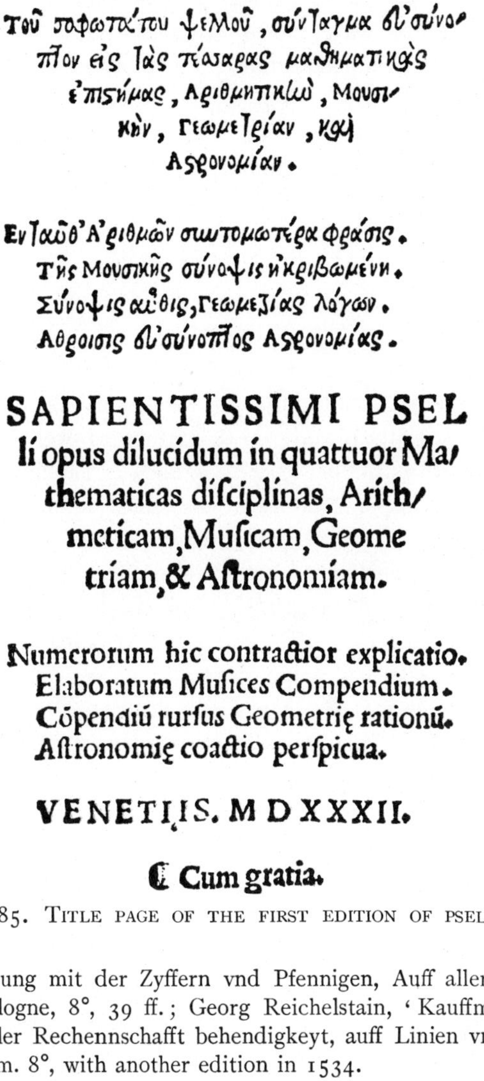

Τοῦ σοφωτάτου Ψελλοῦ, σύνταγμα εὐσύνο-
πτον εἰς τὰς τέσσαρας μαθηματικὰς
ἐπιστήμας, Ἀριθμητικὴν, Μουσι-
κὴν, Γεωμετρίαν, καὶ
Ἀστρονομίαν.

Ἐνταῦθ' Ἀριθμῶν συντομωτέρα φράσις.
Τῆς Μουσικῆς σύνοψις ἠκριβωμένη.
Σύνοψις αὖθις, Γεωμετρίας λόγων.
Ἄθροισις εὐσύνοπτος Ἀστρονομίας.

SAPIENTISSIMI PSEL
li opus dilucidum in quattuor Ma/
thematicas diſciplinas, Arith/
meticam, Muſicam, Geome
triam, & Aſtronomiam.

Numerorum hic contractior explicatio.
Elaboratum Muſices Compendium.
Cōpendiū rurſus Geometrię rationū.
Aſtronomię coactio perſpicua.

VENETIIS. M D XXXII.

¶ Cum gratia.

FIG. 85. TITLE PAGE OF THE FIRST EDITION OF PSELLUS

liche Rechnung mit der Zyffern vnd Pfennigen, Auff allerley handttierung,' Cologne, 8°, 39 ff.; Georg Reichelstain, 'Kauffmans handbüchlin. Aller Rechennschafft behendigkeyt, auff Linien vnd Ziffern,' Frankfort, sm. 8°, with another edition in 1534.

MICHAEL PSELLUS. Ed. pr. 1532. Paris, 1557.

See p. 168.

Title. 'Michael // Psellvs de // Arithmetica,// Mvsica, Geometria:// & Proclus de // Sphæra,// Elia Vineto Santone interprete.// (Woodcut with motto: Inpingvi Gallina.) Parisiis,// Apud Gulielmum Cauellat, in pingui gallina,// ex aduerſo collegij Cameracenſis.// 1557.' (F. 1, r.)

Description. 8°, 10.6 × 16.2 cm., the text being 6.6 × 13.3 cm. 2 ff. unnumb. + 76 numb. + 3 blank = 81 ff. (18 ff. on arithmetic), 22 ll. Paris, 1557.

Editions. See p. 168. This is one of the Latin editions.

See p. 168.

MICHAEL PSELLUS. Ed. pr. 1532. Leipzig, 1590.

See p. 168.

Title. 'Pselli // Philoſophi & Mathemati-//ci clariſsimi // Arithmetica//Edita ſtudio//M. Chriſtophori Meureri,//Mathematum Profeſſoris publici // in Academia // Lipſienſi.// 1590.// Plato interrogatus, cur homo ſit a-//nimal ſapientiſsimum : *ὅτι ἀριθ-//μεῖν ἐπίςαται*, reſpondit.// Lipsiae.' (F. 1, r.)

Colophon. 'Lipſiæ, imprimebat Michaêl // Lantzenberger.// Anno M. D. XC.' (F. 24, v.)

Description. 8°, 9.5 × 15.8 cm., the text being 6.6 × 12.3 cm. 24 ff. unnumb., 30 ll. Leipzig, 1590.

Editions. See p. 168. The Latin dedicatory epistle is dated 'Lipſiæ XV. Cal. Nouemb. Anno poſt Chriſtum natum 1590.' The text is in Greek and Latin.

See p. 168.

MICHAEL PSELLUS. Ed. pr. 1532. Leipzig, 1616.

See p. 168.

Title. 'Pselli // arithmetica // Guilhelmo Xylandro // interprete // Cum Præfatione // Christophori Meureri D.// Mathematum Profeſſoris // in Academia Lipſienſi.// 1616.// Lipsiæ // Typis Abrahami Lambergi.' (F. 1, r.)

Description. 8°, 9 × 14.4 cm., the text being 6.6 × 12.1 cm. 39 pp., 1–8 unnumb., then numb. 1–31; 24–25 ll. Leipzig, 1616.

See p. 168.

ADAM RIESE. Ed. pr. 1533. Leipzig, 1536.

See p. 138.

Title. See Fig. 86.

Description. 4°, 14.6 × 19 cm., the text being 9.6 × 15.1 cm., and the tables 9.5 × 12 cm. 79 ff. unnumb. + 1 blank = 80 ff., 17–24 ll. in the Introduction (3 ff.), the rest of the book consisting of tables. Leipzig, 1536.

Editions. Leipzig, 1533; ib., 1536, 4° (here described).

This is a set of mercantile tables for the multiplication and division of denominate numbers.

GEORGE AGRICOLA. Ed. pr. 1533. Paris, 1533.

Born March 24, 1490, at Glauchau, Saxony; died November 21, 1555, at Chemnitz. He was rector of a school at Zwickau (1518–1522), and in later life a physician. He wrote a number of scientific works.

Title. See Fig. 87.

Description. 8°, 10.2 × 15 cm., the text being 6.8 × 12.4 cm. 7 pp. unnumb. + 3–261 numb. + 6 blank + 1 with woodcut = 273 pp., 28 ll. Paris, 1533.

Editions. Paris, 1533, 8° (here described); Venice, 1533, 8°; ib., 1535, fol.; Basel, 1549, 8°; ib., 1550, fol. There was also an 'Epitome omnium Georgii Agricolae de mensuris et ponderibus per G. Philandrum' published at Lyons in 1552, 8°.

The work can hardly be called an arithmetic, but, like a few others included in this list, it is a valuable book of reference on the history of ancient measures. It consists of five books as follows: 'Liber primus, de menfuris Romanis' (p. 9); 'Liber secundus de Mensuris Græcis' (p. 75); 'Liber tertius, de Pondere rerum quas metimur' (p. 144); 'Liber quartus, de Ponderibus Romanis' (p. 188); 'Liber quintus, de Ponderibus Græcis' (p. 219). The book is also valuable to the student of Roman and Greek numerals, and of the various symbols of measures. Such works explain the origin of certain systems of

Ein Gerechent Bůch=
lein/auff den Schöffel/Eimer/
vnd Pfundtgewicht/ zu eh=
ren einem Erbarn/Weisen
Rathe auff Sanct An=
nenbergk.

Durch Adam Riesen.
1533

Zu Leiptzick/hatt gedruckt diß
gerechent Bůchlein
Melchior Lotter.
Volendet vnd außgangen am abendt
des Newen Jars
1536

FIG. 86. TITLE PAGE OF THE 1536 RIESE

measures employed before the metric system was developed, and of such symbols as are still used by apothecaries.

GEORGII

AGRICOLAE MEDICI LIBRI
quinque de Mensuris & Ponderibus, in quibus plæraque à BVDAEO &
PORTIO parum animaduersa diligenter excutiuntur.

Opus nunc primum in lucem æditum.

PARISIIS,
Excudebat Christianus Wechelus, in uico Iacobæo, sub scuto Basiliensi. Anno
M. D. XXXIII.

FIG. 87. TITLE PAGE OF THE FIRST EDITION OF AGRICOLA

Other works of 1533. Agrippa, p. 167, 1531; Apianus, p. 62; Jordanus, p. 62, 1496; Riese, p. 139, 1522; Schonerus (editor), p. 178, 1534; Vincento, p. 140, 1522; Anonymous, 'Libretto de Abaco,' Venice; Anonymous (sometimes attributed to Regiomontanus), p. 178, 1534.

GIOVANNI MANENTI. Ed. pr. 1534. Venice, 1534.

ZUAN MANENTI. A Venetian mathematician of the sixteenth century.

Title. See Fig. 88.

Colophon. 'In Vinegia per Giouan' Antonio di Ni-//colini da Sabio A Inſtantia de. M.// Zuan Manenti. Nelli Anni del // ſigno-re. MDXXXIIII.//del Meſe di Genaro.// Neſſuno ardiſca Stampar queſte Tariffe// de cambii & de diuerſe coſe ſotto pe//na de excommunicatione late // ſententie come nel Priui-//legio ſe contiene.// M D XXXIIII.' (Large woodcut.) (F. 400, v.)

Description. 12°, 7.5 × 13.9 cm., the text being 5.4 × 10.6 cm. 402 ff. (2 blank) unnumb., 21–26 ll. Venice, 1534.

Editions. There was no other edition.

As the title indicates, this is a set of tables of exchange, and it was intended for the use of Venetian bankers and merchants.

GIOVANNI SFORTUNATI. Ed. pr. 1534. Venice, 1534.

JOHANNES INFORTUNATUS. An Italian arithmetician, born at Siena c. 1500.

Title. See Fig. 89.

Colophon. 'Stampata in Vinegia per Nicolo di Ariſtotile // detto Zoppino.// M.D. XXXIIII.' (F. 129, r.)

Description. 4°, 15.4 × 21 cm., the text being 13.2 × 16.9 cm. 129 ff., 40–41 ll. Venice, 1534.

Editions. Venice, 1534, 4° (here described); ib., 1543, 4°; ib., 1544 (colophon 1545, see p. 177), 4°; ib., 1545, 4° (p. 177); s. l., Venice, c. 1550; ib., 1561, 4° (p. 177); ib., 1568. The privilege is dated 1532.

Sfortunati wrote his treatise along the lines followed by Borghi and Feliciano, and in his preface he acknowledges his indebtedness to them and to 'Maeſtro Luca dal Borgo dell' ordine di ſanto Franceſco' (p. 54) and to the 'operetta di Filippo Calādri Cittadino Fiorentino' (p. 47). Like these authors, he was a popular writer, as the seven editions of his book go to prove. His work is fairly complete as to the operations with integers and fractions, and is satisfactory as to the examples illustrating the Italian business life of the sixteenth century. The treatise closes with some work in practical mensuration and some mercantile tables.

TARIFFA
DE CAMBI E ALTRO
Cōpoſta per Zuā Manenti.

Con priuilegio del Illuſt. Senato veneto
ch'altri che Z.Manēti infra anni X ſtā
par ne far ſtāpar nō la poſſa, ſo'to le pe
ne cōtenute in quello. M D XXXIIII.

FIG. 88. TITLE PAGE OF MANENTI

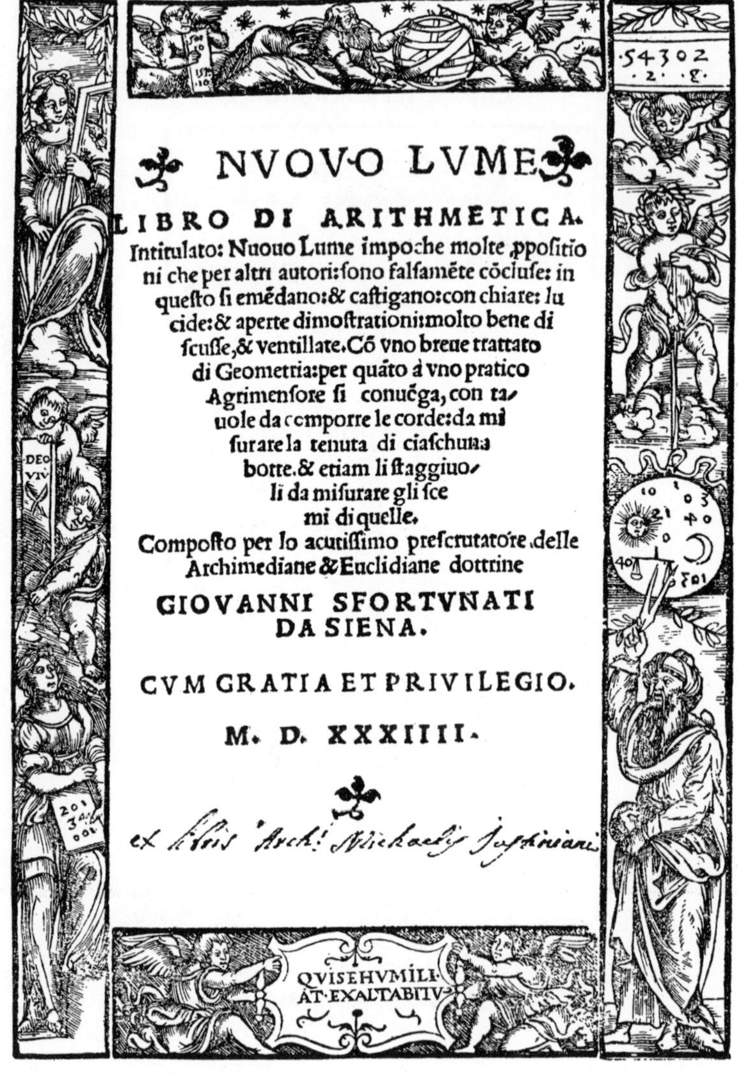

NVOVO LVME

LIBRO DI ARITHMETICA.

Intitulato: Nuouo Lume impoche molte ppositio ni che per altri autori: sono falsamēte cōcluse: in questo si emēdano: & castigano: con chiare: lu cide: & aperte dimostrationi: molto bene di scusse, & ventillate. Cō vno breue trattato di Geometria: per quāto à vno pratico Agrimensore si conuēga, con ta uole da comporre le corde: da mi surare la tenuta di ciaschuna botte. & etiam li staggiuo li da misurare gli sce mi di quelle.

Composto per lo acutissimo prescrutatore delle Archimediane & Euclidiane dottrine

GIOVANNI SFORTVNATI DA SIENA.

CVM GRATIA ET PRIVILEGIO.

M. D. XXXIIII.

ex libris Arch.i Michaelis Justiniani

FIG. 89. TITLE PAGE OF THE FIRST EDITION OF SFORTUNATI

GIOVANNI SFORTUNATI.
Ed. pr. 1534. Venice, 1544–45.

See p. 174.

Title. The title page is practically identical with that of 1534, except for the date: M. D. XLIIII.

Colophon. 'Stampata in Vinegia per Bernardino de Bindoni // Milanefe Anno domini. M. D. XLV.' (F. 129, r.)

Description. 4°, 14.8 × 20 cm., the text being 12.2 × 16.7 cm. 2 ff. unnumb. + 127 numb. = 129 ff., 40 ll. Venice, 1544–45.

Editions. See p. 174. It will be noticed that the dates in the colophon and on the title page do not agree.

See p. 174.

GIOVANNI SFORTUNATI. Ed. pr. 1534. Venice, 1545.

See p. 174.

Title. The title page is practically identical with that of the 1534 edition.

Colophon. 'In Vinegia per Giouan'Antonio & Pietro fratel-//li de Nicolini da Sabio. Ad inftantia di // Giacomo da Coneano libraro a fan // Fantin. M: D. XLV.' (F. 129, r.)

Description. 4°, 15.4 × 21.2 cm., the text being 12.9 × 16.8 cm. The text is practically identical with that of the 1534 edition.

See p. 174.

GIOVANNI SFORTUNATI. Ed. pr. 1534. Venice, 1561.

See p. 174.

Title. The title page is practically identical with that of the 1534 edition.

Colophon. 'In Venetia per Francefco del Leno,// M D LXI.' (F. 129, r.)

Description. 4°, 15.4 × 21.6 cm., the text being 12.9 × 17.3 cm. 5 ff. blank + 129 numb. = 134 ff., 41–43 ll. Venice, 1561.

See p. 174.

ANONYMOUS. (Schonerus editor.)
Ed. pr. 1534. Nürnberg, 1534.

Johannes Schonerus (Schöner) was born at Karlstadt, near Würzburg, January 16, 1477, and died at Nürnberg January 16, 1547. He was a preacher at Bamberg, and later (1526–1546) a teacher of mathematics in the Aegidiengymnasium at Nürnberg, in which Melanchthon took such interest.

Title. See Fig. 90.

Colophon. 'Norimbergæ apud Io. Petreium,// Anno M. D. XXXIIII.' (F. 32, r.)

Description. 4°, 15 × 20.9 cm., the text being 11.6 × 15.6 cm. 32 ff. unnumb., 29–43 ll. Nürnberg, 1534.

Editions. This is merely an edition of the anonymous mediæval 'Algorithmus Demonstratus,' with notes by Schonerus. As might be expected, therefore, it is purely theoretical, being a late variation of the Boethian works. In the preface Schonerus speaks of it as the 'Algorithmus Demonſtratus incerti autoris.' De Morgan thought that it might have been written by Regiomontanus, but he was wrong in asserting that Schonerus attributed it unquestionably to him. As a matter of fact the authorship goes back at least to the fourteenth century. There is said to have been an edition published at Nürnberg in 1533, attributed to Regiomontanus, but I have not seen it. 520

JOHANN ALBERT. Ed. pr. 1534. 520 Wittenberg, 1561.

A Wittenberg Rechenmeister of c. 1500–1565.

Title. 'Rechenbůchlin // Auff der Federn/ Gantz // leicht/ aus rechtem Grund/ In // Gantzen vnd Gebrochen/ Neben // angehefftem vnlangſt ausgelaſſnem Bůch-//lin/ Avff den Linien/ Dem einfel-//tigen gemeinem Man/ vnd anhe-//benden der Arithmetica // zu gut.// Durch Johann. Albert/ // Rechenmeiſter zu Wittembergk/ zuſamen // bracht. Auffs new/ mit allem vleis vber-//ſehen/ gemehrt vnd gebeſſert/ // zum dritten mal.// Wittemberg.// 1561.' (F. 1, r.)

Colophon. 'Gedruckt zu // Wittemberg/ // durch Geor-//gen Lawen // Erben.// 1561.' (F. 118, v.)

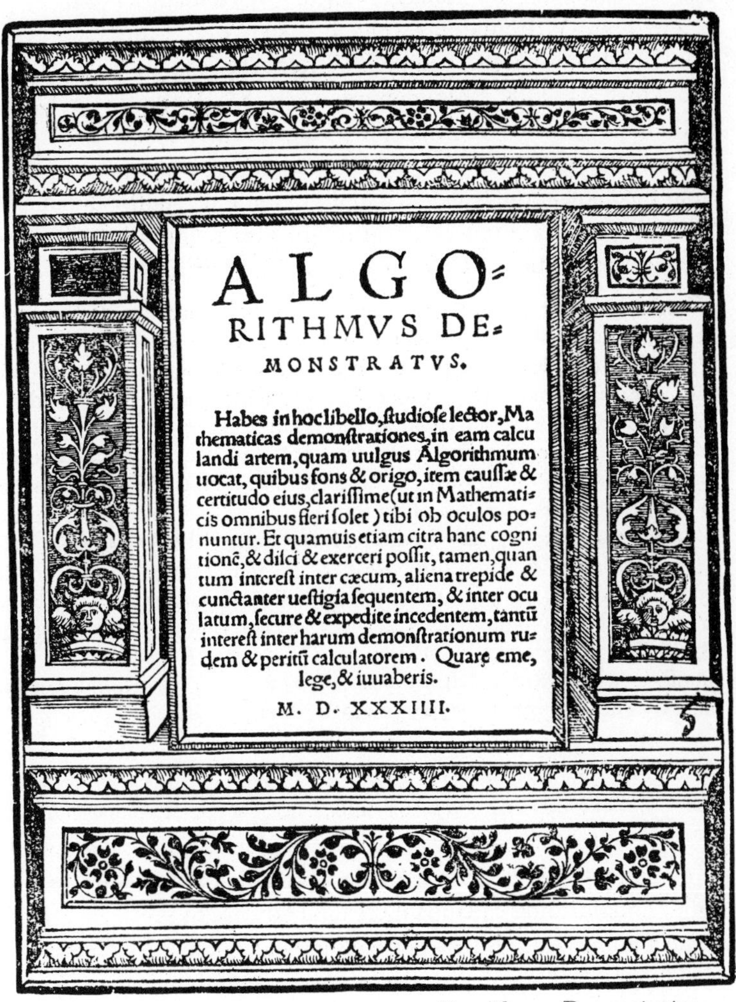

ALGO-
RITHMVS DE-
MONSTRATVS.

Habes in hoc libello, studiose lector, Ma
thematicas demonstrationes, in eam calcu
landi artem, quam uulgus Algorithmum
uocat, quibus fons & origo, item caussæ &
certitudo eius, clarissime (ut in Mathematicis omnibus fieri solet) tibi ob oculos po-
nuntur. Et quamuis etiam citra hanc cogni
tionē, & disci & exerceri possit, tamen, quan
tum interest inter cæcum, aliena trepide &
cunctanter uestigia sequentem, & inter ocu
latum, secure & expedite incedentem, tantū
interest inter harum demonstrationum ru-
dem & peritū calculatorem. Quarę eme,
lege, & iuuaberis.

M. D. XXXIIII.

FIG. 90. TITLE PAGE OF THE 1534 *Algorithmvs Demonstratvs*

Description. 8°, 9.4 × 15.1 cm., the text being 6.7 × 10.8 cm. 120 ff. unnumb. (two blank), 7–24 ll. Wittenberg, 1561.

Editions. Wittenberg, 1534; ib., 1541, 12°; Frankfort, 1541 (the colophon is dated 1542), 8°; Wittenberg, 1553, 8°; ib., 1554; Frankfort, 1558, 8°; Magdeburg, 1559, 8°; Wittenberg, 1561, 8° (here described); ib., 1564, 8°; Magdeburg, 1579, 8°; Wittenberg, 1586, 8°. That this edition was revised in 1541 appears from the dedication, which is dated 'im tauſent/ fünff-//hundert/ ein vnd vier-//tzigſten Jar.' Murhard also mentions an 'Introductio Arithmetices,' Cologne, 1542, 8°. [498; 520]

Although from the title it would seem that algorism ('Auff der Federn') is emphasized, counter reckoning ('die Species auff·den Linien') is first described (ff. A 3–F 2). This is followed by the second part, the algorism: 'Das Ander Rechenbůchlein/ auff der // Feder/ auffs aller kürtzeſt // vnd leichteſt ver-//faſſet' (f. F 3). In each part there are many commercial problems, and the book ranks as one of the most practical of its day. It is a valuable source of information as to the commercial activities of its time.

Other works of 1534. Borgni, p. 16, 1484; Bradwardin, p. 61, 1495; Jordanus, p. 62, 1496; Kolross, p. 164, 1530; Ortega, p. 93, 1512; Peurbach, p. 53, 1492; Reichelstain, p. 169, 1532; Rudolff, p. 151, 1526; Wolphius, p. 154, 1527; Rabbi Elias Misrachi, מלאכת המספר or ספר המספר (perhaps 1533? 1532?), Constantinople, 4° (another edition, with a Latin translation by Schreckenfuchs, and a commentary entitled קצור מלאכת המספר by Münster, Basel, 1546, 4°). [520]

GIOVANNI MARIANI. Ed. pr. 1535. Venice, 1580.

ZUANE MARIANI. A Venetian arithmetician of the sixteenth century.

Title. 'Tariffa perpetva // Con le ragion fatte per ſcontro di // qualunque Mercadante ſi voglia,// che dimoſtra quanto monta ogni // quantità de cadauna mercantia ad // ogni pretio, sì a peſo come a nume//ro. Buona per ogniuno in Venetia,// Dalmatia, & altri luoghi; nelli quali // ſi ragiona, & ſi ſpende a moneda //Venetiana. Et è buona per Verona,// Breſſa, Bergamo, Milan, Cremona,// Mantoua, & altri luoghi dôue ſi // ragiona, & ſi ſpende a moneda Im-//periale, & Breſſana : Con la redutiō // di moneda Venetiana in mone-//da Imperiale, & della imperiale in //

Venetiana. Et è buona a ridurre // ogni forte de ori in moneda cor-//rente, sì Venetiana come Im-//periale: & Breſſana ad // ogni precio.' (F. 1, r.)

Colophon. 'Stampata in Venetia per gli Heredi di // Franceſco Rampazetto. // Ad inſtantia de lAutore Zuane // Mariani.// L'Anno .M. D. LXXX.' (F. 299, r.)

Description. 12°, 8.5 × 15.2 cm., printed in double columns, each being 2.6 × 13.3 cm. 279 ff. numb. + 20 unnumb. = 299 ff., 32 ll. Venice, 1580.

Editions. This is apparently the first of two books by Mariani, the various editions appearing as follows: Venice, 1535; ib., 1553; ib., 1559; ib., 1564, 8°; ib., 1567; ib., 1569, 16°; ib., 1572; ib., 1575; ib., 1579; ib., 1580, 12° (this edition); ib., 1591, 16°. The second book, also a Tariffa, appeared three times at Venice, viz. in 1538, 1555, and 1558. That these were different works I know only from such bibliographers as Riccardi and Libri. **496**

Like other books with the same title, this is simply a set of tables for the use of merchants. It includes both interest and exchange tables, and is adapted to the needs of Northern Italy.

Other works of 1535. Agricola, p. 171, 1533; Angelus Mutinens, p. 140, 1525; Finaeus, p. 160, 1530–32; Grammateus, p. 123, 1518; Köbel, p. 102, 1514; Reisch, p. 82, 1503; Riese, p. 141, 1522; Tonstall, p. 134, 1522; Torrentini, p. 76, 1501; Pedro Melero, 'Compendio de los números y proporciones,' Saragossa, 4°. **520**

HUDALRICH REGIUS. Ed. pr. 1536. Freiburg, 1550.

A German teacher of the first half of the sixteenth century.

Title. 'Vtrivs-//qve arithme-//tices epitome, ex variis // authoribus concinnata, per // Hvdalrichvm // Regium.// Nvnc Tertio omnia // diligenter reuiſa & emendata.// Friburgi Briſgoiæ,// Stephanus Grauius excu-//debat, Anno // M. D. L.' (F. 1, r.)

Colophon. 'Fribvrgi Brisgoiae, // Stephanus Grauius // excudebat,// Anno M. D. L.' (F. 104, v.)

Description. 8°, 10 × 15.3 cm., the text being 7 × 11.1 cm. 1 f. unnumb. + 103 numb. = 104 ff., 17–22 ll. Freiburg, 1550.

Editions. Strasburg, 1536, 8°; Freiburg, 1543, 8°; ib., 1550, 8° (here described). [522]

This work was intended for the Latin schools. It is only slightly practical, and as compared with a book like that of Gemma Frisius it is reactionary. The first part (to f. 48) treats only of Boethian arithmetic, the theory of numbers, closing with the words: 'Hactenus de numerorum Theo-//rijs, nunc de eorundem // Praxi.' The practical part gives the operations in the usual style of the Latin writers of the time, and closes with several pages on the use of counters.

Other works of 1536. Boethius, p. 27, 1488; Bradwardin, p. 61, 1495; Budaeus, p. 99, 1514; Feliciano, p. 148, 1526; Ortega, p. 93, 1512; Peurbach, p. 53, 1492; Riese, p. 139, 1522; Torrentini, p. 76, 1501; Christiernus Torchillus Morsianus, 'Arithmetica practica,' Basel, 8°, with subsequent editions, ib., 1538, 1553, 8° (but see p. 159, 1528); Georg Wålckl, 'Die Wålſch practica/ gezogē auſʒ der kunſt der Proportion,' Strasburg (Nürnberg ?), 8°; Rycharde Benese, 'This boke sheweth the maner of measurynge of all maner of lande, as well of woodlande, as of lande in the felde, and comptynge the true nombre of acres of the same,' London, 4°. (De Morgan includes this book because of its computations and early mathematical tables. Subsequent editions appeared in 1537, 1540, c. 1558, 1562, and 1564.) L. Culman, 'Wie iunge und alte Leut recht petten sollen,' Nürnberg, 8°; ib., 1537. [522; 523; 524; 525]

ABRAHAM BÖSCHENSTEYN. Ed. pr 1536. S. l., 1536.

The son of Johann Böschensteyn (see p. 99).

Title. 'Ein nützlich // Rechenbůch-//lin der Zyffer/ darauß ein // yeder/ durch ſein aygen fleyſz mit // kleyner hüff/ lernen mag anfeng//klich rechenen/ Auſzgangē durch // Abraham Boͤſchenſteyn/ Vnnd // yetzo zům dritten mal mit fleyſz // vberſehen vnnd Corrigiert/ mit // erlichen zůgethanen Exem//plen/ Durch Johann // Boͤſchenſteyn/ den altē.// M. D. XXXVI.' (F. 1, r.)

Description. 8°, 9.5 × 14.5 cm., the text being 6.8 × 11.9 cm. 40 ff. unnumb., 20–23 ll. S. l., 1536.

Editions. There was no other edition, but the book seems to have been written in 1530, for the dedicatory epistle is dated 'Gebē am 19.// tag Aprilis/ An-//no ꝛc. im̄ 30 // jar d' min// derenn // zal.' The work is very rare.

In the epistle the author mentions his father's work (p. 99): 'Wiewol meyn Herr vatter/ herr Johañ Bôſchenſteyn vor 17. jaren auch der gleych zů Augſpurg inn den Truck mitgetheylt hat/ vnnd zům drittenn mal getruckt worden.' Seventeen years before 1530 was 1513, when Johann's book was probably written, since it was published in 1514.

Abraham's work is not much of an improvement on his father's, and resembles it in many respects. It gives seven 'Species,' including 'Duplicatio' and 'Mediatio,' as Johann's work had done. The principal additions are in the applied problems.

GIEL VANDER HOECKE. Ed. pr. 1514. Antwerp, 1537.

A Dutch arithmetician of the first half of the sixteenth century.

Title. See Fig. 91.

Colophon. 'Gheprent Thantwerpen op die Lombaerden veſte // teghen die gulden hant ouer by mi Symon Cock.// Int Jaer ons Heeren; M.CCCCC. ende // XXXVII. den. ix. dach Februarij.' (F. 180, v.)

Description. 8°, 9.4 × 14.7 cm., the text being 7.6 × 12.8 cm. 5 ff. unnumb. + 176 numb. = 181 ff., 26–32 ll. Blackletter, as shown in Figs. 91, 92. Antwerp, 1537.

Editions. So far as I know this edition of 1537 is the first, the date 1514 in the British Museum catalogue being evidently an error for 1544. The book was again published by the same printer in 1544, and there was an edition in 1548. (See *Bibliotheca Mathematica*, 1906, p. 211.)

The author begins with the fundamental processes with integers, considering the subject in a practical way. He then considers the same processes with counters on the line abacus. This preliminary work is followed by chapters on denominate numbers, fractions, the rule of three, roots, and the mediæval proportion. The second part of the treatise is devoted to algebra and the applications of arithmetic. The work closes with a brief treatment of mensuration. It is especially noteworthy on account of the early use of the plus and minus signs, not heretofore noticed by writers on the history of the subject. There is no other Dutch book of this period that makes as much use of these signs (see Fig. 92), and Vander Hoecke should be recognized as among the pioneers in appreciating their value in connection with algebraic quantities. (See Grammateus, p. 125, 1518, who used them in a similar way.)

Een sonderlinghe

boeck in dye edel conste Arithmetica/met veel schoone
perfecte regulen/ als Die numeracie vanden ghetale
metten specië int gheheele/eñ int ghebroken.
Die regule van dryen int gheheele eñ int ghebrokẽ.
Die regule van een valsche positie / eñ twee valsce po=
sitien van diueerschen ghewichte/mate/ eñ ghelde.
Ooc hebdy die edel regule Cos/ die langhe verbor=
ghen heeft gheweest/ welcke regule is die dore van alle
questien. Dese regule hebdy met haer specien/ als Nu=
meratio/ additio/ subtractio/ multiplicatio/ eñ diuisio/
eñ met haer egaliacie oft ghelikinghe/eñ met die regu=
le der quantiteyt/annex der regulen Cos.
Itẽ van alle coopmãscappẽ/als van laken/specerië/eñ
mercerië/diemẽ vercoopt bi gewichte/mate/ eñ nõmer/
eñ dat selue ooc te werckẽ doer die regule van practikẽ
Die regule van gheselscap/met diueerschen inlegge.
Ooc die selue regule met diueerscen inlegge/ met diuer
Die regule van smaeldeelinge oft om (scẽ tide
stellingen bidẽ transpoorten vanden landen/profitelijc
voer alle ontfanghers vã subuencien.
Die regule van mangelinge/van assayen van goude/
eñ van siluer/eñ van mannen van wapenen.
De fabrike vand wijnroedẽ oft visier roedẽ/eñ het vsẽ
De practike van eenen sticke lants te me (rẽ vã diẽ
ten also wel dat onbegangelijc is midts den water/oft
anders/als dat beganghelijc is.
Ghecalculeert eñ versaemt met grooter naersticheyt/
bi Gielis vandẽ hoecke. Eñ gheprent Thantwerpen
op die Lombaerde veste. By mi Symon Cock.

FIG. 91. TITLE PAGE OF VANDER HOECKE

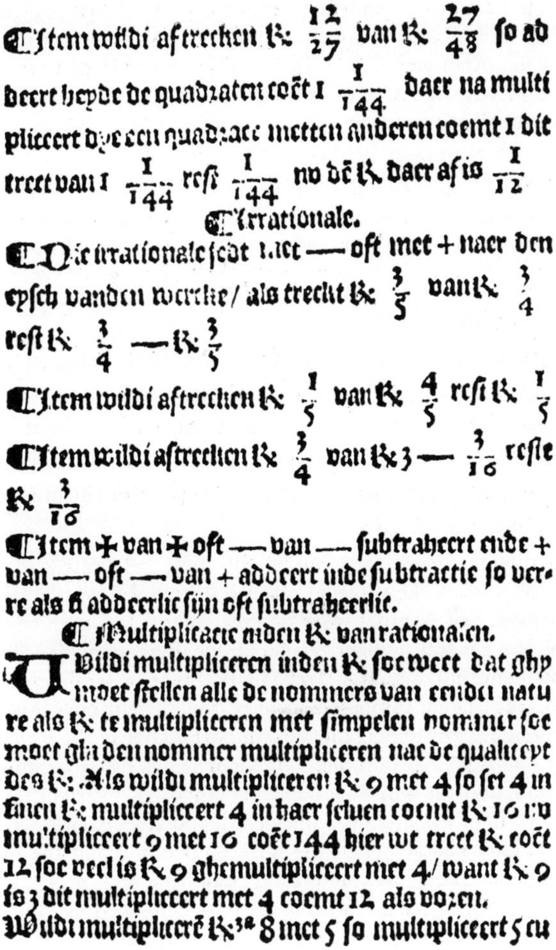

¶ Item wildi aftrecken R $\frac{12}{27}$ van R $\frac{27}{48}$ so ad
deert beyde de quadraten coēt 1 $\frac{1}{144}$ daer na multi
pliceert dye een quadraet metten anderen coemt 1 dit
treet van 1 $\frac{1}{144}$ rest $\frac{1}{144}$ no dē R daer af is $\frac{1}{12}$

¶ Irrationale.

¶ Die irrationale seyt naer — oft met + naer den
eysch vanden wercke / als treckt R $\frac{3}{5}$ van R $\frac{3}{4}$
rest R $\frac{3}{4}$ — R $\frac{3}{5}$

¶ Item wildi aftrecken R $\frac{1}{5}$ van R $\frac{4}{5}$ rest R $\frac{1}{5}$

¶ Item wildi aftrecken R $\frac{3}{4}$ van R 3 — $\frac{3}{16}$ reste
R $\frac{3}{16}$

¶ Item + van + oft — van — subtraheert ende +
van — oft — van + addeert inde subtractie so ver-
re als si addeerlic sijn oft subtraheerlic.

¶ Multiplicatie inden R van rationalen.

Wildi multipliceren inden R soe weet dat ghy
moet stellen alle de nommers van eender natu
re als R te multipliceren met simpelen nommer soe
moet ghi den nommer multipliceren nae de qualiteyt
des R. Als wildi multipliceren R 9 met 4 so set 4 in
sinen R multipliceert 4 in haer seluen coemt R 16 nu
multipliceert 9 met 16 coēt 144 hier wt treet R coēt
12 soe veel is R 9 ghemultipliceert met 4 / want R 9
is 3 dit multipliceert met 4 coemt 12 als voren.
Wildi multiplicerē R 3a 8 met 5 so multipliceert 5 eu

FIG. 92. FROM VANDER HOECKE'S *Arithmetica*

Other works of 1537. Apianus, p. 155, 1527; Bæda, p. 159, 1529; Benese, p. 182, 1536; Köbel, p. 110, 1514; Rudolff, p. 152, 1526; Wolphius, p. 154, 1527; Andrés, p. 122, 1515; B. C. Symphorien Champier, 'Libri VII,' Basel, 8° (one chapter 'De Arithmetica'); Culman, p. 182, 1536; Anonymous, 'An Introduction for to lerne to reckon with the Pen and with the Counters after the true cast of Arsmetyke, or Awgrym,' St. Albans.

NICOMACHUS. Ed. pr. 1538. Paris, 1538.

Born at Gerasa; flourished c. 100 A.D. He was a neo-Pythagorean philosopher and mathematician, and attempted, unsuccessfully, to do for the Greek theory of numbers what Euclid had done for geometry. Two works of his are extant, this treatise and a 'Harmonices Manuale.'

Title. See Fig. 93.

Description. 4°, 15.7 × 23.1 cm., the text being 9.4 × 17.7 cm. 77 pp. numb., 32 ll., Greek. Paris, 1538.

Editions. This is the first edition of the arithmetic of Nicomachus in Greek. A second edition appeared, 'Explicata per Joach. Camerarium,' at Augsburg in 1554 (p. 263). For the commentary of Iamblichus, see p. 188.

The arithmetic of Nicomachus is the most celebrated of the few Greek treatises upon the subject. It was written during the decline of Greek learning, and is not a work of great merit, being chiefly a compilation of the general number theory of the Pythagoreans. There are several commentaries upon the 'Introductio,' that of Iamblichus (c. 325 A.D., see p. 188) being the best known of the ancient ones, and that of Camerarius (see p. 262) being the most important one of the Renaissance.

After a philosophical introduction, Nicomachus classifies numbers as even and odd, and the odd as prime and composite. Perfect, excessive, and defective numbers are also considered, and the elaborate system of ratios which later characterized the work of Boethius and the mediæval writers is given. Polygonal and solid numbers and proportions are treated in the second part, a ratio being loosely defined as 'the relation between two terms,' and proportion as 'the composition of ratios.' The work differs essentially from Euclid in its presentation, being inductive instead of deductive in treatment. It is also a matter of interest that the first multiplication table, the 'mensa Pythagorica' of mediæval writers, to be found in any treatise appears here, although Hilprecht found them on the Babylonian cylinders of about 2000 B.C. The best edition of the works of Nicomachus is that of Hoche (Leipzig, 1866).

ΝΙΚΟΜΑΧΟΥ ΓΕΡΑ-
ΣΙΝΟΥ ἈΡΙΘΜΗΤΙΚΗ͂Σ ΒΙΒΛΙΑ ΔΥΟ.

NICOMACHI GERA-
SINI ARITHMETI-
cæ libri duo.

Nunc primùm typis excusi, in lucem eduntur.

PARISIIS.
In officina Christiani Wecheli.
M. D. XXXVIII.

FIG. 93. TITLE PAGE OF NICOMACHUS

GILLES HUGUETAN. Ed. pr. 1538. Lyons, 1538.

A Lyons arithmetician, born c. 1500.

Title. See Fig. 94.

Colophon. 'Icy finiſſent les tables des comptes compoſees et calculees // par Gilles huguetan : Et imprimees cheux // ledict Gilles et Jacques huguetan // freres. Lan // 1538.' (P. 115.)

Description. Fol., 21.3 × 32.9 cm., the text being 18.1 × 27.3 cm. 25 pp. unnumb. + 90 numb. = 115 pp., 59–64 ll. Lyons, 1538.

Editions. There was no other edition.

Because it is composed largely of multiplication and division tables, and of other tables of use to stationers and merchants, this work is not often included among the arithmetics of the century. It should be so classed, however, since the first eleven folios are devoted to the explanation of the fundamental operations both with written numbers and with counters. The illustrations of counter reckoning are striking, the 'gectz' (counters) being represented full size. The book is one of the earliest Lyons arithmetics in which the line abacus is mentioned.

IAMBLICHUS. Ed. pr. 1538. Arnheim-Deventer, 1668.

Born at Chalcis, in Cœle-Syria, c. 283; died at Alexandria, c. 330. He was a neo-Platonic philosopher and a voluminous writer. Four of his works are extant, this introduction to the arithmetic of Nicomachus being one.

Title. See Fig. 95.

Description. 4°, 15 × 19.8 cm., printed in double columns, Greek on the left and Latin on the right, each column being 5 × 14.8 cm. 12 pp. unnumb. + 181 numb. = 193 pp., 34 ll. Bound with the Camerarius edition of Nicomachus (p. 262, 1554). Arnheim-Deventer, 1668.

Editions. There was no other edition in the sixteenth century than that of 1538.

See Nicomachus, p. 186. This commentary by Iamblichus forms the fourth part of his treatise on the Pythagorean philosophy, the greater part of which is still extant.

Other works of 1538. De la Torre, p. 41, 1489; Glareanus, p. 191, 1539; Mariani, p. 181, 1535; Morsianus, p. 182, 1536; Peurbach, p. 53,

LES TABLES DE DIVERS COMPTES, AVEC LEVRS CANONS, calculees par GILLES HVGVETAN, natif de Lyon, Par lesquelles on pourra facilement trouuer les Comptes tous faictz, tant des achatz que uentes de toutes marchandises, soit en gros, ou en detail, a la Mesure, ou au Poix, a la Charge, ou au Nombre.

Les Tables aussi du fin Dor & Dargent, pour scauoir, scelon que le Marc de billon tiendra de fin, ou daloy, combien il uauldra de poix de fin Or, ou Dargent fin.

Deux Tables seruantz aux Libraires. Et une Table de Despence, a scauoir a tant pour Iour, combien on despend lan & le Moys, & a rayson du Moys combien reuient pour an & pour chascun Iour, & a tant pour An combien on despend le Moys, & chascun Iour.

La maniere de Aualuer, ou Reduyre par icelles Tables toutes Monnoyes, en liures, solz, & deniers.

LART & science de Nombrer, Adiouster, Soustraire, Multiplier, & Partir, par le compte des Gectz.

On les uend a Lyon, a lenseigne de la Sphære, cheux Gilles, & Iaques Huguetan, freres.

1538

FIG. 94. TITLE PAGE OF HUGUETAN

1492; Psellus, p. 168, 1532; Riese, p. 138, 1522; Roche, p. 130, 1520; Rudolff, p. 160, 1530; Tonstall, p. 135, 1522; Tomas Klos, 'Algoritmus:

JAMBLICHUS
CHALCIDENSIS
Ex Cœle-Syria
IN
NICOMACHI GERASENI
Arithmeticam introductionem,
ET DE
FATO.

Nunc primum editus, in Latinum sermonem conversus,
notis perpetuis illustratus
à
SAMUELE TENNULIO.
Accedit
JOACHIMI CAMERARII
Explicatio in duos Libros Nicomachi,
cum Indice rerum & verborum locupletissimo.

ARNHEMIÆ,

Prostant apud JOH. FRIDERICUM HAGIUM,
Daventriæ typis descripsit WILHELMUS WIER,
CIƆ IƆ C LXVIII.

FIG. 95. TITLE PAGE OF THE 1668 IAMBLICHUS

to iesth nauka Liczby, Polska rzecza wydana: Przez Ksiedza Tomasza Klosa,' Cracow (reprinted at Cracow in 1889); Eysenhut, 'Ein künstlich Rechenbuch,' Augsburg, 8°.

HENRICUS LORITUS GLAREANUS.

Ed. pr. 1539. Paris, 1543.

Loriti, Loreti. Born at Mollis, canton of Glarus, Switzerland, in June, 1488; died at Freiburg, Breisgau, May 28, 1563. He was professor of mathematics and philosophy at Basel (1515–1521), and professor in the Collège de France, Paris (1521–1524), and later taught at Basel and Freiburg. He wrote on arithmetic, music, and geometry.

Title. See Fig. 96.

De Vi. Arith-
METICAE PRACTI-
CAE SPECIEBVS, HEN-
RICI GLAREANI
Epitome.

PARISIIS

*Ex officina Iacobi Gazelli, ſub in-
ſigni Inuidiæ, è regione gym-
naſii Cameracenſis.*

1543

Fig. 96. Title page of the 1543 Glareanus

Description. 8°, 10.6 × 17 cm., the text being 6.7 × 12.9 cm. 2 ff. unnumb. + 21 numb. = 23 ff., 29 ll. Paris, 1543.

Editions. Glareanus is sometimes mentioned as the author of two works on arithmetic, the 'Isagoge Arithmetica' and the work here described. Under the former title the following editions are mentioned by various writers: Freiburg, 1539, 8° (unless Tropfke is correct in saying there was an edition of 1538); Paris, 1554, 8°; Lyons, 1554, 8°. Under the above title the following editions appeared: Paris, 1543, 8° (here described. The dedication is dated 'Friburgi Brisgoiae . . . M. D. XXXVIII,' so there may have been an edition as early as 1538); Freiburg, 1543, 8°; Cracow, 1549; Freiburg, 1550, 8° (below); Paris, 1551, 8° (p. 193); Freiburg, 1555; ib., 1558, 8°; Paris, 1558, 8°. The biographer of Glareanus, Schreiber (Freiburg, 1837), gives these as the same work. **526**

Besides this book, Glareanus also published an 'Arithmetica et musica operum Boethii demonstrationibus et figuris auctior,' Basel, 1546, fol., and a 'Commentarius in Arithmeticam et Musicam Boethii,' Basel, 1546, fol.; ib., 1570, 4°. **526**

A handbook for the Latin schools. In it Glareanus first treats of notation, including the Greek, Roman, and Arabic systems; then of the elementary operations with integers; then, briefly, of progressions and proportion. There is nothing in the little book to commend it.

HENRICUS LORITUS GLAREANUS.

Ed. pr. 1539. Freiburg, 1550.

See p. 191.

Title. 'De.VI.Ari//thmeticæ//Practicæ Speciebvs//Henrici Glareani//P. L. Epitome.// Fribvrgi Brisgoiæ.// Cum gratia ac Priuilegio Regio,//ad annos ſex.' (P.)

Colophon. 'Apud Friburgum Briſgoicum//Anno M. D. L.// Stephanus Grauius//excudebat.' (P. 77.)

Description. 8°, 10 × 15.8 cm., the text being 6.4 × 11.7 cm. 2 pp. unnumb. + 2 blank + 74 numb. = 78 pp., 22 ll. Freiburg, 1550.

HENRICUS LORITUS GLAREANUS.

Ed. pr. 1539. Paris, 1551.

See p. 191.

Title. 'De sex Arith-//meticae Practi-//cae Speciebvs,//Henrici Glareani // Epitome.// Parisiis,// Apud Gulielmum Cauellat, in pingui Gallina,// è regione colleij Cameracensis.// 1551.' (F. 1, r.)

Colophon. 'Excudebat Lutetiæ Parisiorum Benedictus // Preuotius Typographus in vico Fre-//mentello, sub Stella Aurea:// III. Non. Ianuarij,// 1551.' (F. 23, r.)

Description. 8°, 10.4 × 16.1 cm., the text being 6.4 × 11.9 cm. 2 ff. unnumb. + 21 numb. + 1 blank = 24 ff., 29 ll. Paris, 1551.

See p. 192.

HIERONYMUS CARDANUS.

Ed. pr. 1539. Milan, 1539.

GERONIMO OR GIROLAMO CARDANO, JEROME CARDAN. Born at Pavia, September 24, 1501; died at Rome, September 21, 1576. He was a physician and professor of mathematics at Milan (1534–1559) and professor of medicine at Pavia and (1562–1570) Bologna. Later he was a papal pensioner at Rome. He was one of the most acute mathematicians of his century, and wrote numerous treatises on mathematics and natural science.

Title. See Fig. 97.

Colophon. 'Anno a Virgineo partu.// M. D. XXXIX.// Io. Antonins Castellioneus Me//diolani Imprimebat Im-//pensis Bernardini // Calusci.' (Printer's mark, with 'B.C.') (F. 304, v.)

Description. 8°, 9.9 × 15.3 cm., the text being 8.2 × 12.7 cm. 304 ff. unnumb., 33 ll. Milan, 1539.

Editions. Milan, 1539, 8° (here described); Nürnberg, 1541; ib., 1542. See also p. 338, 1570. A 1537 edition is given by Villicus, and I have seen it mentioned in a dealer's catalogue; but I think the date a misprint, or that some one has taken the date of the preface instead of looking at the colophon.

This is one of the most pretentious arithmetics of the sixteenth century, and it did much to influence the advanced teaching of the subject. It is in no sense a practical book, having been written by a mathematician for the use of scholars. It opens with a discussion of the kinds of numbers considered in arithmetic, such as integers, fractions, surds, and denominate numbers. This is followed by the fundamental operations

HIERONIMI
C.CARDANI MEDICI MEDIOLA
NENSIS, PRACTICA ARITH
metice, & Menſurandi ſingularis. In qua
que preter alias cōtinentur, verſa
pagina demonſtrabit.

FIG. 97. TITLE PAGE OF CARDAN

with these numbers and a treatment of proportion. The properties of numbers occupies a considerable space and includes much of the ancient theory. The work then runs into algebra, combining this with arithmetic.

There are numerous business applications in the treatise, such as partnership, exchange, profit and loss, and mensuration, but these are treated from the theoretical standpoint rather than from that of the practical needs of the merchant class. The great prominence of the author and the scholarly nature of the work account for the various editions of the book. His well-known 'Ars Magna' (algebra) appeared in 1545.

JOHANN NOVIOMAGUS. Ed. pr. 1539. Paris, 1539.

NEOMAGUS, JAN BRONCKHORST. Born at Nimwegen in 1494; died at Cologne in 1570. He was for a time professor of mathematics at Rostock. He not only wrote on numbers, but edited works of Bæda and Ptolemy.

Title. See Fig. 98.

Description. 8°, 9.7 × 15.4 cm., the text being 6.7 × 12.3 cm. 117 pp. numb. + 2 unnumb. = 119 pp., 26–28 ll. Paris, 1539.

Editions. Paris, 1539, 8° (here described); Cologne, 1544, 8° (below); Deventer, 1551 (p. 197). [526]

The book was intended for the classical schools. It sets forth the Roman and Greek notations, the fundamental operations both with the Hindu numerals and upon the line abacus, the finger notation as found in the works of Bæda, the astrological numerals of the Middle Ages, and the Boethian theory of numbers.

Other works of 1539. Capella, p. 66, 1499; Peurbach, p. 53, 1492; Ringelbergius, p. 166, 1531; Vogelin (see Peurbach, p. 53, 1492); Wolphius, p. 154, 1527; Anonymous, 'Abacho novo con il quale ogni persona puol imparar Abacho senza che alcuno li insegni,' Venice; Anonymous, 'An introduction to algorisme, to learne to reckon with the penne,' London, 8°, with another edition ib., 1581, 8°; Juan Gutiérrez de Gualda, p. 167, 1531. [526]

JOHANN NOVIOMAGUS. Ed. pr. 1539. Cologne, 1544.

See above.

Title. 'De Nvme//ris Libri II. Qvo-//rum prior Logiſticen, & ueterum nu//merandi conſuetudinem: poſterior// Theoremata numerorum complecti-//tur, autore Ioan. Nouiomago.// Nunc recéns ab ipſo autore recogniti.// (Woodcut with motto: 'Discite Ivsticiam moniti.') Coloniæ Ioan. Gymnicus excudebat,// Anno M.D.XLIIII.' (F. 1, r.)

De Numeris li-
BRI DVO, QVORVM PRIOR
Logisticen & veterum numerandi consuetudinem, posterior Theoremata numerorum complectitur, ad doctissimum virum Andream Eggerdem professorem Rostochiensem.
Nunc recens in lucem emissi authore Ioanne Nouiomago.

PARISIIS
Ex officina Christiani Wecheli, sub scuto Basiliensi, in vico Iacobæo: & sub Pegaso, in vico Bellouacensi.
M. D. XXXIX.

FIG. 98. TITLE PAGE OF NOVIOMAGUS

Description. 8°, 10 × 14.8 cm., the text being 6.7 × 11.5 cm. 59 ff., 25 ll. Cologne, 1544.

Editions. See p. 195.

This is merely a reprint of the first edition (p. 195, 1539).

JOHANN NOVIOMAGUS. Ed. pr. 1539. Deventer, 1551.

See p. 195.

Title. 'De nvme//ris libri II. qvo-//rvm prior logisticen, et //ueterum numerandi confuetudinem : pofte-//rior Theoremata numerorum com//plectitur, Autore Ioan.// Nouiomago.// Nvnc recens ab ipso // autore recogniti.// (Woodcut, and 'T. B. Fons Iovis.') Daventriæ,// Theodoricus Bornius excudebat.// Anno M. D. LI.' (F. 1, r.)

Description. 8°, 9.8 × 14.9 cm., the text being 6.8 × 10.7 cm. 50 ff. unnumb., 25 ll. Deventer, 1551.

Editions. See p. 195.

Like the 1544 edition, this is a reprint of that of 1539.

JODOCUS WILLICHIUS. Ed. pr. 1540. Strasburg, 1540.

WILKE, WILCKE, WILD. Born at Resel, East Prussia; died at Lebus, November 12, 1552. He was professor of Greek (1540), and then of medicine, in the university of Frankfort an der Oder.

Title. See Fig. 99.

Colophon. 'Argentorati ex officina // Cratonis Mylii,// mense Sept.// anno // M. D. XL.' (P. 125.)

Description. 8°, 10.2 × 15.4 cm., the text being 8.2 × 11.4 cm. 2 pp. unnumb. + 123 numb. = 125 pp., 26 ll. Strasburg, 1540.

Editions. Strasburg, 1540, 8° (here described); ib., 1545.

This is a book intended for the classical schools. It is written chiefly in Latin, but contains numerous extracts from the Greek. It is based upon Greek models, and contains several quotations from Nicomachus. The plan of treatment is catechetical (see Fig. 100), and it is interesting to note that this work appeared in the same year in which Recorde may have published his 'Ground of Artes' in England (see p. 213), a book in which the author also adopted the catechism form. It is manifestly inspired by Boethius, and is hair-splitting in theory and useless in practice. Willichius begins in a grandiloquent style, 'De Arithmeticæ, quæ

Matheſeos mater eſt, finitione.' The history of arithmetic has few more curious examples than the first chapter of this work, with its learned references to Pythagoras, Augustine, the Platonists, and 'an Arab philosopher named Algebras.' ('Eadem Autem hodie ab authore quodam Arabe Philoſopho, cui nomen erat Algebras, nomine regularum Algebræ

IODOCI
VVILLICHII
RESELLIANI,
Arithmeticæ libri
tres.

ARGENTORATI
M. D. XL.
Cum Gratia & Priuilegio.

FIG. 99. TITLE PAGE OF WILLICHIUS

explicatur.' P. 19.) Willichius follows the ancient Greek plan of dividing arithmetic into two parts, the first being the practical, the logistica of the classical civilization ('vna πρακτικὴ, qua ſupputatio domeſtica fit . . . apud ueteres à ratiocinando λογιςικὴ dicta eſt,' p. 19), and the second being theoretical, the ancient arithmetica. ('Eſt θεωρητικὴ, qua velut ſanctiora myſteria continentur, & hæc ſola intelligentia animi conſtat,' p. 20). In his number mysticism he calls unity Jupiter ('Vnitas eſt Iupiter,' p. 22), saying that others call it Cupid, others Amicitia, and others Concordia, and quotes 'Zarathas the teacher of Pythagoras' as calling one the father and two the mother of numbers. ('Proinde apte Zarathas Pythagoræ præceptor dixit, μόναδα eſſe numerorum patrem, δύαδα

30 ARITHMETICAE

NICOLAVS.

Pariter im= par.

QVis eſt alter numerus par? IVST. Eſt pariter impar, uel à paribus impar, Græcis ἀρτιοπέρισσος, uel ἀρτιάκις πέρισσος. Eſt autem, cum primum diuiditur, mox fit indiuiſibilis, ut 14. 18. 22. NICOL. Quomodo eum finit horum numerorum exquiſitus magiſter Euclides? IVST. Sic: ἀρτιάκις πέρισσος ἐστὶν, ὁ ὑπὸ ἀρτίου ἀριθμοῦ μετρούμενος κατὰ περισσὸν ἀριθμόν. Ita numerus diuidens par eſt, ſed diuiſorius mox impar exurget. NICOL. Cur id nominis illi inditum eſt? IVST. Ideo, quod quilibet eius ordinis numeri pares, facti ſunt per imparem multiplicationem: ut bis ter, ſenarium, bis quinq; denarium conficiunt. Verum ſi cui altius contemplari libet, eundem uocabit imparem in ſua qnantitate, ſed parem in denominatione. Eſto exempli gratia, denarius, cuius altera pars eſt quinarius, qui quantitate, hoc eſt, monadum congregatione eſt impar, ſed quia à binario denominatur, par iudicabitur. Quæ ratio nominis ex Boethio colligitur: Alia autem Euclidi eſſe uidetur. NICOL. Sunt ne huic de illo aliquot theoremata? IVST. Quidni? Vnum eſt. Si numerus dimidium impar habuerit, pariter impar eſt tantum. Nam hic dumtaxat extremum, quod maxi=

Cur uocatur pariter impar

Symbola ex pariter impa= ri.

FIG. 100. FROM THE ARITHMETIC OF WILLICHIUS

autem matrem,' p. 22.) The contrast between this work and that of Gemma, which appeared in the same year, is very marked. It must be admitted, however, that the book has some value in interpreting the ideas of such followers of Nicomachus as Boethius, Jordanus, and Faber Stapulensis, as is shown in Fig. 100. It has at least the merit of having been written for beginners.

GEMMA FRISIUS. Ed. pr. 1540. Wittenberg, 1542.

More properly GEMMA RAINER or REGNIER, the Frisian. Born at Dockum, in East Friesland December 8, 1508; died at Louvain, May 25, 1555. He received the degree of Doctor of Medicine in 1541, when he abandoned his mathematical studies. He wrote upon astronomy and arithmetic, and his son Cornelis was a contributor to the former science.

Title. See Fig. 101.

Colophon. 'Impreſſum Vitebergæ apud // Georgium Rhau.// Anno M.D.XLII.' (F. 80, r.)

Description. 8°, 10 × 15.3 cm., the text being 7.8 × 15.3 cm. 80 ff. unnumb., 23 ll. Wittenberg, 1542.

Editions. Antwerp, 1540, 8°; Wittenberg, 1542, 8° (here described); ib., 1543, 12°; Paris, 1543; 526 Wittenberg, 1544, 8° (p. 202); Paris, 1545 (Peletarius edition); Antwerp, 1547; Wittenberg, 1548; Paris, 1549, 8° (p. 202); Paris, 1550, 8° (p. 203); Wittenberg, 1550, 12°; Antwerp, 1550; Wittenberg, 1551, 8° (p. 203); Paris, 1551; Antwerp, 1552, 8° (p. 203); Paris, 1553, 8° (p. 204); Wittenberg, 1553; ib., 1555; Lugduni, 1556, 8°; Paris, 1557; Leipzig, 1558, 8° (p. 204); s. a., but c. 1558; Paris, 1559; Leipzig, 1559; Paris, 1561, 8°; Wittenberg, 1561, 8° (p. 204); Leipzig, 1562; Antwerp, 1562; Paris, 1562; ib., 1563 (p. 205); Wittenberg, 1563, 8° (p. 205); Cologne, 1565; Leipzig, 1565, 8°; Lugduni, 1566, 8°; Wittenberg, 1567, 4°; Paris, 1567; Venice, 1567, 4° (p. 205); Leipzig, 1568; Paris, 1569, 8°; Wittenberg, 1570, 12°; Cologne, 1571, 8° (p. 206); Paris, 1572; Leipzig, 1572; ib., 1575, 8° (p. 206); Cologne, 1576, 8°; Paris, 1578, 8° (p. 207); Wittenberg, 1579; Leipzig, 1580; Antwerp, 1581, 8° (p. 207); ib., 1582, 8° (the first with Forcadel's notes ?); Wittenberg, 1583, 8° (p. 208); Paris, 1585, 8°; s. l., 1588; Leipzig, 1588, 8° (p. 208); ib., 1591; ib., 1592, 8° (p. 208); Cologne, 1592, 8°; Wittenberg, 1593; Frankfort, 1597 (the only German translation ?). There were numerous editions after 1600. Treutlein (*Abhandlungen*, I, 18), following Murhard, says there were at least twenty-five editions in the sixteenth century. In reality there were more than twice as many; the above list, probably incomplete, names fifty-nine. 526

The editions of Gemma varied but little until Peletarius, several years before the former's death, added his notes. These amplified the text, but they made no changes of importance.

This was the most popular arithmetic of the sixteenth century, at least among those intended for the Latin schools. It combined the older

ARITHME-
TICAE PRACTI-
CAE METHODVS FACILIS,
per Gemmam. Frisium Medi-
cum ac Mathematicum.

VITEBERGAE. M. D. XLII.

FIG. 101. TITLE PAGE OF THE 1542 GEMMA FRISIUS

science of numbers with the commercial arithmetic of the Italian writers in such way as to appeal in a remarkable degree to the teachers of the period. The book opens with a discussion of the various fundamental operations, presented without much explanation and with numbers of relatively small size. These operations include the subjects of duplation and mediation as was customary in the Latin books of that time. The author closes the first part of his work with a treatment of progressions and the rule of three. The second part is devoted to fractions, the sequence being the same as with integers. The third part includes such common rules of business as partnership, alligation, and rule of false, together with roots and a little algebra. The fourth part treats of proportion, and has a few pages on arithmetical recreations.

GEMMA FRISIUS. Ed. pr. 1540. Wittenberg, 1544.

See p. 200.

Title. This edition is substantially the same as that of 1542 with the following exceptions: F. 1, r., title page: 'Vitebergæ Anno M,D,XLIIII.'

Colophon. 'Anno M,D,XLIIII.' (F. 88, r.)

Description. 8°, 9.5 × 15 cm., the text being 7.9 × 11.3 cm. 88 ff. unnumb., 23 ll. Wittenberg, 1544.

See p. 201.

GEMMA FRISIUS. Ed. pr. 1540. Paris, 1549.

See p. 200.

Title. 'Arithmeticæ practicæ // Methodvs facilis per // Gemmam Frisivm, Medicvm // ac Mathematicum, Iam recens ab ipſo // authore emendata & multis in lo-//cis inſigniter aucta.// Hvc accesservnt Iacobi Pe-//letarii Cenomani annotationes: Eiuſdem item de Fractio-//nibus Aſtronomicis compendium: Et de cognoſcēdis per // memoriam Calendis, Idib. Nonis, Feſtis mobilibus, & loco // Solis & Lunæ in zodiaco.// (Woodcut with motto: 'In pingvi. gallina.') Parisiis.// Apud Gulielmum Cauellat, in pingui gallina, ex // aduerſo collegij Cameracenſis.// 1549.' (F. 1, r.)

Description. 8°, 10.5 × 16.6 cm., the text being 7 × 12.7 cm. 1 f. unnumb. + 95 numb. = 96 ff., 26 ll. Paris, 1549.

Editions. This is the ninth edition (see p. 200), and the first which I have seen with the notes of Peletarius, although these notes bear the date 1545 (f. 77, r.) They are here given in an appendix ('Iacobvs Peletarius Lectori,' f. 77, r.), but later, as in the 1571 edition (see p. 206), they are introduced in the body of the text.

See p. 201.

GEMMA FRISIUS. Ed. pr. 1540. Paris, 1550.

See p. 200.

Title. Except for the date (1550) the title page is the same as that of the 1549 edition (p. 202).

Description. 8°, 10.3 × 16.1 cm., the text being 7.3 × 13 cm. 1 f. unnumb. + 95 numb. = 96 ff., 26 ll. Paris, 1550.

See p. 201.

GEMMA FRISIUS. Ed. pr. 1540. Wittenberg, 1551.

See p. 200.

Title. 'Arithme//ticae practicae // methodvs facilis, per // Gemmam Frifium Medi-//cum ac Mathe-//maticum.// (Woodcut showing computing with counters.) Vvitebergae.// Anno M. D. LI.' (F. 1, r.)

Description. 8°, 9 × 14.2 cm., the text being 6.4 × 11.2 cm. 87 ff. unnumb. + 1 blank = 88 ff., 23 ll. Wittenberg, 1551.

See p. 201.

GEMMA FRISIUS. Ed. pr. 1540. Antwerp, 1552.

See p. 200.

Title. 'Arithme-//ticæ practicæ metho-//dus facilis, per Gemmam Frifium Me-//dicum ac Mathematicum, iam re-//cens ab ipfo auctore emen-//data, & multis in locis // infigniter aucta.// Cor. Graphevs.// Si numerandi artem, cunctis ex artibus illam // Vel primam, exactè difcere lector amas,// Hanc gemmā, ingenio fummus quam Gemma libello // Hoc paruo includit, carpito, doctus eris.// Sunt conati alij prolixis tradere chartis // Hanc artem, at multis non placet ille labor.// Porrò hic Gemma ſuam gemmam ſic temperat, ipſa // Vt placeat cunctis commoditate breui.// ¶Antuerpiæ, apud Gregorium Bontium // à Cæſ. Maieſt. librarium admiſſum.// Cum gratia & priuilegio.// 1552.' (F. 1, r.)

Description. 8°, 9.5 × 13.9 cm., the text being 7.4 × 11.8 cm. 4 ff. unnumb. + 76 numb. = 80 ff., 28 ll. Antwerp, 1552.

See p. 201.

GEMMA FRISIUS. Ed. pr. 1540. Paris, 1553.

See p. 200.

Title. Except for the date (1553) the title page is the same as that of the 1549 edition (p. 202).

Description. 8°, 9.8 × 14.8 cm., the text being 6.8 × 12.7 cm. 1 f. unnumb. + 95 numb. = 96 ff., 26 ll. Paris, 1553.

See p. 201.

GEMMA FRISIUS. Ed. pr. 1540. S. l. (Leipzig), 1558.

See p. 200.

Title. 'Arithme-//ticæ practicæ // methodvs facilis, per // Gemmam Friſium Medicum ac // Mathematicum // (Woodcut as in the 1542 edition.) Anno M. D. LVIII.' (F. 1, r.)

Description. 8°, 10.1 × 15.4 cm., the text being 7.8 × 11 cm. 87 ff. unnumb., 22–23 ll. S. l. (Leipzig), 1558.

Editions. See p. 200.

See p. 201. A set of manuscript notes in the back, in a sixteenth-century hand, gives an interesting synopsis of arithmetic as taught in the universities of that time.

GEMMA FRISIUS. Ed. pr. 1540. Wittenberg, 1561.

See p. 200.

Title. 'Arithme-//ticæ practicæ me-//thodvs facilis, per Gem-//mam Frisivm Medi-//cum ac Mathematicum.// (Woodcut as in the 1542 edition.) Vvitebergæ // ex officina Hæredvm // Georgii Rhavv // M.D.LXI.' (F. 1, r.)

Colophon. 'Vvitebergæ // ex officina Hæredvm // Georgii Rhavv // M.D.LXI.' (F. 88, v.)

Description. 8°, 9.8 × 14.9 cm., the text being 7.7 × 11.2 cm. 88 ff. unnumb., 23 ll. Wittenberg, 1561.

See p. 201.

GEMMA FRISIUS. Ed. pr. 1540. Paris, 1563.

See p. 200.

Title. The title page is substantially the same as that of 1549 (p. 202), with the following addition: 10th line, 'Quibus demum ab eodem Peletario additæ ſunt Radicis // vtriuſque demonſtrationes.'

Description. 8°, 10.4 × 16.3 cm., the text being 7.1 × 12.5 cm. 2 ff. unnumb. + 102 numb. = 104 ff., 26 ll. Paris, 1563.

See p. 201.

GEMMA FRISIUS. Ed. pr. 1540. Wittenberg, 1563.

See p. 200.

Title. 'Arithme-//ticæ practicæ me-//thodvs facilis, per // Gemmam Friſium Medicum ac // Mathematicum.// VVitebergæ // M.D.LXIII.' (F. 1, r.)

Colophon. 'VVitebergæ // ex officina Hæredvm.// Georgii Rhavv.// M.D.LXIII.' (F. 88, r.)

Description. 8°, 10.5 × 15.6 cm., the text being 8.1 × 11.2 cm. 88 ff. unnumb., 23 ll. Wittenberg, 1563.

Editions. See p. 200.

GEMMA FRISIUS. Ed. pr. 1540. Venice, 1567.

See p. 200.

Title. 'Aritmetica//prattica facilissima,//composta da Gemma Frisio Medico,// et Matematico; // Con l'aggiunta dell' Abbreuiamento de i Rotti Aſtronomici di // Giacomo Pelletario; & del conoſcere à mente le Calende,// gl' Idi, le None, le Feſte Mobili, il luoco del Sole, & della Luna nel // Zodiaco; & la dimoſtratione della Radice Cubica: lequali tutte // coſe dal latino, ha in queſta lingua ridotte Oratio Tosca-//nella della famiglia di Maeſtro Luca Fiorentino; & halle de-//dicate // allo illvstre ſignore, il ſignor//Ettore Podocataro.// (Woodcut.) In Venetia, Appreſſo Giouanni Bariletto.// MDLXVII.' (F. 1, r.)

Colophon. 'Registro.// *A B C D E F G H I K L M.// Tutti ſono Quaderni; eccetto M,// che è Quinterno // In Venetia,// Appreſſo Giouanni Bariletto.//MDLXVII.'

Description. 4°, 14.9 × 20.7 cm., the text being 10.6 × 17.4 cm. 5 ff. unnumb. + 51 numb. = 56 ff., 40–42 ll. Venice, 1567.

Editions. See p. 200. This is the only Italian edition of the sixteenth century.

See p. 201. The 'Maeſtro Luca Fiorentino' mentioned in the title page was a well-known arithmetician of Florence. A manuscript of his of c. 1425 is described in the second part of this work.

GEMMA FRISIUS. Ed. pr. 1540. Cologne, 1571.

See p. 200.

Title. 'Arithme-//ticae practicae // methodvs facilis, per Gemmam // Friſium, Medicum ac Mathematicum, iam re-//cèns ab ipſo authore emendata, & multis // in locis inſigniter aucta.// Hvc accesservnt Iacobi Pe-//letarij Cenomani annotationes: Eiuſdem item de // Fractionibus Aſtronomicis compendium: Et // de cognoſcendis per memoriam Calendis,// Idibus, Nonis, Feſtis mobilibus, // & loco Solis & Lunæ in // Zodiaco.// Nunc verò a Ioanne Stein recognita, & no-//uis aucta additionibus.// (Woodcut with motto: Benedices // Coronæ Anni // Benignitatis // Tvae. Psal. 64 //) Coloniæ,// Apud Maternum Cholinum.// M. D. LXXI.// Cum gratia & priuilegio Cæſ. Maieſt.' (P. 1.)

Description. 8°, 9.5 × 15.8 cm., the text being 6.5 × 12.1 cm. 3 pp. unnumb. + 281 numb. = 284 pp., 28–30 ll. Cologne, 1571.

See p. 201.

GEMMA FRISIUS. Ed. pr. 1540. Leipzig, 1575.

See p. 200.

Title. 'Arithme-//ticæ practicæ // methodvs facilis,// per // Gemmam Frisivm // Medicum ac Mathematicum.// (Woodcut representing a counting house, with line reckoning.) Lipsiæ // Iohannes Rhamba excudebat.// M. D. LXXV.' (F. 1, r.)

Description. 8°, 9.5 × 15.3 cm., the text being 7.8 × 11 cm. 87 ff. unnumb., 23–24 ll. Leipzig, 1575.

See p. 201. The book has been rebound with added pages which have been used for manuscript. These notes are in Latin, in a German sixteenth-century hand, and include both work on the fundamental operations and numerous commercial problems.

GEMMA FRISIUS. Ed. pr. 1540. Paris, 1578.

See p. 200.

Title. The title page is substantially the same as that of 1542 (p. 201), except for the following: 'In me Mors,// In me Vita.// Parisiis,// Apud Hieronymum de Marnef, & viduam // Guliel. Cauellat, ſub Pelicano monte D. Hilarij.// 1578.' (F. 1, r.)

Description. 8°, 10.9 × 17.3 cm., the text being 7.4 × 14.1 cm. 3 ff. unnumb. + 93 numb. = 96 ff., 29 ll. Paris, 1578.

See p. 201.

GEMMA FRISIUS. Ed. pr. 1540. Antwerp, 1581.

See p. 200.

Title. 'Arithmeticæ // practicæ methodvs fa-//cilis, per Gemmam Frisivm, Medi-//cum, ac Mathematicum conſcripta: iam recens ab Auctore // pluribus locis aucta & recognita.// In eandem Ioannis Steinii & Iacobi // Peletarii Annotationes.// Eiuſdem de Fractionibus Aſtronomicis Compendium,// & de cognoſcendis per memoriā Kalendis, Idibus, No-//nis, Feſtis mobilibus, loco'qȝ Solis & Lunę in Zodiaco.// (Woodcut of astronomer.) // Antverpiæ.// Apud Ioannem Bellerum ad inſigne Aquilæ // aureæ. Anno 1581.' (P. 1.)

Description. 8°, 9.7 × 15.7 cm., the text being 6.5 × 13 cm. 5 pp. unnumb. + 178 numb. + 1 blank = 184 pp., 31 ll. Antwerp, 1581. With this copy is bound: 'Ivl. Pacii // a Beriga // Institvtiones // Logicæ,// in vsvm scholarvm // Bernensivm // editæ. // Bernæ. M.D.C.'

See p. 201.

GEMMA FRISIUS. Ed. pr. 1540. Wittenberg, 1583.

See p. 200.

Title. 'Arithme-//ticæ Practicæ // Methodus facilis,// per // Gemmam Frisivm // Medicum ac Mathematicum.// (Woodcut as in the 1542 edition.) Vvitebergæ //Ex officina Matthæi VVelacij. // M. D. LXXXIII.' (F. 1, r.)

Description. 8°, 9.5 × 15 cm., the text being 7.9 × 11.2 cm. 87 ff. unnumb., 23 ll. Wittenberg, 1583.

See p. 201.

GEMMA FRISIUS. Ed. pr. 1540. Leipzig, 1588.

See p. 200.

Title. 'Arithme-//ticæ practicæ // methodvs facilis,// per // Gemmam Frisivm // Medicum ac Mathematicum.// Lipsiæ,// Anno M. D. LXXXVIII.' (F. 1, r.)

Colophon. 'Lipsiæ,// ex officina typo-//graphica Abrahami // Lambergi.// Anno // M. D. LXXXVIII.' (F. 87, r.)

Description. 8°, 9.4 × 15 cm., the text being 7.8 × 11.5 cm. 87 ff. unnumb., 24 ll. Leipzig, 1588.

See p. 201.

GEMMA FRISIUS. Ed. pr. 1540. Leipzig, 1592.

See p. 200.

Title. 'Arithme//ticæ Practicæ // Methodvs facilis,// per // Gemmam Frifium Medi-//cum ac Mathematicum.// (Woodcut as in the 1542 edition.) Lipsiæ,// Anno // M. D. XCII.' (F. 1, r.)

Colophon. (Woodcut of Pegasus.) 'Lipsiæ,// ex officina typo-// graphia Abrahami // Lambergi.// Anno // M. D. XCII.' (F. 88, r.)

Description. 8°, 9.2 × 15.5 cm., the text being 8.1 × 11.4 cm. 88 ff. unnumb., 24 ll. Leipzig, 1592.

See p. 201.

HENRICUS URANIUS. Ed. pr. 1540. Solingen, 1540.

A German classicist, born at Reesz, Prussia. He lived at Emmerich (Emrich) when he wrote this work.

Title. See Fig. 102.

Description. 8°, 10 × 15 cm., the text being 7.9 × 11.7 cm. 20 ff. unnumb., 28 ll. Solingen, 1540.

Editions. There was no other edition.

This is a semi-historical discussion of the various measures which the sixteenth century received from the Roman civilization, or which were

DE RE

NVMARIA, MENSVRIS ET PON-
deribus Epitome ex Budæo, Portio, Alciato
& Georgio Agricola concinnata
per Henricum Vranium
Ressensem.

IN ZOILVM.

Zoile quid properas? nihil hic quod rodere poßis,
Omnia sunt doctis ante probata uiris.
Ocyus hinc fugito, si nescis quo sit eundum:
En monstrabo, malam Zoile adito crucem.

SALINGIACI, Ioannes Soter
excudebat. AN. M. D. XL.

FIG. 102. TITLE PAGE OF URANIUS

mentioned in the most commonly read classics of the Renaissance period. No arithmetical operations are given in the book. It begins with the fractional parts of the *as*, the twelfth being called the uncia (the Troy ounce), the sixth the sextans, the fourth the quadrans, and so on.

PHILIP MELANCHTHON. Ed. pr. 1540. Leyden, 1540.

Born at Bretten, Baden, February 6, 1497; died April 19, 1560. His family name was Schwarzerd; he assumed the Greek equivalent, Melanchthon, when he entered the university of Heidelberg (1509). He was one of the most famous classicists of the Renaissance, and a leader in the Reformation.

Title. See Fig. 103.

MATHEMA-
TICARVM DISCI-
PLINARVM, TVM
ETIAM ASTROLO-
GIAE ENCO-
MIA,
*
PER PHIL. MELANCHT.

ITEM
Phænomena Ioachimi Camerarij,
elegantiſsimo carmine
deſcripta.

APVD SEB. GRYPHIVM
LVGDVNI,
1540.

FIG. 103. TITLE PAGE OF MELANCHTHON

Description. 8°, 9.9 × 15.2 cm., the text being 7.2 × 12.2 cm. 40 pp. numb., 30 ll. Leyden, 1540.

Editions. There was no other separate edition.

The work of Melanchthon consists of three letters: (1) to Simon Grynaeus, dated 'Vitebergæ, mēſe Auguſto. M. D. XXXI'; (2) to Johann Reiffenstein, dated 'Menſe Augusto. Anno M. D. XXXVI'; (3) to Johann Schonerus (see p. 178), dated 'Vuittebergæ, Meſe Augusto, Anno M. D. XXXVI.' These epistles are all upon the value and the nature of mathematical thought, and are replete with classical and religious references. As mathematics they have no value.

The verses of Camerarius, 'Ioachimi Camerarii Phaenomena ad clarissimum ivvenem Danielum Stibarum,' relate chiefly to astronomy.

MAGNUS AURELIUS CASSIODORUS.

Ed. pr. 1540. Paris, 1550.

CASSIODORIUS. Born at Scylaceum, c. 470; died, probably at Rome, c. 564. A Roman statesman and historian.

Title. 'Aurelij Caſsiodori Se-//natoris Cos.qve Romani // de quatuor Mathematicis diſciplinis // Compendium. // Parisiis // Apud Vaſcoſanum, uia Iacobæa ad inſigne Fontis. // M. D. L.' (F. 1, r.)

Description. 4°, 15.9 × 20.8 cm., the text being 9.7 × 17.2 cm. 1 f. unnumb. + 7 numb. = 8 ff., 25–29 ll. Paris, 1550.

Editions. Paris, 1540; ib., 1550, 4° (here described); ib., 1580. The 'Opera' appeared in Paris in 1598 (and 1584?). The Compendium is embodied in the 'Disciplinarum liberalium orbis, ex P. Consentio et Magno Aurelio Cassiodoro,' published at Basel in 1528.

This brief treatise on the nature of arithmetic, music, geometry, and astronomy, the four mathematical disciplines, was held in high esteem in the Middle Ages.

Other works of 1540. Anianus, p. 32, 1488; Beldamandi, p. 15, 1483; Benese, p. 182, 1536; Borghi, p. 21, 1484; Ortega, p. 93, 1512; Paxi, p. 79, 1503; Rudolff, p. 152, 1526; p. 160, 1530; Scheubel, p. 233, 1545; Wolphius, p. 154, 1527; Anonymous (Ortega?, p. 91, 1512), 'Œuvre tres subtile y profitable de l'art y science de arismeticque y geometrie translate nouvellement d'Espaignol en Francoys,' Paris, 8°.

Works of 1541. Agrippa, p. 167, 1531; Albert, p. 180, 1534; Cardan, p. 193, 1539; Riese, p. 139, 1522; Ringelbergius, p. 167, 1531; Tagliente, p. 115, 1515; Georg Rheticus, Arithmetic, Strasburg.

ANONYMOUS. Various authors.

Ed. pr. 1542. Cologne, 1542.

Title. See Fig. 104.

ARITHME
TICES INTRODV-
ctio ex uarijs authoribus con-
cinnata.

Coloniæ excudebat Ioannes Gymnicus
Anno M. D. XLII.

FIG. 104. TITLE PAGE OF THE 1542 ANONYMOUS WORK

Description. 8°, 9.7 × 15.1 cm., the text being 6.9 × 12.3 cm. 20 ff. unnumb., 25–29 ll. Cologne, 1542.

Editions. Cologne, 1542, 8° (here described); ib., 1546; Dortmund, 1549, 8° (see below).

This is one of several anonymous compilations made in the sixteenth century for use in the Latin schools. It has no merit, save that of brevity. It contains a brief treatment of the fundamental operations, followed by a chapter 'De Progreſione,' the 'Regvla mercatorvm ſeu de tribus,' and 7 pages 'De minutijs.' I notice that folio B 6, v., is an exact copy of Wolphius (see p. 154) folio B 1, v. This is the same as the work next mentioned, published at Dortmund in 1549.

ANONYMOUS. Ed pr. 1542. Dortmund, 1549.

Title. 'Brevis // Arithme//tices Intro-//dvctio ex Variis // Authoribus con-//cinnata. Tremoniae excud. Melch. Soter./ Anno M.D.LXIX.' (P. 1.)

Description. 8°, 9.8 × 14.9 cm., the text being 6.1 × 12.3 cm. 48 pp. unnumb. (1 blank), 25 ll. Dortmund, 1549.

See above.

ROBERT RECORDE. Ed. pr. c. 1542. London, 1558.

Born at Tenby, Pembroke, c. 1510; died in Southwark prison, probably soon after June 28, 1558 (the date of his will). He was educated at Oxford and Cambridge, and taught mathematics at the former and probably at the latter university. He became royal physician and wrote on medicine as well as mathematics.

Title. See Fig. 105.

Colophon. 'Imprinted at London in Paules churchyard // at the ſigne of the Braſen Serpent // by Reginalde VVolfe.// Anno Domini M. D. LVIII.' (F. 205, v.)

Description. 8°, 8.9 × 13. 1 cm., the text being 8 × 11.8 cm. 205 ff. unnumb., 31 ll. London, 1558.

Editions. There is considerable uncertainty as to the date of the first edition of this work of Recorde's. It appeared, however, between 1540 and 1542. For a discussion of the question see the *Dictionary of National Biography*, and De Morgan, p. 22. The former says there were twenty-seven editions of the book, but there were at least twenty-eight (see p. 214). On account of the influence of the work on English education, the bibliography has been extended through the seventeenth century. London, c. 1542; ib., 1543, 8°; ib., 1549, 8°; ib., 1551, 8°; ib., 1556; 1552; London, 1558, 8° (here described); ib., 1561, 8° (the earliest seen by De Morgan); 1570; London, 1571, 8°; ib., 1573; ib., 1577; ib. 1579, 8° (p. 217); ib., 1582, 8° (the Mellis edition);

THE GROVND OF
ARTES:
Teaching the worke and practise of
Arithmetike, both in whole numbres
and Fractions, after a more easyer
and exacter sorte, then anye lyke
hath hytherto beene
set forth: with di-
uers new ad-
ditions.
Made by M. ROBERTE
RECORDE
Doctor of Physike.

FIG. 105. TITLE PAGE OF THE 1558 RECORDE

ib., 1586, 8°; ib., 1590, 8° (the Dee and Mellis edition); ib., 1594, 8° (p. 207); ib., 1596, 8° (p. 219); ib., 1618, 8°; ib., 1623, 8°; 1636; London, 1646, 8° (p. 219); ib., c. 1646 (p. 220); 1652; 1654; London, 1662, 8° (p. 220); ib., 1668, 8° (p. 221); 1673; 1699. Presumably all of these were published at London. 526

ADDITION.

none other examples for to learne the nume
ration of this forme.

But this shall you marke, that as you dyd in the other kynd of Arithmetik set a pricke in the places of thousandes, in this worke you shall set a starre, as you see before.

S. Then I perceave Numeration: but I praye you, how shall I do in this art to adde two summes or more togither?

ADDITION.

Master.

The easiest way in this arte, is to adde but two summes at ones togyther: how be it, you maye adde more, as I wil tel you anone. therefore whenne you wylle adde two summes, you shall fyrste set downe one of them, it forceth not whiche, and then by it draw a lyne crosse the other lynes. And afterwarde sette downe the other summe, so that that lyne maye be betwene them: as if you woulde adde 2659 to 8342, you must set your summes as you see here.

Addition two summes.

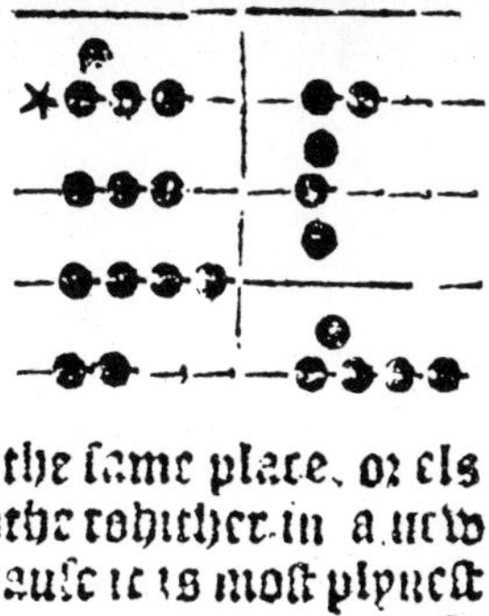

And then if you lyst, you maye adde the one to the other in the same place, or els you may adde them bothe togither in a new place: which way, bycause it is most playnest

I

Fig. 106. Counter reckoning from the 1558 Recorde

The first arithmetic to be published in England was that of Tonstall (p. 132). There must have been several other books published between that date and the first appearance of Recorde's, because in the preface to the latter the author says : 'And if any man obiect, that other books haue bene written of Arithmetike alreadie ſo ſufficiently, that I needed not now to put penne to the booke, except I wil cōdemne other mens writings: to them I anſwere. That as I cōdeme no mans diligence, ſo I know that no man can ſatisfie euery man, and therefore like as many do eſteeme greatly other bookes, ſo I doubt not but ſome will like this my booke aboue any other Engliſh Arithmetike hitherto written, & namely ſuch as ſhal lacke inſtructers, for whoſe ſake I haue plain-ly ſet forth the exāples, as no book (that I haue ſeene) hath hitherto: which thing ſhall be great caſe to the rude readers.' (From the 1594 edition.)

This is, however, the first commercial arithmetic of any note used in the English schools. It is written in the form of a dialogue between the master and his pupil, and the language is so formal that it seems strange that the book should have been so successful. The first part is devoted to integers, the fundamental operations being followed by a section on denominate numbers. This is followed by proportion and the 'golden rule' of three, the backer rule of three (inverse proportion), the double rule of three (compound proportion), the rule of three composed of five numbers, and the rule of fellowship (partnership). The second part relates to fractions, and includes the same general topics as the first, together with alligation and the rule of false. Counter reckoning is given (Fig. 106) as well as computation with Arabic numerals.

Among other works, Recorde wrote 'The Castle of Knowledge' (p. 253) and 'The Whetstone of Witte' (p. 286), the latter being chiefly on algebra. His geometry, 'The Pathway to Knowledge,' appeared in London in 1551, 4°; ib., 1574, 4°.

Recorde's works were the most influential English mathematical publications of the sixteenth century. Thomas Willsford, in his 1662 edition of the 'Ground of Artes,' was able to say with much truth that this book was 'entail'd upon the People, ratified and ſign'd by the approbation of Time.'

Other works of 1542. Albert, p. 180, 1534; Cardan, p. 193, 1539; Finaeus, p. 163, 1530–32; Gemma, p. 200, 1540; Ortega, p. 93, 1512; Diego el Castillo, 'Tratado de quentos,' Salamanca, 4°; Giambattista Verini, 'Spechio del mercatanti,' Milan, 8° (Brunet says, 'Libro de Abaco e gioco de memorie,' Milan, sm. 8°); Han vander Wehn, 'Exempelrechenschaft der Regel de Tri, die man nennt die Kaufmanns güldene Regel ganz und gebrochen,' s. l., 8°. 526

ROBERT RECORDE. Ed. pr. c. 1542. London, 1579.

See p. 213.

Title. 'The // Grounde of Artes :// teaching the work and pra-//ctife of Arithmetike, bothe in whole numbers // and Fractions, after a more eafyer and // exacter forte than any like hath hither-//to bin fet foorthe :// Made by Mayfter Roberte Re-// cord, Doctor in Phyfike, and now of late// diligently ouerfeene and augmented // with new and neceffarie //Additions.//

I. D.

That vvhich my freende hath vvell begonne,
For very loue to common vveale,
Neede not all vvhole to be nevv done,
But nevv encreafe I do reueale.

Some thyng heerein, I once redreft,
And novve agayne for thy behoofe,
Of zeale I doe, and at requeft,
Both mend and adde, fitte for all proofe.

Of Numbers vfe, the endleffe might,
No vvitte nor language can expreffe,
Applie and Trie, both day and night,
And then this truth thou vvilt confeffe.

Printed at London by H. Bynneman.// Anno Domini. 1579.' (F. 1, r.)

Colophon. 'Imprinted at London by Henry // Binneman, and John Harifon.// Anno Domini M.D.LXXVII.' (F. 261, r.)

Description. 8°, 8.7 × 13.9 cm., the text being 6.5 × 11.7 cm. 261 ff. unnumb., 27 ll. London, 1579.

It will be noticed that the colophon is dated 1577. This is therefore one of the cases where a large edition was printed, and a new title page was added from year to year as necessary. The edition is more rare than its date would suggest.

ROBERT RECORDE. Ed. pr. c. 1542. London, 1594.

See p. 213.

Title. 'The // Grovnd of // Artes, teaching // the perfect worke and practife // of Arithmeticke, both in whole numbers // and Fractions, after a more eafie and exact // fort, than hitherto hath been ſet foorthe.// Made by M. Robert Record,// D. in Phiſicke.// And now lately diligently corrected and beauti-//fied with ſundry new Rules and neceſſary Addi-//tions: And further endowed with a third part, of // Rules of Practiſe, abridged into a briefer method // than hitherto hath bene publiſhed: with di-//uerſe ſuch neceſſairie Rules as are // incident to the trade of // Merchandiſe.// Whereunto are alſo added diuerſe Tables and In-//ſtructions that will bring great profite and delight vnto Merchants, Gentlemen, and others,// as by the Contents of this Trea-//tiſe ſhall appeare.// By Iohn Mellis.// At London,// Imprinted by T.D. for Iohn Hariſon, at the Greyhound in // Paules Churchyard.// 1594.'// (P. 1.)

Colophon. 'Imprinted at London by Thomas Dawſon, for Iohn // Harriſon, dwelling in Paules // Churchyard, at the ſigne of // the Greyhound,// 1594.' (P. 558.)

Description. 8°, 10.1 × 15.5 cm., the text being 7.6 × 13.1 cm. 34 pp. unnumb. + 493 numb. 33–525 + 1 blank = 528 pp., 30–31 ll. London, 1594.

Editions. See p. 213. This is probably the third edition by Mellis. In the dedication he says: 'And ſeeing that within this 8. yeares, two impreſſions of theſe my labors dedicated to your Worſhip are already worne out . . .', although not much reliance can be placed on the statement, since it also appears in the 1596 edition.

See p. 216. The first two parts, covering 404 pp. of this book, are substantially identical with the 1558 edition (p. 213). The third part (pp. 405–557), the work of John Mellis, also appeared in the 1591 edition. It includes the 'Rules of Practiſe,' 'The order & worke of the Rule of three in broken Numbers, after the trade of Marchants, digreſſ-ing ſomething from M. Recordes,' 'Loſſe and Gaine,' 'Rules of Payment' (equation of payments), barter, exchange, interest, and other business applications, together with a chapter on 'Sportes and Paſtimes done by Number.' These mathematical recreations had already appeared in printed textbooks, and they played an interesting rôle until the latter half of the nineteenth century. Those in this treatise related to number guessing, the rules being easily developed by our present algebra, but rather mysterious by sixteenth-century arithmetic.

ROBERT RECORDE. Ed. pr. c. 1542. London, 1596.

See p. 213.

Title. The title page is missing.

Colophon. 'Imprinted at London by Richard // Field, for Iohn Harrifon, dwelling // in Pater nofter Row at the // figne of the Greyhound.// 1596.' (P. 559.)

Description. 8°, 9.8 × 15.5 cm., the text being 8 × 13.3 cm. 559 pp., 30–31 ll. London, 1596.

Editions. See p. 213. A note on the last page says that this copy was bought in 1686 for 1s. 6d., not a very low price at that time for a book only ninety years old. This edition is practically identical with that of 1594 (p. 217).

ROBERT RECORDE. Ed. pr. c. 1542. London, 1646.

See p. 213.

Title. 'Records Arithmetick :// or,// The Grovnd // of Arts : // Teaching // The perfect work and Practice of Arithmetick,// both in whole Numbers and Fractions, after a more // eafie and exact form than in former time hath been fet forth :// Made by M. Robert Record, D. in Phyfick. // Afterward, augmented by M. John Dee.// And fince enlarged with a third part of Rules of Pra-//ctife, abridged into a briefer method than hitherto hath been // publifhed, with divers neceffary Rules incident to the Trade // of Merchandife : with Tables of the valuation of all Coyns,// as they are currant at this prefent time.// By John Mellis.// And now diligently perufed, corrected, illuftrated and en-// larged ; with an Appendix of figurate Numbers, and the Extraction // of their Roots, according to the method of Chriftian Vrftitius : with // Tables of Board and Timber meafure ; and new Tables of Intereft // upon Intereft, after 10 and 8 per 100 ; with the true value of // Annuities to be bought or fold prefent, Refpited, or in Rever-//fion : the firft calculated by R. C. but corrected, and the // latter diligently calculated by Rob : Hart-well, Philomathemat.// Scientia non habet inimicum nifi ignoran-tem.// Fide. ------- fed ------ Vide.// London,// Printed by

M. F. for John Harifon, and are to be fold by // Geo: Whittington, and Nath: Brooks, at the fign of // the Angell in Corn-hill. 1646.' (P. 1.)

Description. 8°, 10 × 15.5 cm., the text being 8.6 × 13.7 cm. 27 pp. unnumb. + 629 numb. = 656 pp., 31 ll. London, 1646.

Editions. See p. 213.

Like the 1594 edition this has the additional 'third part' by John Mellis. The Hartwell chapter on roots begins on p. 573, and is based, as the title says, on the work of Urstisius.

ROBERT RECORDE. Ed. pr. c. 1542. London, 1646 (?).

See p. 213.

Title. This edition is said to have been published in 1646, but no date appears in the book itself. It has the same title page as the one dated 1646, except as follows: The words 'Records Arithmetick' (l. 1) are here omitted; the last lines read 'Printed by M. F. for John Harifon, and are to be // fold at his Shop in Pauls-Church-yard.' The number of words on some of the lines in the title page is different, but with the above exceptions the page is the same. The body of the book is from the same setting of type as in the other edition of 1646.

Description. 8°, 10.2 × 15.7 cm., the text being 8.6 × 13.7 cm. 27 pp. unnumb. + 629 numb. = 656 pp., 32 ll. London, s. a. (1646 ?).

See p. 216.

ROBERT RECORDE. Ed. pr. c. 1542. London, 1662.

See p. 213.

Title. The title page is practically the same as that of 1646, except as to the imprint: 'Printed by James Flefher, and are to be fold by Jofeph // Cranford, at the figne of the Gunn in St. Pauls // Church-yard. 1662.'

Description. 8°, 10.7 × 16.5 cm., the text being 9.2 × 13.9 cm. 22 pp. unnumb. + 536 numb. = 558 pp., 33 ll. London, 1662.

See p. 216.

ROBERT RECORDE. Ed. pr. c. 1542. London, 1668.

See p. 213.

Title. The title page of this edition is practically the same as that of 1646, except as to the imprint: 'Printed by James Flesher, and are to be sold by // Robert Boulter, at the Turks-head in Bishopsgate-//street, next the great James. 1668.'

Description. 8°, 10.4 × 16.4 cm., the text being 9.2 × 13.7 cm. 22 pp. unnumb. + 536 numb. = 558 pp., 33 ll. London, 1668.

JOHANN FREY. Ed. pr. 1543. Nürnberg, s. a. (1543).

A Nürnberg gauger of the middle of the sixteenth century.

Title. See Fig. 107.

Colophon. 'Gedruckt zů Nůrnberg durch // Georg Wachter.' (F. 36, r.)

Description. 8°, 9.5 × 15.5 cm., the text being 7.1 × 10.9 cm. 36 ff. unnumb., 21–26 ll. Nürnberg, s. a. (1543).

Editions. There was no other edition. The book bears no date except in the dedicatory epistle to the reader, which closes with the words, 'Anno 1543.' This epistle gives, also, the only reference to the author, who there speaks of himself as 'ich Johañ Frey/ burger zu Nůrmberg.'

The subject of gauging occupied a great deal of attention on the part of German writers on arithmetic in the sixteenth century, and occasionally, as in this instance, separate books were prepared. It was not so common in England as on the Continent, not appearing, for example, in as extensive a work as the Dee and Mellis edition of Recorde mentioned above. It had, however, some standing in the early American arithmetics, and is found as a separate chapter as late as the middle of the nineteenth century. The title page gives some idea of the work of the gauger before the days of standardization of casks. The American rule for gauging a cask was substantially as follows: Add to the head diameter 0.7, 0.65, 0.6, or 0.55 of the difference between the head and bung diameters (according to the degree of curvature of the staves), and multiply the square of this sum by the length; divide by 359 for ale or beer gallons, and by 294 for wine. Thus a cask with bung diameter 36 in., head diameter 30 in., and length 48 in., contains 153.65 ale gallons.

This particular work is interesting because, although it was printed in 1543, some of the numerals are quite like those of a hundred years earlier. The mediæval 4, 5, and 7 are exclusively used in the engraved

Ein new Visier bůchlein/
welches innhelt/wie man durch den Quadraten
auff eynes yeden lands Eich/ein Růtten zůbe
reyten/vñ damit yetlichs vnbekants vaß
Visieren/vnd solches innhalt erken-
nen sol/Auffs new gebessert
vnd gemert.

FIG. 107. TITLE PAGE OF FREY

figures, although the types used in the body of the book give the later forms. These mediæval figures may be seen in several illustrations in the second part of this bibliography.

ANONYMOUS. Ed. pr. 1543. Paris, 1543.

The author of the *Theologoumena* is unknown. He lived after Anatolius (Bishop of Laodicea, 270 A.D.) and probably after Iamblichus (fourth century).

Title. See Fig. 108.

Description. 4°, 15.2 × 23 cm., the text being 9.1 × 18.1 cm. 7 pp. unnumb. + 61 numb. (5–65) + 2 blank = 70 pp., 30 ll. Paris, 1543. Greek text, except the dedicatory epistle, which, being dated 'Lutitiæ Parifiorum 1543. 6. Calendas Iulias,' shows this to be the first edition.

Editions. This edition was unknown to De Morgan. The best edition is that of Fr. Ast, Leipzig, 1817.

A work of little importance, on the Greek theory of numbers. Gow (p. 88) describes it as a 'curious farrago.' Cantor (*Geschichte der Mathematik*, Kap. 22) says that the author may have drawn on Iamblichus. The most valuable feature of the work is the light which it throws on an earlier work by Speusippus, nephew of Plato.

Other works of 1543. Archimedes, p. 228, 1544; Gemma, p. 200, 1540; Glareanus, p. 192, 1539; Recorde, p. 214, c. 1542; Regius, p. 182, 1536; Sfortunati, p. 174, 1534; Tonstall, p. 134, 1522; Nicolaus Medlerus, 'Rudimenta Arithmeticæ practicæ,' s. l., 8° (with subsequent editions: Wittenberg, 1550; ib., 1558; Leipzig, 1556; Weissenfels, 1564, 8°). 496;526

MICHAEL STIFEL. Ed. pr. 1544. Nürnberg, 1544.

STIEFEL, STYFEL. Born at Esslingen, April 19, 1487; died at Jena, April 19, 1567. He was a priest, a reformer, and a fanatic, but was one of the most skillful arithmeticians of his time.

Title. See Fig. 109.

Colophon. 'Excudebatur Norimbergæ // apud Ioh. Petreium.' (F. 325, r.)

Description. 4°, 15.5 × 20.2 cm., the text being 10.2 × 15.1 cm. 326 ff. (1 blank, 6 unnumb.), 33 ll. Nürnberg, 1544.

Editions. Stifel wrote five works on mathematics, all but one appearing after his period of religious fanaticism. These works, with their various editions, are as follows:

1. 'Ein Rechen Büchlein vom End Christ. Apocalysis in Apocalysim,' Wittenberg, 1532. A little-known work on the theory and mysticism of numbers.

ΤΑ ΘΕΟΛΟ-
ΓΟΥΜΕΝΑ ΤῆΣ ἈΡΙΘΜΗΤΙΚῆΣ.

Habes hic o ſtudio-
SE LECTOR, NOVVM OPVS-
culum antehac nuſquàm excuſum, in quo ita Numerorum ratio explicatur, vt non ſit obſcurum intelligere hanc arithmetica ád interioré illá de philoſophia diſputationem, quam Theologiam veteres vocabant, conferre plurimum.

PARISIIS.
Apud Chriſtianum wechelum ſub ſcuto Baſilienſi, in vico Iacobæo: & ſub Pegaſo, in vico Bellouacenſi. M. D. XLIII.

FIG. 108. TITLE PAGE OF THE *Theologoumena*

ARITHMETI
CA INTEGRA.

Authore Michaele Stifelio.

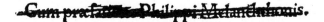

Norimbergæ apud Iohan. Petreium.
Anno Christi M. D. XLIIII.

Cum gratia & priuilegio Cæsareo
atq; Regio ad Sexennium.

FIG. 109. TITLE PAGE OF STIFEL'S SECOND ARITHMETIC

2. 'Arithmetica Integra,' Nürnberg, 1544, 4° (here described); ib., 1545, 4°; 1546; Nürnberg, 1548; ib., 1586, 4°.

3. 'Deutsche Arithmetica' (p. 231), Nürnberg, 1545, 4°.

4. 'Rechenbuch von der Welſchen vnd Deutſchen Practick/ auff allerley vorteyl vnd behendigkeit/ mit erklerung viler Exempeln/ . . .', Nürnberg, 1546.

5. 'Die Coss', Königsberg, 1553–54, 4° (p. 259); 1571; 1615.

The dedication and Melanchthon's preface, both of which have been removed from this copy, are dated 'Vuitebergæ 1543.' This copy has evidently been owned by some one unsympathetic with Melanchthon, because not only has the preface been removed but the reformer's name has been crossed out of the title page (Fig. 109). The work is one of the most scholarly arithmetics and algebras that came out in Germany in the sixteenth century. It is divided into four books, the first being an arithmetic and giving not only the theory of the subject, but a considerable amount of practical work. The second book is on irrational numbers, a chapter which we now insert in algebra, and the third is on algebra itself, the name of which subject is said to be 'à Gebro Aſtronomo, auctore eius,' a common opinion at that time. This work did for Germany what Cardan's and Tartaglia's did for Italy. It was a storehouse from which subsequent writers drew, and, although not a practical mercantile book, it materially influenced even the elementary textbook makers. Stifel himself recognized the demand for such a work, for he says: 'Quanqȝ autem plurimi de Arithmetica libelli extent, & quotidie plures noui gignunter, ego tamen adhuc nullum uidi qui integrā artem traderet.' Stifel makes much use of the plus and minus signs.

ARCHIMEDES. Ed. pr. 1544. Basel, 1544.

Born at Syracuse, c. 287 B.C.; died there in 212 B.C. The greatest of Greek mathematical physicists.

Title. See Fig. 110.

Colophon. 'Basileae, per Ioannem // Hervagivm, anno ab orbe re-//dempto, M.D. XLIIII. menſe Martio.' (P. 69 of the last part.)

Description. Fol., 21.8 × 31.7 cm., the text being 13.2 × 23.5 cm. 455 pp. (7 blank and 12 unnumb.), 51 ll. The work is made up of four parts, all of the same date, separately paged, and bound together. The 'Ψαμμίτης, De harenæ numero,' is the sixth work in the collection. Basel, 1544.

ΑΡΧΙΜΗΔΟΥΣ

ΤΟΥ ΣΥΡΑΚΟΥΣΙΟΥ, ΤΑ ΜΕΧΡΙ
νῦν σωζόμενα, ἅπαντα.

ARCHIMEDIS SYRACVSANI
PHILOSOPHI AC GEOMETRAE EX-
cellentiſſimi Opera, quæ quidem extant, omnia, multis iam ſeculis deſi-
derata, atq; à quàm pauciſſimis hactenus uiſa, nuncq;
primùm & Græcè & Latinè in lu-
cem edita.

Quorum Catalogum uerſa pagina reperies.

Adiecta quoq; ſunt
EVTOCII ASCALONITAE
IN EOSDEM ARCHIMEDIS LI-
bros Commentaria, item Græcè & Latinè,
nunquam antea excuſa.

Cum Cæſ. Maieſt. gratia & priuilegio
ad quinquennium.

Ioannes Bapt. Grillinzonij

BASILEAE.
Ioannes Heruagius excudi fecit.
An. M D XLIIII.

FIG. 110. TITLE PAGE OF THE 1544 ARCHIMEDES

Editions. This seems to be the first edition of the works of Archimedes to contain the 'De arenae numero.' This chapter also appears in the 1558 edition (below), and separately in Paris in 1557, 8° (the Hamellius edition, below). It was not in the Tartaglia edition of 1543. It appeared in the elaborate editions of Barrow (1675), Torelli (1792), Peyrard (1807–8), and Heiberg (1880). Riccardi says that it 'fu illustrato dal Clavio ne' suoi commenti alla sfera del Sacrobosco.'

There are several works of Archimedes extant. The 'De arenae numero,' included in this edition, is not an arithmetic, but it treats of the numeration of large numbers. It is addressed to Gelon, King of Syracuse, and proposes to show, 'by geometric proofs which you can follow, that the numbers which have been named by us and are included in my letter to Zeuxippus are sufficient to exceed not only the number of a sand-heap as large as the whole earth, but one as large as the universe.' Archimedes then proceeds to develop a system of numeration by octads. In this work he incidentally refers to a fact which would now be expressed by the symbols $x^m \cdot x^n = x^{m+n}$.

ARCHIMEDES. Ed. pr. 1544. Paris, 1557.

See p. 226.

Title. 'Paschasii // Hamellii Regii // Mathematici // Commentarius // in // Archimedis // Syracuſani præclari Mathematici librū // de numero arenæ, multis locis per // eundem Hamellium // emendatum.// Lvtetiae // Apud Gulielmum Cauellat, ſub pingui Gallina,// ex aduerſo collegij Cameracenſis.// 1557.' (F. 1, r.)

Description. 8°, 11 × 17.1 cm., the text being 7.1 × 12.6 cm. 48 ff. numb., 23–30 ll. Paris, 1557.

Editions. See above.

This commentary of Hamellius is quite as satisfactory as any of the earlier ones.

ARCHIMEDES. Ed. pr. 1544. Venice, 1558.

See p. 226.

Title. 'Archimedis // opera non nvlla // à Federico Commandino // Vrbinate // nvper in Latinvm conversa, // et commentariis // illvstrata.// Quorum nomina in ſequenti pagina leguntur.

// Cvm privilegio in annos X.// Venetiis,// apud Paulum Manutium, Aldi F.// M D LVIII.' (F. 2, r.)

Description. Fol., 21.1 × 30 cm., the text being 13.2 × 22.8 cm. 128 ff. (2 blank, 8 unnumb.), 39 ll. Venice, 1558. With this is bound the 'Commentarii // in opera non Nvlla // Archimedis.// Venetiis,// apud Paulum Manutium, Aldi F.// M D LVIII.' (F. 60, r.) It contains the book entitled 'Archimedis // Liber de Arenæ // nvmero.'

See p. 228.

JUAN SARAVIA, de la calle Beronese.

Ed. pr. 1544. Medina, 1544.

A Spanish arithmetician, of Medina, of the middle of the sixteenth century.

Title. See Fig. 111.

Colophon. '℄Fue impreſſa la preſente:// obrallamada Inſtruciō de mercaderes enla muy // noble villa de medina d'l campo por Pedro de // Caſtro ĩpreſſor. Acoſta d'Antoño de vrueña // mercader d'libros. Acaboſe atreynta dias // del mes de Julio. Anō de mil ꝛ quiniē-//tos ꝛ quarenta ꝛ quatro anōs.' (F. cvij, r.)

Description. 8°, 14 × 20.1 cm., the text being 12.1 × 16.9 cm. 2 ff. unnumb. + 105 numb. (Roman numerals) = 107 ff., 30 ll. Medina, 1544.

Editions. Medina, 1544, 4° (here described); ib., 1547, 4°; Venice, 1561, 8° (p. 231.)

This rare work was evidently written about 1542 (see f. xcviij, r.). It is not a textbook on arithmetic, but it relates to commercial problems, the topics being curiously interspersed with biblical illustrations to show the justice of the customs involved.

Other works of 1544. Apianus, p. 155, 1527; Bæda, p. 131, 1521; Finaeus, p. 163, 1530–32; Gemma, p. 200, 1540; Grammateus, p. 123, 1518; Köbel, p. 102, 1514; Noviomagus, p. 195, 1539; Peurbach, p. 53, 1492; Riese, p. 139, 1522; Sfortunati, p. 177, 1534; Tonstall, p. 136, 1522; Vander Hoecke, p. 183, 1537; H. Bock, 'Ein new Rechenbuchlein,' Nürnberg, 8°; Johannes Bogardus, a work on finger reckoning based on Aventinus (p. 136), Paris; Leonard Hegelin, 'Ein künstlich Rechenbüchlin auff Zyffer vnnd andern hüpschen Regeln,' Ulm(?);

Lib de cam°. tiamo 2 Cason 7 Lucy 6

¶ Inſtrucion de mercaderes
muy prouechoſa. En la qual ſe enſeña como deuen los
mercaderes tractar. y de que manera ſe han de eui-
tar las vſuras de todos los tractos de ventas ⁊ com
pras. Aſſi alo contado como alo adelantado: y alo fia
do. y delas compras del cenſo al quitar: y tractos de
compañia: y otros muchos contratos. Particular
mente ſe habla del tracto delas lanas. Tambien ay
otro tractado de cambios. Enel qual ſe tracta delos
cambios licitos y reprouados. Nueuamente compueſto por el doctor Sarauia dela calle Beronēſe.
Año. M.D.xliiij   NIHIL NIMIS

FIG. III. TITLE PAGE OF SARAVIA

Antonio Martín, 'Tractado de Arithmética y Geometría,' Alcalá, 4°; H. Vuelpius, 'De minutiis phyſicis et practicis aſtronomicae arithmeticae regulis,' Cologne, 4°, and 'Libellus de communibus et uſitatis arithmeticae practicae regulis,' ib., 4°. 528

JUAN SARAVIA. Ed. pr. 1544. Venice, 1561.

See p. 229.

Title. 'Institvtione//de' Mercanti//che tratta del comparare //et vendere,//et della vsvra chepvo//occorrere nella Mercantia//insieme con vn trattato//de' Cambi.//Et in somma ſi ragiona di//tutto quello che al Mercante Christiano//ſi conuiene. //Composta per il Dottor Sarava,//& nuoamente tradotta di lingua spagnuola//dal S. Alfonso D'Vlloa.//Cvm Privilegio.//In Venetia//Apreſſo Bolognino Zaltieri.//M D LXI.' (F. 1, r.)

Colophon. 'In Venetia apreſſo Bolognino Zaltieri // 1561.' (F. 135, v.)

Description. 8°, 10.2 × 15.1 cm., the text being 7.6 × 12.2 cm. 135 ff. (3 unnumb.), 29 ll. Venice, 1561.

Editions. See p. 229.

This is an Italian translation of the Spanish edition of 1544 (p. 229). The book was evidently written in 1542 as already stated, for the author speaks (f. 109) 'del cãpo queſto anno M D XLII.'

MICHAEL STIFEL. Ed. pr. 1545. Nürnberg, 1545.

See p. 223.

Title. See Fig. 112.

Colophon. 'Zu Nůrnberg Truckts Johan Petreius.// 1545.' (F. 96, r.)

Description. 4°, 15.5 × 20.7 cm., the text being 10.8 × 15.5 cm. 4 ff. unnumb. + 92 numb. = 96 ff., 32–34 ll. Nürnberg, 1545.

Editions. See p. 226.

The 'Deutsche Arithmetica' is divided into three parts. In the 'Hauſsrechnung' there are 12 chapters, all relating to the arithmetic of the common people, there being 'no household so narrow and poor that common arithmetic is not both serviceable and necessary to its welfare.' ('Es iſt kein Hauſzhaltung/niendart ſo gering vnnd ſchlecht/

Deutsche Arithmetica.
Inhaltend.

Die { Haußrechnung. / Deutsche Coß. / Kirchrechnung. }

Mein lieber Leser/Nach dem die Coß (welche ist ein Kunstrechnung der gantzen Arithmetick) bißher den Deutschen/mit vil frembden worten/vermengt vnd verblend/schwer ist gewesen/So wirt sie hie mit new erfundnem vortheil vnnd Regeln / sehr leicht vnd kurtz herfur bracht vnd gelehrt/ vnd mit guten Deutschen bekantlichen worten vñ Exempeln erweyset. Das ander so hierin gelert wirt von der Haußrechnung vnd Kirchenrechnung/ bringt seinen bericht gnugsam mit sich. Alles durch Herr Michael Stifel/auff ein besondere newe vnd leichte weis gestellet.

Zu Nürnberg truckts Johan Petreius.
1545.
Cum Priuilegio ad Quinquennium.

FIG. 112. TITLE PAGE OF STIFEL'S *Deutsche Arithmetica*

das jr die Hauſzrechnung oder gemeine rechnung/ nicht nützlich vnnd dieſtlich sey.') Stifel first treats 'Vom Algorithmo der Rechenpfenning,' recommending the operations as being 'wunderleichtlich durch die Rechenpfenning gelernet vnd gelehret.' The coss (algebra) relates largely to arithmetic, although touching upon the most common algebraic operations (Fig. 113). In it Stifel gives the rule for dividing one fraction by another by using the inverted divisor as a multiplier. The third part relates to the ecclesiastical calendar ('Von der Kirchenrechnung/die man nennet Computum Eccleſiaſticum'), a subject in which he acknowledges his indebtedness to 'Johannes de Sacro buſto.' With his usual commendable but effusive piety Stifel closes with praise not only to the Prince but to 'vnſerm Vater im Hymel vnd ſeinem eynigen naturlichen Son/ vnſerm Herren Ieſu Chriſto.'

JOHANN SCHEUBEL. Ed. pr. 1545. Leipzig, 1545.

SCHEYBL, SCHEUBELIUS. Born at Kirchheim, Württemberg, August 18, 1494; died February 20, 1570. Professor of Mathematics in the University of Tübingen, to which institution he bequeathed most of his manuscripts. He wrote on arithmetic and algebra, and edited part of Euclid.

Title. See Fig. 114.

Colophon. 'Lipſiæ ex Officina Michaëlis // Blum, a reſtituta ſalute.// Anno M. D. XLV.// Idib: Maij.' (F. 255, r.)

Description. 8°, 10 × 15.9 cm., the text being 6.9 × 12 cm. 255 ff. unnumb. + 1 blank = 256 ff., 20–26 ll. Leipzig, 1545.

Editions. It is sometimes stated that this work appeared at Strasburg in 1540, but I cannot verify the statement. I know of no edition other than this of 1545. Scheubel also wrote a 'Compendium arithmeticae' (p. 246, 1549). Murhard mentions an 'Arithmetica sive de Arte supputandi Liber,' Lipsiae, 1545, 8°, but he may refer to the 'De Numeris.' Scheubel also published an algebra (Paris, 1551), and the seventh, eighth, and ninth books of Euclid (1558). 528

This work is the production of a scholar rather than a man conversant with the demands of business. While Scheubel tried to write a mercantile arithmetic, the result was far removed from the needs of the common people. It carries the work in subjects like the roots so far that the ordinary Rechenmeister could not have used it. Moreover, it is written in Latin and is much more extended than the work of Gemma Frisius, so that it appealed neither to the business school nor

Der Ander theyl

Von disen zweyen zeichen/ + vnd —. VII.

WO ich von zeychen reden werde/ soltu mich verstehn von disen zeichen + vnd —/ Deñ solliche verzeichnis/ Sum: oder Sum: A. oder ſ. ꝛc. Werde ich nicht zeychen nennen/ sondern/ namen/ oder benennung der zalen. Wa ich nu rede von gleichen zeichē/ soltu es verstehn von + vnd + / oder von — vnd —. Also auch/ wa ich von vngleichen zeichen rede / so verstehe es/ von + vnd —.

So haben nu dise zwey zeichen + vnd — / ein sonderlichen Algorithmum/ welchē ich hie stellen will auff 4 Regeln. Denn er gehöret zum Algorithmo der vngerechneten zalen wie du woll sehen wirst/ vnd alles was vorhin gesagt ist von disen namen sum: sum: A. ꝛc. das gehöret alles hie her / als vnter ein einigen Algorithmum.

Die erst Regel von dem Addiren vnd Subtrahiren. VIII.

Wey gleiche zeichen / machen eben das selbig zeichen/ im Addiren vñ Subtrahiren/ ohn allein so du im subtrahiren die zal / die du soltest subtrahirē/ nicht kanst subtrahirē.

Exempla vom Addiren.

8 Sum: + 7.	8 Sum: — 18.
12 Sum: + 11.	3 Sum: — 6.
20 Sum: + 18.	11 Sum: — 24.

Hie sihest nu vor augen/ wie + vnd + mache im ersten exemplo

FIG. 113. FROM STIFEL'S *Deutsche Arithmetica*

to the ordinary classical school. A great deal of attention is given to exchange, the rule of three, and the extracting of roots of high order. Attention is also given to problems which would now form part of algebra, and there is a little treatment of geometry from the standpoint of mensuration. The nature of the work can be somewhat understood

DE NVME-
RIS ET DIVERSIS RATIONIBVS
ſeu regulis computationum opuſculum,
a Ioanne Scheubelio compoſitum.
Non ſolum ad uſum quendam uul
garem, ſed etiam cognitionem
& ſcientiam exquiſitiorem
arithmeticæ acco-
modatum.

M. D. XLV.

Fig. 114. Title page of Scheubel

from the titles of the five 'tractati': 1. 'De numeris integris'; 2. 'De proportionibus, proportionalitatibus & alijs'; 3. 'De minutijs uulgaribus'; 4. 'De minutijs phyſicalibus'; 5. 'Aliquot regulæ.'

While Scheubel is not much appreciated to-day, he was really ahead of his time. He tried to banish the expression 'rule of three' and to substitute 'rule of proportion.' His explanation of square root is in some respects the best of the century, and he dismisses with mere mention the 'duplatio' and 'mediatio' of his contemporaries. He extracts various roots as far as the 24th, finding the binomial coefficients by means of the Pascal triangle a century before Pascal made the device famous.

SACROBOSCO. Ed. pr. 1545. Wittenberg, 1550.

See p. 31.

Title. 'Ioannis // de Sacrobvsto // Libellus de Sphæra.// Accessit eivsdem // avctoris Compvtvs // Ecclefiafticus, Et alia quædam // in ftudioforum gra-//tiam edita.// Cum Præfatione Philippi // Melanthonis.' On f. 67, r., begins 'Libellvs // Ioannis de Sacro//bvsto, de Anni Ratione, // sev vt vocatvr vvl-//go Compvtvs Ec-//clesiasti-//cvs.// Cvm Praefatione // Philippi Melan-//thonis.// Anno M. D. XLV.'

Colophon. 'Impreffum Vuitebergæ apud // Iohannem Cratonem.// Anno // M. D. L.' (F. 134, r.)

Description. 8°, 10.1 × 15.5 cm., the text being 6.5 × 10.8 cm. 134 ff. unnumb. + 2 blank (2 plans) = 136 ff., 26 ll. Wittenberg, 1550.

Editions. The preface by Melanchthon is dated 'Menfe Auguf-to.// Anno // M. D. XXXVIII,' so that his edition of the book could not have appeared earlier than that year. The fact that this volume was printed in 1550, as shown by the colophon, makes it probable that the date 1545 on the title page is that of the first edition. There are several editions of the 'Sphæra' in Mr. Plimpton's library, but this is the only one containing the Computus. 528

This is less properly included in a list of arithmetics than many of the other computi. Since several others have been included, this, which is one of the most celebrated, is given place; but the arithmetical work is practically nil.

EUCLID. Ed. pr. (arith. books) 1545. Rome, 1545.

See p. 11.

Title. See Fig. 115.

Colophon. 'Stampata in Roma per Antonio Blado Afolano.// M D XLV.' (P. 12.)

Description. 8°, 9.8 × 15.1 cm., the text being 6.7 × 10.9 cm. 112 pp. (3 blank, 6 unnumb.), 24–26 ll. Rome, 1545.

Editions. There were several editions of one or more of those books of Euclid that relate to some part of the theory of arith-

I QVINDICI
LIBRI DEGLI ELEMEN
TI DI EVCLIDE, DI GRE
CO TRADOTTI IN
LINGVA THO.
SCANA.

IN ROMA. M D XXXXV.

Con gratia & priuilegio del S. N. S. *Paulo Terzo,*
& della Serenissima republica Venetiana
per cinque anni.

FIG. 115. TITLE PAGE OF THE 1545 EUCLID

metic. Of these separate books, this is the rare first Italian edition. Among others may be mentioned those of Wittenberg, 1546; 1549; Paris, 1551, 4° (p. 238); ib., 1554, 4° (p. 238); 1555 (by Scheubel); Paris, 1557, 8° (p. 240); Wittenberg, 1564,

8° (p. 240). There were very many editions of Euclid's work published before 1600, practically all including Book V, on proportion, and some including the other arithmetical books.

Euclid's treatment of arithmetic was purely theoretical, no work on computation being included. This copy has the following note by De Morgan, who once owned it: "This book was also printed in Greek — same year, place, size, printer, and dedication. There is a Greek copy in the British Museum. A. De Morgan, Feb. 29, 1852."

Other works of 1545. Bæda, p. 131, 1521; Feliciano, p. 149, 1526; Psellus, p. 168, 1532; Stifel, p. 226, 1544; Sfortunati, p. 177, 1534; Willichius, p. 197, 1540; Antonius de Barres, 'Arithmeticae practicae libri IV,' Louvain, 4°; Johann Obers, 'Newgestelt Rechenpüchlin,' Augsburg; Petro a Spinosa, 'Tractatus proportionum,' Salamanca, 1545, fol. 528

EUCLID.[11] Ed. pr. (arith. books) 1545. Paris, 1551.

Title. See Fig. 116. (Note the signature of Giuliano de Medici.)

Description. 4°, 16.3 × 23.4 cm., the text being 9.5 × 17.3 cm. 162 ff. (2 blank, 20 unnumb.), 29 ll. Paris, 1551.

Editions. See p. 237.

This is not the same as the 1545 edition already described, but, like it, this relates to the numerical side of mathematics.

EUCLID. Ed. pr. (arith. books) 1545. Paris, 1554.

See p. 11.

Title. 'Evclidis // Elementa qvædam // Arithmetica.// Lvtetiæ,// Apud Vascosanum, uia Iacobæa, ad insigne Fontis.// M. D. LIIII.// Cvm privilegio regis.' (F. 1, r.)

Description. 4°, 15.4 × 19.4 cm., the text being 11.2 × 17.8 cm. 18 ff. numb., 29 ll. Paris, 1554.

Editions. See p. 237.

This work is made up of certain extracts from the various books of Euclid relating to Arithmetic. It is in Greek with a Latin translation following each definition or theorem. It consists of such standard old definitions as 'Numerus autem, ex unitatibus composita multitudo,' and such theorems as 'Omnis primus numerus, ad omnem numerum quem non metitur, primus est.' There are no discussions, illustrations, or proofs of the propositions.

Euclidis elementorum

LIBER DECIMVS, PETRO

Montaureo interprete.

Ad Ioannem Bellaïum Cardinalem.

LVTETIAE,

Apud Vascosanum, uia Iacobæa ad insigne Fontis.

M. D. LI.

CVM PRIVILEGIO.

Di Giuliano de Medicj.

FIG. 116. TITLE PAGE OF THE 1551 EUCLID

EUCLID. Ed. pr. (arith. books) 1545. Paris, 1557.

See p. 11.

Title. 'Evclidis // Elementorvm // Libri XV. Grae-//cè & Latiné,// Quibus, cùm ad omnem Mathematicæ ſcientiæ // partem, tùm ad quamlibet Geometriæ tra-//ctationem, facilis comparatur aditus.// Επίγραμμα παλαιόν.// Σχήματα πέντε Πλάτωνος, ἃ Πυθαγόρας σο-//φὸς εὗρε.// Πυθαγόρας σοφὸς εὗρε, Πλάτων δ' ἀρίδηλ' ἐδί-//δαξεν // Εὐκλείδης ἐπὶ τοῖσι κλέος περικαλλὲς ἔτευξεν. Lvtetiae,// Apud Gulielmum Cauellat, in pingui Gallina, // ex aduerſo collegij Cameracenſis.// 1557.' (F. 1, r.)

Description. 8°, 10.5 × 16.9 cm., the text being 7 × 12.8 cm. 16 ff. unnumb. + 130 numb. = 146 ff., 25 ll. Paris, 1557.

See p. 238.

EUCLID. Ed. pr. (arith. books) 1545. Wittenberg, 1564.

See p. 11.

Title. 'Arithmetices // Evclideae // Liber Primvs.// Aliâs in ordine reliquorum // Septimvs : Qui citra // præcedentium Sex librorum // Geometricorum opem eruditè//perſequitur, cum reliquis duobus // ſequentibus, uera principia ac // ſolidiora fundamenta Logi-//ſtices, id eſt, ut uocant,// Arithmetices Pra-//cticæ. // Per // Ioan. Sthen. Luneb.// In scholarvm vsvm κατὰ τὸ ὅτι tractatus ἐρωτηματικῶς, diſquiſitione nimi-//rum Dialectica quæ Dialogorum // est propria.// 1564.' (F. 1, r.)

Colophon. 'VVittenbergæ.// Anno // 1564.' (F. 106, v.)

Description. 8°, 9.2 × 14.8 cm., the text being 6.3 × 11.5 cm. 107 ff. unnumb., 22–24 ll. Wittenberg, 1564.

Editions. See p. 237.

GASPARD DE TEXEDA. Ed. pr. 1546. Valladolid, 1546.

A Valladolid arithmetician of the first half of the sixteenth century.

Title. See Fig. 117.

Colophon. '℄Fue impreſſala preſente // obra d'Arithmetica En la muy noble // y felice villa de Valladolid (Pincia // otro tiempollamada) En la offici-//na de Franciſco Fernandez // de

Suma
De Arithmetica pratica
Y de todas Mercadrias
Con la Sorden de contadores.
Hecho por
Gaspar de Texeda
Con Priuilegio
Imperial

FIG. 117. TITLE PAGE OF TEXEDA

cordoua/ junto alas // escuelas ma//yores // Acabose a quatro dias del mes // de Henero deste año del // señor de mill ꝛ quini-//entos ꝛ quaren//ta ꝛ seys // Años.' (F. lxiiij, r.)

Description. 8°, 13.5 × 19.5 cm., the text being 10.3 × 16.5 cm. 64 ff., numb. in Roman, 32–34 ll. Title page engraved on wood. Valladolid, 1546.

Editions. There was no other edition. De Morgan (p. 103) gives the date as 1545, which is that of the privilege, the colophon of his copy having been torn out.

This rare Spanish arithmetic gives the fundamental processes with integers, fractions, and denominate numbers, introduces some practical mensuration under the title 'De Geometria,' and gives a rather extended treatment of the business rules. It is interesting because of the treatment of Spanish and Arab (algoristic) notation, or, as the author says, 'de numerar en castellano y en guarismo.' For example, his two methods of writing 160,462,009,621, are:

c. lx. U462 q̊s. . ix U621
160 U462 q̊s 009 U 621. (F. iiij, r.)

PIETRO CATANEO. Ed. pr. 1546. Venice, 1546.

A sixteenth-century arithmetician, of Siena.

Title. See Fig. 118.

Colophon. 'Stampato in Venetia per Niccolo Bascarini.// M D XLVI.' (F. 64, r.)

Description. 4°, 15.3 × 20.8 cm., the text being 12 × 16 cm. 1 f. unnumb. + 63 numb. = 64 ff., 32–35 ll. Venice, 1546.

Editions. A very rare edition, and probably the first in spite of the words 'nvovamente stampate,' for the dedication is dated M. D. XLVI. (F. 1, v.) There were two later editions, Venice, 1559, 4° (p. 244), and Venice, 1567, 4° (p. 244).

The work is fairly practical, and in many respects is in advance of its time. Unlike most Venetian books it uses the Florentine name 'biricvocolo' for the common form of multiplication, and gives the 'a danda' division before the galley form, recommending it as 'molto necessario.' The applications, while not numerous, are practical, and throw some light upon the business customs of Siena and Venice. Cataneo was not, however, an original writer. His arithmetic is composed quite largely of didactic statements to be found in the works of his predecessors, and the fact that he gives four methods of multiplication shows that he could not escape the influence of writers like Paciuolo.

LE PRATICHE DELLE DVE
PRIME
MATHEMATICHE
DI PIETRO DE CATANI
DA SIENA

LIBRO D'ALBACO
E GEOMETRIA

NVOVAMENTE STAMPATE,
IN VENETIA M D XLVI

FIG. 118. TITLE PAGE OF THE 1546 CATANEO

Other works of 1546. Anonymous, p. 213, 1542; Boethius, p. 27, 1488; Euclid, p. 237, 1545; Glareanus, p. 192, 1539; Helmreich, p. 303, 1561; Manzoni, p. 257, 1553; Misrachi, p. 180, 1534;

Rudolff, p. 152, 1526; Sole, p. 143, 1526; Stifel, p. 226, 1544; Anonymous, 'An introduction for to lerne to recken with the pen, or with the counters accordyng to the trewe cast of Algorisme, in hole numbers, or in broken, newly corrected. And certayne notable and goodly rules of false positions thereunto added, not before sene in our Englysche Tonge,' London, 8°, with another edition at London in 1574, sm. 8°; Anonymous, 'Ein new kurtz Rechenbüchlein auff der Linien und Federn,' Frankfort, 8°, possibly by Gülfferich (p. 269, 1555); Alfonzo López de Corella, 'Secretos de las cuatro mathemáticas ciencias,' Valladolid.

Works of 1547. Gemma, p. 200, 1540; Saravia, p. 229, 1544; Tagliente, p. 114, 1515.

Works of 1548. Gemma, p. 200, 1540; Ghaligai, p. 132, 1521; Riese, p. 139, 1522; Stifel, p. 226, 1544; Tagliente, p. 114, 1515; Uberti, see Tagliente, p. 114, 1515; Wolphius, p. 154, 1527; Anonymous, 'Specie principali, et primi principii del' Arithmetica di C. de C. P.,' Bologna. 528

PIETRO CATANEO. Ed. pr. 1546. Venice, 1559.

See p. 242.

Title. 'Le // pratiche // delle dve prime // Matematiche // di Pietro Cataneo // con la aggionta,// libro d'Albaco e Geometria con il // pratico e uero modo di mifurar la Terra.// Non piv mostra da altri.// (Woodcut of griffin with motto : 'Virtute dvce // comite Fortvna.') In Venetia, apreffo Giouanni Griffio, MD LIX.' (F. 1, r.)

Colophon. 'In Venetia, apreffo Giouan Griffio, ad inftantia // di M. Pietro Cataneo, M D LIX.' (F. 83, v.)

Description. 4°, 15.3 × 20.3 cm., the text being 11.4 × 15.1 cm. 1 f. unnumb. + 82 numb. + 1 blank = 84 ff., 32–35 ll. Venice, 1559.

See p. 242.

PIETRO CATANEO. Ed. pr. 1546. Venice, 1567.

See p. 242.

Title. 'Le // pratiche // delle dve prime // Matematiche // di Pietro Cataneo Senese,// ricorrette, & meglio ordinate, con alcune ag-//giontioni de lo steffo Autore.// Diuife in libri quattro.//

(Woodcut of griffin, with motto: 'Virtvte dvce,// comite Fortvna.') In Venetia, appreſſo Giouanni Griffio,// M D LXVII.' (F. 1, r.)

Colophon. 'In Venetia, appreſſo Giouan Griffo, ad inſtantia di // M. Pietro Cataneo, M D LXVII.' (F. 88, r.)

Description. 4°, 15.5 × 20.7 cm., the text being 11.7 × 15.7 cm. 88 ff. numb., 32–35 ll. Venice, 1567.

Editions. See p. 242. This differs from the first (1546) edition only in the fact that the part relating to geometry contains considerable additional matter.

See p. 242.

JACQUES PELETIER. Ed. pr. 1549. S. l., 1607.

Born at Mans in 1517; died at Paris in July, 1582. He became principal of a college, traveled extensively, and contributed both to literature and to elementary mathematics.

Title. 'L'Arithmeti-//qve de Iacqves // Peletier dv // Mans, // Departie en quatre liures.// Troiſieme edition, reucuë et augmentee.// Par Iean de Tovrnes.// M. DC. VII.' (P. 3.)

Description. 8°, 10.3 × 16 cm., the text being 7.7 × 13 cm. 297 pp. (6 blank, 1 unnumb.), 26 ll. S. l., 1607.

Editions. Poitiers, 1549, 4° (Graesse, Sup.); ib., 1552, 8° (Graesse says 1551); Lyons, 1554, 8°. Graesse mentions all of these, but the above title shows this edition of 1607 to be the third, possibly the third revision. I have also seen mentioned an edition of 1567, 4°, a Lyons edition of 1570, and a Latin edition at Paris in 1563 and 1578.

The work is quite practical, although it contains a considerable amount of mediæval matter. The first book treats of the fundamental operations with integers, the second of fractions, the third of roots and proportion, and the fourth of the applications of arithmetic. It contains a number of such traditional problems as the hare and hound. Peletier also wrote a chapter 'De fractionibus astronomicis compendium de cognoscendis per memoriam calendis,' that was published in his editions of Gemma Frisius. Graesse mentions an 'Arithmeticae modus,' Paris, 1563, 8°, probably the Paris edition referred to above. Peletier also wrote one of the first practical textbooks on algebra.

JOHANN SCHEUBEL. Ed. pr. 1549. Basel, 1549.

See p. 233.

Title. The title page is missing in this copy, but the running headline is 'Compendium Arithmeticæ.' (See the 1560 edition.)

Colophon. 'Basileæ, per Iacobvm // Parcvm, expensis // Ioannis Oporini,// Anno 1549.' (F. 87, r.)

Description. Sm. 8°, 8.7 × 13.7 cm., the text being 6.3 × 11.8 cm. 87 ff. unnumb., 27 ll. Basel, 1549.

Editions. Basel, 1549, sm. 8° (here described); ib., 1560, 8° (below). That this is the first edition is seen in the 'Epiſtola Dedicatoria,' which bears date 'Tubingæ, idibus Martij // annni ſeſqui milleſimi // quadrageſimi noni.' (P. 7.)

While open to some of the criticism mentioned in connection with Scheubel's 'De numeris' (p. 233, 1545), this book is more practical than his earlier one, and was enough in demand to warrant two editions. It is not, however, a commercial textbook.

JOHANN SCHEUBEL. Ed. pr. 1549. Basel, 1560.

See p. 233.

Title. 'Compen-//divm Arithme-//ticæ Artis, vt bre-//uiſsum ita longè utiliſsimum eru//diendis tyronibus, non ſolùm pro-//pter ordinem, quo paucis perſtrin-//guntur omnia huius artis capita: ſed // etiam cauſa perſpicuitatis, quæ plu-//rimùm delectat & iuuat diſcentes,// ſummoperè expetēdum: per Ioan-// nem Scheubeliū adornatum // & conſcriptum.// Iam denuò ab ipſo autore recognitum // & emandatum.// Continent autem utrunqȝ hoc Compendiū,// numerorum ſcilicet & calculorum, ſeu // proiectilium (ut uocant) ra-//tiocincationem.// Baſilæ, anno 1560.' (P. 1.)

Colophon. 'Basiliæ // excudebat Iacobus Parcus,// expenſis Ioannis Opo-//rini, anno M.D.LX.// menſe Martio.' (P. 205.)

Description. 8°, 9.5 × 15 cm., the text being 6.1 × 12 cm. 14 pp. unnumb. + 191 numb. (3–193) + 1 blank = 206 pp., 23–24 ll. Basel, 1560.

See above.

JOHANN FISCHER (Piscator).

Ed. pr. 1545. Stettin, s. a. (1565 ?).

A German Rechenmeister of the second half of the sixteenth century.

Title. 'Ein kurtz Rechenbůchlein fůr die anfahende Schůler gemacht //Durch Johann Fiſcher. Gedruckt zu Alten Stettin// in Johan Eichorns Druckerey.' (F. 1, r.)

Description. 8°, 9.8 × 15.4 cm., the text being 6.7 × 11.2 cm. 16 ff. unnumb., Stettin, s. a. (1565 ?).

Editions. This work appeared first in Latin under the title 'Arithmeticae Compendium, pro Studiosis hujus artis tyronibus recognitum,' Leipzig, 1549, with subsequent Latin editions, ib., 1554, 1559, 1582, 1592, 1598, and Wittenberg, 1545, ib., 1592, all 8° The editions of the German translation were: Stettin, s. a. (1565 ?), 8° (here described); Frankfort an der Oder, 1566; Leipzig, 1581, 8°; 1592, 8°. Fischer also published a work, said to be different from the 'Compendium,' entitled 'Ein künstlich Rechenbüchlein,' Wittenberg, 1559, with four editions from 1559 to 1592. 528

As the title suggests, this is merely a compendium, designed to serve as an introduction to practical arithmetic. It has no more merit than any brief primer.

JUAN DE YCIAR, Vizcayno.

Ed. pr. 1549. Saragossa, 1549.

ICIAR. A Basque arithmetician, born at Durango in 1525. On f. 3, r., is a large portrait with the inscription 'Ioannes de Yciar ætatis sve anno xxv.' He lived in Saragossa, and was well known as a calligrapher.

Title. See Fig. 119.

Colophon. '¶Fue impreſſo el preſente libro en la muy noble //y leal ciudad // d'çaragoça en caſe de Pedro Bernuz/a coſta //del auctor y de Miguel de çapila mercader d'libros,// Acabo ſe a .xvj. de Febrero del año de mil y // quinientos y quarenta y //nueue.' (F. 61, v.)

Description. Fol., 19 × 28.9 cm., the text being 14.3 × 21.5 cm. 4 ff. unnumb. + 56 numb. + 3 blank = 63 ff., 27–38 ll. Saragossa, 1549.

Libro intitulado Arithmetica practica/ muy util y prouechoso para toda persona que quisiere exercitar se en aprender a contar/ agora nueuamente hecho por Juan de Yciar Vizcayno.

1 2 3 4 5 6 7 8 9 0

Cæsaraugustæ. M.D.XLIX.

ANNO 1548

FIG. 119. TITLE PAGE OF YCIAR

Editions. Saragossa, 1549, fol. (here described); ib. 1555, 4°; ib., 1564, 4°. The book probably had several other editions, for Heredia (I, 154) says that this is 'une des plus anciennes éditions de ce traité.'

There are numerous interesting features in this book. Among these is Yciar's fanciful explanation of the origin of the Roman numerals, part of it traditional, as that V stands for five because it was the fifth Latin vowel, and part more recent, as that L was half of the old form for C. U is used instead of M, as with several Spanish writers, and *cuento* is used for million as was their general custom. (See p. 60, Ciruelo.) The treatment of the fundamental operations is followed by progressions, compound numbers, roots, mensuration, and such common applications as 'las compañias ſin tiempo' and 'con tiempo,' and 'las reglas de teſtamentos.' [528]

Other works of 1549. Anianus, p. 32, 1488; Anonymous, p. 213, 1542; Boethius, p. 27, 1488; Euclid, p. 237, 1545; Gemma, p. 202, 1540; Glareanus, p. 192, 1539; Köbel, p. 102, 1514; Recorde, p. 214, c. 1542; Hans Bock, 'Ein new Rechenbüchlein auff der Linien und Federn,' Nürnberg (probably a second edition of the 1544 book, p. 229); Joannes Stigelius, 'Arithmetica,' Leipzig, 8°, with a second edition, s. l., 1554 (Victorinus Strigelius of 1563? see p. 311); Juan Vejar, 'Arithmetica practica,' Saragossa, 4°. (Yciar's work?) [528]

VALENTIN MENHER de Kempten.

Ed. pr. 1550. Antwerp, 1565.

MENNHER. A German-Dutch arithmetician of the sixteenth century. See also p. 281, 1556.

Title. 'Practicqve // pour brievement // apprendre à Ciffrer, & tenir Liure // de Comptes, auec la Regle de // Coſs, & Geometrie. // Par M. V. Menher Alleman. // (Woodcut of counting house.) A Anvers, l'an M. D. LXV. // Auec priuilege du Roy pour 4 ans.' (F. 1, r.) Bound with this in the third part is 'La Regle d'Algebra,' or 'Coſs.' Also the 'Practicqve // des Triangles // Spheriqves. // . . . Anvers . . . M. D. LXIIII.'

Colophon. 'Imprimé en Anuers par Ægidius Dieſt, // l'An de noſtre Seigneur Ieſu Chriſt. // M. D. LXV. // 19. Ianuarij.' (F. 113, v.)

Description. 8°, 9.6 × 15.6 cm., the text being 7.3 × 12.6 cm. 113 ff., 26 ll. The algebra contains 120 ff. unnumb. (part 3); the geometry, 102 ff. (part 4), besides the 'Practique des triangles sphériques.' Antwerp, 1565.

Editions. Menher wrote three or four arithmetics, as follows:

1. 'Practique briesve pour cyfrer et tenir Livres de compte,' Antwerp, 1550, 8°; 1556 (probably the one mentioned on p. 281); Antwerp, 1565, 8° (here described). Unlike the De Morgan copy this does not have 1564 for 1565 in the colophon. Indeed I think De Morgan probably looked at the colophon of the geometry instead of the arithmetic.

2. 'Arithmetique seconde,' Antwerp, 1556 (p. 281). A comparison of this with the 'Practique' shows it to be substantially the same work. In his epistle to the reader, Menher speaks of the 1565 edition of the 'Practique' as merely a revision of 'noſtre ſeconde Arithmetique de l'an M. D. LVI.'

3. 'Livre d'Arithmetiqve,' Antwerp, 8°, 1573 (p. 347). I know of no other sixteenth-century edition of this work, although there was a Rotterdam edition in 1609, 8°. The work is entirely different from the 'Practique.' 529

4. 'Arithmetica Practice,' Autorff, 1560, 8°. I know nothing of this work.

The 'Practique' and the 'Arithmetiqve seconde,' essentially the same work, are mercantile textbooks, possessed of the spirit of the 'Livre d'Arithmetiqve' (p. 347), but not as successfully written. Menher was one of the pioneers among the Dutch arithmeticians, and his successors, particularly in the period from 1600 to 1650, produced some very practical textbooks.

ADAM RIESE. Ed. pr. 1550. Leipzig, 1550.

See p. 138.

Title. See Fig. 120.

Colophon. 'Gedruckt zu Leipzig durch // Jacobum Berwalt.' (F. 196, r.)

Description. 4°, 15.5 × 17.8 cm., the text being 10.6 × 14.4 cm. 4 ff. unnumb. + 196 numb. = 200 ff., 29–31 ll. Leipzig, 1550.

Rechenung nach der lenge/ auff den Linihen vnd Feder.

Darzu forteil vnd behendigkeit durch die Proportiones/Practica genant/ Mit grüntlichem vnterricht des visierens.

Durch Adam Riesen.

im 1550. Jar.

Cum gratia & priuilegio Cæsareo.

FIG. 120. TITLE PAGE OF THE 1550 RIESE

Editions. See p. 140. This is the fourth of Riese's books (p. 139). Bound with it is Isaac Riese's arithmetic of 1580 (p. 365). The date is also given in the dedicatory epistle, 'im 1550 jhar.'

The first forty six folios contain the treatise 'auff den Linihen,' the counter reckoning. This is followed (ff. 47–105) by that 'auff der Feder,' the common algorism. The third part is the 'Practica,' and the fourth the 'Vifieren' or gauging. The book represents the culmination of Riese's work, and is the best exponent of the practical arithmetic of the middle of the century in Germany.

Other works of 1550. Agricola, p. 171, 1533; Agrippa, p. 167, 1531; Anianus, p. 32, 1488; Borghi, p. 22, 1484; Cassiodorus, p. 211, 1540; Feliciano, p. 149, 1526; Gemma, p. 200, 1540; Glareanus, p. 192, 1539; Lonicerus, p. 253, 1551; Medlerus, p. 223, 1543; Regius, p. 181, 1536; Riese, p. 139, 1522; Tagliente, p. 114, 1515; Torrentini, p. 76, 1501. There was also an edition of Sfortunati, c. 1550 (p. 174, 1534). Two other works published c. 1550 should be mentioned: Anonymous, 'Opera che insegna a tener conto de libro secondo lo cōsueto di tutti li lochi della Italia al modo mercantile,' s. l. a., with some mercantile arithmetic; William Buckley, 'Arithmetica memorativa sive compendaria Arithmeticæ tractatio,' 8°, s. l. a., but later in Seton's Logic, London, 1572, 1574, 1577, 1584, 8°. 529

JOHANN SCHEUBEL. Ed. pr. 1550. Paris, 1551.

See p. 233.

Title. 'Algebrae//compendiosa//facilisqve descri-//ptio, qua depromuntur magna // Arithmetices miracula.// Authore Ioanne Scheubelio Mathematicarum//profeffore in academia Tubingenfi. // Parisiis, // Apud Gulielmum Cauellat, in Pingui Gallina,// ex aduerfo Collegii Cameracenfis.//1551.//Cvm privilegio." (F. 1, r.)

Colophon. 'Excudebat Lutetiæ Parifiorum, Benedictus Preuotius Typo-//graphus, in vico Frementello, fub infigni ftellæ aureæ. //1551.' (F. numb. 52, v.)

Description. 4°, 13.2 × 18.5 cm., the text being 9.4 × 15.2 cm. 52 ff. numb., 32–37 ll. Paris, 1551.

Editions. There was no other edition. 529

I have included this algebra because it contains some work in the extracting of roots by the galley method, and therefore shows the persistence of this mediæval plan.

ROBERT RECORDE. Ed. pr. 1551. London, 1596.

See p. 213.

Title. 'The Castle // of // Knowledge.// To Knowledge is this Castle ſet,// All Learnings friends wil it ſupport,// So ſhall their name great honour get,// And gaine great fame with good report. // Though ſpitefull Fortune turn'd her wheele,// To ſtay the Sphere of Vranie,// Yet doth the ſphere reſiſt that wheele,// And flee'th all Fortunes villanie // Though earth do honour Fortunes ball,// And beetles blinde her wheele aduance, // The heauens to Fortune are not thrall,// The ſpheres ſurmount all Fortunes chance.// London // printed by Valentine Sims, aſsigned // by Bonham Norton.// 1596.' (P. 1.)

Colophon. 'Imprinted at London by Valentine // Simmes. 1596.' (P. 236.)

Description. 4°, 14.4 × 19.2 cm., the text being 12.5 × 17.5 cm. 3 pp. unnumb. + 1 blank + 232 numb. = 236 pp., 41 ll. London, 1596.

Editions. London, 1551; ib., 1556, fol.; ib., 1596, 4° (here described).

Recorde's name does not appear on the title page, but he signs the letter of dedication to 'Princesse Marie,' 'Robert Record Phyſicion.' The work is on astronomy, and is of interest in the history of arithmetic only in the operations involving sexagesimal fractions. Division is performed by the galley method, and there are no symbols for degrees, minutes, and seconds. The arithmetical part includes the rule of three. Like the author's other works (pp. 213, 286), this is in the catechism form.

Other works of 1551. Borghi, p. 16, 1484; Euclid, p. 238, 1545; Gemma, p. 200, 1540; Glareanus, p. 193, 1539; Noviomagus, p. 197, 1539; Peletier, p. 245, 1549; Recorde, p. 214, c. 1542; Tonstall, p. 134, 1522; Adam Lonicerus, 'Arithmeticae brevis Introductio,' Frankfort, 8°, with subsequent editions, ib., 1568, 8°; 1570, 12°; 1581; 1585; 1600, 8° (Tropfke puts the first edition as 1550); Andrés García de Lovas, 'Tratado del cómputo,' Salamanca, 8°; Innocenzo Ringhieri, 'Centi givochi liberali . . . in dieci libri descritti,' Bologna, 4°, with subsequent editions, Venice, 1553, 4°; Bologna, 1580; Lyons (French translation), 1555, 4° (contains some number games).

MARCO AUREL. Ed. pr. 1552. Valencia, 1552.

Aurel was, as he states, a German. He lived, however, for several years in Valencia, and published a work there in 1541.

Title. See Fig. 121.

Description. 4°, 14.5 × 19.3 cm., the text being 9.9 × 16.6 cm. 4 ff. unnumb. + 140 numb. = 144 ff., 31–36 ll. Valencia, 1552.

Editions. There was no other edition.

Aurel, in his letter to the reader, dwells upon the unfortunate state of mathematics in Spain, and says that he feels called upon to assist in making known a science so necessary to humanity. Of the twenty-four chapters in the book, the first six may be said to relate to arithmetic as we ordinarily consider it, the rest referring entirely to algebra. The arithmetical chapters present the subject in a fairly practical way, but are deficient in genuine problems. To subtract is called 'Restar,' as at present in Spanish, the same root appearing occasionally in other languages, and our word 'rest' (for remainder) being a relic of this name. Division is performed entirely by the galley method, but the figures are not canceled as is generally the case. 'Proporcion' is used for ratio, and 'proporcionalidad' for proportion, as was generally the custom in the early arithmetics of all Latin countries, a custom derived from the Boethian books. The 'proporciones' (ratios) are treated at considerable length after the fashion set by the mediæval writers. The applications are almost entirely under the 'Regla de tres' (rule of three).

In the part devoted to algebra, surd numbers are first treated, the root symbols showing the German influence. The plus and minus signs are also used as extensively as in the works of writers like Stifel and Scheubel, and the symbols for the various powers of the unknown quantity are such as are found in the works of contemporary writers in other countries.

LILIUS GREGORIUS GYRALDUS.

Ed. pr. 1552. Venice, 1553.

A philosopher of Ferrara, of the middle of the sixteenth century.

Title. See Fig. 122.

Description. 8°, 9.3 × 15.5 cm., the text being 6.1 × 12.4 cm. 184 pp. (2 blank, 16 unnumb.), 29 ll. Venice, 1553.

Editions. Venice, 1552, 8°; ib., 1553, 8° (here described). This treatise also appeared in his 'Opera,' Basel, 1580, sm. fol., and Leyden, 1696.

LIBRO PRIMERO, DE ARITHMETICA Algebratica, enel qual ſe contiene el arte Mercantiuol, con otras muchas Reglas del arte menor, y la Regla del Algebra, vulgarmente llamada Arte mayor, o Regla de la coſa: ſin la qual no ſe podra entender el decimo de Euclides, ni otros muchos primores, aſsi en Arithmetica como en Geometria: compueſto, ordenado, y hecho Imprimir por Marco Aurel, natural Aleman: Intitulado, Deſpertador de ingenios. Va dirigido al muy magnifico ſeñor moſſen Bernardo Cimon, Ciudadano dela muy inſigne y coronada Ciudad de Valencia.

de Hieronymo de Nogeras Frco. gilberte

Con Priuilegio de ſu Mageſtad, por tiempo de diez años.

EN VALENCIA,
En caſa de Ioan de Mey, Flandro.
Año. 1552.

FIG. 121. TITLE PAGE OF AUREL

LILII GREGORII
GYRALDI FERRARIEN. SVA=
RVM QVARVNDAM ANNO=
tationum Dialogiſmi XXX. *ad Am=*
pliß. Card. *Saluiatum.*

Item *Laurentij Frizzolij Solianenſis Dialogiſmus unicus*
de ipſius Lilij uita & operibus.

VENETIIS, *Apud Gualterum Scottum.*

M D L I I I.

FIG. 122. TITLE PAGE OF THE 1553 GYRALDUS

This set of dialogues is of interest in the history of mathematics in that the second and third parts deal with notation and finger symbolism. 'Dialogismvs secvndvs de manus & digitorum nominibus déq; numerandi per eos antiquorum ratione' (p. 10) is an almost unknown sketch of finger symbolism. 'Dialogismvs tertivs ad Baptistam Lucarinum FR. filium optimæ ſpei ac indolis puerum, de notis & finguris numerorum, quibus Latini ac Græci utebantur' (p. 20) is an equally interesting sketch of the Greek and Latin numerals. Gyraldus also published a 'Brevis instructio de Grecorꝝ numerali supputatione' in 1513, fol.

Other works of 1552. Agricola, p. 171, 1533; Gemma, p. 203, 1540; Ghaligai, p. 132, 1521; Herman Gülfferich, 'Ein new kurtz Rechenbüchlein,' Frankfort, 8°, with editions ib., 1555, 12°, 1568, 8° (see also p. 244, 1546, and p. 292, 1559); Ortega, p. 94, 1512; Peletier, p. 245, 1549; Recorde, p. 214, c. 1542; Riese, p. 139, 1522; Joachim Camerarius, 'Arithmologia,' 12° (there was also an edition at Basel, s. a., 16°); Dunkel, 'Arithmetica,' Leipzig, 8°; Christopher Falconius, 'Rechenbuch,' Königsberg in Preußen, 4° (Murhard cites as a different work of the same date 'Rechenbuch auff die Preusche müntz mass und Gewicht,' ib., 4°). 530

DOMENICO MANZONI. Ed. pr. 1553. S. l. (Venice), 1553.

A sixteenth-century arithmetician, born at Oderzo.

Title. See Fig. 123.

Description. 8°, 10.5 × 15.8 cm., the text being 8.1 × 13.4 cm. 16 ff. unnumb., 26 ll. except where arranged in sections. S. l. (Venice), 1553.

Editions. There was no other edition of this work. Manzoni had already published a textbook entitled 'Libretto molto utile per imparar a leggere, scrivere et Abaco, con alcuni Fondamenti della Dottrina Christiana,' Venice, 1546, 8°, 31 pp. of which are devoted to arithmetic. He also published in 1553 a more elaborate treatise, 'La Brieve Risolvtione di Aritmetica universale in qualsi voglia negotio, doue interuenga numero, peso, & misura,' Venice, 1553, 8°, 246 ff.

A beautifully printed little manual of the fundamental operations of arithmetic. The 'Abbreviatvre delle monete, Peſi, & miſure, chi ſi uſano in Vinegia' (f. 2) is helpful to students of the history of sixteenth-century arithmetic and to all who are interested in the metrology of the countries with which Venice traded.

ABACHETTO NVOVO
MOLTO COPIOSO, ET ARTI-
ficioſamente ordinato à modo
di tariffa d'alcune moltipli-
cationi & partimenti di
numeri rappreſentanti
diuerſe nature, &
qualità di mo-
nete, peſi, et
miſure.

OPERETTA A CIASCVNO
Vtiliſſima, & uniuerſalmente in
tutte le Città & Prouintie
molto neceſſaria.

Di Dominico manzoni.

MDLIII.

FIG. 123. TITLE PAGE OF MANZONI

MICHAEL STIFEL. Ed. pr. 1553. Königsberg i. Pr., 1553.

See p. 223.

Title. See Fig. 124.

Colophon. 'Gedrůckt zu Ko̊nigs-//berg in Preuſsē durch Alexandrum // Behm von Luthomiſl/ Voll // endet am dritten tag defs Herbft-//monats/ Als mann zalt nach // der geburt vnfers lieben // herrn Jefu Chrifti.// 1554.' (P. 505.)

Die Coss
Christoffs Rudolffs
Die schönen Exempeln der Coss
Durch
Michael Stifel
Gebessert vnd sehr gemehrt.

Den Innhalt des gantzen Buchs
such nach der Vorred.

Zu Königsperg in Preussen
Gedrückt/ durch Alexandrum
Lutomyslensem im jar.

1553.

FIG. 124. TITLE PAGE OF STIFEL'S RUDOLFF

Description. 4°, 14.7 × 19.6 cm., the text being 9.9 × 16 cm. 507 pp. (4 blank, 12 unnumb.), 21–26 ll. Königsberg in Preußen, 1553.

Editions. Königsberg in Preußen, 1553, 4° (here described); 1571. See also p. 226.

Although the book is nominally an algebra (the word 'cofs' coming from the Italian *cosa*, 'thing,' meaning the unknown quantity), the first part of the book is devoted entirely to arithmetic. This part is intended as an introduction to the algebra that comes later, and gives the fundamental operations as treated by Rudolff, with Stifel's commentary. It is entirely theoretical, and it naturally leads into the theory of irrational numbers, which constitutes the first part of the algebra.

Other works of 1553. Albert, p. 180, 1534; Boethius, p. 27, 1488; Gemma, p. 204, 1540; Gyraldus, p. 254, 1552; Mariani, p. 181, 1535; Morsianus, p. 182, 1536; Ringhieri, p. 253, 1551; Rudolff, p. 152, 1526; Bernardu Wojewódki, 'Algorithm, to jest nauka liczby, po polsku na linijoch uczyniony,' Cracow. **530**

CLAUDE DE BOISSIÈRE. Ed. pr. 1554. Paris, 1554.

CLAUDIUS BUXERIUS. Born in the diocese of Grenoble, probably c. 1500. He also wrote on poetry, music, and astronomy. For his *Rythmomachia* see p. 271.

Title. See Fig. 125.

Colophon. 'Acheué d'Imprimer le xiij. // iour d'Octobre,// 1554.' (F. 74, 1.)

Description. 4°, 10 × 14.8 cm., the text being 7.2 × 12.9 cm. 2 ff. unnumb. + 73 numb. = 75 ff., 32 ll. Paris, 1554.

Editions. Paris, 1554, 8° (here described); ib., 1563, 8° (p. 262).

This is a theoretical work in two 'livres,' probably written for students in the University of Paris. It is the second printed work which I have noticed that carries the system of numeration as high as thousands of quintillions, 'Mille de Quintillions,' although Chuquet, in his manuscript of 1484, carries it to 'nonyllions.' De la Roche (p. 128) followed Chuquet in this as in other respects. Boissière's plan starts with the names of millions, bimillions (for million millions), trimillions, etc. He then says that to avoid ambiguity, as of bimillions for two millions, these names are abridged to billion, trillion, etc. He moreover numerates in periods of six figures each, as in England at present, and not in periods of three figures as is now the custom in France and America.

He first treats of the classes of number according to the old plan, distinguishing between digits (1–9), articles (multiples of 10), and composites (articles + digits), of which names we still use the digit. Instead of extending the fundamental operations to six, or even nine, as his contemporaries so often did, he limits them to four, as we do. His work in addition and multiplication is substantially like ours, but his 'substraction' is as follows (1563 edition):

Reſte	939973901	C.
Debte ou ſomme	1840006503	A.
Paye ou à Subſtraire	900032602	B.

L'ART
D'ARYTHME-
TIQVE CONTENANT
TOVTE DIMENTION, TRES-
SINGVLIER ET COMMODE,
tant pour l'art militaire que
autres calculations.

Auec priuilege
du Roy.

Imprimé à Paris, par Annet Briere, à l'enſeigne ſainct Sebaſtian, rue des Porées.

1554.

FIG. 125. TITLE PAGE OF BOISSIÈRE

In division he uses the galley method only. Boissière is one of the first writers, I believe the first in France, to invert the divisor in the division of fractions, as Stifel had done a few years before in Germany. Book I closes with a somewhat extended treatment of the rule of three.

The second book relates largely to mensuration, figurate numbers, roots, and the mediæval proportions.

Some of the rules are in verse. Of military matters, referred to in the title page, there is little mention. 530

CLAUDE DE BOISSIÈRE. Ed. pr. 1554. Paris, 1563.

See p. 260.

Title. This is practically identical with that of the 1554 edition, except for the following: 'Reueu & augmenté par Lucas Trembley Parifien,// profeffeur des Mathematiques.// A Paris,// Pour Guillaume Cauellat, à l'enfeigne de la Poulle//graffe, deuant le college de Cambray.// 1563.' (F. 1, r.)

Description. 8°, 10.2 × 16.8 cm., the text being 6.8 × 13 cm. 3 ff. blank + 1 unnumb. + 71 numb. = 75 ff., 25–30 ll. Paris, 1563.

See p. 260. Two other books are bound with this: 'La Declara// tion et Vsage de // L'inftrument nommé Canomettre,// Par G. des Bordes, Gentilhō-// me bordelois, profeffeur ez // Mathematiques.// . . . 1570.' and 'Vsage // Dv // Compas // de // Proportion.// Par D. Henrion, Mathem.// . . . M. DC. XVIII.'

JOACHIM CAMERARIUS. Ed. pr. 1554. Deventer, 1667.

Born at Bamberg, April 12, 1500; died at Leipzig, April 17, 1574. The office of chamberlain (Kammermeister) to the Prince-Bishop of Bamberg being hereditary in the family of Liebhard, he took the Latin name of Camerarius. He was a distinguished classicist, a friend of Melanchthon, and a professor at Tübingen and Leipzig. Of the various commentaries on Nicomachus, his was the most important of the Renaissance.

Title. 'Explicatio // Ioachimi Camerarii // Papebergenfis// in dvos libros //Nicomachi Geraseni // Pythagorei // Deductionis // Ad Scientiam Numerorum.// Et Notæ // Samuelis Tennulii // in //Arithmeticam // Jamblichi Chalcidensis.// Daventriæ.// Typis Wilhelmi Wier, CIↃ IↃCLXVII.' (P. 1.)

Description. 4°, 14.7 × 19.8 cm., the text being 10.3 × 14 cm. 2 pp. unnumb. + 239 numb. = 241 pp., 34 ll. Bound with the commentary of Iamblichus on Nicomachus (p. 188, 1538). Deventer, 1667.

Editions. The first edition was Augsburg, 1554, 8°. There was also an edition published in 1569.

Camerarius also wrote a work entitled 'De logistica,' published at Augsburg in 1554; ib., 1557, 8°; Leipzig (?), 1569, 8°. See also p. 257, 1552. 530

A commentary on the theoretical work of Nicomachus (p. 186).

Other works of 1554. Albert, p. 180, 1534; Bæda, p. 131, 1521; Buteo, p. 292, 1559; Euclid, p. 238, 1545; Finaeus, p. 160, 1530–32; Fischer (Piscator) p. 247, 1549; Glareanus, p. 192, 1539; Huswirt, p. 74, 1501; Nicomachus, p. 186, 1538; Peletier, p. 245, 1549; Psellus, p. 168, 1532; Riese, p. 139, 1522; Stigelius, p. 249, 1549; Tagliente, p. 114, 1515; Barth. Barchi, 'Tariffe della valuta di tutte le monete,' Mantua, 4° (hardly an arithmetic); Claudio Bertholio, 'De numerandi ratione aphorismi,' Paris, 8°; Ian Gentil (Vander Schuere, in his 1634 edition, f. 201 — see p. 424 — refers to his arithmetic as published at Paris, 1554); Caspar Hützler, 'Eyn behende und Kunstrik Rekensboeck op. der Linien und Tziferen,' Lübeck, 8°.

PETRUS RAMUS. Ed. pr. 1555. Paris, 1555.

PIERRE DE LA RAMÉE. Born at Cuth, Vermandois, 1515; killed at Paris the night of August 24–25, 1572, in the Massacre of St. Bartholomew. He lectured on philosophy at Paris, and for a short time at Heidelberg. He wrote on arithmetic, geometry, optics, and mathematics in general.

Title. See Fig. 126.

Colophon. 'P. Rami Eloqventiæ et Phi-//losophiæ professoris //Regij Arithmeticæ,// Finis.' (P. 111.)

Description. 4°, 13.8 × 19.1 cm., the text being 9.5 × 17.2 cm. 128 pp. (2 blank, 16 unnumb.), 26–27 ll. Paris, 1555.

Editions. Paris, 1555, 4° (here described); ib., 1557, 8°; Basel, 1567; Paris, 1584. See also Gleitsman, 1600 (p. 427), and the 'Libri Duo' of Ramus, 1569 (p. 330).

This arithmetic was popular in the Latin schools for half a century. It is theoretical, consisting largely of definitions, extracts from the Greek writers, a little work on the fundamental operations, and the mediæval theory of ratios. Ramus had not the faculty of putting together a textbook that should be a rival to that of Gemma Frisius.

P. Rami, eloquentiæ
ET PHILOSOPHIÆ PROFES-
SORIS REGII, ARITH-
meticæ libri tres,

Ad

Carolum Lotharingum Cardinalem.

PARISIIS,
Apud Andream Wechelum, ſub Pegaſo, in
vico Bellouaco, *Anno Salutis,*
1555.

Cum priuilegio Regis.

FIG. 126. TITLE PAGE OF RAMUS

JACOBUS MICYLLUS. Ed. pr. 1555. Basel, 1555.

MOLTZER. Born at Strasburg, April 6, 1503; died at Heidelberg (?), January 28, 1558. He was a well-known classical scholar.

Title. See Fig. 127.

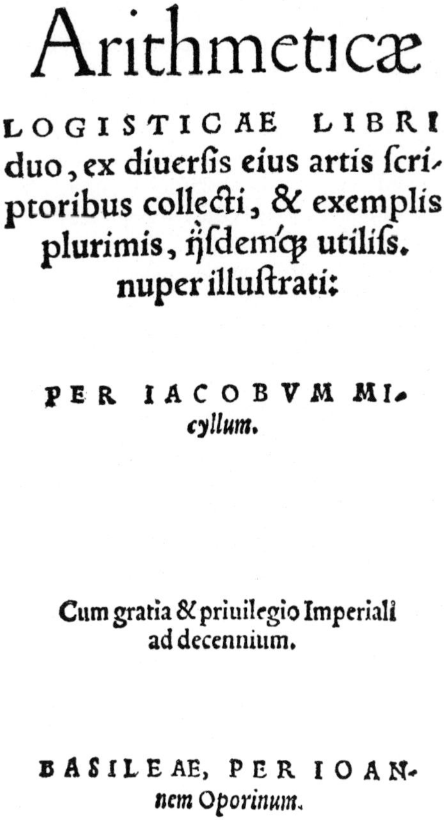

Arithmeticæ

LOGISTICAE LIBRI duo, ex diuerſis eius artis ſcriptoribus collecti, & exemplis plurimis, ijſdemq; utiliſs. nuper illuſtrati:

PER IACOBVM MIcyllum.

Cum gratia & priuilegio Imperiali ad decennium.

BASILEAE, PER IOANnem Oporinum.

FIG. 127. TITLE PAGE OF MICYLLUS

Colophon. 'Basileæ, ex officina // Ioannis Oporini, Anno Salutis humanæ // M. D. LV. Menſe Martio.' (P. 319.)

Description. 8°, 9.3 × 15.2 cm., the text being 7.8 × 11.3 cm. 320 pp. (3 blank, 23 unnumb.), 25–27 ll. Basel, 1555.

Editions. There was no other edition. Treutlein's statement (*Abhandlungen*, I, 15) that the book was published in Heidelberg is incorrect. It probably came from the fact that the 'Epistola Nvncvpatoria' is dated 'Heidelbergæ, 10 Calend. Nouemb. Anno Domini 1553.'

The book was written for the classical schools. Although Micyllus gives the fundamental operations in a practical manner, the latter part of his book is theoretical, presenting some of the ancient arithmetic in the Greek language. The work includes an unusually complete treatment of sexagesimals, 'De partibvs Aſtronomicis, & earum ſupputatione' (p. 201), and a chapter on the computus (see p. 7), 'De Temporvm svppvtatione, qva Eccleſiaſtici utuntur.' The latter shows that the name 'Computus' had become unpopular, for it begins with the following statement: 'Est & temporum ſupputatio quædam, quem Computum Eccleſiaſticum, barbaro nomine, uocant.' (P. 269.) The following is an example of his applied problems: 'Scribit Plinius, Alexandriam à Rhodo diſtare 583000 paſsuum. Cupio autem ſcire, quot ſtadia ijdem paſsus efficiant. Diuido igitur 583000 per 125.'

MICHAEL NEANDER. Ed. pr. 1555. Basel, 1555.

Born in the Joachimsthal, April 3, 1529; died at Jena, October 23, 1581. He was professor of mathematics, Greek, and medicine, in the university of Jena. He wrote on physics and cosmography.

Title. See Fig. 128.

Colophon. 'Basileæ, ex Officina // Ioannis Oporini, Anno // Salutis humanæ M. D. LV. Men-//ſe Ianuario.' (P. 119.)

Description. 4°, 13.9 × 19.5 cm., the text being 9.1 × 15.7 cm. 104 pp. numb. + 15 unnumb. + 1 blank = 120 pp., 25–29 ll. Basel, 1555.

Editions. There was no other edition. The 'Epistola Nvncvpatoria' is dated 'Die Paſchalis, Anno LIIII,' but the book was printed, as the colophon shows, in 1555.

This is a historical treatise on Greek, Egyptian, Roman, Arabic, and mediæval European weights and measures, and is particularly interesting because of the symbolism which it contains. The origin of our present apothecary's symbols of measure is seen in the ancient Roman abbreviations. The text is mostly in Latin, but eight pages of 'Γαληνου περι Μετρων καὶ Σταθμῶν' are in Greek.

ΣΥΝΟΨΙΣ

MENSVRARVM ET PONDERVM, PONDERATIONISque mensurabilium secundum Romanos, Athenienses, γεωργοὺς, καὶ ἱπποιάτρους, ex præstantissimis authoribus huius generis cōtracta, opera MICHAELIS NEANDRI ex Valle Ioachimica,

Anno M D LIIII.

ACCESSERVNT ETIAM, QVAE APVD *Galenum hactenus extabant de ponderum & mensurarum ratione uehementer deprauata, nunc Græcè & Latinè multo correctiora, eiusdem* MICHAELIS NEANDRI *opera.*

Item rerum & uerborum in his omnibus memorabilium Index.

LEVITICI XIX.

Μέτρα δίκαια, καὶ χοῦς δίκαιος ἔσται ἐν ὑμῖν.

BASILEAE, PER IOANnem *Oporinum.*

Fig. 128. Title page of Neander

ANTOINE CATHALAN. Ed. pr. 1555. Paris, 1556.

A French arithmetician. The author's name does not appear in this edition. A work with the same title is assigned to Cathalan, Lyons, 1555.

Title. See Fig. 129.

L'ARITHMETIQVE

ET MANIERE DE APPRENDRE
a Chiffrer & compter par la plume & par les gestz en nombre entier & rompu, facile a apprendre, & tresutile a toutes gens.

De nouueau reueue & corrigee.

A laquelle sont adioustees plusieurs questions & exemples pour faire la science plus facile, & plus legiere a comprendre.

A PARIS.

Par Iehan Ruelle, demourant en la Rue sainct Iacques, a l'enseigne de la queue de Regard.

1556.

FIG. 129. TITLE PAGE OF THE 1556 CATHALAN

Description. Small 8°, 7.3 × 11.4 cm., the text being 5.3 × 9.5 cm. 79 ff. numb. + 1 unnumb. = 80 ff., 27 ll. Paris, 1556.

Editions. Lyons, 1555, 16°; Paris, 1556, sm. 8° (here described); 1559.

This is a very good little primer of algorism for the time. Concerning this work the author says: 'Lequel art trouua premierement vn philofophe d'Arabie, nommé Algus. Dont cefte fcience prent fon nom d'Algorifme' (f. 3). In multiplication the gelosia arrangement is given as a third method, under the name 'Mvltiplication per quarreaux.' In division only the galley form appears. The 'premier liure' covers the four fundamental operations with integers, and a brief treatment of progressions. The 'fecond liure' relates to the use of counters ('gectz'), which the author esteems so highly as to say: 'Et note que cefte efpece de addition eft plus vtile & facile aux gectz que aux chifres;' 'il eft facile par les gectz, & difficile par les chifres.' The last part of the book is devoted to such standard problems as the testament, the pipes filling the cistern, and the broken eggs, and to applications like exchange and pasturage.

LODOICO BAËZA. Ed. pr. 1555. Paris, 1555.

A Spanish scholar of the sixteenth century.

Title. See Fig. 130.

Colophon. 'Excudebat Benedictus Preuotius, via // Frementella, ad infigne ftellæ aureæ.' (F. 68, r.)

Description. 8°, 10.3 × 15.6 cm., the text being 7.8 × 13 cm. 2 ff. unnumb. + 66 numb. = 68 ff., 26–27 ll. Paris, 1555.

Editions. This rare first edition of Baëza's arithmetic is seldom found in libraries. The second edition (1556), 8°, is, however, the same impression with the title page changed.

The book is entirely theoretical, making much of the classification of numbers and of the ancient ratio systems. It is in Latin with numerous Greek quotations.

Other works of 1555. Euclid, p. 237, 1545; Finaeus, p. 163, 1530–32; Gemma, p. 200, 1540; Glareanus, p. 192, 1539; Herman Gülfferich, p. 257, 1552; Mariani, p. 181, 1535; Ringhieri, p. 253, 1551; Valturius, p. 10, 1472; Yciar, p. 249, 1549; Jacob Cuno, 'Arithmetica,' Wittenberg, 8°; Taddeo Duni, 'Liber de arithmetica,' Basel, 4°; Melchior Goldammer, 'Arithmetica Pratica,' Wittenberg, 8°; Jerónimo de Valencia, 'Arte de computo,' in Santaella's 'Vocabularium ecclesiasticum,' a separate edition appearing at Saragossa in 1601. (See also p. 140, 1523.)

NVMERANDI DOCTRINA

PRAECLARA METHODO

exposita, in qua breuiter continentur, & exponuntur apertè ea, quæ ex vniuersa Arithmetica sunt ad vsum potiora.

Authore Lodoico Baëza.

LVTETIAE,

Apud Gulielmum Cauellat, sub pingui Gallina, ex aduerso collegij Cameracensis.

1555.

FIG. 130. TITLE PAGE OF BAËZA

CLAUDE DE BOISSIÈRE. Ed. pr. 1556. Paris, 1556.

See p. 260.

Title. See Fig. 131.

Description. 8°, 10.4 × 16.6 cm., the text being 6.7 × 13.1 cm. 1 f. unnumb. + 52 numb. + 1 blank = 54 ff., 26–27 ll. Paris, 1556.

Editions. Paris, 1556, 8° (here described); French edition, ib., 1556, 8°.

Of the three standard treatises on the ancient number game of Rythmomachia mentioned in this list, the others being the one of 1496 of uncertain authorship and Barozzi's work of 1572, this is the clearest. It describes very carefully the checkerboard on which the game is played, the nature of the *calculi* used, and the general mode of procedure. Moreover, it is profusely illustrated (Fig. 132), which adds much to the value of the book. The game was connected with the mediæval number classifications and ratios, and could never have been understood by any save those who were well educated in the ancient theoretical arithmetic.

GALLUS SPÄNLIN. Ed. pr. 1556. Nürnberg, 1566.

An Ulm Rechenmeister, as he describes himself on the title page.

Title. See Fig. 133.

Colophon. 'Gedruckt zu Nůrmberg//durch Chriſtoff // Heuſzler.' (P. 382.)

Description. 8°, 9.2 × 14.7 cm., the text being 7.1 × 11.5 cm. 15 pp. unnumb. + 4 blank + 365 numb. = 384 pp., 21–25 ll. Nürnberg, 1566.

Editions. The dedication to 'Den Edlen Ehrnveſten/ Fůrſichtigen/ Erſamen vnd Weyſen Herren/ Eltern Burgermeiſtern vnd Rath/ des heyligen Reichs Stadt Ulm' is dated '3 Julij/ Anno Chriſti/ 1556,' so that this is possibly the date of the first edition. I know of no other edition except this of 1566.

The author devotes twenty-one pages to 'Rechnung auff der Linien,' this counter reckoning being still the popular method at the time he wrote. He then (p. 24) takes up the processes with Arabic numerals, at first using a few abstract numbers, but soon, as was the custom, introducing many practical applications. On the whole the book may be said to be a rather good exponent of Riese's school.

NOBILISSI-
MVS ET ANTIQVISSI-
mus ludus Pythagoreus (qui Rythmo
machia nominatur) in vtilitatem & re-
laxationem ſtudioſorum comparatus
ad veram & facilem proprietatem &
rationem numerorum aſſequendam,
nunc tãdem per Claudium Buxerium
Delphinatem illuſtratus.

LVTETIAE,

*Apud Gulielmum Cauellat, ſub pingui Gal-
lina, ex aduerſo collegij Cameracenſis.*

*Abacus & calculi vęneunt in Palatio,
apud Ioannem Gentil.*

1556

CVM PRIVILEGIO REGIS.

FIG. 131. TITLE PAGE OF THE 1556 BOISSIÈRE

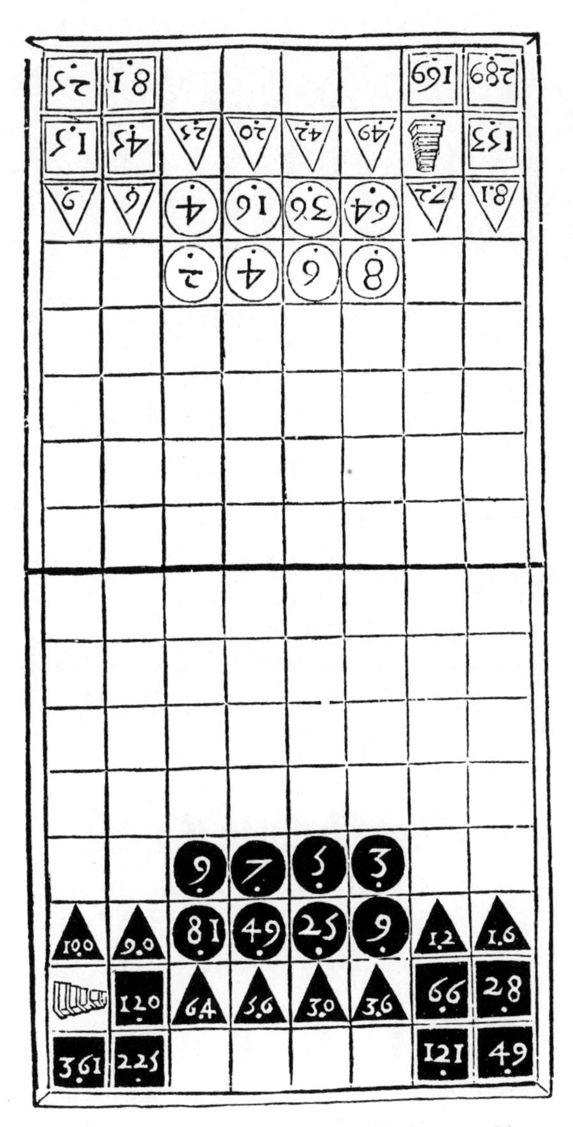

FIG. 132. FROM BOISSIÈRE'S *Rythmomachia*

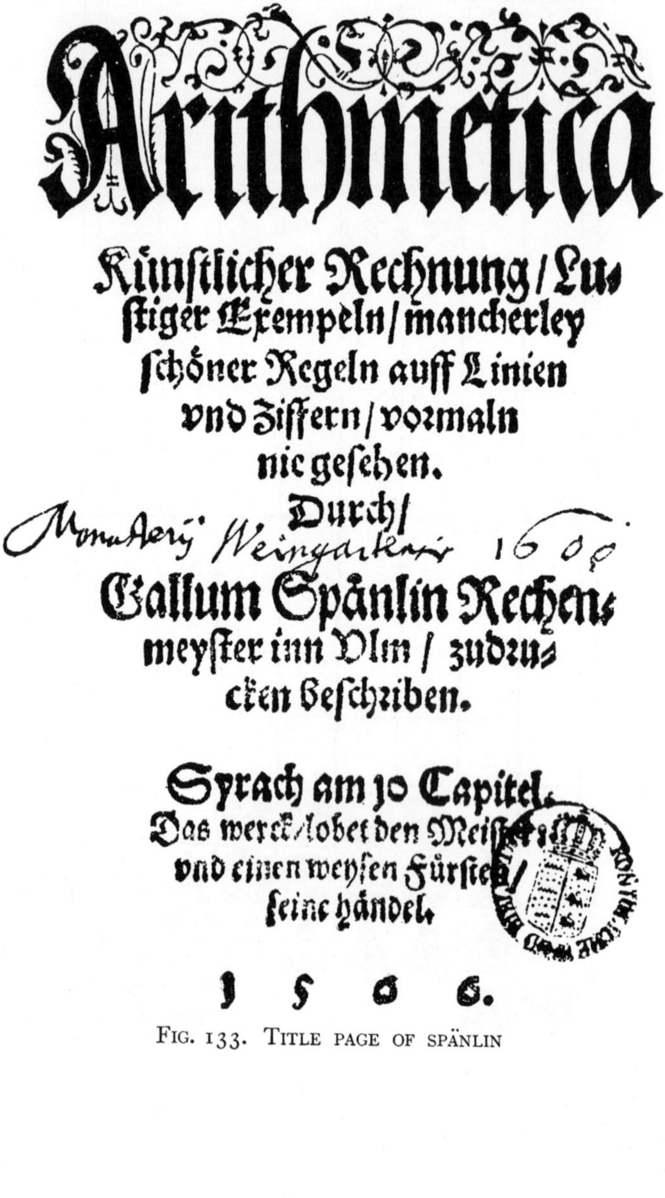

Arithmetica

Künstlicher Rechnung / Lustiger Exempeln / mancherley schöner Regeln auff Linien vnd Ziffern / vormaln nie gesehen.

Durch /

Monasterii Weingartensis 1600

Gallum Spänlin Rechenmeyster inn Vlm / zudrucken beschriben.

Syrach am 10 Capitel.
Das werck lobet den Meister / vnd einen weysen Fürsten seine händel.

1 5 6 6.

FIG. 133. TITLE PAGE OF SPÄNLIN

DOMENICO DELFINO. Ed. pr. 1556. Venice, 1565.

A sixteenth-century Venetian. He was born of a noble family, which had produced a doge, a general of the Camaldoli, and other men of prominence, and which was later to produce a cardinal and a captain-general of the Venetian fleet.

Title. See Fig. 134.

Description. 8°, 9.8 × 14.4 cm., the text being 8.5 × 12.7 cm. 416 pp. (56 unnumb.), 26 ll. Venice, 1565.

Editions. Venice, 1556, 4°; ib., 1565, 8° (here described); ib., 1568, 8°; ib., 1584, 8°, and later. The title page bears the date 1565 as stated, and the dedicatory letter 'Al Mag.co et Ecc.mo Sig. Nicolò Crasso,' signed by Ludouico Dolce, is dated 'In Venegia a VII. di Febraio. M D LXIIII,' 1565 new style. The author's preface, 'Intentione dell' avtore,' which follows, bears no date, but succeeding this there is a letter 'All'Illvstrissimo et Reverendissimo Signor, Il S. Cristoforo Madrvccio, Cardinal di Trento, Governator Meritissimo di Milano,' signed by F. Nicolo Croce, and dated 'Di Vinetia il dì primo di Settembre. M D L VI.' There is no colophon to fix the date more accurately, unless it appeared on some page after 360, with which this copy terminates.

The book, a 'summary of all the sciences,' devotes a brief chapter to the Boethian arithmetic (pp. 43–48): 'Dell' Aritmetica, de' svio // inuentori, utilità, modo, & altri ſecre // ti Cap. III.' The author is highly lauded in Croce's letter as of noble family, 'Illuſtre per lo splendor del ſangue, non dimeno uia piu per l'ornamento e per la gloria della dottrina.' The Delfino family was at that period very prominent in Venice.

NICOLÒ TARTAGLIA. Ed. pr. 1556. Venice, 1556.

TARTALEA. Born at Brescia, c. 1506; died at Venice in 1559. He was one of the best mathematicians of his time, and was the first to give a general solution of the cubic equation.

Title. See Fig. 135.

Description. Fol., 21.1 × 29.1 cm., the text being 13.8 × 23.6 cm. 285 ff. (3 blank, 6 unnumb.), 52–56 ll. Venice, 1556.

Editions. Tartaglia's works include the 'Nova Scientia,' chiefly algebraic, editions of Euclid, Archimedes, and Jordanus,

SOMMARIO

DI TVTTE LE

SCIENZE,

DEL MAGNIFICO MESSER

Domenico Delfino, nobile Vinitiano,

DAL QVALE SI POSSONO IMPARAR

molte cose appartenenti al uiuere humano,

& alla cognition di DIO.

Con la Tauola, & le Postille delle cose piu notabili.

CON PRIVILEGIO.

IN VINEGIA APPRESSO GABRIEL

GIOLITO DE' FERRARI.

M D L X V

FIG. 134. TITLE PAGE OF DELFINO

LA PRIMA PARTE DEL GENERAL TRATTATO DI NVMERI, ET MISVRE DI NICOLO TARTAGLIA,

NELLAQVALE IN DIECISETTE LIBRI SI DICHIARA TVTTI GLI ATTI OPERATIVI, PRATICHE, ET REGOLE NECESSARIE NON SOLAmente in tutta l'arte negotiaria, & mercantile, ma anchor in ogni altra arte, ſcientia, ouer diſciplina, doue interuenghi il calculo.

MALIGNITA'

NOIAR NON PVO
A' FORTEZZA

CON LI SVOI PRIVILEGII.

In Vinegia per Curtio Troiano de i Nauò.
M D LVI.

FIG. 135. TITLE PAGE OF TARTAGLIA

the 'Quesiti et inventioni' and 'Regola generale,' chiefly physical, and the 'General Trattato' containing the arithmetic. Of the arithmetic there were editions in Venice, 1556 and 1560, 4° (?); ib., 1592–93, 4° (p. 279); Paris, 1578, 8° (below); Antwerp, 1578, 8°. There was also a 'Scelta di Abbaco ridotto dal famosissimo Nicolò Tartaglia,' published at Venice in 1596.

The entire work consists of six volumes, bound usually in three, and the publication extended over a period of five years, from 1556 to 1560 inclusive. This volume contains the arithmetic, the most scholarly contribution to the subject that appeared in the sixteenth century. It is more elaborate than the treatise of Paciuolo (p. 54, 1494), and like that work it enters into the various minute details of the operations and commercial rules of the Italian arithmeticians. For example, Tartaglia gives seven methods for the multiplication of integers, and four for division; he enters very fully into the discussion of denominate numbers and exchange; and his treatment of such rules as 'Regula de tri' is unusually elaborate. Indeed, there is no other treatise that gives as much information concerning the arithmetic of the sixteenth century, either as to theory or application. The life of the people, the customs of the merchants, the struggles to improve arithmetic, are all set forth by Tartaglia in an extended but interesting fashion. 531

NICOLÒ TARTAGLIA. Ed. pr. 1556. Paris, 1578.

See p. 275.

Title. 'L'Arithmetiqve // de Nicolas // Tartaglia Brescian,// grand mathematicien,// et prince des practiciens.// Diuiſée en deux parties.// La declaration ſe verra en la page ſuyuante.// Recueillie, & traduite d'Italien en François, par // Gvillavme Gosselin de Cæn.// Auec toutes les demonſtrations Mathematiques : & pluſieurs in-//uentions dudit Gosselin, eſparſes chacune // en ſon lieu.// A tres-Illustre & Vertueuſe Princeſſe Mar-// gverite de France, Royne de Nauarre.// Premiere Partie.// A Paris,// Chez Gilles Beys, rue S. Iacques, au Lis blanc.// 1578. // Avec Privilege dv Roy.' (F. 1, r.)

Description. 8°, 10.6 × 16.6 cm., the text being 7.5 × 13.4 cm. 286 ff. (152 in part I, 134 in part II, 28 being unnumb.), 27–32 ll. Paris, 1578.

See p. 278. The 'Seconde Partie' follows f. numb. 136 of the first part, and is of the same date. The Privilege follows f. numb. 122 of the second part, and is dated 'Paris le 17. Septembre. 1577.'

This is merely a French translation of the work already described with notes by Gosselin. These notes are, rather naïvely, printed in more prominent type than the original text, but are generally of little value. They cover such points as multiplying by or dividing numbers ending in zero, they amplify certain demonstrations (as in the division of fractions), and they adapt the commercial chapters to French usage.

NICOLÒ TARTAGLIA. Ed. pr. 1556. Venice, 1592.

See p. 275.

Title. 'Tvtte l'opere // d'arithmetica // del famosissimo // Nicolò Tartaglia.// Nelle qvale in XVII. libri con varie prove,// & ragioni, moſtraſi ogni prattica naturale, & artificiale; i modo, & le regole da // gli Antichi, & Moderni vſate nell' arte mercantile; & oue interuiene calcolo,// peſi, denari, tariffe, calmeri, baratti, cambi di banchieri, e di fiere, ſaldi, ſconti,// giuochi, traffico di compagnie, compre, vendite, portar mercantie da un paeſe // all'altro, conuertir monete, congiungimento di metalli, & opere de' zecchieri.// Sopra le qual coſe tutte, formanſi belliſſimi queſiti, & ſi ſciolgono le diffi-//coltà, con vgual chiarezza, & diligenza, per vtile rileuato de i mercanti, & te-//ſorieri, à Capitani, e Matematici, & Aſtrologhi, &c. // Parte Prima.// Con Privilegio. // In Venetia,// All'Inſegna del Leone. M. D. XCII.' (F. 1, r.)

Description. 4°, 13.6 × 18.8 cm., the text being 11.5 × 18 cm. 479 ff. (204 in part I, 275 in part II, 8 unnumb.), 40 ll. Venice, 1592. Riccardi says that it was not completed until 1593.

Editions. See p. 278.

This is substantially the same as the first volume of the 'General Trattato' of 1556 (p. 275), except that it is in quarto instead of folio.

ORONTIUS FINAEUS. Ed. pr. 1556. Paris, 1556.

See p. 160.

Title. See Fig. 136.

Description. Fol., 19.6 × 28.5 cm., the text being 13 × 21.4 cm. 6 ff. unnumb. + 136 numb. = 142 ff., 15–34 ll. Paris, 1556.

ORONTII FINAEI, DELPHINATIS, REGII MATHEMATICARVM PROFESSORIS,

De rebus mathematicis, hactenus desideratis, Libri IIII.

¶ Quibus inter cætera, Circuli quadratura Centum modis, & suprà, per eundem Orontium recenter excogitatis, demonstratur.

LVTETIAE PARISIORVM,
Anno Christi Seruatoris,
M. D. LVI.

Ex officina Michaëlis Vascosani, uia Iacobæa ad insigne Fontis.

Cum Priuilegio Regis.

FIG. 136. TITLE PAGE OF THE 1556 FINAEUS

Editions. There was no other edition. For his arithmetic, see p. 160.

Although chiefly on geometry, this work is included in the list of arithmetics because its treatment of proportion is more arithmetical than that of Euclid and his followers. (Ff. 25–29.) There was, however, nothing original in the work of Finaeus.

VALENTIN NABOD. Ed. pr. 1556. Cologne, 1556.

NAIBOD, NAIBODA. Born at Cologne; died at Padua, March 3, 1593. He was for a time professor of mathematics at Cologne. He wrote on astronomy as well as arithmetic.

Title. See Fig. 137.

Colophon. 'Coloniae, Typis Iohannis // Bathenij.' (F. 100, r.)

Description. 8°, 9.5 × 14.9 cm., the text being 6.3 × 11.9 cm. 8 ff. unnumb. + 92 numb. = 100 ff., 28 ll. Cologne, 1556.

Editions. There was no other edition.

This is a Latin work, with occasional Greek passages, and was written for the classical schools of Germany. Although assuming to be a practical treatise on calculating, it so lacks the merit of brevity as to be unfitted for use as a textbook. It has few problems, and these are of no commercial value. In his desire to exalt the classical learning, Nabod, like other Latin writers of his time, assigns the Arab-Hindu numerals to the Pythagoreans.

VALENTIN MENHER de Kempten.
Ed. pr. 1556. Antwerp, 1556.

See p. 249.

Title. See Fig. 138.

Colophon. 'Imprimé en Anuers par Ian Loë // l'An de noſtre Seigneur 1556 // le 20 iour d'Auril.' (F. 184, v.)

Description. 8°, 9.8 × 14.9 cm., the text being 7.8 × 13.2 cm. 184 ff. unnumb., 27–28 ll. Antwerp, 1556.

Editions. See p. 250.

This is usually mentioned as the second of Menher's arithmetics. It is, however, the same as the work described on p. 249, 1550, save for a few minor changes.

VALENTINI NABODI DE CALCVLATORIA NVMERORUMQUE NATURA SECTIONES QUATUOR.

A D

CLARISSIMVM VIRVM

GASPARVM DOVCIVM FLORENTINVM, CAESAREAE Maiestatis Consiliarium.

COLONIAE AGRIPPINAE,

Apud hæredes Arnoldi Birckmanni. 1556.

FIG. 137. TITLE PAGE OF NABOD

ARITHMETI-
QVE SECONDE PAR
M. VALENTIN MENNHER
de Kempten.

Auec grace & priuilege de l'Em-
pereur pour quatre ans.

FIG. 138. TITLE PAGE OF MENHER

PIERRE FORCADEL. Ed. pr. 1556–57. Paris, 1556–57.

Born at Béziers; died at Paris in 1574. He was (1560) professor of mathematics in the Collège Royal Paris. He also wrote on astronomy, and translated the works of several Greek mathematicians.

Title. The title page of this copy of Book I is missing. The title page of Book II reads as follows: 531

'Le // Second Livre de // l'Arithmetiqve de P. For-//cadel de Beziers.// Av qvel seront declarees les // fractions vulgaires, auec leurs demonſtrations, par les // quantitez continues, & premieres cauſes des egaliſſemens // de l'Algebre.// Le tout nouuellement inuenté par l'auteur.// (Woodcut with motto: In Pingvi Gallina.) A Paris,// Chez Guillaume Cauellat, à l'enſeigne de la // poulle graſſe, deuant le college de Cambray.// 1556.// Avec Privilege.' (F. 94.)

The title page of Book III reads as follows:

'Le // Troysieme Li-//vre de l'Arithmetiqve // de P. Forcadel de Beziers.// Avqvel sont traictees les de-//monſtrations de toutes ſortes de racines, auecques l'entiere pra-//ctique de l'extraction d'icelles, enſemble pluſieurs queſtions, reigles,// & demonſtrations Mathematiques, auecques le propre ſubiect de l'Algebre.// Le tout de l'inuention de l'Autheur.// A Paris,// Chez Guillaume Cauellat, a l'enſeigne de la // Poulle graſſe, deuant le college de Cambray.// 1557.// Avec Privilege.' (F. 1, r., of Part III.)

Description. 4°, 13.2 × 19.2 cm., the text being 10.1 × 16 cm. 323 ff. (5 + 93, 110, 4 + 111, in the above three books); 25–31 ll. Paris 1556–57.

Editions. Forcadel wrote four arithmetics, as follows:

1. 'L'arithmetiqve,' Paris, 1556–57, here described.
2. 'L'Arithmétique par les gects,' Paris, 1558, 8°.
3. 'Arithmetiqve entiere et abregee,' Paris, 1565 (p. 316).
4. 'Arithmétique demonſtrée,' Paris, 1570, 4°.

This is perhaps the most elaborate French treatise on arithmetic published in the sixteenth century. Its three books of about one hundred pages each form a work of the nature of the great contemporary Italian arithmetic of Tartaglia, or rather of Tonstall's Latin treatise. Of

L'ARITHMETICQVE DE P. Forcadel de Beziers.

(1556. voir le tome 2

De la cognoiſſance & diſpoſition des figures, autrement des nombres ſimples.

TOVTE l'Arithmeticque conſiſte à ſçauoir cognoiſtre dix figures, ou caracteres, propres guidons de toute computation Mathematicque: qui ſont cy deſſous rengez ſelon leur propre nature.

	vn.	deux.	trois.	quatre.	cinq.	ſix.	ſept.	huit.	neuf.	zero ou nulle.	
vn.	1.	2.	3.	4.	5.	6.	7.	8.	9.	0.	zero.
deux.	2.	3.	4.	5.	6.	7.	8.	9.	0.	1.	premieres.
trois.	3.	4.	5.	6.	7.	8.	9.	0.	1.	2.	ſecondes.
quatre.	4.	5.	6.	7.	8.	9.	0.	1.	2.	3.	tierces.
cinq.	5.	6.	7.	8.	9.	0.	1.	2.	3.	4.	quartes.
ſix.	6.	7.	8.	9.	0.	1.	2.	3.	4.	5.	quintes.
ſept.	7.	8.	9.	0.	1.	2.	3.	4.	5.	6.	ſixieſmes.
huit.	8.	9.	0.	1.	2.	3.	4.	5.	6.	7.	ſeptieſmes.
neuf,&c.	9.	0.	1.	2.	3.	4.	5.	6.	7.	8.	huitieſmes.
zero ou nulle.	0.	1.	2.	3.	4.	5.	6.	7.	8.	9.	neufieſmes.
	zero.	premieres.	ſecondes.	tierces.	quartes.	cinquieſmes.	ſixieſmes.	ſeptieſmes.	huitieſmes.	neufieſmes,&c.	

FIG. 139. FIRST PAGE OF TEXT OF THE 1556–57 FORCADEL

practical problems it has very few, and it is equally deficient in the theories of the ancient arithmeticians. It is simply a ponderous work on the theory of arithmetical calculations and rules, valuable for a scholar but useless as a practical textbook.

For a biographical study of Forcadel see Boncompagni's *Bulletino*, II, 424.

Other works of 1556. Baëza, p. 269, 1555; Cathalan (Anonymous), p. 269, 1555; Gemma, p. 200, 1540; Medlerus, p. 223, 1543; Medicus, p. 290, 1557; Psellus, p. 168, 1532; Recorde, p. 214, c. 1542; Riese, p. 139, 1522; Ringelbergius, p. 168, 1531; Xylander, p. 356, 1577; Anonymous, 'Arithmetices Epitome,' Freiburg, 12°; A. Lottini, 'Calculi e conti per quelli che hanno denari,' Lyons, 8°; Casp. Pauerus, 'Logistice astronomica,' Wittenberg, 8°; Joannes Pierius Valerianus, 'Hieroglyphica sive de Sacris Aegyptiorvm literis Commentarii' (ancient finger reckoning), Basel, fol., with editions ib., 1567, fol.; ib., 1575, fol.; Lugduni, 1579, fol.; Juan Díaz Freyle, 'Sumario compendioso . . . con algunas reglas tocantes al Aritmética,' Mexico (the first arithmetic printed in America). 531; 532; 533

ROBERT RECORDE. Ed. pr. 1557. London, 1557.

See p. 213.

Title. See Fig. 140.

Colophon. '¶Imprinted at London,//by Jhon Kyngſton.// Anno domini. 1557.' (F. 164, v.)

Description. 4°, 13 × 18 cm., the text being 8.1 × 14.7 cm. 164 ff. unnumb., 36 ll. London, 1557.

Editions. There was no other edition.

Recorde speaks of this work as 'The ſeconde parte of Arithmetike, containyng the extraction of Rootes in diuerſe kindes, with the Arte of Coſſike nombers, and of Surdes nombers also, in ſondrie ſortes.' It is not, however, purely algebraic, for the first half of the book is a treatise on Boethian arithmetic. For example, the following is one of the definitions: 'A Diametralle nomber, is ſuche a number as hath twoo partes of that nature: that if thei bee multiplied together, thei will make the ſaied diametralle nomber.' The treatise on 'Cossike nombers' begins on f. S 1, and 'The rule of equation, commonly called Algebers Rule' on f. Gg 4. The sign of equality, 'a paire of paralleles, or Gemowe lines of one lengthe, thus: ═══, bicauſe noe .2. thynges, can be moare equalle,' is found for the first time in print on f. Ff 1 (see Fig. 141).

The whetſtone of witte,

whiche is the ſeconde parte of Arithmetike: containyng thextraction of Rootes: The *Coßike* practiſe, with the rule of *Equation*: and the woorkes of *Surde Nombers.*

Though many ſtones doe beare greate price,
The whetſtone *is for exerſice*
As neadefull, and in woorke as ſtraunge:
Dulle thinges and harde it will ſo chaunge,
And make them ſharpe, to right good vſe:
All arteſmen knowe, thei can not chuſe,
But vſe his helpe: yet as men ſee,
Noe ſharpeneſſe ſemeth in it to bee.
The grounde of artes *did brede this ſtone:*
His vſe is greate, and moare then one.
Here if you liſt your wittes to whette,
Moche ſharpeneſſe therby ſhall you gette.
Dulle wittes hereby doe greately mende,
Sharpe wittes are fined to their fulle ende.
Now proue, and praiſe, as you doe finde,
And to your ſelf be not vnkinde.

These Bookes are to bee ſolde, at the Weſte doore of Poules, by Jhon Kyngſtone.

FIG. 140. TITLE PAGE OF RECORDE'S *Whetſtone of witte*

The Arte

as their workes doe extende) to distincte it onely into twoo partes. Whereof the firste is, *when one nomber is equalle vnto one other.* And the seconde is, *when one nomber is compared as equalle vnto .2. other nombers.*

Alwaies willyng you to remēber, that you reduce your nombers, to their leaste denominations, and smalleste formes, before you procede any farther.

And again, if your *equation* be soche, that the greateste denomination *Cossike*, be ioined to any parte of a compounde nomber, you shall tourne it so, that the nomber of the greateste signe alone, maie stande as equalle to the reste.

And this is all that neadeth to be taughte, concernyng this woorke.

Howbeit, for easie alteratiō of *equations*. I will propounde a fewe exāples, bicause the extraction of their rootes, maie the more aptly bee wroughte. And to auoide the tediouse repetition of these woordes: is equalle to: I will sette as I doe often in woorke vse, a paire of paralleles, or Gemowe lines of one lengthe, thus: =====, bicause noe .2. thynges, can be moare equalle. And now marke these nombers.

1. 14.ze. —+— .15.ꝭ ===== 71.ꝭ.
2. 20.ze. —— .18.ꝭ ===== .102.ꝭ.
3. 26.ʒ —+— 10ze ===== 9.ʒ —— 10ze —+— 213.ꝭ.
4. 19.ze —+— 192.ꝭ ===== 10ʒ —+— 108ꝭ —— 19ze
5. 18.ze —+— 24.ꝭ. ===== 8.ʒ. —+— 2.ze.
6. 34ʒ —— 12ze ===== 40ze —+— 480ꝭ —— 9.ʒ.

1. In the firste there appeareth. 2. nombers, that is 14.ze.

Fig. 141. From recorde's *Whetstone of witte*

LUCAS LOSSIUS. Ed. pr. 1557. Frankfort a. d. Oder, 1557.

A German arithmetician of the sixteenth century, born at Lüneberg.

Title. See Fig. 142.

ARITHME-
TICES EROTE-
MATA PVE-
RILIA.

IN QVIBVS SEX SPECIES
huius vtilissimæ artis, & Regula, quam
vocant, Detri, breuiter & per-
spicuè traduntur.

IN GRATIAM ET VSVM SCHOLA-
*rum puerilium Latinarum collecta, & in
lucem iam recens edita.*

A

*LVCA LOSSIO
Luneburgensi.*

FRANCOFORDIÆ AD ODERAM
IN OFFICINA IOHANNIS
EICHORNI.

FIG. 142. TITLE PAGE OF LOSSIUS

Description. 8°, 9.2 × 14 cm., the text being 6.8 × 11.9 cm. 4 ff. unnumb. + 27 numb. = 31 ff., 22–24 ll. Frankfort. For date see f. 4, v.: 'Datæ Lunebar, Anno 1557.// 7. Februarij.'

Editions. Frankfort an der Oder, 1557, 8° (here described); s. l., 1562, 8°; Leipzig, 1568, 8°; Frankfort, 1569; Magdeburg, 1585, 8°. 533

A small book, intended, as the title states, for beginners in the Latin schools. It is arranged on the catechism plan, a feature not common with arithmetics printed in Germany at this time, although extensively used by Recorde in England (see p. 210). It begins: 'Quid eſt Arithmetica? Est benè & artificioſè numerandi & computandi ſcientia.' (F. 5.) The 'species' are treated in a practical way, but the problems are all traditional, part of them being based upon biblical incidents.

Other works of 1557. Archimedes, p. 228, 1544; Camerarius, p. 263, 1554; Gemma, p. 200, 1540; Gutiérrez de Gualda, p. 167, 1531; Jacob, p. 298, 1560; Paxi, p. 80, 1503; Psellus, p. 170, 1532; Ramus, p. 263, 1555; Rudolff, p. 153, 1526; Tagliente, p. 115, 1515; Martinus Carolus Cressfelt, 'Arithmetica, Reeckeninge op den Linien end Cyfferen na allerley Hantieringe,' Deventer (second edition 1577); Sixtus Medicus, 'De Latinis numerorum notis,' Venice, 4° (colophon date, 1556; hardly an arithmetic). 533

GIOVANNI FRANCESCO PEVERONE.

Ed. pr. 1558. Lyons, 1581.

An Italian arithmetician of the sixteenth century, born at Cuneo, in Piedmont.

Title. See Fig. 143.

Description. 4°, 15.3 × 21.7 cm., the text being 10.2 × 17.6 cm. 136 pp. (60 on arithmetic), 35–37 ll. The dedicatory epistle is dated 1556. Bound with this is 'Il breve Trattato// di Geometria.' Lyons, 1581.

Editions. Although the dedicatory epistle is dated 1556, and the portrait of the author bears the date 1550, I know of no edition before 1558, Lyons, 8°. This edition of 1581 seems to have been the second. 533

The first part of the work treats of the operations with integers. It is not a particularly progressive textbook, as is seen in the fact that it includes the 'gelosia' multiplication of the early Venetian writers. The second book relates to fractions, the third to business operations, and the fourth to roots. The work is in no sense a scholarly production.

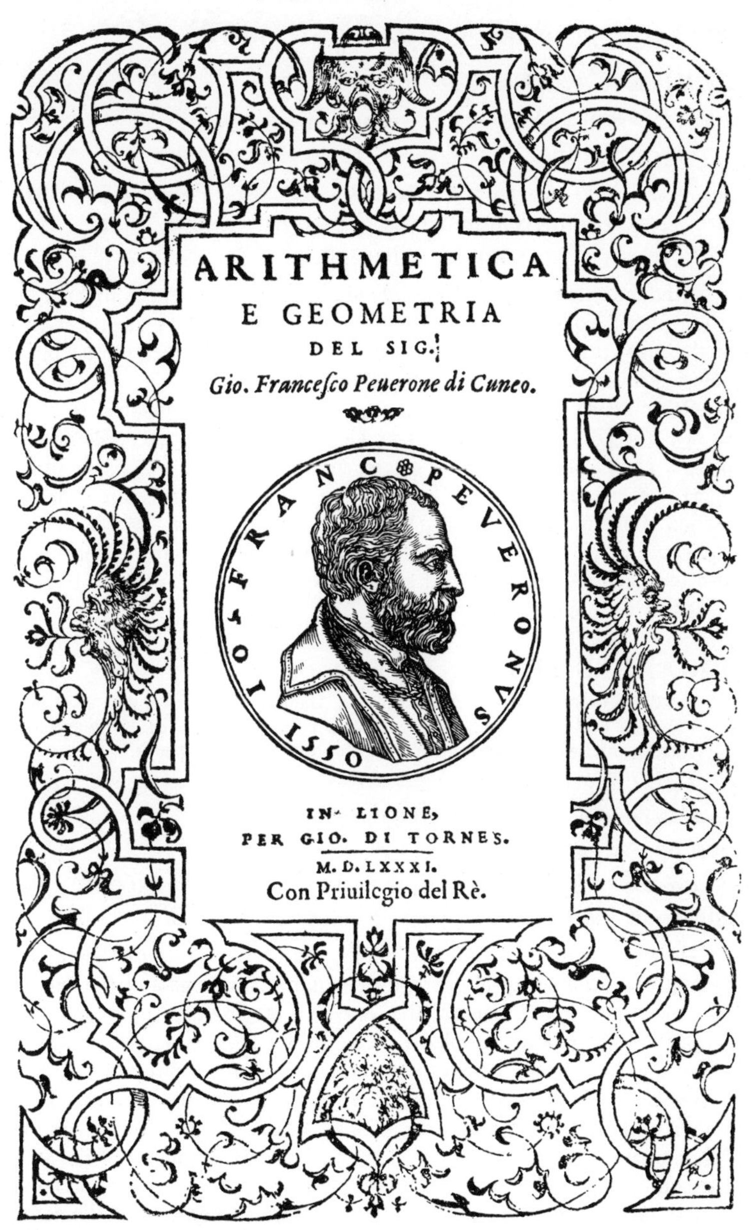

ARITHMETICA
E GEOMETRIA
DEL SIG.
Gio. Francesco Peuerone di Cuneo.
IO. FRANC. PEVERONVS
1550
IN LIONE,
PER GIO. DI TORNES.
M. D. LXXXI.
Con Priuilegio del Rè.

FIG. 143. TITLE PAGE OF PEVERONE

ALVISE CASANOVA. Ed. pr. 1558. Venice, 1558.

A Venetian teacher of the sixteenth century.

Title. See Fig. 144.

Description. 4°, 15.5 × 20.8 cm., the text being 12.5 × 16.1 cm. 136 ff. (partly numbered), 33–38 ll. Venice, 1558.

Editions. There was no other edition.

Although this rare work is usually classed as a commercial arithmetic, it is rather a collection of bookkeeping problems. As such it was one of the most prominent of the century, and it gives an idea of the business questions of the Venetian merchants of its time.

Other works of 1558. Anianus, p. 32, 1488; Archimedes, p. 228, 1544; Forcadel, p. 284, 1556–57; Gemma, p. 204, 1540; Glareanus, p. 192, 1539; Mariani, p. 181, 1535; Medlerus, p. 223, 1543; Recorde, p. 213, c. 1542; Riese, p. 141, 1522; Andreas Clatovenus, 'Arithmetica Bohemice,' Prag, 8° (Wydra, in his *Historia Matheseos in Bohemia*, p. 18, mentions it as printed in Nürnberg, 1530). There were also two arithmetics published s. a., but c. 1558, viz.: Benese, p. 182, c. 1536; Gemma, p. 200, 1540.

JOHANNES BUTEO. Ed. pr. 1559. Lyons, 1559.

BOTEO, BUTÉON, BATEON. Born in Dauphiné, c. 1485–1489; died in a cloister in 1560 or 1564. He belonged to the order of St. Anthony, and wrote chiefly on geometry, exposing the pretenses of Finaeus.

Title. See Fig. 145.

Description. 8°, 10.4 × 16.9 cm., the text being 7.2 × 13.2 cm. 400 pp., 23–27 ll. Lyons, 1559.

Editions. Lyons, 1559, 8° (here described); ib., 1560, 8°. An edition of his 'Opera' appeared in 1554.

The work is divided into five books, the first treating of the fundamental operations with integers, the second of fractions and the rule of position, the third of algebra, the fourth of arithmetical problems, and the fifth of algebraic problems. The problems are not of practical value, and hence the arithmetic never attained any popularity.

Other works of 1559. Albert, p. 180, 1534; Anianus, p. 32, 1488; Cataneo, p. 244, 1546; Gemma, p. 200, 1540; Mariani, p. 181, 1535; Johann Fischer (Piscator), p. 247, 1549; Herman Gülfferich, 'Ein new Rechenbüchlin auff der Linien und Federn' (perhaps the work mentioned on p. 257, 1552); Pedro Juan Monzó (Monzoni), 'Elementa Arithmeticae,' Valencia, 8°, and s. l., 1566, 1569.

SPECCHIO LVCIDISSIMO,
nel quale ſi uedeno eſſere diffinito
tutti i modi, & ordini de
ſcrittura, che ſi deue
menare nelli nego-
tiamenti della
Mercan-
tia,

*CAMBII, RECAMBII, CON
li loro corriſpondentie, diſgarbu-
gliando, & illuminando
l'intelletto a nego-
tianti.*

OPERA NON PIV VEDVTA,
compoſta per Aluiſe Caſanoua,
Cittadin Venetiano. Anno
Domini. M.D.LVIII.

CON GRATIA ET
Priuilegio dell'Illuſtriſsimo
Senato di Venetia per
anni diece.

FIG. 144. TITLE PAGE OF CASANOVA

IOAN.

BVTEONIS

LOGISTICA, QVÆ

& Arithmetica vulgò dicitur in libros quinque digesta: quorum index summatim habetur in tergo.

EIVSDEM,

Ad locum Vitruuij corruptum restitutio, qui est de proportione lapidum mittendorum ad balistæ foramen, Libro Decimo,

LVGDVNI,

APVD GVLIELMVM ROVILLIVM,

SVB SCVTO VENETO.

M. D. LIX.

Cum priuilegio Regis.

FIG. 145. TITLE PAGE OF BUTEO

FRANCISCUS BAROCIUS. Ed. pr. 1560. Pavia, 1560.

FRANCESCO BAROZZI. Born at Venice, c. 1538; died after 1587. He wrote on cosmography, and edited Proclus.

Title. 'Francisci Barocii // patritii Veneti // opvscvlvm,// in quo vna Oratio, & duæ Quęſtiones:// altera de certitudine, & altera // de medietate // Mathematicarum continentur.// Ad Reuerendisſimum Danielem Barbarum Patriarcham // Aquileienſem deſignatum Virum Clariſſ.// Patavii, E. G. P.// M. D. LX.' (F. 1, r.)

Description. 4°, 14.1 × 19.6 cm., the text being 10.1 × 15.3 cm. 40 ff. numb., 26 ll. Pavia, 1560.

Editions. There was no other edition.

This philosophical discussion has been mentioned because it includes some reference to the old Boethian arithmetic. No arithmetical processes are discussed.

OLIVIERO FONDULI. Ed. pr. 1560. Bologna, 1560.

A Bolognese teacher of the middle of the sixteenth century.

Title. See Fig. 146.

Description. 8°, 9.9 × 14.6 cm., the text being 6.9 × 11.1 cm. 23 ff. unnumb. (possibly one missing at the end), 24 ll. The date is given on f. 1, v.: 'Dat. Bonoiæ . die xxv. Septemb. 1560.' Bologna, 1560.

Editions. There was no other edition.

This rare little handbook of commercial customs, from the press of Pellegrino Bonardo of Bologna, is hardly an arithmetic, although it explains certain arithmetical processes, and gives a considerable number of business problems.

SIMON JACOB. Ed. pr. 1560. Frankfort, 1565.

Born at Coburg; died at Frankfort am Main, June 24, 1564. He was one of the best-known Rechenmeisters of his time.

Title. See Fig. 147.

Colophon. 'Getruckt zu Franck-//furt am Main/ bey Georg Raben/ //in verlegung Sigmund Feyerabends/ //vnd Simon Hůters. Sigmund Feyerabent // Simon Hůtter // M. D. LXV.' (F. 363, v.)

PRATICHE
DE FIORETTI MERCHANTILI.
vtiliſsime a ciaſcheduna perſona, di mandare a
memoria le breue Inuētioni fabrichate ſopra
il Valutar de peſi, & miſure, & altre in
ſtruttioni neccessarie da ſapere.
Et anchora a quadrare Muraglie, Taſſelli, & Coperti a
Tuade, Fieno, & Legne: Con la decchiaratrone, &
Exempli loro come legendo intenderai.

FIG. 146. TITLE PAGE OF FONDULI

Ein New vnd Wol-
gegründt Rechenbuch/auff den Linien vñ Ziffern/
sampt der Welschen Practic vnd allerley vortheilen/neben der
extraction Radicum, vñ von den Proportionen/mit vilen lustigen Fragen vñ
Auffgaben/ꝛc. ¶ Deßgleichen ein vollkomner Bericht der Regel Falsi/mit neuwen Inuentio-
nibus/Demonstrationibus/ vnd vortheilen/so biß anher für vnmüglich geschetzt/gebessert/der-
gleichen noch nie an tag kommen. ¶ Vnd dann von der Geometria/ wie man mancherley Fel-
der vnd ebne/auch allerley Corpora/Regularia vnd Irregularia/messen/Aream finden vñ rechnen sol. Alles durch Simon Jacob von Coburg/ Bürger vnd Rechenmeister zu
Franckfurt am Main/mit fleiß zusammen getragen/vnd jetzt
erstmals getruckt.

Mit Röm. Keyser. Mt. Gnad vnd Freiheit nicht nachzutrucken.
Getruckt zu Franckfurt am Mayn/1565.

FIG. 147. TITLE PAGE OF JACOB

Description. 4°, 15.2 × 18.6 cm., the text being 10.3 × 15 cm. 354 ff. (349 numb.), 30 ll. Frankfort, 1565.

Editions. Jacob published two works, as follows:

1. 'Rechenbüchlein auf den Linien und mit Ziffern,' Frankfort, 1557, 12°; 1559; 1574, 12°; 1589; 1590; 1599.

2. 'Ein new vnd wolgegrůndt Rechenbuch,' 1560; Frankfort, 1565, 4° (here described); 1569; 1600, 4° (below), and in the seventeenth century.

Jacob's arithmetics followed the general plan of the popular books of Riese, and were deservedly well received in the second half of the century. They were commercial textbooks, and although they do not show any mathematical advance they are historically valuable for their applied problems. The title page (fig. 147) is interesting because of the variety of mathematical instruments illustrated.

Other works of 1560. Borghi, p. 16, 1484; Buteo, p. 292, 1559; Feliciano, p. 149, 1526; Psellus, p. 168, 1532; Scheubel, p. 246, 1549; Tartaglia, p. 278, 1556; Gasparo Rizzo, 'Abbaco nvovo molto copioso et artificiosamente ordinato,' Venice, 8°; Juan Ventallol, 'Aritmética,' a rare work, s. l. a., which appeared in Spain about this time. 533

SIMON JACOB. Ed. pr. 1560. Frankfort, 1600.

See p. 295.

Title. This is practically the same as in the 1565 edition described above.

Colophon. 'Getruckt zu Franckfurt am Mayn/ bey // Matthes Beckern/ In Verlegung Chris-//ſtian Egenolphs Erben.// Anno 1600.' (F. 360, v.)

Description. 4°, 15.5 × 19.2 cm., the text being 10.5 × 14.8 cm. 11 ff. unnumb. + 349 numb. = 360 ff., 28 ll. Frankfort, 1600.

See above. 533

JOSEPHUS UNICORNUS. Ed. pr. 1561. Venice, 1561.

GIUSEPPE UNICORNO. A Bergomese arithmetician, born in 1523; died in 1610.

Title. See Fig. 148.

Description. 8°, 9.4 × 13.9 cm., the text being 7.4 × 12.7 cm. 1 f. unnumb. + 78 numb. = 79 ff., 30 ll. Venice, 1561.

IOSEPHI VNICORNI
BERGOMATIS
LIBER
DE VTILITATE MATHEMATICA-
RVM ARTIVM,

*In quo candide Lector multa quidem scitu periucun-
da, admiratione digna, & humano usui
necessaria passim reperies.*

VENETIIS, *Apud Dominicum de Nicolinis,*
Anno Domini, M D L X I.

FIG. 148. TITLE PAGE OF UNICORNUS

Editions. There was no other edition.

This work is not an arithmetic, Unicorn's textbook on this subject appearing much later (p. 412, 1598). It is a prolix dissertation on the uses of mathematics, and of arithmetic in particular, with extracts from the ancient and mediæval writers. Like Agrippa's 'De vanitate scientiarum,' it is interesting but profitless.

NICOLAUS WERNER. Ed. pr. 1561. Nürnberg, 1561.

A Nürnberg Rechenmeister, born c. 1520.

Title. See Fig. 149.

Colophon. 'Gedrückt zu Nürnberg/ //durch Johann vom Berg/ //vnd Ulrich Newber.' (F. 156, r.)

Description. 4°, 14.2 × 19.3 cm., the text being 10.1 × 15.5 cm. 156 ff. unnumb., 23–31 ll. Nürnberg, 1561.

Editions. There was no other edition.

This is a purely mercantile arithmetic, and like many of the Italian textbooks it assumes some preliminary knowledge of the fundamental operations. It begins with a treatment of Welsch (Italian) practice, and is composed almost entirely of practical problems of the day, solved by this method. Welsch practice differs from the rule of three only in having, ordinarily, unity for the first term, as in the following example: If 1 book cost $2, how much will 7 books cost? This method was very popular with all German arithmeticians of the sixteenth century, and survived in the chapter on Practice still to be found in English arithmetics. It appeared also in the early American textbooks. The book gives a very good idea of the business requirements of the second half of the sixteenth century. The value placed upon exchange and 'profit and loss' may be seen in the fact that twenty-six pages are devoted to the former and forty-seven to the latter, while partnership also has twenty-six pages for its share.

JOHANNES MONHEMIUS.

Ed. pr. 1561. Düsseldorf, 1561.

A German teacher of the second half of the sixteenth century.

Title. See Fig. 150.

Description. 8°, 9.4 × 15 cm., the text being 6.5 × 12 cm. 22 ff. unnumb., 25–27 ll. Düsseldorf, 1561.

Editions. There was no other edition.

Rechenbüch

Von der welschen prac-
tick auff allerley Kauffmans hendel/ vnd
sonderlich souiel der Nürnbergischen Lands art/ vnd
gebrauch belangt/ mit sonderm vorteyl/ vnd behendigkeit auff mancherley art practicē

Durch

Nicolaum Werner
Rechenmeyster/ vnd Burger zu
Nürnberg/ Seinen Discipeln/ Schülern
vnd sonst menigklichen zu nutz in
Druck verfertigt.

Nürnberg M. D. LXI.

FIG. 149. TITLE PAGE OF WERNER

This work is almost unknown to students of arithmetic. It seems to be the only edition, although the introduction bears the date 1542. It is a small work and is based somewhat on the mediæval texts. It gives

METHO-
DVS ARITHMETICES
COMPVTATORIAE,OMNEM SVP-
putandi artem tradens:tam illam,quæ notis
numerorum,quàm quæ olim calculis,
nunc nummis ſuper æquè diſtan
tes lineas ſurſum ac de-
orſum poſitis,
perficitur.

Auctore Io. Monhemio.

DVSSELDORPII
Anno 1 5 6 1.

FIG. 150. TITLE PAGE OF MONHEMIUS

the fundamental operations with integers and a brief treatment of the rules of three, partnership, and false. There is also a brief explanation of counter reckoning.

BENEDICTUS HERBESTUS. Ed. pr. 1561. Cracow, 1577.

Born at Novomiasti, Poland, in 1531; died at Jaroslaw, March 4, 1593. He was a Jesuit priest at Cracow.

Title. See Fig. 151.

Description. 8°, 9.5 × 15.3 cm., the text being 8 × 11.5 cm. 42 ff. unnumb., 27–28 ll. Cracow, 1577.

Editions. 1561; Cracow, 1564, 8°; ib., 1566; ib., 1569; ib., 1577, 8° (here described).

A Latin book intended for the Church schools, and revised, as the 'Proæmivm auctoris' states, 'Mariæburgi in Prufsia, fub finem Anni Domini 1576.' It consists of two parts, the first being devoted to the arithmetical operations, chiefly with counters (and this as late as 1577), and the second to the calendar. Besides the operations, the first part also includes a chapter on progressions ('De Progressione. Capvt VII.'), and one on the rule of three ('De tribvs nvmeris integris. Capvt VIII.').

ANDREAS HELMREICH. Ed. pr. 1561. Eisleben, 1561.

A Halle Rechenmeister of the latter half of the sixteenth century. The dedication of the 1561 edition is signed by 'Andreas Helmreich Rechenmeifter vnd Vifierer zu Halle,' and the 1588 edition by 'Andreas Helmreich von Eifzfeldt/ Notarius publicus/ Rechenmeifter vnd Vifierer zu Halle.'

Title. See Fig. 152.

Colophon. 'Gedruckt zu Eisleben/ bey // Vrban Ganbifch.' (F. 111, r.)

Description. 4°, 14.2 × 19.3 cm., the text being 9.3 × 14.9 cm. 111 ff. unnumb. + 1 blank = 112 ff., 27–33 ll. Eisleben, 1561.

Editions. This is probably the first edition of Helmreich's arithmetic, the dedicatory epistle being dated 'zu Halle in Sachffen den Sontag Oculi/ welcher war der 9. tag des Monats Marcij/ nach Chrifti Ihefu vnfers lieben Herrn vnd Seligmachers Geburt 1561.' Murhard (I, 155) mentions, however, an edition appearing at Halle in 1546, 4°. The book was republished at Leipzig in 1588 (p. 306), and again in 1595, 4° (p. 306), and possibly ib., 1596, 4°.

The work is an unsuccessful attempt to combine the old and new arithmetics. The author begins with a semi-Boethian treatment of ratios, assuming a knowledge of the fundamental operations with integers, and then introduces a course in mercantile arithmetic. He closes with a considerable amount of work on mensuration, including

ARITHME-
TICA LINEA-
RIS,
eiq; adiuncta
FIGVRATA,
cum quibusdam ex
COMPVTO
necessarijs:
auctore BENEDICTO
HERBesto, Societatis
IESV Presbytero,
cum facultate Superiorum.

CRACOVIAE,
CVM GRATIA ET PRIuilegio S. R. M.
In Officina Matthaei Siebeney-
cher: Anno DOMini
1577.

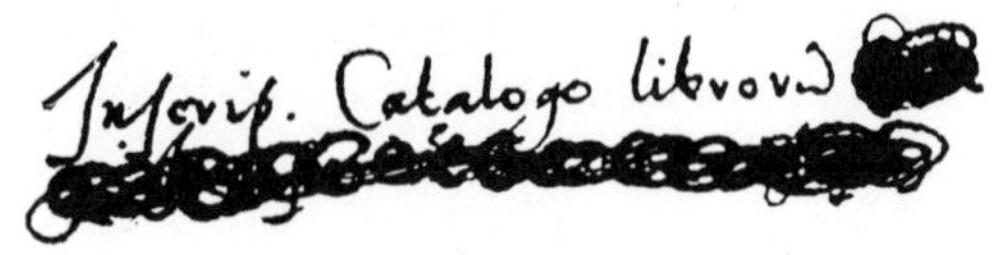

FIG. 151. TITLE PAGE OF THE 1577 HERBESTUS

Rechembüch

von vorteil vnd behendig-
keit nach der Welſchen Practica / mit jren
eigentlichen / deutlichen vnd vnterſchiedlichen Proportzen /
ſampt gewiſſer vnd behender art / mancherley Viſier-
ruthen vnd Schnüre / auff alle Ohme vnd Eich
zu machen / zuſamen bracht durch

Andream Helmreich Rechenmeiſter
vnd Viſierer zu Halle in Sachſſen.

FIG. 152. TITLE PAGE OF THE 1561 HELMREICH

gauging (Visierrechnung). It is somewhat surprising that such a wearisome treatise ever went through three editions.

Other works of 1561. Albert, p. 178, 1534; Borghi, p. 16, 1484; Feliciano, p. 148, 1526; Gemma, p. 204, 1540; Recorde, p. 214, c. 1542; Rudolff, p. 152, 1526; Sfortunati, p. 177, 1534; Tagliente, p. 115, 1515; Wolphius, p. 154, 1527; A. Citolini, 'Tipocosmia,' Venice, 8° (a synopsis of science, including arithmetic); Ognibene de Castellano, 'Opera,' Venice, 4° (principally on geometry, but including some arithmetic; see also p. 375, 1582); Antonio Maria Venusti, 'Compendio utilissimo,' Milan, 8° (includes some treatment of exchange).

ANDREAS HELMREICH. Ed. pr. 1561. Leipzig, 1588.

See p. 303.

Title. 'Rechenbüch//Von vortheil vnd behendig-//keit/ nach der Welſchen Practica/ mit // jhren vnterſchiedlichen Proportionen.// Grund vnd vrſach der Regel Detri/ // Sampt gewiſſer vnd behender art/ mancherley Vi-//ſier Ruthen vnd Schnůre/ nach dem Quadrat vnd // Cubo, auff alle Ohme vnd Eiche zu machen/ alles mit war-//hafftigen perſpectiuiſchen/ Geometriſchen vnd Arithmetiſchen demon-//ſtrationen, auſz den Bůchern Euclidis fundirt/ mit fleiſz // zuſammen gebracht vnd in Druck auſzgangen///Durch//Andream Helmreich/Rechenmeiſter vnd Viſierer // zu Hall in Sachſſen.// (Cut showing dimensions of a cask.) Im Jahr 1588.' (P. 3.)

Colophon. 'Gedruckt zu Leipzig/ //Durch Abraham Lamberg //Anno // M. D. LXXXVIII.' (P. 336.)

Description. 4°, 15.5 × 19.2 cm., the text being 10.1 × 14.5 cm. 9 pp. blank + 16 unnumb. + 317 numb. = 342 pp., 31 ll. Leipzig, 1588.

See p. 303.

ANDREAS HELMREICH. Ed. pr. 1561. Leipzig, 1595.

See p. 303.

Title. 'Rechenbuch/ //Erſtlich/Von // Vortheil vnnd Behen- // digkeit/ nach der Welſchen Practica/ mit ih-//ren vnterſchiedlichen Proportionibus, Grund vnd vr-//ſach der Regel Detri oder proportionum.//II. Von zubereitung mancherley Viſier Ruthen/

// vnnd Schnůren/ damit alle Våhſſer vnnd andere Corpora nach dem // Quadrat vnd Cubo auff alle Ohme vnd Eiche-Viſiere werden.// III. Wie man kůnſtlich das Feld vnd Erdreich/ auff man-//cherley art/ mit gewiſſer Meßruthen vnd Schnuren/ nach eines jeden Landes // oder Stadt gebrauch/ Geometriſcher vnd Jdiotiſcher weiß/ vnd was hier-//innen der Vnterſcheid ſey/ recht meſſen ſol.// IIII. De Diſtantijs Locorum, das iſt/ Wie man wunderbar-//licher Geographiſcher vnd Coſmographiſcher/ nach Geometriſcher weiſe/ durch // Arithmeticam, in der gantzen Welt/ zweyer Stedte oder Wonung Diſtantiam, oder // wie weit die von einander gelegen/ nach ihren Longitudinibus, vnd Latitudinibus, // ſol Rechnen vnd finden/ Sampt einer Landtaffel daraus zu // machen vnd zu beſchreiben.// V. Vnnd wie man die Fůnff Horologia communia, oder ge-//meine Sonnen Vhrn/ als Horizontale, Meridionale, Septentrionale, Ori-//entale, vnd Occidentale, auff einen Cubum, auſzwendig/ oder ſonſten an die Wen-//de/ Mawer vnd ebenen/ Vnd das Fundamentum Horologiorum abreiffen. Item/ // der Sonnen vnd des Monds am Himmel geſchwinden Lauff rechnen. Vnd auch eine ſonderliche gar kůnſt-//liche Sonnen Vhr/ der Cylinder genandt/ darinne alle Stunde nach der Sonnen ſchein deß Tages grůnd-//lich zu erſehen vnd zu befinden/ machen/ vnd ſampt andern mehr ſchőnen Kůnſten vnd Vbun-//gen/ nůtzlich gebrauchen ſol.// Alles mit warhafftigen Demonſtrationibus vnd Figuren/ aus den Bůchern Euclidis vnd andern fundirt. Durch // Andream Helmreich von Eißfeldt/ Rechenmeiſter vnd // Viſierer zu Halle in Sachſen an der Sahle. // Cum Priuilegio.' (P. 1.)

Colophon. 'Gedruckt zu Leipzig/durch Zacha-//riam Berwald. // Im Iahr/ // M. D. XCV.' (P. 645.)

Description. 4°, 14.5 × 18.7 cm., the text being 10.9 × 15.4 cm. 17 pp. unnumb. + 3 blank + 627 numb. = 647 pp., 26–34 ll. Leipzig, 1595.

Editions. See p. 303.

This edition contains the Rechenbuch of 1561, together with much additional matter as summarized on the title page, viz. books III, IV (pp. 306–631). This includes a considerable amount of work on gauging, surveying, cosmography, and the 'Sphere.' The first 304 pages are practically identical with the first edition, except for the addition of numerous woodcuts.

GUGLIELMO PAGNINI. Ed. pr. 1562. Lucca, 1562.

A Lucca mathematician of the sixteenth century.

Title. 'Practica // Mercantile // Moderna. // Di Guglielmo Pagini // Lucchese. // In Lucca per il Bus-//dragho. MDLXII.' (P. 5.)

Description. 4°, 14.5 × 20.6 cm., the text being 9.1 × 15.3 cm. 14 pp. unnumb. + 161 numb. = 175 pp., 19–28 ll. Lucca, 1562.

Editions. There was no other edition.

This is a commercial arithmetic, the only one I have seen that appeared in Lucca in the sixteenth century. The author begins with the operations on denominate numbers, but places division immediately after addition. The galley method of dividing is not given at all, in which Pagnini shows a more advanced spirit than most of his contemporaries. There is first the 'Modo di partire ditti partiri piccoli,' or short division; then the 'Modo di partir per ripiegho,' 'Modo di partire per colonna,' and finally the 'Modo di partire, à danda per altro nome partire grande.' Following division is the work in multiplication, including 'regoletto,' 'biricuocholo,' 'crocetta,' 'per colonna,' and 'per ripiegho.' Subtraction or 'traction' ('Modo di trare, o sottrare') follows multiplication. The author then takes up fractions in the same order. This work in the fundamental operations is followed by chapters on exchange and mercantile problems, the 'Regola del tre,' profit and loss ('Gvadagni e perdite'), partnership ('Compagnie'), barter, interest and discount ('Meriti e sconti simplici e capo d'anno'), and a further treatment of exchange. Altogether the arithmetic is one of the most interesting of the smaller books of the time issued under Florentine influence.

JUAN PEREZ DE MOYA. Ed. pr. 1562. Salamanca, 1562.

Born in San Stefano (Santisteban del Puerto), in the Sierra Morena, in the first third of the sixteenth century. He studied at Alcalá and Salamanca. He was canon at Granada in the latter half of the century.

Title. See Fig. 153.

Colophon. 'En Salamanca. //Por Mathias Gast. // Año de 1562.' (P. 765.)

ARITHMETICA PRACTICA, Y SPECV-latiua del Bachiller Iuan Perez de Moya.

Agora nueuamente corregida, y añadidas por el mismo author muchas cosas, con otros dos libros, y vna Tabla muy copiosa de las cosas mas notables de todo lo que en este libro se contiene.

Va dirigida al muy alto y muy poderoso señor don Carlos Principe de España nuestro señor.

Con licencia y priuilegio Real.

EN SALAMANCA.
Por Mathias Gast.
1562

Esta tassado à cinco blancas el pliego.

FIG. 153. TITLE PAGE OF THE 1562 MOYA

Description. 8°, 9.3 × 14.9 cm., the text being 7 × 12.1 cm. 811 pp. (765 numb.), 29 ll. Salamanca, 1562.

Editions. This is the first edition of this work that I have been able to find, in spite of the words 'agora nueuamente corregida' in the title. It appeared again at Alcalá in 1573, and at Madrid in 1598, 8°, and there were at least thirteen editions between 1609 and 1706. For the 1703 edition see below. 534

Moya also published three other works: 'Reglas para cõtar sin pluma y de reduzir unas monedas castellanas en otras,' in 1563, 4°; 'Manval de contadores,' Alcalá, 1582, 8°, and Madrid, 1589, 8°; and 'Tratado de matemáticas,' Alcalá, 1573, fol., containing a section on arithmetic less complete than the 'Arithmetica practica.' 534

This is an elaborate treatise of 765 pages of text, covering the ordinary calculating by algorism, the use of counters, business arithmetic, the elements of algebra (of which word he gives the etymological meaning, adding the rest of the ancient title, 'almucabala'), practical geometry ('Trata algvnas reglas de Geometria pratica neceſſarias para el medir de las heredades,' p. 304), and the calendar. Besides all this, Moya gives a large amount of information concerning matters of historical interest. For example, he treats the subject of notation very fully, giving the Greek, Roman, Hebrew, digital, and astrological systems, together with a brief mention of other systems. Altogether it is the most noteworthy book on mathematics published in Spain in the sixteenth century.

JUAN PEREZ DE MOYA.

Ed. pr. 1562. Barcelona, 1703.

See p. 308.

Title. 'Arithmetica // Practica,// y Especvlativa,// del Bachiller // Jvan Perez // de Moya.// Aora nvevamente corregida, // y añadidas por el miſmo Autor mu-//chas coſas.// Con otros dos Libros,// y vna Tabla muy copioſa de las coſas mas // notables de todo lo que en eſte Libro // ſe contiene.// Año 1703. Con Licencia: En Barcelona, en la Imprenta // de Rafael Figueró.' (F. 1, r.)

Description. 8°, 15 × 20 cm., the text being 10.7 × 17.9 cm. 396 pp. (380 numb.), 39–41 ll. Barcelona, 1703.

Editions. See p. 310.

See p. 310. It shows the popularity of this Spanish treatise that this edition should appear 141 years after the first one, and that other editions were published as late as 1761, 1776, and 1784.

ANONYMOUS. Ed. pr. 1562. Paris, 1562.

Title. See Fig. 154.

Description. 8°, 10.5 × 15.9 cm., the text being 6.9 × 12.5 cm. 98 pp. numb., 24–30 ll. Paris, 1562.

Editions. There was no other edition.

The work is largely theoretical, but it contains a few applied problems. It was without merit.

Other works of 1562. Benese, p. 182, 1536; Budaeus, p. 99, 1514; Euclid, p. 11, 1482; Gemma, p. 200, 1540; Ghaligai, p. 132, 1521; Lossius, p. 290, 1557; Ramus, p. 330, 1569; Riese, p. 139, 1522; Xylander, p. 356, 1577; Francesco Spinola, 'De Intercalandi ratione corrigenda, & tabellis quadratorum numerorum, à Pythagoreis dispositorum,' Venice, 8°. **534**

VICTORINUS STRIGELIUS.

Ed. pr. 1563. Leipzig, 1563.

STRIGEL. Born at Kaufbeurn, December 26, 1524; died at Heidelberg, June 26, 1569. He was professor of theology at Jena, Leipzig, and Heidelberg, and, aside from the work here mentioned, wrote entirely on that subject.

Title. See Fig. 155.

Description. 8°, 9.1 × 14.9 cm., the text being 6.3 × 11.8 cm. 83 ff. unnumb. + 1 blank = 84 ff., 21–27 ll. Leipzig, 1563.

Editions. There was no other edition. It seems by the preface that the work was written in 1551. See p. 249, 1549. For the date of the printing of this work, 1563, see f. 6, v., of the preface. F. 10, v., gives the year in which it was written, MDLI. **535**

This comparatively unknown work presents the subject of arithmetic from the classical standpoint. The author speaks of the dignity of arithmetic ('De dignitate arithmeticae'), and follows this discussion by

ARITHME-
TICA.

PARISIIS,

Apud Andream Wechelum.

1562.

Cum privilegio Regis.

J. M. Thomasij Δ

FIG. 154. TITLE PAGE OF THE ANONYMOUS 1562 ARITHMETIC

Arithmeticus

LIBELLVS

CONTINENS

NON MODO PRAECEPTA

NOTA ET VSITATA, SED

etiam demonstrationes

præceptorum,

EDITVS A

Victorino Strigelio.

Lipsiæ,

IN OFFICINA

VOEGELIANA.

FIG. 155. TITLE PAGE OF STRIGELIUS

a series of definitions from Euclid, a brief treatment of the operations, the theory of proportion, the operations with fractions, and some of the Greek theory of numbers. One curious feature of the book is the notation of Greek fractions after the Arabic manner.

PIERRE SAVONNE. Ed. pr. 1563. Lyons, 1571.

A French arithmetician, born at Avignon c. 1525.

Title. 'L'Arithmetiqve // de Pierre Savonne,// dict Talon, natif d'Auignon // comté de Veniſſe. // En laquelle ſont contenues pluſieurs reigles briefues & ſubtiles, pour les traffiques de pluſieurs pays, mentionnez // en la table dudit liure : avec la difference des poids, aunages // & monnoyes de chacun deſdits lieux, alliage de metaux :// neceſſaire pour tous Maistres de monnoyes, Orfeures & // Changeurs, avec le fait & maniement des Changes & Ban-//ques qui ſe font iournellement à Lyon, & par les places ac-//coustumees : comme Flandres, Angleterre, Heſpagne,// Italie, & autres lieux.// A Lyon, par Benoist Rigavd. // M. D. LXXI.' (F. 1, r.)

Colophon. 'A Lyon, de l'Imprimerie de // Pierre Rouſsin.// 1571.' (F. 148, r.)

Description. 8°, 10.8 × 17 cm., the text being 7.4 × 13.6 cm. 156 ff. (147 numb.), 28–30 ll. Lyons, 1571.

Editions. Paris, 1563, 4°; ib., 1565, 4°; Lyons, 1571, 8° (here described); ib., 1585; ib., 1588, 8°.

This is one of the early French commercial arithmetics, well arranged but with no marked peculiarities. It was so popular that an edition appeared as late as 1672. Like all arithmetics appearing in Lyons it devotes much attention to banking and exchange, this city being at that time the commercial center of France, and the seat of one of the great international fairs.

Other works of 1563. Bæda, p. 131, 1521; Boissière, p. 262, 1554; Feliciano, p. 148, 1526; Gemma, p. 205, 1540; Moya, p. 310, 1562; Ortega, p. 93, 1512; Peletier, p. 245, 1549; Wolffgang Hobel, 'Ein nützlich Rechenbüchlein mit viel ſchönen Regeln und Fragſtücken,' Nürnberg, with editions s. l. (Nürnberg?), 1565 and 1577, 8°; Simon Schweder, 'Rechenbuch von alles kauffmanschaft der Landt auff der Feder und Linien,' Königsberg, 8°. 535; 536

COSIMO BARTOLI. Ed. pr. 1564. Venice, 1589.

A Florentine geometer, born in 1503; died in 1572. He also translated the works of Finaeus (see pp. 160, 164).

Title. 'Cosimo Bartoli//gentil'hvomo, et//Accademico Fiorentino,//Del Modo di Misvrare//le diſtantie, le ſuperficie, i corpi, le//piante, le prouincie, le proſpettiue, &//tutte le altre coſe terrene, che poſſo-//no occorrere a gli huomini,//Secondo le vere regole d' Euclide, & de gli altri//piu lodato ſcrittori.// (Elaborate woodcut.) In Venetia, Per Franceſco Franceſchi Saneſe. 1589.' (F. 1, r.)

Colophon. 'In Venetia,//Per Franceſco Franceſchi Saneſe. //M. D. LXXXIX.' (F. 148, r.)

Description. 4°, 15.5 × 21.6 cm., the text being 11.2 × 15.5 cm. 148 ff. (6 unnumb.), 29 ll. Venice, 1589.

Editions. Venice, 1564, 4°; ib., 1589, 4° (here described), and one edition after 1601. The dedicatory epistle is dated 'il di 10. di Agoſto del 1559,' so there may have been an earlier edition than that of 1564.

Although the book is on practical mensuration, the 'libro sesto' (f. 130, r.) is upon square and cube root. The galley method is used, and the common sixteenth-century device of annexing $2n$ ciphers in square root and dividing the root by 10^n (and similarly for cube root) is employed. Bartoli also gives a table of squares to 662^2. The chapter on roots is followed by one on the 'Regola delle tre coſe, ouero quattro proportionali.'

Works of 1564. Albert, p. 180, 1534; Apianus, p. 155, 1527; Benese, p. 182, 1536; Euclid, p. 240, 1545; Gutiérrez de Gualda, p. 167, 1531; Herbestus, p. 303, 1561; Köbel, p. 111, 1514; Mariani, p. 181, 1535; Medlerus, p. 223, 1543; Riese, p. 139, 1522; Sole, p. 146, 1526; Tagliente, p. 115, 1515; Thierfelder, p. 391, 1587; Ulman, p. 391, 1587 (Thierfelder); Yciar, p. 249, 1549; Barlaamo, 'Arithmetica demonstratio eorum quæ in secundo libro Elementorum (Euclidis) sunt,' Strasburg (see also p. 343, 1572); Giovanni Camilla, 'Enthosiasmo,' Venice, 8° (containing a little work on arithmetic); Manuel Fernández Lagasa, 'Libro de quentas,' Salamanca, 4°, with a little work on arithmetic; Petrus Nonius (Nuñez), 'Libro de Algebra en Arithmética y Geometría,' with two editions at Antwerp in 1567, 8°; the 'Opera' of Nonius appeared at Basel in 1592. 535

PIERRE FORCADEL. Ed. pr. 1565. Paris, 1565.

See p. 284.

Title. See Fig. 156.

Description. 4°, 14.8 × 20 cm., the text being 9.9 × 15.9 cm. 192 pp. numb., 32–39 ll. Paris, 1565.

Editions. See p. 284. This is the third work on arithmetic published by Forcadel. The dedicatory epistle is dated 'De Paris, ce 19. de May. 1565.' There was only one edition, and this is not often found in dealers' catalogues.

In this book Forcadel does not, as in his other works, take up the subject of counter reckoning, but he gives a very satisfactory treatment of the fundamental operations, the rule of three, partnership, alligation, and the common applications of the day. It is not, however, as practical as the arithmetics of Savonne and Trenchant.

ANTICH ROCHA de Gerona.

Ed. pr. 1565. Barcelona, 1565.

A Spanish arithmetician of the second half of the sixteenth century, born in Gerona, lecturer at Barcelona.

Title. See Fig. 157.

Description. 8°, 9.7 × 14.7 cm., the text being 7.5 × 12 cm. 314 ff. (267 numb.), 28 ll. The dedication by Rocha bears the date 'Hecha en Barcelona, a. 23. de Nouiembre. 1564.' Barcelona, 1565.

Editions. There was no other edition.

This rare compilation is based upon several Italian books, and the writer claims to have consulted a large number of authors. His list includes the names of Feliciano, Faber Stapulensis, Buteo, Scheubel, Finaeus, Ramus, Gemma, and various other writers of the time. Rocha has, however, omitted some of the best textbook-makers who preceded him. The book is a fairly complete elementary treatise, the writer having taken up the fundamental operations with various kinds of numbers, and treated each rule in a rather scientific way. Although a considerable number of practical problems relating to mercantile affairs appear in the last half of the work, the style of the writer is so prolix that the book could never have been well received by the mercantile classes. Bound with this work is another with the following title: 'Compendio

ARITHMETIQVE
ENTIERE ET ABREGEE
DE PIERRE FORCADEL,
LECTEVR DV ROY ES
MATHEMATIQVES.

A PARIS.
Chez Charles Perier, rue S.Iean de Beauuais,
au Bellerophon.

1565.

FIG. 156. TITLE PAGE OF THE 1565 FORCADEL

y breue // inſtruction por tener Libros de Cuen//ta, Deudas, y de Mercaduria: muy prouechoſo // para Mercaderes, y toda gente de negocio. // traduzido de Frances en Caſtellano. En Barcelona.// En caſe de

ARITHMETICA por Antich Rocha de Gerona compueſta, y de varios Auctores recopilada: prouechoſa para todos eſtados de gentes.

Va anadido vn Compendio, para tener y regir los libros de Cuenta: traduzido de lengua Franceſa en Romance Caſtellano.

EN BARCELONA
En caſa de Claudio Bornat, a la Aguila fuerte.

1565.

Con priuilegio por diez años.

FIG. 157. TITLE PAGE OF ROCHA

Claudio Bornat // al Aguila fuerte. 1565.// Con priuilegio por diez años.' (F. 286.) This relates, as the title suggests, entirely to bookkeeping, and is one of the earliest treatises upon the subject in the Spanish language. In the fourth book of this second part is a treatise on algebra, one of the first to appear in Spain.

Rocha speaks of an Aritmética by Juan Ventallol, of which we do not know the date, but which must have appeared before 1565 (p. 298).

ERHART HELM. Ed. pr. 1565. Frankfort, 1592.

A Frankfort arithmetician and gauger of the middle of the sixteenth century.

Title. '1592. Erhart Helm/ // Mathematicus // zu Franckfurt am Mayn/ // von // Geometrifcher Abmeffung der Erden.// Item :// Kůnftliche Vifier vnnd // Wechfelruthen/ auß dem Quadrat/ // durch die Arithmeticam vnnd Geometri-//am/ gerecht zumachen/ fampt einer luftigen behen-//digkeit in Weinrechnung/ // alles durch obgedachten // Authorem befchrieben/ vnd jetzt von neuwem // widerumb fleiffig erfehen vnd cor-//rigiert.// Franckf. bey Chrift. Egen. Erben. 1592.' (F.1, r.)

Description. 8°, 9.5 × 15.4 cm., the text being 6.5 × 12.2 cm. 23 ff., 26–27 ll. Bound with Adam Riese's arithmetic of 1592 (p. 143). Frankfort, 1592.

Editions. The work appears in the 1565 edition of Adam Riese (p. 142), but not with a separate title page. I have seen no separate edition before this one of 1592.

This is a brief treatise on mensuration, and in particular on gauging, but it contains some explanation of arithmetical processes, including the extraction of roots, and a table of square roots to the equivalent of three decimal places.

Other works of 1565. Agrippa, p. 167, 1531; Apianus, p. 62, 1496 (Jordanus); Belli, p. 343, 1573; Cusa, p. 43, c. 1490; Delfino, p. 275, 1556; Fischer (Piscator), p. 247, 1549; Gemma, p. 200, 1540; Hobel, p. 314, 1563; Jacob, p. 295, 1560; Jordanus, p. 62, 1496; Menher, p. 249, 1550; Padovanius, p. 389, 1587; Riese, p. 142, 1522; Savonne, p. 314, 1563; Priscian, Rhemnius, Fanius, Bæda, Metianus, 'Liber de nummis, ponderibus, mensuris, numeris, eorumque notis, et de vetere computandi ratione, ab Elia Vineto emendati,' Paris, 8°. (Priscian's 'De figuris et nominibus numerorum' had already appeared in his works published at Venice in 1470, fol., with later editions in 1488, 1492, 1495, 1496, 1519, 1525. Such books, of which Mr. Plimpton has several, are not generally included in this list.) 535

HIERONYMUS MUNYOS. Ed. pr. 1566. Valencia, 1566.

Born at Valencia; died in 1584. He was professor of mathematics and Hebrew at Ancona, and later at Valencia. He also wrote an astronomical work. The family name is more strictly Muñoz.

Title. See Fig. 158.

Description. 4°, 15 × 20.4 cm., the text being 9.3 × 14.6 cm. 4 ff. unnumb. + 77 numb. + 1 blank = 82 ff., 27–31 ll. Valencia, 1566.

Editions. There was no other edition. The 'epistola' is dated 'Calendis Aprilis, anni M. D. Lxvj.'

The work consists of three books. The first treats of the fundamental operations, including proportion and some work in the Greek theory of numbers. The second book treats of fractions, including sexagesimals, these being needed by the astronomers for whom Munyos was writing. The third book relates to ratio and proportion. Altogether the work is too theoretical to have much influence upon the development of arithmetic.

IAN TRENCHANT. Ed. pr. 1566. Lyons, 1578.

A Lyons arithmetician, born c. 1525.

Title. 'L'Arithme-//tiqve de Ian // Trenchant,// Departie en trois // liures.// Enſemble vn petit diſcours des Changes.// Avec // L'art de calculer aux Getons.// Reueüe & augmentée pour la quatriême edition,// de pluſieurs regles & articles,// par l'Autheur. // A Lyon,// par Michel Iove,// et Iean Pillehotte.// à l'enſeigne du Ieſus.// 1578.// Auec priuilege du Roy.' (P. 1.)

Description. 8°, 10.4 × 15.5 cm., the text being 7 × 13.7 cm. 375 pp. numb. + 5 unnumb. = 380 pp., 30–31 ll. Lyons, 1578.

Editions. Lyons, 1566; ib., 1571; ib., 1578, 8° (here described). The dedication of the 1578 edition bears the date 'De Lyon ce 9. de Iuillet 1571,' and this is often given as the date of the first edition. Although this is described on the title page as the fourth edition, I know of only two earlier. Cantor says that not less than six editions were published at Lyons from 1588 to 1602. I have seen mentioned editions of 1608, 1610, 1632, and 1643. 537

Trenchant was one of the best of the sixteenth-century textbook-makers of commercial arithmetic in France. His work is divided into

INSTITVTIONES
ARITHMETICAE AD PERCIPIENDAM ASTROLOGIAM ET
Mathematicas facultates necessariæ.
AUCTORE
Hieronymo Munyos Valentino Hebraicæ linguæ pariter atq; Mathematum in Gymnasio Valentino publico professore.

VALENTIAE.
Ex typographia Ioannis Mey.
Anno 1566.

FIG. 158. TITLE PAGE OF MUNYOS

three books, the first dealing with the fundamental operations with integers and fractions, and containing a considerable number of applied problems. The second book treats of the rule of three in its various forms together with such applications as barter, partnership, commission, and alloys. The third book treats of the properties of numbers, including figurate numbers, roots, and progressions, and has some work on discount, together with a few recreations. In the 1578 edition the third book is followed by a chapter on exchange, and an explanation of the method of calculating with counters.

GEORGIO LAPAZZAIA. Ed. pr. 1566. Naples, 1569.

LAPEZAJA, LAPIZAYA, LAPAZAIA. Probably a resident of Naples, but born at Monopoli, a town in the province of Bari, on the Adriatic. He was a priest, and wrote only the work here described.

Title. See Fig. 159.

Colophon. 'In Napoli // Apreffo Mattio Cancer. M.D.LXIX. // Con Priuilegio per anni diece. // Marius Carrafa Archiepifcopus Neapolitanus.' (P. 262.)

Description. 4°, 14.1 × 21 cm., the text being 10.8 × 16.9 cm. 13 pp. unnumb. + 250 numb. = 263 pp., 29 ll. Naples, 1569.

Editions. Naples, 1566, Latin edition with the title 'De familiarite arithmeticæ et geometriæ,' this date appearing also in the privilege, 'Datum Neapoli die vltimo Iulii M.D.LXVI;' ib., 1566, Italian edition; ib., 1569, 4° (here described); 1575; Naples, 1590, 4° (p. 324). There were several editions after 1600, one as late as 1784. For the 1601 edition see p. 324. **496;537**

The book has nothing to commend it except its popular style. Lapazzaia begins with the fundamental operations with integers, and then treats of ratio, fractions, and progressions. He considers also the rule of three, the rule of five, interest, exchange, partnership, alligation, rule of false, and the extraction of roots. The last part of the work is on mensuration.

Other works of 1566. Belli, p. 343, 1573; Fischer (Piscator), p. 247, 1549; Gemma, p. 200, 1540; Herbestus, p. 303, 1561; Monzó, p. 292, 1559; Georg Meyer, 'Rechenbüchlein defs Silberkauffs und gemachter Arbeit,' Augsburg, 16°; Mathäus Nessen (Nesse), 'Zwei neue Rechenbücher,' Breslau, 8°; Johannes de Segura, 'Mathematicae quaedam selectae propositiones,' Alcalá, 4°, and 'Compendium Arithmeticae et Geographiae partis,' Alcalá, 4°. **537**

D'ARITMETICA E GEO·
METRIA DELL'ABBATE GEOR·
GIO LAPAZZAIA MONOPOLITANO.

FIG. 159. TITLE PAGE OF THE 1569 LAPAZZAIA

GEORGIO LAPAZZAIA. Ed. pr. 1566. Naples, 1590.

See p. 322.

Title. 'Opera // terza // de Aritmetica // et Geometria.// Dell'abbate Georgio Lapazaia//da Monopoli.//Intitolata il Ramaglietto.// In Napoli,//Appresso gli Eredi di Mattio Cancer.// M. D. LXXXX.' (P. 1.)

Colophon. 'In Napoli // Appresso gl'Eredi di Mattio Cancer.// M. D. LXXXX.' (P. 176.)

Description. 4°, 13.9 × 18.7 cm., the text being 10.8 × 16.5 cm. 1 p. unnumb. + 1 blank + 174 numb. = 176 pp., 29 ll. Naples, 1590.

Editions. See p. 322.

Although bearing a different title, this is merely a revision of the 1566 work, with a slight variation in the problems. The preface is dated 1569, when the second edition of the Italian version appeared, but it is not found in that edition. The title 'Opera terza' means simply the second revision, or the third writing of the book. Lapazzaia's work shows the increasing attention given by the Church schools to the business needs of the people.

GEORGIO LAPAZZAIA. Ed. pr. 1566. Naples, 1601.

See p. 322.

Title. 'Libro // d'Aritmetica // e Geometria, // dell'Abbate Giorgio Lapazzaia // Canonico Monopolitano, e Protonotario Apostolico.// Nouamente in quest'vltima impressione espurgato da molti errori, & arric-//chite d'vna Prattica d'Abbaco, non meno vtilissima, che necessaria.// Al Signor Diego d'Aldana,// Presidente della Regia Camera della Summ. per Sua Maiestà.// (Large woodcut, coat of arms.) In Napoli,// Appresso Tarquinio Longo. MDCI.' (P. 1.)

Colophon. 'In Napoli,// Appresso Tarquinio Longo. MDCI.' (P. 215.)

Description. 4°, 14.9 × 20 cm., the text being 10.7 × 15.8 cm. 216 pp. (3 blank, 8 unnumb.), 30–38 ll. Naples, 1601.

See p. 322.

NICOLAUS PETRI. Ed. pr. 1567. Amsterdam, 1635.

Born at Deventer. He taught at Amsterdam from 1567 to 1588. He also wrote on algebra and astronomy.

Title. 'Practicque //Om te leeren//Reeckenen/ Cypheren // ende Boeckhouwen/ met die regel Cofs/ //ende Geometrie/ feer profijtelijcken voor allen // koop-luyden. Van nieus gecorrigeert // ende vermeerdert/ //Deur Nicolaum Petri Daventrienfem.// L'homme propofe, Et dieu dispofe. A° 1603.// (Woodcut of author.) t'Amstelredam,//Voor Hendrick Laurentfz. Boeckvercooper op het // water int Schrijf-boeck, Anno 1635.' (F. 1, r.)

Description. 8°, 10.3 × 16.3 cm., the text being 8.9 × 14.4 cm. 6 ff. unnumb. + 281 numb. = 287 ff., 23–32 ll. Bound with this is '1596.// Iournael-Boeck // gheteeckent met die // Letter // Anno M. DC. XXXV.,' 10 ff. Amsterdam, 1635.

Editions. From the preface it appears that the work was first published in 1567. The other sixteenth-century editions were Amsterdam, 1576, 8°; 1583; 1591; and Alkmaar, 1596, 8°. 537

The arithmetic is of the ordinary Dutch type. It uses only the galley form of division, and its chief value to the student of history lies in its business problems.

Other works of 1567. Agrippa, p. 167, 1531; Borghi, p. 16, 1484; Cataneo, p. 244, 1546; Gemma, p. 205, 1540; Mariani, p. 181, 1567; Nonius (Nuñez), p. 315, 1564; Ramus, p. 263, 1555; Tagliente, p. 115, 1515; Conradus Dasypodius, 'Logistica,' Strasburg, 8°; Guill. de la Toissonière, 'Compost arithmétical,' Lyons.

STEFANO GHEBELINO. Ed. pr. 1568. Brescia, 1568.

An arithmetician of Brescia, of the second half of the sixteenth century.

Title. See Fig. 160.

Colophon. 'In Brescia,// Apresso Vincenzo // di Sabbio.// M. D. LXVIII.' (F. 34, r.)

Description. 4°, 13.5 × 19.5 cm., the text being 10.1 × 16.5 cm. 11 ff. unnumb. + 33 numb. = 44 ff., 37 ll. Brescia, 1568.

Editions. There was no other edition.

The dedicatory epistle shows that this work was written at Brescia in October, 1568. It is composed almost entirely of tables for the use

TAVOLE BREVISSIME ARITHMETICHE, Con lequali ciascuno ageuolißimamente potrà far' in un tratto. quasi ogni sorte de conti, Composte per Stefano Ghebelino Bresciano, & della Scola di M. Hieronimo Cataneo Nouarese Matematico raro.

FIG. 160. TITLE PAGE OF GHEBELINO

of merchants and bankers in the north of Italy. There is some explanatory matter in the beginning relating to the arithmetic of exchange, but the work can hardly be called a school textbook. Riccardi speaks of it as 'uno dei primi esempi di tavole di conti fatti.'

HUMPHREY BAKER. Ed. pr. 1562. London, 1580.

Born at London; died after 1587.

Title. 'The Well ſpring of // Sciences.// Which teacheth the perfect // worke and practiſe of Arith-//meticke, both in whole Num-//bers and Fractions: ſet // forthe by // Humfrey Baker // Londoner, 1562.// And nowe once agayne peruſed // augmented and amended in all // the three partes, by the sayde // Aucthour: where unto he // hath alſo added certein // tables of the agree-// ment of meaſures // and waightes // of diuers places in Europe, // the one with the other, as // by the table following // it may appeare.// 1580.' (F. 1, r.)

Description. 8°, 9.9 × 14 cm., the text being 5.6 × 11.2 cm. 227 ff. (28 unnumb., last folio missing), 24–26 ll. London, 1580.

Editions. London, 1568, 8° (written in 1562, which explains the date on the title page); ib., 1574, 8°; ib., 1580, 8° (here described); ib., 1583, 8°; ib., 1591, 8°. There were several editions after 1601 (see pp. 328, 329). **537**

For a long time Baker's arithmetic was the only English rival to Recorde's 'Ground of Artes' (see p. 213), and it was in many respects better than that popular work. This edition is more complete than that of 1568, the book having, as the author states, been rewritten. In 'The Prologue to the gentle Rerder' he says: 'Hauing ſometime now twelue yeres ſithence (gentle Reader) publiſhed in print one Engliſhe boke of Arithmetick, conteyning as I suppose, ſundry necſsarie and profitable documentes for ſuch as are vvilling to attayne any knowlege therein. I have bene often ſince that time, and of very late alſo, requeſted by ſundry of my friendes to peruſe the ſame vvorke, and as I ſhold nowe iudge it expedient, to adde ſomething more therevnto, and to amplifye the ſame.' He complains of the criticism of foreigners that English arithmetic is not as advanced as that on the continent: 'For vvhen I perceyued the importunitie of certayne ſtraungers not borne within this lande, at this preſent, and of late dayes ſo farre proceeding, that they aduaunced and extolied them ſelues in open talke and writinges, that they had attayned ſuch knowledge and perfection in Arithmetike, as no engliſh man the like: Truly me thought that the ſame reporte not only tended to the (disprayſe) diſpraiſe our Countreymen in general: But touched eſpetially ſome others & me, that had trauailed & written publiquely in the ſame facultie. For vnto this ſame effecte they haue of late paynted the corners and poſtes in euery place within this citie with their peeuiſhe billes, making promiſe and bearinge men in hande that they coulde teache the ſumme of that Science in breefe Methode and compendious rules ſuch as before their arriuall hath not bene taughte within this Realme.' These words, and others in the same strain, give an interesting picture of English arithmetic in 1580, and of the work of the teacher at that time. The criticism was a just one, for the Dutch, French, Germans, Spanish, and Italians were much ahead of the English at that period in the matter of arithmetic.

Baker follows the continental models, giving the usual operations and the applications to 'Marchandiſe,' 'Felowſhip,' barter, alligation, false position, and the like. He closes his text with 'Quſtions of Paſtime.' He still uses duplation, generally uses the form 'substraction' (following the Dutch books of the time), and makes relatively little of 'Deuiſion,' which he treats by the usual galley method, but he succeeds in producing a fairly practical mercantile book.

Other works of 1568. Anianus, p. 32, 1488; Delfino, p. 275, 1556; Gemma, p. 200, 1540; Gülfferich, p. 257, 1552; Lonicerus, p. 253, 1551; Lossius, p. 290, 1557; Riese, p. 139, 1522; Sfortunati, p. 174, 1534; Mauritius Steinmetz, 'Aritmeticae praecepta in quaeſtiones redacta,' Leipzig, sm. 8°.

HUMPHREY BAKER. Ed. pr. 1562. London, 1659.

See p. 327.

Title. 'The // Wel-spring // of // Sciences: // teaching The perfect Work and Practice of // Arithmetick,// both in Numbers and Fractions.// Set forth by // Humphrey Baker // Londoner.// And now again Peruſed, Augmented, and // Amended in all three Parts, by // the ſaid Authour.// Whereunto are added certain Tables of // the agreement of Meaſures & Weights // of divers places in Europe,// the one with the other, as by the Table appeareth.// London,// Printed for A. Kemb, at St. Margarets Hill in // Southwark, to bee ſold by Tho. Brewſter,// at the three Bibles in Pauls Church-yard: 1659.' (P. 1.)

Description. 8°, 8.4 × 13.6 cm., the text being 6.3 × 11.2 cm. 366 pp. (54 unnumb.), 26–27 ll. London, 1659.

Editions. See p. 327.

This is not materially different from the 1580 edition (p. 327). The publisher says that 'The friendly Reader may pleaſe to take notice that in this Impreſſion of 1659, the whole Book hath been reviſed, every Queſtion therein examined, the Faults that were committed in former Impreſſions, Corrected, the whole reſtored to its first integrity.' In spite of this statement, there is little improvement in the book either in methods of operating or in symbolism. It is interesting to read that 'The Fractions [in the tables at the end of the book] which before were in the common way (and ſo the figure being ſmall in many not diſcerned) are put into the decimal parts, and ſo the ſame with the integral, but farre more true than the Common Fractions can expreſs it in one figure, and if in the common it be expreſt in many, (as it must be, if true) then the decimal is far more eaſie, becauſe the Denominator is one and the ſame to all, whereas the other is differing.' In fact, very slight knowledge of decimals is shown, and when they are employed the bar is generally used instead of the point.

HUMPHREY BAKER. Ed. pr. 1562. London, 1687.

See p. 327.

Title. 'Licenſed,// Feb. 28, 168$\frac{6}{7}$.// Rob. Midgley.' (P. 4, first page of print.) 'Baker's // Arithmetick :// Teaching // The perfect Work and Practice of // Arithmetick both in // Whole Numbers & Fractions.// Whereunto are Added // Many Rules and Tables of // Intereſt, Rebate, and Purchaſes, & c.// Also // The Art of Decimal Fractions,// intermixed with Common Fractions, for the // better Underſtanding thereof.// Newly Corrected and Contracted, and // made more plain and eaſie // By Henry Phillippes.// London.// Printed by J. Richardſon for William Thackery at the // Angel in Duck-Lane, and Matthew Wotton at the Three // Daggers in Fleet ſtreet, and George Conyers at the // Ring without Ludgate, 1687.' (P. 5.)

Description. 12°, 8.3 × 14.3 cm., the text being 6.8 × 12.8 cm. 8 pp. blank + 10 unnumb. + 228 numb. = 246 pp., 32–39 ll. London, 1687.

Editions. See p. 327.

Phillippes, the editor, pays a deserved tribute to Baker in his letter 'To the Reader.' He begins as follows: 'This little Book, as it was one of the firſt, ſo it is one of the beſt of this Subject, and hath had as good Acceptance, as any other; which may appear by the often Impreſſions of it. Indeed as long as the Author lived, he was careful to be ſtill adding and correcting it: and though he be dead, yet his Book is thought worthy to live, and not only to live, but to flouriſh.' It certainly speaks well of the book that this edition should have been published 121 years after the first one appeared. The treatment of decimal fractions is very satisfactory, and, of course, is not found in the original edition. These fractions, the necessity for which became apparent in the sixteenth century, were first scientifically treated at any length in a work by Stevin, published in 1585. (See p. 386.)

PETRUS RAMUS. Ed. pr. 1569. Basel, 1569.

See p. 263.

Title. 'P. Rami Arith-//meticæ libri // dvo: Geometriæ // septem et viginti.// (Woodcut.) Basiliæ, per Evsebivm // Epiſcopium, & Nicolai fratris hæredes.// Anno M. D. LXIX.' (P. 1.)

Description. 4°, 17.5 × 23.3 cm., the text being 12.3 × 18.2 cm. 198 pp. (2 blank, 6 unnumb.), 42 ll. Basel, 1569.

Editions. Basel, 1569, 4° (here described); Paris, 1577, 8° (p. 331); Basel, 1580, 4° (p. 331); Paris (Stadius edition), 1581, 12°; Frankfort (Schonerus edition), 1586, 8° (p. 331); ib., 1591, 8° (Stegerus edition); ib., 1592, 8° (Schonerus); ib., 1596, 8° (the Snellius and Schonerus 'Explicationes,' p. 333); ib., 1599, fol. (Schonerus, p. 333); Lemgo, 1599, 4°. There was also published at Paris in 1562, and in two editions the same year, an 'Arithmetica,' in two books, without the author's name, attributed to Ramus, but I do not know whether it is the same as this work. There was also an English edition of 'The Art of Arithmeticke in whole numbers and fractions . . . by P. Ramus . . . translated . . . by William Kempe,' London, 1592, 8°. 537

This is a better book than the 'Libri Tres' of 1555. Although it is too theoretical to have met the commercial needs, it is a nearer approach to a practical work than its predecessor.

PETRUS RAMUS. Ed. pr. 1569. Paris, 1577.

See p. 263.

Title. 'Petri Rami // Professoris Regii,// Arithmeticæ // libri dvo.// Parisiis,// Apud Dionyſium Vallenſem, ſub // Pegaſo, in vico Bellouaco.// 1577.' (F. 1, r.)

Description. 8°, 10.3 × 16 cm., the text being 7.1 × 13.1 cm. 97 ff. (2 blank, 1 unnumb.) with chart; 32 ll. Paris, 1577.

See p. 330.

PETRUS RAMUS. Ed. pr. 1569. Basel, 1580.

See p. 263.

Title. 'P. Rami // Arithmeticae // libro dvo // Geometriae // septem et viginti.// (Woodcut.) Basileæ, per Evsebivm // Epiſcopium, & Nicolai fratris hæredes.// M D LXXX.' (P. 1.)

Colophon. 'Basileæ, per Evsebivm Episco-//pium, & Nicolai fratris hæredes. Anno // M. D. LXXX.' (P. 192.)

Description. 4°, 15.9 × 21.2 cm., the text being 11.1 × 17 cm. 200 pp. (9 unnumb.), 28–35 ll. Basel, 1580.

See p. 330.

PETRUS RAMUS and LAZARUS SCHONERUS.
Ed. pr. 1569. Frankfort, 1586.

See p. 263.

Title. See Fig. 161.

Description. 8°, 10.5 × 17.3 cm., the text being 7 × 13.1 cm. 16 pp. unnumb. + 406 numb. = 422 pp., 26–32 ll. Frankfort, 1586.

Bound with this is 'P. Rami, Regii // Eloquentiæ et // Philosophiæ Pro-//feſſoris, liber de moribus // veterum Gallorum,// ad // Carolum Lotharingum // Cardinalem.// Parisiis,// Apud Andream Wechelum.// 1562.// Cum privilegio Regis.'

Editions. See p. 330. The commentary of Schonerus also appeared with that of Snellius in 1596 (p. 333), and without the latter in 1599 (p. 333). There were also editions by Steger published at Leipzig in 1591 and at Frankfort in 1592.

PETRI RAMI
ARITHMETICES LI-
BRI DVO, ET ALGEBRÆ
totidem: á LAZARO SCHONERO
emendati & explicati.

Eiusdem SCHONERI *libri duo: alter, De Numeris figuratis; alter, De Logistica sexagenaria.*

FRANCOFURDI
Apud heredes Andreæ Wecheli,
MDLXXXVI.
Cum S. Cæs. Maiestatis priuilegio ad sexennium.

FIG. 161. TITLE PAGE OF THE 1586 RAMUS

The first of these works is one of several commentaries on the theoretical arithmetic of Ramus (p. 263). It is more practical than that of Snellius (mentioned below), giving the various operations and making an attempt at introducing some commercial problems.

In the first part are included two works by Schonerus, 'De numeris figuratis Lazari Schoneri liber' and 'Lazari Schoneri De logistica sexagenaria liber.' The former is, as the title suggests, a treatise on the Greek theory of numbers, and the second is on the sexagesimal fractions used by the astronomers. Schonerus writes his sexagesimals thus: $\begin{matrix} \text{IIæ} & \text{Iæ} & \text{o} & \text{I} & \text{II} \\ 3. & 39. & 40. & 20. & 40., \end{matrix}$ for $3 \cdot 60^2 + 39 \cdot 60 + 40 + \frac{20}{60} + \frac{40}{60^2}$.
This is one of the early approaches to our symbols °, ′, ″.

PETRUS RAMUS and RUDOLPHUS SNELLIUS.

Ed. pr. 1569. Frankfort, 1596.

See p. 263. SNELLIUS, born at Oudewater, October 8, 1546; died at Leyden, March 2, 1613. **537**

Title. See Fig. 162.

Description. 8°, 9.6 × 16 cm., the text being 6.6 × 12.8 cm. 3 pp. unnumb. + 154 numb. = 157 pp., 29–30 ll. Frankfort, 1596. Bound with this are 'Rvdolphi // Snellii in // P. Rami Geome-//triam Præle-//ctiones,' and 'Rudolphi // Snellii in Sphæram Cor-//nelii Valerii // prælectiones,' both of 1596.

Editions. See p. 330. The first edition of Snell's commentary.

Like the arithmetic of Ramus, this work is theoretical rather than practical. Only the prominence of Ramus could have justified such efforts as these of Snellius, Salignacus, and Urstisius.

PETRUS RAMUS. Ed. pr. 1569. Frankfort, 1599.

See p. 263.

Title. 'Petri Rami // Arithmeticae // libri dvo: Geometriae // septem et viginti.// A Lazaro Schonero recogniti & aucti.// Francofvrti,// Apud Andreæ Wecheli heredes,// Claudium Marnium, & Ioannem Aubrium.// M. D. XCIX.' (P. 3.)

Description. Fol., 17 × 22.5 cm., the text being 11.9 × 17.6 cm. 244 pp. in the arithmetic, 184 pp. in the geometry, 39 ll. Frankfort, 1599. Bound with this is the geometry of Ramus.

IN P. RAMI ARITH-
METICAM

RVDOLPHI
SNELLII

Explicationes lectissimæ:

LAZARI SCHONERI, BERNH. Salignaci, & Christiani Vrstisii, commentationibus passim locupletatæ.

FRANCOFVRTI

Ex Officina Typographica Ioannis Saurii, impensis hæredum Petri Fischeri.

M. D. XCVI.

FIG. 162. TITLE PAGE OF THE 1596 RAMUS AND SNELLIUS

PETRUS RAMUS. Ed. pr. 1569. Basel, 1569.

See p. 263.

Title. See Fig. 163.

Colophon. 'Basileæ, per Evsebivm Episco-//pium, & Nicolai ſratris hæredes. Anno Salutis humanæ//M. D. LXIX.' (P. 190 of the geometry, bound with the above, or 534 of the entire book.)

Description. 4°, 17.5 × 23.3 cm., the text being 12.3 × 18.2 cm. 16 pp. unnumb. + 320 numb. = 336 pp. of the above (not including the rest of the work), 42 ll. Basel, 1569.

Editions. Basel, 1569, 4° (here described); ib., 1578 (Schonerus edition); Frankfort (also Schonerus edition), 1599, 4°. The first three books also appeared at Paris in 1567, 8°, under the title 'Præmium Mathematicarum.' 538

An extensive and tiresome treatise on the philosophy of elementary mathematics in general.

THOMAS DE MERCADO.
Ed. pr. 1569. Salamanca, 1569.

A Spanish priest of the middle of the sixteenth century.

Title. See Fig. 164.

Description. 8°, 13.5 × 19.4 cm., the text being 11.5 × 16 cm. 277 ff. (29 unnumb.), 32 ll. Salamanca, 1569.

Editions. There was no other edition.

This is not a textbook on arithmetic, but a treatise on the applications of the subject to mercantile affairs. It is so prolix and theoretical that it was never republished. That it is the first edition appears from 'La Tassa' and from the dedication to the king, the former being dated October 6, 1569, and the latter May 6 of the same year. The license is, however, dated August 13, 1568, and one of the decrees May 9, 1568. Although the work professes to be of a mercantile character, it is too ponderous in style for the purpose for which it was intended. It is interesting historically because in several chapters the author has considered the development of arithmetic and of mercantile customs. It is also interesting because of its reference to the recently awakened commerce. For example, chapter 13 has the title 'De los tratos de Indias, y tratantes en ellos.' Chapter 16 is also suggestive of the methods of trade of the period, the title being 'De los baratas y

Nobilissimo iuueni Nicolao Burckhusio hunc librum Ioannes Sturmius [illegible] [illegible] [illegible] d. d.

D. 15 Ap 158[illegible]

P. RAMI SCHOLARVM MATHEMATICARVM, LIBRI VNVS ET TRIGINTA.

BASILEAE, PER EVSEBIVM EPISCOPIVM, & Nicolai Fratris hæredes.

ANNO M. D. LXIX.

P. Ramus Jo. Sturmio Argentin. academiae nostrae dono dedit

Fig. 163. Title page of the *Libri vnvs et triginta* of Ramus

TRATOS Y CON-
TRATOS DE MERCADERES
y tratantes diſcididos y determinados, por
el Padre Preſentado Fray Thomas
de Mercado, de la orden de los
Predicadores.

Con licencia y priuilegio real.

EN SALAMANCA.
Por Mathias Gaſt. Año de
1569.

Eſta taſſado en cinco reales.

FIG. 164. TITLE PAGE OF MERCADO

de la nauegacion de las Indias.' The treatment of interest and exchange is more extensive than usual, the former having been by no means popular in Spain. The last part of the book is devoted entirely to legal questions.

Other works of 1569. Belli, p. 343, 1573; Camerarius, p. 263, 1554; Gemma, p. 200, 1540; Herbestus, p. 303, 1561; Jacob, p. 298, 1560; Lapazzaia, p. 322, 1566; Lossius, p. 290, 1557; Mariani, p. 181, 1535; Monzó, p. 292, 1559; Urstisius, p. 361, 1579; Jacob Frey, 'Exempelbüchlein allerley Kaufmannshändel,' Nürnberg (there was also an Augsburg edition of 1603, 16°); Adriaen van der Gucht, 'Cyferbouck,' Bruges, 4°; James Peele, 'The patheway to perfectnes in th' accomptes of debitour and creditour,' London, fol. (second edition). 538

HIERONYMUS CARDANUS.

Ed. pr. 1570. Basel, 1570.

See p. 193.

Title. See Fig. 165.

Colophon. 'Basileæ,// ex officina Henricpetrina, Anno // Salvtis M. D. LXX. Mense // Martio.' (P. 111 of the third part.)

Description. Fol., 20.4 × 30 cm., the text being 13.2 × 24 cm. 291 pp. (4 blank, 16 unnumb.) in this book. Bound with this is the 'Ars Magna' (second edition), 163 pp. numb., and the 'De aliza regvla liber, hoc est, algebraicæ logiſticæ ſuæ . . .,' 120 pp. (111 numb.), 41 ll. Basel, 1570.

Editions. There was no other separate edition of the 'Opus Novum.'

This work is particularly interesting in its application to physical problems, these being well illustrated. The only reason for including it in a list of arithmetics is that it contains some work on proportion less geometric than that given in Euclid. See also p. 193, 1539.

Other works of 1570. Belli, p. 343, 1573; Boethius, p. 27, 1488; Feliciano, p. 148, 1526; Forcadel, p. 284, 1556–57; Gemma, p. 200, 1540; Glareanus, p. 192, 1539; Lonicerus, p. 253, 1551; Recorde, p. 214, c. 1542; Riese, p. 139, 1522; Tagliente, p. 115, 1515; Anonymous, 'Briefue arithmétique fort facile à comprendre'; Johann Weber, 'Gerechnet Rechenbüchlein auf Erfurtischen Wein- und Tranks-Kauff,' Erfurt, with a second edition in 1583.

Works of 1571. Digges, p. 343, 1572; Gemma, p. 206, 1540; Riese, p. 139, 1522; Savonne, p. 314, 1563; Stifel, p. 260, 1553;

HIERONYMI CARDANI MEDIO LANENSIS, CIVISQVE BONO- NIENSIS, PHILOSOPHI, MEDICI ET Mathematici clarissimi,

OPVS NOVVM DE PROPORTIONIBVS NVMERORVM, MOTVVM, PONDERVM, SONORVM, ALIARVMQVE RERVM

mensurandarum, non solùm Geometrico more stabilitum, sed etiam uarijs experimentis & obseruationibus rerum in natura, solerti demonstratione illustratum, ad multiplices usus accommodatum, & in V libros digestum.

PRAETEREA.

ARTIS MAGNÆ, SIVE DE REGVLIS ALGEBRAICIS, LIBER VNVS, ABSTRVSISSIMVS & inexhaustus plane totius Arithmeticæ thesaurus, ab authore recens multis in locis recognitus & auctus.

ITEM.

DE ALIZA REGVLA LIBER, HOC EST, ALGEBRAICAE logisticæ suæ, numeros recondita numerandi subtilitate, secundum Geometricas quantitates inquirentis, necessaria Coronis, nunc demum in lucem edita.

Opus Physicis & Mathematicis inprimis utile & necessarium.

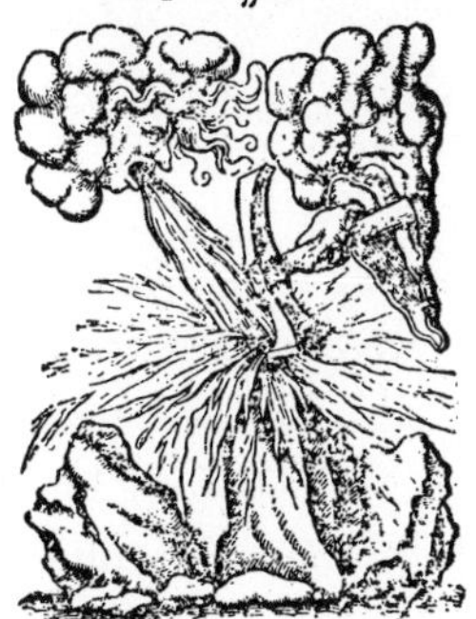

Cum Cæs. Maiest. Gratia & Priuilegio.

BASILEÆ.

FIG. 165. TITLE PAGE OF THE 1570 CARDAN

Trenchant, p. 320, 1566; Anonymous, 'Les principaux fondemens d'arithmétique'; Nicolaus Eschenburg, 'Arithmetica logiftica,' Frankfort; Alex. Vandenbussche, 'Arithmétique militaire,' Paris, 4°. 497

FRANCISCUS BAROCIUS. Ed. pr. 1572. Venice, 1572.

See p. 295.

Title. See Fig. 166.

Description. 4°, 13.9 × 19.3 cm., the text being 10.3 × 17 cm. 3 ff. unnumb. + 23 numb. = 26 ff., 17–36 ll. Venice, 1572.

Edition. There was no other edition. A German translation was published in Leipzig in 1616.

This is an attempt to popularize the mediæval number game of Rithmomachia (Rithmimachia, Rythmomachia), set forth in Latin possibly by Shirwode or by Faber Stapulensis, in an edition of Boethius in 1496 (see p. 63), and afterwards amplified by Claude Boissière (see p. 271, 1556). The game was often, with no authority, attributed to Pythagoras. Barocius (or Barozzi) amplified the treatment attributed to Faber Stapulensis, and his discussion of the subject is clearer than that of the latter, although hardly equal to that of Boissière already described. He had already published a philosophical discussion of arithmetic as stated on p. 295.

LEONARD AND THOMAS DIGGES.
Ed. pr. 1572. London, 1579.

LEONARD DIGGES came of an ancient family whose seat was Digges Court, Barham, Kent. He studied at Oxford, and was an expert mathematician for the time. He died c. 1571.

THOMAS was a son of Leonard, and was born in Kent. He was educated at Oxford, and died in London, August 24, 1595.

Title. See Fig. 167.

Colophon. 'Imprinted at Lon-//don, by Henrie Bynneman, dwel-//ling in Thames Street, neere vnto // Baynardes Caftle.// Anno 1579.' (P. 192.)

Description. 4°, 12.8 × 18 cm., the text being 8.9 × 14.7 cm. 16 pp. unnumb. + 191 numb. = 207 pp., and one plan (p. 176), 35 ll. London, 1579.

Editions. London, 1572, 4°; ib., 1579, 4° (here described); ib., 1585, 4°; ib., 1590, 4°. The 1579 edition was a revision,

IL NOBILISSIMO ET ANTIQVISSIMO GIVOCO PYTHAGOREO NOMINATO Rythmomachia CIOE BATTAGLIA DE CONSONANTIE DE NVMERI.

Ritrouato per vtilità, & ſolazzo delli Studioſi.

Et al preſente per Franceſco Barozzi Gentil'huomo Venetiano in lingua volgare in modo di Paraphraſi compoſto.

IN VENETIA.

Appreſſo Gratioſo Perchacino. 1572.

FIG. 166. TITLE PAGE OF BAROZZI

¶ An Arithmeticall Militare Treatise, named
STRATIOTICOS:
Compendiously teaching the Science of Nũbers,
as vvell in Fractions as Integers, and so much of the Rules and Æquations Algebraicall and Arte of Numbers Cossicall, as are requisite for the Profession of a Soldiour.

Together with the Moderne Militare Discipline, Offices, Lawes and Dueties in euery wel gouerned Campe and Armie to be obserued: Long since attẽpted by LEONARD DIGGES Gentleman, Augmented, digested, and lately finished, by THOMAS DIGGES, his Sonne.

Whereto he hath also adioyned certaine Questions of great Ordinaunce, resolued in his other Treatize of Pyrotechny *and great Artillerie, hereafter to bee published.*

VIVET POST FVNERA VIRTVS.

AT LONDON:
Printed by Henrie Bynneman.
Anno Domini. 1579.

FIG. 167. TITLE PAGE OF THE 1579 DIGGES

for the preface 'To the Reader' (f. a 2) states that it was 'finifhed the 13. of October. 1579.'

Considering its date, this work is a very good introduction to the study of arithmetic. The arithmetic proper extends, however, only to page 32. Then follows a brief treatment of algebra (pp. 33-51), after which are certain problems (to p. 70) relating to military matters. Pp. 81-191 are devoted entirely to military affairs.

The father and son wrote several mathematical treatises, but none directly on arithmetic. One was the 'Pantometria' of 1571, which De Morgan includes, and which is in Mr. Plimpton's library, but which I have omitted because it is in no sense an arithmetic.

Other works of 1572. Buckley, p. 252, 1550; Gemma, p. 200, 1540; Grammateus, p. 123, 1518; Mariani, p. 181, 1535; Barlaamo, 'Λογιστικη, sive arithmeticæ, algebraicæ libri VI' (in a work on spherics), Strasburg, with later editions at Paris, 1594, 4°; 1599, 4°; 1600, 4° (see also p. 315, 1564); John Seton (see Buckley, p. 252, c. 1550).

SILVIO BELLI. Ed. pr. 1573. Venice, 1573.

Born at Vincenza. He died in 1575. He was an architect at Rome and Ferrara, and wrote on practical geometry.

Title. See Fig. 168.

Description. 4°, 15 × 20.6 cm., the text being 9 × 15.1 cm. 46 ff. (40 numb.), 19-21 ll. Venice, 1573.

Editions. This is the only separate edition of this semi-geometric work. Belli published a 'Libro del misurar con la vista,' which passed through the following editions: Venice, 1565, 4°; ib., 1566, 4°; 1569, 4°; Venice, 1570, 4°; ib., 1573, 4°; ib., 1595, 4°. This was united with the 'Della Proportione' in 1595 (Venice) to form the 'Quattro libri geometrici.'

This work is included in this list because of the treatment of mediæval proportion which it contains.

LUCAS PAETUS. Ed. pr. 1573. Venice, 1573.

A Venetian jurist of the sixteenth century.

Title. See Fig. 169.

Description. Fol., 20.5 × 29.5 cm., the text being 15 × 24.5 cm. 8 pp. unnumb + 93 numb. + 1 blank = 102 pp., 50 ll. Venice, 1573.

SILVIO BELLI
VICENTINO
DELLA
PROPORTIONE, ET PROPORTIONALITA'
Communi Paſſioni del Quanto.

LIBRI TRE.

Utili, & neceſſarij alla vera, & facile intelligentia dell' Arithmetica, della Geometria, & di tutte le ſcientie & arti.

Al Magnanimo Aleſſandro Farneſe Card.le

CON PRIVILEGIO.

IN VENETIA, *Appreſſo Franceſco de' Franceſchi Saneſe.* 1573.

FIG. 168. TITLE PAGE OF BELLI

LVCAE PAETI

IVRISCONSVL.

DE MENSVRIS, ET

PONDERIBVS ROMANIS,

ET GRAECIS,

CVM HIS QVAE HODIE ROMAE SVNT COLLATIS

LIBRI QVINQVE.

EIVSDEM VARIARVM LECTIONVM LIBER VNVS

AD SANCTISSIMVM OPTIMVMQVE PRINCIPEM

PIVM QVINCTVM PONT. MAX.

MAXIMILIANI.II

VENETIIS MDLXXIII.

Fig. 169. Title page of paetus

Editions. There were two editions of this work published at Venice in 1573. (See next title.)

Although not an arithmetic, this work is a scholarly and interesting contribution to the history of the weights and measures of Greece and Rome, and the symbols inherited by the Middle Ages. It also contains several illustrations of ancient measures.

LUCAS PAETUS. Ed. pr. 1573. Venice, 1573.

See p. 343.

This is a different edition from that just described. The title page is, however, substantially the same.

Description. Fol., 17.5 × 23.2 cm., the text being 11.6 × 18.5 cm. 144 pp. (127 numb.), 38 ll. Venice, 1573.

See above.

VALENTIN MENHER. Ed. pr. 1573. Antwerp, 1573.

See p. 249.

Edited by Michiel Cognet, born c. 1549 at Antwerp; died at Antwerp.

Title. See Fig. 170.

Colophon. 'Antverpiæ // Typis Ant. Dieſt. 1573.'

Description. 8°, 9.3 × 13.6 cm., the text being 6.7 × 11.8 cm. 141 ff. unnumb., 24 ll. Antwerp, 1573.

Editions. I have no doubt there was an earlier edition, although Cognet may have edited this from a manuscript left by Menher.

This is one of the best of the purely business arithmetics of its time. It shows, better than most works of the kind, the state of commerce in the second half of the sixteenth century in Antwerp, then the most progressive of the mercantile cities of the North. In it may be studied the merchandise, the trade routes, the customs of merchants and bankers, and the prices prevailing in that period. It was to the Netherlands what Riese's book had been to Germany and Borghi's to Italy. As the title page shows, it also took a progressive attitude with reference to practical geometry. (On Cognet see p. 365.)

Other works of 1573. Köbel, p. 102, 1514; Moya, p. 310, 1562; Recorde, p. 214, c. 1542; Peter Beausard, 'Arithmetices praxis,' Louvain, 8°; Simon Köpfer, 'Grundbüchlein der Regel Detri,' Nürnberg; Bartolomeo Piccini, 'Trattato de' Cambi,' Florence, 4°.

Works of 1574. Anonymous, p. 244, 1546; Baker, p. 327, 1568; Buckley, p. 252, c. 1550; Jacob, p. 298, 1560; Riese, p. 139, 1522; Rudolff, p. 152, 1526; Seton (see Buckley, p. 252, c. 1550); Lorenzo

LIVRE

D'ARITHMETIQVE,

contenant pluſieurs belles queſtions & demandes, propres & vtiles à tous ceux qui hantent la Trafique de Marchandiſe.

Compoſé par feu Valentin Mennher Allemand: reueu, corrigé, & augmenté en pluſieurs endroits

par Michiel Cognet.

ENSEMBLE

Vne ample declaration ſur le fait des Changes.

ITEM

Vn petit diſcours de bien & deuëment diſconter, auec la Solution ſur diuerſes opinions y propoſées.

AVEC

La Solution des queſtions Mathematiques par la ſupputation de Sinus, illuſtrées & amplifiees par les demonſtrations Geometriques neceſſaires à icelles.

A ANVERS

Chéz Iean waesberghe, à l'eſcu de Flandres.

AVEC PRIVILEGE.

1573.

FIG. 170. TITLE PAGE OF THE 1573 MENHER

Bonocchio, 'Breve et universale risolutione d'aritmetica,' Brescia (Venice ?), 4°; ib., 1597; Milles de Norry, 'Arithmétique,' Paris, 4°; Johann Sekgerwitz, 'Rechenbüchlein auff allerley Handthierung,' Breslau.[497;539]

DIOPHANTUS. Ed. pr. 1575. Toulouse, 1670.

A Greek mathematician, c. 300 A.D. He was the first great writer upon algebra.

Title. 'Diophanti // Alexandrini // Arithmeticorvm // libri sex,// et de nvmeris mvltangvlis // liber vnvs.// Cvm commentariis C. G. Bacheti V.C.// & obſeruationibus D. P. de Fermat senatoris Toloſani // Acceſſit Doctrinæ Analyticæ inuentum nouum, collectum // ex varijs eiuſdem D. de Fermat Epiſtolis.// (Engraving 'Rabault Facit,' with motto: 'Obloqvitvr nvmeris septem discrimina vocvm.') Tolosæ, // Excudebat Bernardvs Bosc, ē Regione Collegij Societatis Ieſu.//M. DC. LXX.' (P. 1.)

Description. Fol., 23.4 × 36.3 cm., the text being 15.1 × 23.9 cm. 5 pp. blank + 6 unnumb. + 341 numb. + 48 of notes = 400 pp., 50–55 ll. Toulouse, 1670.

Editions. Basel, 1575. There was no other sixteenth-century edition.

Athough entitled an arithmetic this is really a treatise on algebra, the first systematic one ever written. It contains, however, a good deal of matter upon the Greek theory of numbers, notably the 'Clavdii Gasparis Bacheti Sebusiani, in Diophantvm Porismatvm, Liber Primus,' 'Liber Secundus,' and 'Liber Tertius.' A certain amount of this work also enters into the treatise itself, but this is generally algebraic in character, the standard problem requiring the finding of a number satisfying given conditions. This leads to numerous indeterminate (Diophantine) equations. This edition, by Bachet and Fermat, is one of the best that has been published.

FRANCISCUS MAUROLYCUS.

Ed. pr. 1575. Venice, 1575.

FRANCESCO MAUROLICO. Born at Messina, September 16, 1494; died near there, July 21, 1575. He entered the priesthood and later became professor of mathematics at Messina. He wrote chiefly on astronomy, and edited several works of the Greek mathematicians.

Title. See Fig. 171.

Colophon. 'In monaſterio S. Maria // à parte 19. Iulij die,// ☿. 11. Indictionis,// 1553.' (P. 305.)

Description. 8°, 15.7 × 20.9 cm., the text being 10.6 × 17.5 cm. 20 pp. unnumb. + 285 numb. + 1 blank = 306 pp., 39–40 ll. Venice, 1575.

D. FRANCISCI

MAVROLYCI,

ABBATIS MESSANENSIS,

Opuſcula Mathematica;

Nunc primùm in lucem ædita, cum rerum omnium notatu dignarum.

INDICE LOCVPLETISSIMO.

PAGELLA HVIC PROXIME CONTIGVA,

eorum Catalogus eſt.

CVM PRIVILEGIO.

Venetijs, Apud Franciſcum Franciſcium Senenſem.

M D L X X V.

FIG. 171. TITLE PAGE OF THE *Opuscula* OF MAUROLYCUS

Editions. There was no other edition. (See below.)

It appears from the colophon that this work was composed in 1553, although it was not published until 1575. It includes (pp. 26–47) a 'Compvtvs ecclesiasticvs in svmmam collectvs.' The rest of the treatise is chiefly astronomical. It forms the first of two volumes on mathematics, the second being the 'Arithmeticorum libro duo' (see below).

FRANCISCUS MAUROLYCUS.

Ed. pr. 1575. Venice, 1575.

See p. 348.

Title. See Fig. 172.

Colophon. 'Libri fecundi Arithmeticorum Maurolyci finis: hora //decimaoctaua, diei Sabbati, qui fuit Iulij 24[9]. Cùm // Meffanæ cum multo pontis & arcus // apparatu expectaretur Io. Cerda,// Methynenfium Dux,// Prorex. Indict. 15.// M. D. LVII.// Venetiis, M D LXXV.// Apud Francifcum Francifcium Senenfem.' (P. 183.)

Description. 4°, 15.9 × 20.9 cm., the text being 13.3 × 17.2 cm. 200 pp. (175 numb.), 40 ll. Venice, 1575.

Editions. This is the first edition, and from the colophon it appears that it lay in manuscript from 1557 to 1575. A second edition appeared in Venice in 1580. The 'Arithmeticorum libri dvo' formed the second volume of the 'Opuscula Mathematica' (Venice, 1575; p. 348).

The work is mediæval, dealing solely with the Boethian theory of numbers. It was one of the last of the extensive sixteenth-century Italian works of this nature, and shows considerable originality in the treatment of figurate numbers. Maurolycus was by no means a mere compiler, but a man of creative power.

HENRICUS BRUCÆUS. Ed. pr. 1575. Rostock, 1575.

Born at Alost, Flanders, c. 1531; died at Rostock, December 31, 1593. He was professor of mathematics at Rome, and later professor of medicine at Rostock.

Title. See Fig. 173.

Description. 8°, 9.5 × 15.1 cm., the text being 6.5 × 11.6 cm. 76 ff. (1 blank), 25 ll. Rostock, 1575.

D. FRANCISCI

MAVROLYCI,

ABBATIS MESSANENSIS,

Mathematici celeberrimi,

ARITHMETICORVM LIBRI DVO,

NVNC PRIMVM IN LVCEM EDITI,

Cum rerum omnium notabilium.

INDICE COPIOSISSIMO.

CVM PRIVILEGIO.

Venetijs, Apud Franciſcum Franciſcium Senenſem.

M D LX X V.

Leonard Flammel

FIG. 172. TITLE PAGE OF THE *Libri duo* OF MAUROLYCUS

Editions. There was no other edition.

This is a Latin-school manual, in two 'books.' The first book treats of arithmetic, and chiefly of mediæval ratios. The second treats of algebra, the equations being considered from the standpoint of geometry,

HENRICI
BRVCÆI
BELGÆ,

MATHEMATICARVM
EXERCITATIONVM
LIBRI DVO.

ROSTOCHII
EXCVDEBAT IACOBVS LVCIVS
TRANSYLVANVS.
Anno M. D. LXXV.

FIG. 173. TITLE PAGE OF BRUCÆUS

and some attention being given to surd numbers, roots, and the rule of false.

Other works of 1575. Gemma, p. 206, 1540; Köbel, p. 102, 1514; Lapazzaia, p. 322, 1566; Mariani, p. 181, 1535; Salignac, p. 359,

1577: Xylander, p. 356, 1577: Mauricius Steinmetz Gersbach, 'Arithmetices praecepta,' Leipzig, 8°. 539; 540

Works of 1576. Gemma, p. 200, 1540; Petri, p. 325, 1567; Tagliente, p. 115, 1515; Joseph Lange, 'Arithmetica,' Copenhagen, 8°.

GIR̊JKA GO̊RLA Z GO̊RLSSTEYNA. 539

Ed. pr. 1577. Czerny, 1577.

A Polish Rechenmeister of the latter half of the sixteenth century.

Title. See Fig. 174.

Colophon. 'Wytiſſtēnow // Starém Héſté Praʒ kém // v Girijka Çʒerneho.// Letha Páně/ // M. D. LXXVII.' (F. XC, r.)

Description. 8°, 9.5 × 15.9 cm., the text being 7.2 × 11.7 cm. 9 ff. unnumb. + 89 numb. (Roman) = 98 ff., 21 ll. Czerny, 1577.

Editions. There was no other edition.

This is one of the few arithmetics in the Polish language published in the sixteenth century, and is very rare. There was a copy in the Boncompagni sale, but the book is seldom mentioned by bibliographers. It consists of five parts, the first dealing chiefly with the fundamental operations with counters, the second with written operations, the third with fractions, the fourth with business arithmetic, and the fifth with the rule of false and allied topics.

DIONIS GRAY. Ed. pr. 1577. London, 1577.

A London goldsmith of the second half of the sixteenth century.

Title. See Fig. 175.

Description. Sm. 8°, 8.8 × 13.8 cm., the text being 6.5 × 12 cm. 126 ff. (8 unnumb.). London, 1577.

Editions. London, 1577, 8° (here described); ib., 1586, 8°.

This is a practical arithmetic, consisting of four parts. 'The firſt containeth ſoundrie partes of Arithmetique, that is to ſay,' the fundamental operations with integers, including progressions. 'The ſeconde parte, containeth the ſaid partes ſeruyng for practiſe of broken numbers or fraccions.' 'The third part containeth the ſondrie Rules of proportion, furthered by vſe of the foreſaid partes,' and includes alligation and the rule of false. 'The fourth parte containeth ſondrie Rules of breuetie,' or short processes. It is one of the earliest English arithmetics to contain rules and definitions in rhyme. For example, in

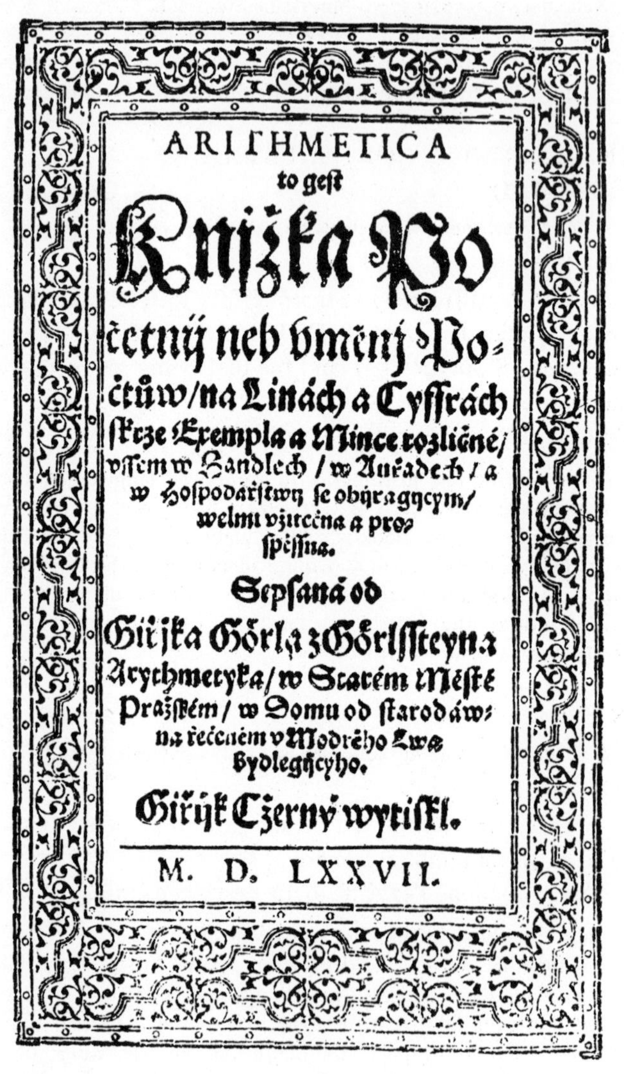

ARITHMETICA
to gest
Knjžka Po-
četnij neb vměnj Po-
čtůw/na Linách a Cyffrách
skrze Exempla a Mince rozličné/
vssem w Handlech / w Auřadech / a
w Hospodářstwj se obýragycym/
welmi vžitečna a pro-
spěssna.
Sepsaná od
Giřjka Görla z Görlssteyna
Arythmetyka/ w Starém Městě
Pražském / w Domu od starodáw-
na řečeném v Modrého Lwa
Bydlegycyho.
Giřijk Czerný wytiskl.
M. D. LXXVII.

FIG. 174. TITLE PAGE OF GI$\overset{e}{\text{R}}$JKA G$\overset{e}{\text{O}}$RLA Z G$\overset{e}{\text{O}}$RLSSTEYNA

speaking of addition Gray says: ‘And for to amplifie the effecte, take here a fewe lines in verſe:

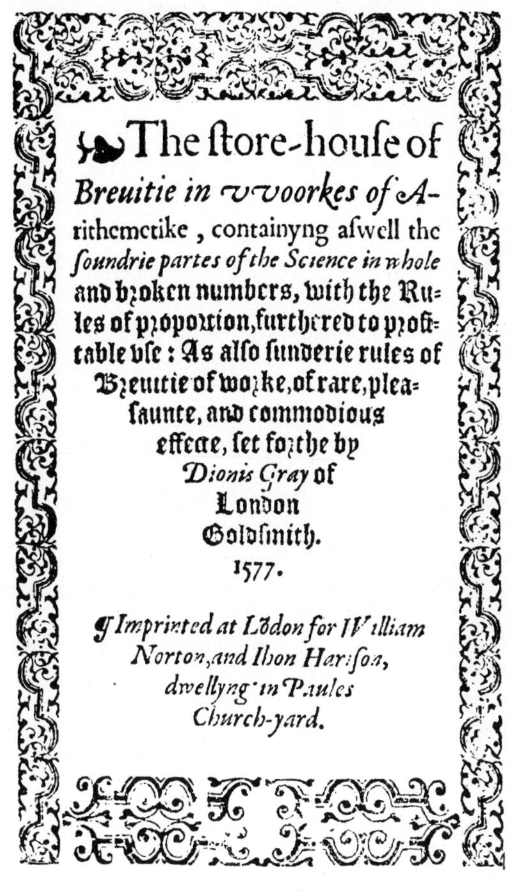

The ſtore-houſe of
Breuitie in vvoorkes of A-
rithemetike, containyng aſwell the
ſoundrie partes of the Science in whole
and broken numbers, with the Rules of proportion, furthered to profitable uſe: As alſo ſunderie rules of
Breuitie of worke, of rare, plea-
ſaunte, and commodious
effecte, ſet forthe by
Dionis Gray of
London
Goldſmith.
1577.

¶ *Imprinted at Lōdon for William*
Norton, and Ihon Hariſon,
dwellyng in Paules
Church-yard.

FIG. 175. TITLE PAGE OF GRAY

‘¶ Of ſondrie ſommes perticulars, one totall for to frame,
Set them doune right orderly, as worke doeth beſt require:
What place ye giue to any one, the reſt let haue the ſame,
So maie you well performe the ’ffecte, of what you doe deſire.’

The rule is then continued in a series of verses.

GUILIELMUS XYLANDER.

Ed. pr. 1577. Heidelberg, 1577.

WILHELM HOLTZMANN. Born at Augsburg, December 26, 1532; died at Heidelberg, February 10, 1576. He was professor of Greek at Heidelberg.

Title. See Fig. 176.

Colophon. 'Excudebat Iacobus Mylius, impenſis // Matthæj Harniſch.// M. D. LXXVII.' (F. 52, v.)

Description. 4°, 14.9 × 19.2 cm., the text being 10.5 × 16.1 cm. 2 ff. unnumb. + 50 numb. = 52 ff., 31 ll. Heidelberg, 1577.

Editions. There was no other edition of this work. Xylander translated several Greek works into Latin, among them Euclid (Basel, 1562) and Diophantus (Basel, 1575, fol.). The 'Opuscula' appeared a year after his death. Xylander also edited Psellus (Basel, 1556).

As the title states, this work is divided into four parts. The first and most extensive (ff. 1–22) relates to astronomy. The second is purely arithmetical (ff. 22–36), and treats of common fractions, giving the usual operations and the rule of three. The arrangement of this part is peculiar, addition and subtraction being treated together, after which division is explained, multiplication coming last. This order would be justified if Xylander had reduced his fractions to fractions having a common denominator, but he merely follows the usual rule of cross multiplication (f. 29, v.). The third section is 'De svrdis, qvos vocant, nvmeriis iis, qvi a qvadratis primò naſcuntur, Inſtitutio docendo explicanda,' a chapter now conventionally placed in our algebras. The fourth section relates to the celestial globe and the astrolabe (ff. 46–50).

PIETRO ANTONIO CATTALDI.

Ed. pr. 1577. Bologna, 1577.

CATALDI, CATALDO. Born at Bologna in 1548; died at Bologna, February 11, 1626. Professor of mathematics and astronomy at Florence (1563), Perugia (1572), and Bologna (1584). He wrote several mathematical works, and to him is due the beginning of the theory of continued fractions (1613).

Title. See Fig. 177.

Description. 4°, 14.3 × 19.6 cm., the text being 10.9 × 16.7 cm. 8 ff. unnumb., 40 ll. Bologna, 1577.

OPVSCVLA
MATHEMATICA
DOMINI
GVILIELMI XYLANDRI
AVGVSTANI.

Aphorismi Cosmographici liber I.

De Minutiis liber I.

De Surdorum Numerorum natura & tractatione liber.

De vsu Globi & Planisphærij tractatus.

HEIDELBERGÆ

Excudebat Iacobus Mylius, impensis
Matthæj Harnisch.

M. D. LXXVII.

FIG. 176. TITLE PAGE OF XYLANDER

DVE LETTIONI
DI PIETR'ANTONIO
CATTALDI BOLOGNESE

FATTE NELL'ACADEMIA
dei Dissegno di Perugia,

ALLI GENEROSI, ET VIRTVOSISSIMI
Signori Academici, il Signor Caualier Paciotto,
& il Signor Caualiero Anastagi.

IN BOLOGNA.

Per Giouanni Rossi MDLXXVII.

Con licentia de' Superiori.

FIG. 177. TITLE PAGE OF CATTALDI

Editions. There was no other edition. Cattaldi wrote, under the pseudonym Perito Annotio, a 'Prima Parte della Pratica Aritmetica' (Bologna, 1602) and a 'Trattato dei numeri perfetti' (Bologna, 1603), but like all of his strictly mathematical works they appeared after 1600. The second part of the 'Pratica' appeared under his own name in 1606.

This book hardly deserves to be ranked in a list of arithmetics. I have included it, however, because in his first address Cattaldi treats somewhat of numbers, and in the second address he applies arithmetic to mensuration. The treatment is mediæval, and the chief interest in the book is a typographical one; for, the printer not being able to set such numbers, all of the fractions have been written in by hand.

Other works of 1577. Borghi, p. 16, 1484; Buckley, p. 252, c. 1550; Capella, p. 68, 1499; Cressfelt, p. 290, 1557; Herbestus, p. 303, 1561; Hobel, p. 314, 1563; Ramus, p. 331, 1569; Seton, p. 252 (see Buckley, c. 1552); Anonymous, 'Arithmetica,' Debreczin; Miguel Berenguer, 'De numerorum antiquorum notis,' Saragossa; Georg Gehrl, 'Ein nutzlich und künſtlich Rechenbuch auff der Federn,' Prag, 8°; Johann Jung, an arithmetic, Lübeck, c. 1557 (no copy extant?); Bernhard Salignacus, 'Tractatus de Arithmetica Partium et Alligationis,' Frankfort (Peacock says 1575); also 'Regula veri,' Heidelberg, 1578, and 'Arithmeticae Libri II, et Algebrae totidem,' Frankfort, 1580, 4°; ib., 1593, 4°. 498

JACQUES CHAUVET CHAMPENOIS.

Ed. pr. 1578. Paris, 1578.

Professor of mathematics in the University of Paris, in the second half of the sixteenth century.

Title. See Fig. 178.

Description. 8°, 10.1 × 17 cm., the text being 7.5 × 14 cm. 9 pp. unnumb. + 383 numb. = 392 pp., 28–29 ll. Paris, 1578.

Editions. There was no other edition. The privilege is dated 'a Paris le huictieſme iour de Septembre, M. D. LXXVII.', and the dedicatory epistle 'De Paris ce 28. de Nouembre 1577.' 539

The chief interest in the book lies in the fact that it was written by a man who was so interested in military matters as to take a large number of his applied problems from army life. The arrangement of the book is not peculiar, but the great array of military problems is unique, and, having been prepared especially for this work, would form an

LES
INSTITVTIONS
DE L'ARITHMETIQVE DE IACQVES CHAVVET CHAMPENOIS, Professeur és Mathematiques en l'Vniuersité de Paris, diuisees en quatre parties: auec vn petit Traicté des fractions Astronomiques.

A PARIS,

Chez Hierosme de Marnef, au mont S.Hilaire, à l'enseigne du Pelican.

1578.

AVEC PRIVILEGE DV ROY.

FIG. 178. TITLE PAGE OF CHAUVET CHAMPENOIS

interesting source for the study of army conditions in France in the sixteenth century.

Other works of 1578. Capella, p. 66, 1499; Gemma, p. 200, 1540; Ramus, p. 355, 1569; Salignacus, p. 359, 1577; Tartaglia, p. 278, 1556; Trenchant, p. 320, 1566. 539

JOHANN OTTO. Ed. pr. 1579. Leipzig, 1579.

A Freiburg Rechenmeister, born c. 1529. The dedicatory epistle is dated 1579, and is signed 'Johan Otto. Ætatis fuæ 50,' which approximately fixes the date of Otto's birth.

Title. See Fig. 179.

Colophon. 'Getruckt zu Leipzig/ bey Johan // Rhambawes Seligers hinderlaſ-//nen Erben/ 1579.' (P. 423.)

Description. 4°, 14.5 × 19.1 cm., the text being 11.7 × 17.5 cm. 424 pp. (381 numb.), 28–38 ll. Leipzig, 1579.

Editions. There was no other edition.

The book is composed almost entirely of tables, although the first few pages give a brief treatment of counter reckoning.

CHRISTIAN URSTISIUS. Ed. pr. 1579. Basel, 1579.

Allassiderus, Allassisiderus, Wursteisen, Urstis. Born at Basel in 1544; died at Basel March 30, 1588. He was educated at Basel and became professor of mathematics (1565) and afterwards (1585) of theology in that university.

Title. See Fig. 180.

Colophon. 'Basileæ Helve-//tiorvm,// per Sebastianvm Hen-//ricpetri, An. hvmanitatis // Filii Dei, CIↃ. IↃ. LXXIX.// Menſe Auguſto.' (P. 192.)

Description. 8°, 10.3 × 16.6 cm., the text being 6.4 × 11.2 cm. 192 pp., 25–28 ll. Basel, 1579.

Editions. On p. 192 is a woodcut with the date 1569, evidently used from some earlier work of the printer's. Murhard (vol. I, p. 173) mentions a work by Urstisius, 'Zwey Bücher von der Rechenkunſt, deſsgleichen in der deutſchen Sprache nie ausgegangen,' Basel, 1569, 4°, in which this woodcut may have been used. He also mentions an edition at Basel in 1595. An English translation by T. Hood appeared in 1596. 539

Calculator.

Ein newes / liebliches / vnd nützliches ausgerechnetes Rechenbuch für alle / so Arithmeticam lieb haben / Insonderheit aber für Kauffleut / Amptspersonen / Händler / Krämer vnd allesampt (sie seien hie oder außländische Personen) so sich des Einkauffens vnd verkauffens zu Leipzig gebrauchen / auff die Wag vnd Gewicht obgedachter löblichen handel Stadt Leipzig gestelt : Auch andern Stedten gleichsfals dienstlich / als zu Franckfort an der Oder / zu Freiberg / Dresden / Kemptnitz vnd Annaberg / etc. vnd in summa allenthalben / wo der Centner auff 110. pfund ausgeteilet ist. Darinnen alle keuff auff Centner gut / vnd anderer fürnembsten wharen austragende Summa gar behend one Multipliciren vnd Diuidiren , allein durch die allerleichteste species der Addition oder Summirens / vnd auch an einem blat etzlich hundert tausent Exempla vnd Fragen zufinden / vnd auffzulösen sein. Was aber mehrers zu diesem Tittel gehörig / ist auff der andern seiten dieses blats vermeldet

Durch

Johan Otthen Notarien, der Rechenkunste Studiosum, zu Freiberg in der alten berhümbten Churfürstlichen Sächsischen Bergstadt / Mitbürger.

Psalm. 90.

Mandauit DOMINUS angelis suis de te, vt custodiant te in omnibus vijs tuis.

Leipzig /

1 5 7 9.

FIG. 179. TITLE PAGE OF OTTO

This is a book written by a gymnasium teacher, who was filled with a love of the classical learning, and yet who recognized that the old Boethian arithmetic must give way to the practical treatment demanded

ELEMENTA
A·RITH-
METICÆ, LOGI-
CIS. LEGIBVS.
DEDVCTA,
In usum Academiæ Basil.
Opera & studio
CHRISTIANI VRSTISII,
Mathematicarum profes-
soris.

BASILEÆ,
PER SEBASTIANVM HEN-
RICPETRI.

FIG. 180. TITLE PAGE OF URSTISIUS

by modern conditions. As a result it is a somewhat heavy treatment of part of the old theory of numbers, together with the fundamental operations, roots, and certain commercial applications of the rule of three, of partnership, and of alligation. Urstisius expresses his indebtedness chiefly to Euclid, Ramus, Salignacus, Gemma Frisius, and Scheubel, not a very extensive list for 1579.

Other works of 1579. Digges, p. 340, 1572; Gemma, p. 200, 1540; Mariani, p. 181, 1535; Riese, p. 139, 1522; Tagliente, p. 115, 1515; Anonymous, 'Nouvelle et facile méthode d'arithmétique,' Lyons, 16°; Miguel de Eleyzalde, 'Guia de contadores,' Madrid, 4°.

GIAMBATTISTA BENEDETTI.

Ed. pr. 1580. Turin, 1585.

JOHANNES BAPTISTA BENEDICTUS. Born at Venice, August 14, 1530; died at Turin, January 20, 1590. Philosopher and mathematician of the Duke of Savoy.

Title. 'Io. Baptistae // Benedicti // Patritij Veneti Philoſophi. //Diversarvm Specvlationvm // Mathematicarum, & Phyſicarum // Liber.// Quarum ſeriem ſequens pagina indicabit.// Ad Sereniſsimvm Carolvm Emanvelem // Allobrogvm, et Svbalpinorvm //Dvcem Invictissimvm.// (Large woodcut.) Tavrini, Apud Hæredem Nicolai Beuilaquæ, MDLXXXV.// Superioribus permiſſum.' (F. 3.)

Description. Fol., 20.7 × 30 cm., the text being 14.5 × 23 cm. 425 pp. numb. + 10 unnumb. + 1 blank = 436 pp. (118 in the part devoted to arithmetic), 42–46 ll. Turin, 1585.

Editions. Turin, 1580, fol.; ib., 1585, fol. (here described); Venice, 1599, fol.

This work is composed of six parts, of which the first is entitled 'Theoremata Arithmetica.' The other five parts relate respectively to perspective, mechanics, Aristotle, Euclid's book on proportion, and physics. The arithmetic is a scientific consideration of various matters of theory, and is best illustrated by the following theorems, which, as is usually the case, are stated in the form of questions: 'Theorema IIII. Cvr multiplicaturi fractos cum integris, rectè multiplicent numerantem fracti per numerum integrorum, partianturque productum per denominantē fracti, ex quo numerus quæſitus colligitur.' 'Theorema XXIX. Qvid cauſæ eſt, cur ſubtracto duplo producti duorum numerorum ad inuicem multiplicatorū ex ſumma ſuorum quadratorum, ſemper quod ſuper eſt duorum numerorum quadratum differentiæ ſit?' All such questions are answered by the aid of diagrams, quite as Euclid would have done. The graphic treatment is even applied to such problems as that of the couriers ('Theorema CXIIII'). There is an interesting

'Appendix de specvlatione regvlae falsi,' which closes the arithmetic, and sets forth an elaborate explanation, with graphic aids, of the rule of false which was then so common.

As a specimen of graphic arithmetic, combining the Euclidean theory with the Renaissance practice, Benedetti's book is worthy of more attention than it has received. It may be inferred from some of his statements that, although purely a theorist himself, he recognized the obsolete nature of much that the practical arithmetics had to offer. It is possible that the common partnership problems were already considered too traditional, for he says: 'Svpponunt *antiqui* aliquot mercatores dantes pecunias lucro in diuerſis vnius anni temporibus,' etc.

ISAAC RIESE. Ed. pr. 1580. Leipzig, 1580.

One of the five sons of Adam Riese (p. 138).

Title. See Fig. 181.

Colophon. 'Gedruckt zu Leipzig // ben Hans Rhambaw/ //im Jar // M. D. LXXX.' (P. 402.)

Description. 4°, 14.1 × 17.9 cm., the text being 11 × 16.5 cm. 36 pp. unnumb. + 366 numb. = 402 pp. Leipzig, 1580.

Editions. There was no other sixteenth-century edition.

Although this book contains a brief treatment of the operations, it is largely made up of tables for the use of merchants. It is therefore not a textbook, and I believe it went through only one other edition (Leipzig, 1619).

WILLEM RAETS. Ed. pr. 1580. Antwerp, 1580.

A Dutch arithmetician of Maastricht. The privilege is dated May 22, 1576, and mentions only the name of Raets. Coignet (Cognet) in his preface, however, speaks of his particular friend ('mijn zōderlinge goet vrient') Raets as dead ('meynen ouerledē vrient'), so that he very likely died between 1576 and 1580. Coignet was also the editor of Menher's arithmetic (see p. 346).

Title. See Fig. 182.

Description. 8°, 9.8 × 14.7 cm., the text being 7.4 × 11.9 cm. 88 ff. unnumb., 27–31 ll. Antwerp, 1580.

Editions. There was no other edition.

This work is one of several elementary business arithmetics appearing in the Low Countries about this time. It is a small book of no special merit save as it shows the style of commercial problems of the period. The similarity of the title page to that of Menher's 'Arithmetiqve seconde' of 1556 (p. 283) is interesting.

Ein newes Nütz-
bar gerechnetes Rechenbuch/

auff allerley Handtirung/ nach dem Centner vnd Pfundt gewicht/ alda der Centner
für 110. Pfundt gewirdiget/ darinnen die Bezalunge
in einkauffen vnd verkauffen/ one sondere weitleufftige Rechnung
gar behend zu finden: Auch von allerley Maß/ Elen/ vnd Gewicht kauff. Sampt mehr angehengten Taffeln auff die Ma-
terialische Specereien/ Neben vergleichung etlicher Müntze vnd
Gewicht gerichtet. Desgleichen eine Wechsel Rechnung auff
Müntz vnd Goldt/ ꝛc. Allen Kauffleuten/ Hendlern/
Apoteckern/ Schössern/ ꝛc. Vnd dem gemeinen Mann sehr dienstlich zugebrauchen.

Alles mit fleis auff die Meißnische Müntze vnd
Wehrunge gestellet vnd gerechnet
Durch
Isaac Riesen/ Burger vnd Viesirer
zu Leipzig/ Vormals der gestalt in
Druck nie außgangen.

Mit Churfürstlicher Sechsischer Begnadung vnd
Freiheit in zehen Jaren nicht nachzudrucken.
Anno 1580.

FIG. 181. TITLE PAGE OF ISAAC RIESE

Other works of 1580. Apianus, p. 155, 1527; Baker, p. 327, 1568; Cassiodorus, p. 211, 1540; Gemma, p. 200, 1540; Gyraldus, p. 254, 1552; Mariani, p. 180, 1535; Maurolycus, p. 350, 1575; Ramus, p. 331, 1569; Ringhieri, p. 253, 1551; Salignacus, p. 359, 1577; J. Ammo-

ARITHMETICA
Oft
Een niew Cijfferboeck/ van
Willem Raets/ Maestrichter.

VVaer in die Fondamenten seer grondelijck verclaert eñ met veel schoone questien gheillustreert vvorden, tot nut ende oorbaer van alle Coopliedẽ ende lief hebbers der seluer Consten.

Met noch een Tractaet vande VVisselroede, met Annotatien verciert, door Michiel Coignet.

T'hantwerpen,
Ten huyse van Hendrick Hendricsen/ inde Lelieblome. 1580.
Met Priuilegie van thien Iaeren.

FIG. 182. TITLE PAGE OF RAETS

nius, 'Isagoge Arithmetica,' Wittenberg, 8°; H. Flicker, 'Arithmetices introductio,' Cologne, 8°, and 'Compendium calculorum, seu projectilium ratiocinationis,' ib., 8°; C. Zuccantini, 'Libro d'Albaco,' Siena, 12°. 539

ANTONIO MARIA VISCONTI.

Ed. pr. 1581. Brescia, 1581.

A mathematician of Piacenza, of the latter part of the sixteenth century.

Title. See Fig. 183.

Colophon. 'Brixiae.// apud Iacob. & Policretum de Turlinis.// M D L XXXI.// Svperiorvm Premissv.' (P. 301.)

Description. 4°, 14.8 × 20.2 cm., the text being 9.4 × 15.6 cm. 304 pp. (289 numb.), 37 ll. Brescia, 1581.

Editions. There was no other edition. Riccardi believes the date 1551 in Murhard is a misprint. I find no such edition.

This rare and curious book is a combination of algebra, advanced arithmetic, geometry, and the mensuration of river lands. The arithmetic is designed to be an application of the algebra, and includes roots, equation of payments, proportion, and a little bookkeeping.

Other works of 1581. Anonymous, p. 195, 1539; Fischer (Piscator), p. 247, 1549; Gemma, p. 200, 1540; Lonicerus, p. 253, 1551; Peverone, p. 290, 1558; Ramus, p. 330, 1569; Riese, p. 139, 1522; Johann Kandleon, 'Arithmetica,' Regensburg (referred to in the carelessly prepared Boncompagni sale catalogue, but probably the Kaudler book of 1591). There was also 'A short Introduction to Arithmetic' published anonymously in London c. 1581–90, 8°.

JULIUS CAESAR of Padua.

Ed. pr. 1582. Frankfort, 1678.

A German-Italian teacher of the sixteenth century.

Title. 'Julii Cæsaris // von Padua // Arithmetiſche // Practick/ // Welche in allen Låndern // ſehr nůtzlich kan gebraucht // werden/ bey Kauff- und Ver-//kauffung allerley Wahren/ auch // die groſſen Můntz-Sorten in // kleine/ und die kleine in groſſe // zu verwandeln:// Samt der Erklårung/ wor-//inn ein Jedweder/ der nur die // Ziffern keñet/ alſobald ſehen kan/ // wie diß Bůchlein zu verſtehen iſt.// Nebenſt Morgen- und Abend- //Gebehten und Geſången/ den // reyſenden Perſonen gar bequem // bey ſich zu fůhren.// Und dann letzlich/ eine kurtze Be-// ſchreibung/ der denckwůrdigſten // Sachen/ ſo von Anfang der Welt/ biß // zu dieſer Zeit/ vorgangen.// Franckfurt am Mayn/ // Druckts Blaſius Jlßner/ im Jahr 1678.' (P. 1.)

ANTONII MARIAE
VICECOMITIS
CIVIS PLACENTINI,

Practica Numerorum, & Menſurarum, ac Alluuionis partitionem, inueſtigandi, & vt in Indice ſequenti.

BRIXIAE.

APVD IACOBVM, ET POLYCRETVM
de Turlinis Fratres. 1581.

Fig. 183. Title page of Visconti

Description. 12°, 4.7 × 11 cm., the text being 3.9 × 9.8 cm. 504 pp. (239 and 258 numb.); tables 28 ll., other pages 22 ll. Frankfort, 1678.

Editions. Strasburg, 1582, 16°; ib., 1583; ib., 1585; ib., 1592, and after 1600 as late as 1679.

The first part of the work is devoted to multiplication tables. This is followed by a chapter on chronology. The last part of this edition is a separate book of prayer, 'Chriſtliche Morgen- Und Abend Gebeht,' of the same date (1678). There is nothing arithmetical in the work except the tables.

GASPARO SCARUFFI. Ed. pr. 1582. Reggio, 1582.

An Italian jurist of the sixteenth century.

Title. See Fig. 184.

Colophon. 'In Reggio,// Per Hercoliano Bartoli. // M.D.-LXXXII.' (F. 65, v.)

Description. Fol., 21 × 30.2 cm., the text being 12.7 × 19.5 cm. 65 ff., 32 ll. Reggio, 1582.

Editions. This is the only sixteenth-century edition of Scaruffi's work, the privilege being dated July 15, 1582. The dedicatory epistle is, however, dated at Reggio, May 16, 1579.

The work is called in the running headlines a 'Discorso sopra le monete,' and is a historical treatise on money and coinage, touching slightly on exchange.

GASPARO SCARUFFI. Ed. pr. 1582. Reggio, 1582.

See above.

Title. 'Breve Instrvttione // sopra il discorso // fatto dal Mag. M. // Gasparo Scarvffi,// per regolare le cose delli // danari.// (Woodcut representing a bishop, surrounded by these words: S. Prosper // Episcopvs // Regii //.) In Reggio,// per Hercoliano Bartoli.// M. D. LXXXII.' (F. 1, r.)

Colophon. 'Di Reggio il xvij. Aprile. M.D.LXXXI.' (F. 9, r.) The colophon and title page do not agree as to date.

Description. Fol., 21 × 30.2 cm., the text being 12.8 × 20 cm. 9 ff., 27–30 ll. Bound with the preceding work. Reggio, 1582.

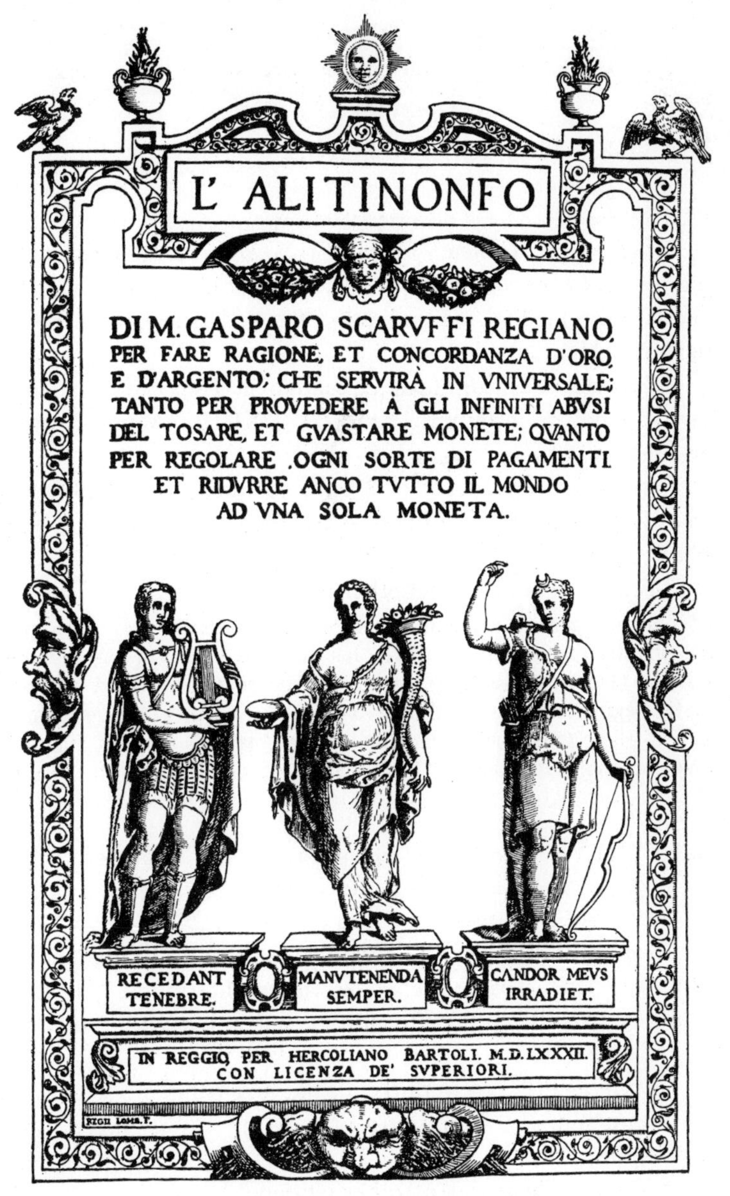

L' ALITINONFO

DI M. GASPARO SCARVFFI REGIANO,
PER FARE RAGIONE, ET CONCORDANZA D'ORO,
E D'ARGENTO; CHE SERVIRÀ IN VNIVERSALE;
TANTO PER PROVEDERE À GLI INFINITI ABVSI
DEL TOSARE, ET GVASTARE MONETE; QVANTO
PER REGOLARE .OGNI SORTE DI PAGAMENTI
ET RIDVRRE ANCO TVTTO IL MONDO
AD VNA SOLA MONETA.

RECEDANT TENEBRE.

MANVTENENDA SEMPER.

CANDOR MEVS IRRADIET.

IN REGGIO, PER HERCOLIANO BARTOLI. M.D.LXXXII.
CON LICENZA DE' SVPERIORI.

FIG. 184. TITLE PAGE OF SCARUFFI

Editions. There was no other edition.

A verbose commentary on certain parts of Scaruffi's work. See p. 370.

JOANNES THOMAS FREIGIUS.

Ed. pr. 1582. Basel, 1582.

A Swiss educator of the sixteenth century.

Title. 'Ioan. Thomæ // Freigii I. V. D.// Pædagogvs.// Hoc est, libellvs // ostendens, qva ratio-//ne prima artivm ini-//tia pueris quàm facilli-//mè tradi poſ-//ſint.// Basileæ,// per Sebastianvm // Henricipetri.' (P. 1.)

Colophon. 'Basileæ,// per Sebastianvm Hen-//ricpetri, anno salvtis // noſtræ inſtauratæ CIↃ. IↃ. XXCII.// Menſe Septembri.' (P. 383.)

Description. 8°, 10.1 × 15.8 cm., the text being 6.8 × 12.4 cm. 17 pp. unnumb. + 1 blank + 366 numb. = 384 pp., 30–31 ll. Basel, 1582.

Editions. There was no other edition.

This is a general summary of the subject-matter of education, published, after the author's death, by his two sons John Thomas and John Oswald (Joannes Osualdus) Freigius. The section devoted to arithmetic begins on p. 144 and ends on p. 156, and p. 145 is reproduced in Fig. 185. Only the fundamental operations with integers, fractions, and compound numbers are given, save for nine lines on the 'Aurea Regula . . . uulgô uocatur regula Detri.'

MATTHEW HOSTUS. Ed. pr. 1582. Antwerp, 1582.

A German educator of the sixteenth century.

Title. See Fig. 186.

Colophon. 'Matthæus Hoſtus Francofordiae ad Oderam haec obſeruata congerebat & edebat elegantioris literaturae // ſtudioſis gratificaturus, Anno Chriſto nato CIↃ. IↃ. LXXXI.' (On map at end.)

Description. 8°, 10 × 14.6 cm., the text being 7.6 × 13.3 cm. 61 pp. numb. + 3 blank = 64 pp., 20–32 ll. Antwerp, 1582 (colophon 1581).

Editions. There was no other edition.

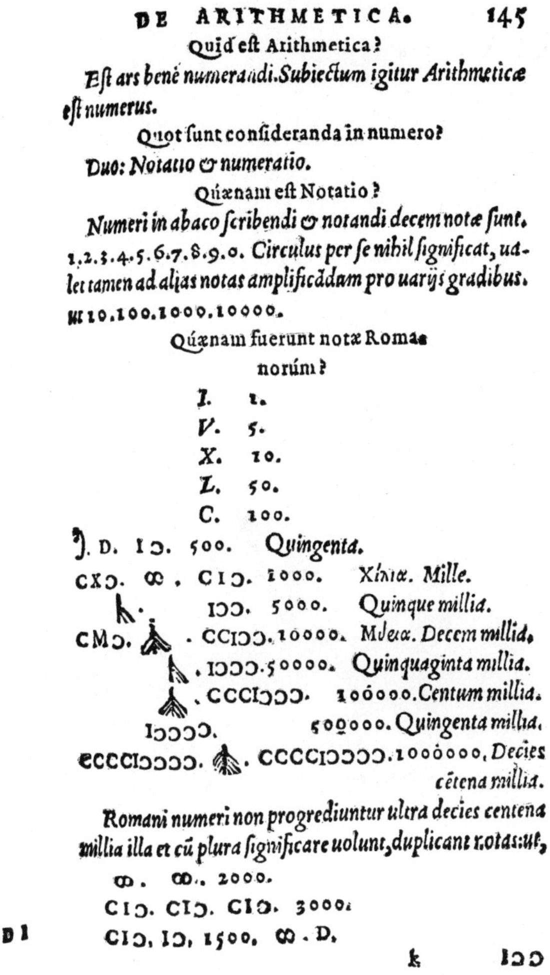

DE ARITHMETICA. 145

Quid est Arithmetica?

Est ars benè numerandi. Subiectum igitur Arithmeticæ est numerus.

Quot sunt consideranda in numero?

Duo: Notatio & numeratio.

Quænam est Notatio?

Numeri in abaco scribendi & notandi decem notæ sunt. 1.2.3.4.5.6.7.8.9.0. *Circulus per se nihil significat, ualet tamen ad alias notas amplificandam pro uarijs gradibus.* ut 10.100.1000.10000.

Quænam fuerunt notæ Romanorum?

I. 1.
V. 5.
X. 10.
L. 50.
C. 100.

Ↄ. D. IↃ. 500. *Quingenta.*

CXↃ. ∞. CIↃ. 1000. Χίλια. *Mille.*

ↁ. IↃↃ. 5000. *Quinque millia.*

CMↃ. ↂ. CCIↃↃ. 10000. Μύρια. *Decem millia.*

ↇ. IↃↃↃ. 50000. *Quinquaginta millia.*

ↈ. CCCIↃↃↃ. 100000. *Centum millia.*

IↃↃↃↃ. 500000. *Quingenta millia.*

CCCCIↃↃↃↃ. ↈ. CCCCIↃↃↃↃ. 1000000. *Decies cētena millia.*

Romani numeri non progrediuntur ultra decies centena millia illa et cū plura significare uolunt, duplicant notas: ut,

∞. ∞. 2000.

CIↃ. CIↃ. CIↃ. 3000.

DI CIↃ. IↃ. 1500. ∞. D.

ↁ IↃↃ

FIG. 185. FROM THE *Pædagogvs* OF FREIGIUS

DE

NVMERATIONE

EMENDATA,

VETERIBVS LATINIS

ET GRÆCIS VSITATA,

Matthæo Hosto auctore.

ANTVERPIÆ,

Ex officina Christophori Plantini.

M. D. LXXXII.

Fig. 186. Title page of Hostus

This is a semi-historical treatise on the various numeral systems found in Renaissance literature. It includes the Arabic system, and the Greek, Latin, and Hebrew ('gens Iudaica,' as Hostus speaks of it) systems, and a chapter 'De notis Numerorum Aſtronomicis quibuſdam vſitatis,' a set of mediæval astrological numerals also fully described by Noviomagus.

MAFFEO POVEIANO. Ed. pr. 1582. Bergamo, 1582.

A Veronese arithmetician of the sixteenth century.

Title. See Fig. 187.

Description. 4°, 15.1 × 16.5 cm., the text being 10.3 × 15.6 cm. 92 ff. (83 numb.), 22–26 ll. Bergamo, 1582.

Editions. There was no other edition.

An ordinary treatment of the fundamental operations, with a few applications to mercantile affairs. The book had not enough merit to warrant a second edition. The second part of the work treats of elementary mensuration. The book is little known, and, like many others in this list, is not mentioned by De Morgan.

Other works of 1582. Gemma, p. 200, 1540; Moya, p. 310, 1562; Sacrobosco, p. 32, 1488; Stevin, p. 386, 1585; Clement, 'Summa del arte arithmetica, de Fr. de Sant Clement,' Barcelona, 4°; Ognibene de Castellano, 'Il lineamento pertinente all' intendere facilmente quello, che Euclide & altri Eccellentiss. Mathematici hà trattato oscuramente,' Vincenza, 8° (contains some theory of numbers; see also p. 306, 1561); Mellema, 'Arithmétique composée de plusieurs inventions et problèmes nouveaux,' Antwerp, 2 vol., 1582 and 1586.

CHRISTOPHER CLAVIUS. Ed. pr. 1583. Rome, 1583.

Christoph Klau. Born at Bamberg in 1537; died at Rome, February 6, 1612. He was a Jesuit priest, and taught mathematics in the Jesuit college at Rome. He wrote a number of treatises on mathematics.

Title. See Fig. 188.

Description. 8°, 10.5 × 16.6 cm., the text being 8 × 13.3 cm. 219 pp. numb. + 13 unnumb. = 232 pp., 38 ll. Rome, 1583.

Editions. Rome, 1583, 8° (here described); Cologne, 1584; Rome, 1585, 8° (p. 378); Cologne, 1592. There were also editions after 1600 (see p. 378 for the 1602 edition). The Italian translation of 1586 is mentioned on p. 378. The collected works of Clavius in five volumes appeared at Basel in 1612, fol. 497;541

FATTORE

LIBRO D'ARITHMETICA,
ET GEOMETRIA PRATTICALE.

DI M. MAFFEO POVEIANO
VERONESE.

Opera noua, & vtilissima

Delle più breui, & generali prattiche, che usar si possano, necessarie ad ogn'uno.

Con licenza de' Superiori.

IN BERGAMO L'ANNO DI N. SIG.
M D LXXXII.
Per Comin Ventura, Stampatore in essa Città.

FIG. 187. TITLE PAGE OF POVEIANO

CHRISTOPHORI
CLAVII
BAMBERGENSIS
E SOCIETATE
IESV

EPITOME ARITHME-
ticæ Practicæ.

PERMISSV SVPERIORVM.
ROMAE Ex Typographia Dominici Basæ. 1583.

FIG. 188. TITLE PAGE OF CLAVIUS

Clavius was an excellent teacher of mathematics, and his textbooks were models of good arrangement. This work is an attempt at a practical arithmetic. It is conservative in treatment, the applications being confined, as was the custom, largely to the rule of three. It was too scholarly to be popular in schools under the mercantile influence, but it was influential in the classical schools.

CHRISTOPHER CLAVIUS. Ed. pr. 1583. Rome, 1585.

See p. 375.

Title. 'Christophori // Clavii // Bambergensis // e Societate // Iesv // Epitome Arithmeticæ // Practicæ nunc denuo ab ipſo auctore // recognita. // (Woodcut with I. H. S. in center.) Permissv Svperiorvm // Romæ Ex Typographia Dominici Baſæ. 1585.' (P. 1.)

Description. 8°, 10.3 × 16 cm., the text being 8 × 13 cm. 337 pp. (321 numb.), 31 ll. Rome, 1585.

See above.

CHRISTOPHER CLAVIUS. Ed. pr. 1583. Rome, 1586.

See p. 375.

Title. 'Aritmetica // Prattica // composta dal Molto // Reuer. Padre Chriſtoforo Clauio // Bambergenſe della Com-//pagnia di I E S V. // Et tradotta da Latino in Italiano dal Signor // Lorenzo Caſtellano Patritio // Romano. // Con Licentia dei Svperiori. // (Woodcut with I H S in center.) In Roma, // Nella Stamperia di Domenico Baſa. // M. D. LXXXVI.' (P. 1.)

Description. 8°, 10.3 × 15.6 cm., the text being 8.2 × 13.3 cm. 302 pp. (275 numb.), 39 ll. Rome, 1586.

Editions. See p. 375.

This is merely an Italian translation of the 1583 edition (see p. 375). Clavius was unable, however, to popularize the book in the mercantile schools of Italy, although several editions appeared after 1600.

CHRISTOPHER CLAVIUS. Ed. pr. 1583. Rome, 1602.

See p. 375.

Title. 'Arithmetica // Prattica // composta dal Molto // Reuer. Padre Chriſtoforo Clauio // Bambergenſe della Com-//pagnia di

IESV.// Et tradotta da Latino in Italiano dal Signor Lorenzo // Caſtellano Patritio Romano.// Reuiſta dal medemo Padre Clauio //con alcune aggiunte.//Con Licentia de i Svperiori.//In Roma, //(woodcut with I. H. S. in center.) Per li Heredi di Nicolò Mutij // M. DC. II.' (P. 1.)

Colophon. 'In Roma,// Per li Heredi di Nicolò Mutij M D C II.' (P. 312.)

Description. 4°, 10.4 × 15.7 cm., the text being 8.3 × 13.3 cm. 312 pp. (281 numb.), 39 ll. Rome, 1602.

Editions. This is the second Italian edition (p. 375).

See p. 378.

NICOLAUS REYMERS. Ed. pr. 1583. Leipzig, 1583.

A German surveyor, born at 'Henstede in Dietmarschen.' The 'Beschluss' is dated 'zu Hattſtede in Diethmarchen,' September 14, 1583.

Title. 'Geodæsia // Ranzoviana.// Landt Rechnen/ // vnd Feldmeſſen/ ſampt meſſen aller-//hand grőſſe. Alles auff eine leichte/ behende/ // vnd vormals vnbekandte newe art/ kűnſt-//lich/ grűndlich vnd deutlich // beſchrieben/ // Zu Ehren // Dem Edlen/ Beſtrengen // vnd Ehrnuehſten Herrn/ Heinrichen // Rantzouen/ Herrn Johans ſeligen Sohne/ der // Kőn. Mayſt. zu Dennemarcken/ etc. In den // Fűrſtenthumben Schleſewick/ Holſtein/ vnd Diethmar-//ſchen/ Stadthaltern/ Rhat vnd Ambtman auff // Segeberge/ Erbgeſeſſen zum // Breitenberge/ etc.// Durch // Nicolaum Reymers/ von Henſtede/ //in Dietmarſchen. // Cvm Privilegio.' (F. 1, r.)

Colophon. 'Gedruckt zu Leipzig bey // Georg Defner/ // Im Iahr // M. D. LXXXIII.' (F. 44, v.)

Description. 4°, 15.1 × 18.6 cm., the text being 9.7 × 14.5 cm. 44 ff. unnumb., 25 ll. Leipzig, 1583.

Editions. There was no other edition.

Although nominally a book on surveying, this work may properly rank as an arithmetic, the 'Erſte Buch' being entirely devoted to that subject. Of this book the first chapter is entitled 'von zahlen'; the second, 'von brűchen'; the third, 'von ſummieren'; the fourth, 'von vielfeltigen,' multiplication thus directly following addition; the fifth, 'von abziehen'; the sixth, 'von Theilen' (division); the seventh, 'von den Wurtzel'; the eighth, 'von der gevierten Wurtzel.' The rest of the work is devoted to mensuration, and in particular to surveying. The chief interest in the first book is in the use made of the compound numbers then needed in surveying. Like most such manuals, it shows no insight into educational problems, and the treatment is very unsatisfactory.

Other works of 1583. Baker, p. 327, 1568; Caesar, p. 370, 1582; Gemma, p. 200, 1540; Petri, p. 325, 1567; Reisch, p. 82, 1503; Anonymous, 'Rechenbüchlein auf Erfurtifche weifs,' Erfurt, 8°; Johnn Weber, 'Ein new Kunstlich Rechenbuch auff den linien und ziffern,' Leipzig, 4° (see also p. 338, 1570).

PETRUS BUNGUS. Ed. pr. 1583–84. Bergamo, 1584–85.

Born at Bergamo; died September 24, 1601, at Bergamo. He was a canon of the cathedral in that city. He wrote only on the mystery of numbers.

Title. See Fig. 189, which gives the title page of the first part. Bound with this is the second part with the title: 'Mysticae // nvmerorvm // significationis // pars altera, // Io. Petro Bongo Canonico Bergomate // avctore,// In qua de Numeris in Sacris libris potiffimum repertis, ex Theo-//logorum maxime fententia, & probatorum aliorum cuiufuis // facultatis Scriptorum, ita exacte, dilucide, & accurate differi-//tur, vt ferme nil addi, aut detrahi poffe videatur: // Opus varia fane, et multiiuga adeo refertum doctrina, vt non Theo-//logis folum; fed etiam Philofophis, Mathematicis, atque alijs ftu-//diofis omnibus, tam vtile, quam iucundum fit futurum.// De Svperiorvm licentia.// Bergomi CIƆ IƆ XXCIV.// Typis Commini Venturæ, eiufdem Vrbis typographi.'

Description. Fol., 20.6 × 30.4 cm., the text being 17 × 28.9 cm. 276 pp. (245 numb.) in the first part; 198 pp. (177 numb.) in the second part; 36 ll. Bergamo, first part 1585, second part 1584.

Editions. The first edition appeared in 1583–84, 'Bergomi, typis Comini Venturæ.' The second part of this work, as the title page shows, is therefore of the first edition. The first edition of the first part was evidently exhausted before that of the

MYSTICAE

NVMERORVM

SIGNIFICATIONIS

LIBER

IN DVAS DIVISVS PARTES,

R. D. PETRO BONGO

CANONICO BERGOMATE

AVCTORE:

Opus maximarum rerum, & plurimarum doctrinæ, suauitate, copia, & uarietate refertum,

Theologis, Philosophis, Mathematicis, atque alijs studiosis omnibus, tam vtilitatem, quam iucunditatem allaturum.

DE SVPERIORVM LICENTIA.

BERGOMI CIↃ IↃ XXCV.

Typis Comini Venturæ, & Socij.

FIG. 189. TITLE PAGE OF THE 1585 BUNGUS

second part was sold, and hence the first part of the present work is the second edition.

There were also editions as follows: Venice, 1585, 8°; Bergamo, 1590; ib., 1591, 4° (which De Morgan incorrectly calls the second edition); ib., 1614 (p. 384). It appeared under the

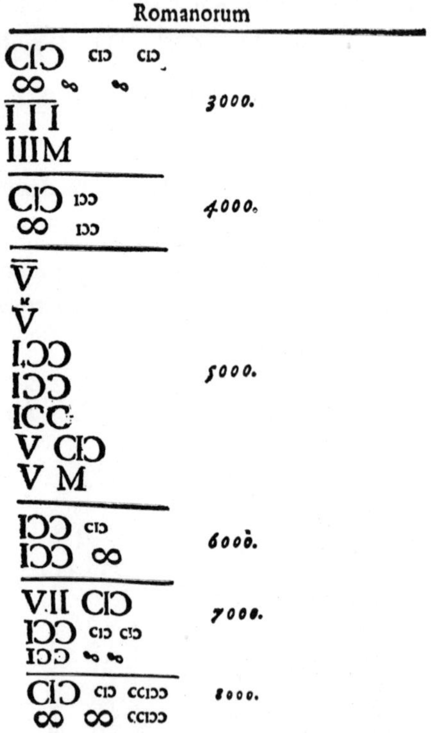

Romanorum	
CIↃ CIↃ CIↃ ∞ ∞ ∞ $\overline{\text{III}}$ IIIM	3000.
CIↃ IↃↃ ∞ IↃↃ	4000.
$\overline{\text{V}}$ $\overset{M}{\text{V}}$ I.ↃↃ IↃↃ ICC V CIↃ V M	5000.
IↃↃ CIↃ IↃↃ ∞	6000.
V.II CIↃ IↃↃ CIↃ CIↃ IↃↃ ∞ ∞	7000.
CIↃ CIↃ CCIↃↃ ∞ ∞ CCIↃↃ	8000.

Fig. 190. From the 1614 edition of Bungus

title 'Numerorum mysteria,' Bergamo, 1599, 4° (p. 384). I have an edition under this title, published at Paris in 1618 (1617 in the colophon), 4°. For the 'Praecipuae numerorum notae et earum valor,' Parma, 1689, see p. 384.

This is a mass of erudition, prolix and unscientific, relating to the mystery of numbers. It was written by a priest for the use of preachers, and it includes all of the allusions to such matters as the mystic three that Bungus could find in ancient literature. He takes up the various numbers from one to ten in the same way, together with a few of the more interesting larger numbers. For students interested in popular number mysticism the book still remains the classic in its way. It is also of much value in showing the nature of the Roman numerals in use in the sixteenth century. (See Figs. 190, 191.)

Numeratio.

CIↃ CCIↃↃ ∞ CCIↃↃ	9000.
CC·I·ↃↃ C·C·I·Ↄ·Ↄ $\overline{\mathrm{X}}$ $\overline{\overline{\mathrm{X}}}$ CC·I·CC ↃMC ↃMↃ IMI	10000.
CCIↃↃ CIↃ CC·I·ↃↃ ∞	11000.
CCIↃↃ CIↃ CIↃ CC·I·ↃↃ ∞ ∞	12000.
CC·IↃↃ CIↃ CIↃ CIↃ CCIↃↃ ∞ ∞ ∞	13000.
CCIↃↃ CIↃ IↃↃ CCIↃↃ ∞ IↃↃ	14000.
CCIↃↃ IↃↃ	15000.

FIG. 191. FROM THE 1614 EDITION OF BUNGUS

Other works of 1584. Buckley, p. 252, 1550; Cassiodorus, p. 211, 1540; Clavius, p. 375, 1583; Delfino, p. 275, 1556; Köbel, p. 102, 1514; Ramus, p. 263, 1555; Anton Schulze, 'Arithmetica oder Rechenbüchlein,' ('Arithmetica oder Rechenbuch neben einer dienlichen Anleitung zum Buchhalten'?) s. l., with subsequent editions at Liegnitz, 1600, 4°, and Frankfort, 1600, 4°. 497;541

PETRUS BUNGUS. Ed. pr. 1583–84. Bergamo, 1599.

See p. 380.

Title. 'Petri Bongi // Bergomatis // Numerorum myſteria.// Opvs maximarvm rervm // Doctrina, et copia refertvm,// In quo mirus in primis, idemq́ perpetuus Arithmeticæ Pythagoricæ cum // Diuinæ Paginæ Nvmeris conſenſis, multiplici ratione probatur.// Poſtrema hac editione ab Auctore ipſo copioſo Indice, & ingenti // Appendice avctvm.// Cum Superiorum approbatione.// Bergomi, Typis Comini Venturæ, eiuſdem vrbis Typographi.// ∞ IↃ XCIX.' (P. 1.)

Description. 4°, 17.5 × 24.3 cm., the text being 12.6 × 18.8 cm. 770 pp. (676 numb.), 32 ll. Bergamo, 1599.

See p. 382. Although the title is slightly changed, this is the work already described.

PETRUS BUNGUS. Ed. pr. 1583–84. Bergamo, 1614.

See p. 380.

Title. 'Petri Bvngi // Bergomatis // Nvmerorvm myſteria // Ex abditis plurimarum diſciplinarũ fontibus hauſta: // Opvs maximarvm rervm // Doctrina, & copia refertum: In quo mirus in primis, idemq́; perpe-//tuus Arithmeticæ Pythagoricę cum Diuinæ Paginæ Nu-//meris conſenſus, multiplici ratione probatur.// Poſtrema hac editione ab Auctore ipſo copioſo Indice, & ingenti // Appendice avctvm.// Illuſtriſſimo viro, Virtutum omnium, ac diſciplinarum //genere ornatiſſimo // Ranvtio Gambaræ // Comiti Virolæ. & c. // Bergomi, Typis Comini Venturæ. 1614.' (P. 1.)

Description. 4°, 17.5 × 23.4 cm., the text being 12.6 × 18.8 cm. 970 pp. (753 numb.), 32 ll. Bergamo, 1614.

See p. 382.

PETRUS BUNGUS. Ed. pr. 1583–84. Parma, 1689.

See p. 386.

Title. 'Præcipuæ // Numerorum // Notæ,// Et earum valor,// Secundum // Petrum // Bungum.// Parmae, Ex Typographia Ducali.// CLↃ. IↃC. LXXXIX.' (P. 1.)

Description. 4°, 11.1 × 17.7 cm., the text being 7.9 × 12.6 cm. 16 pp. (12 numb.), 13–19 ll. Parma, 1689.

Editions. See p. 380. This is simply an extract from the later editions of the 'Numerorum Mysteria.'

The treatment of Roman numerals by Bungus is the most elaborate and interesting to be found in any of the works of Renaissance writers.

MONTE REGAL PIEDMONTOIS.

Ed. pr. 1585. Lyons, 1585.

The title states that he was professor of mathematics in the University of Paris; evidently in the second half of the sixteenth century.

Title. See Fig. 192.

Description. 32°, 5.3 × 9.8 cm., printed on vellum in double columns, each 1.8 × 9.5 cm. 144 pp. (100 numb.), 27–28 ll. Lyons, 1585.

Editions. There was no other edition. The privilege is dated August 6, 1581, but the work does not seem to have been published until 1585. The book is exceedingly rare, and is unknown to most bibliographers. The author speaks of having published part of the tables in Venice in 1575.

This is a collection of tables, largely for multiplication, beautifully printed on split vellum. These tables give the products of numbers to 100 times 1000. The last ten pages contain a table of arrangement of soldiers in order of battle: 'Le moyen et ordre qui ſe doit tenir pour mettre en ordonnance les batailles de dix soldats iuſques au nôbre de quarante mille.' This is dedicated 'A pvissant et illustre Seigneur le Baron de Mont-clar.'

INVENTION
NOVVELLE ET
admirable, pour faire toute
ſorte de cõpte, tant de marchandiſe, comme de chãger
monnoyes, poids, meſures, de
diuerſe maniere, pour vendre
& acheter, laquelle ſeruira en
tout le monde auec grande
facilité, ſans getõs ne plume.

*Et a la fin le moyen de mettre vn
exercit en bataille, de cent
iuſques à quarante mille.*

Le tout nouuellement composé &
mis en lumiere par le Monte Regal Piedmontois, Profeſſeur de
Mathematique en l'Vniuerſité de
Paris.

A LYON,
Il ſe vend en rue Merciere à
l'enſeigne de la Sphere.
1585.
Auec priuilege du Roy.

FIG. 192. TITLE PAGE OF MONTE REGAL PIEDMONTOIS

SIMON STEVIN. Ed. pr. 1582. 541 Leyden, 1585.

Born at Bruges in 1548; died at the Hague in 1620. He was a merchant, a soldier, and an officer in the civil service.

Title. See Fig. 193.

Description. 8°, 10.5 × 16 cm., the text being 7.6 × 12.7 cm. 31 pp. unnumb. + 1 blank + 642 numb. + (203 numb. + 12 unnumb. + 1 blank in 'La Pratiqve d'Arithmetiqve') = 890 pp., 29 ll. Leyden, 1585.

Editions. Stevin's first work was an interest table, Antwerp, 1582. His arithmetic first appeared in Flemish, at Leyden, 1585, and was reprinted in that language at Gouda in 1626 and in 1630. The French translation, which is here described, appeared at Leyden in 1585 and again in 1586. The first edition under the editorship of Girard appeared at Leyden in 1625. For the edition of 1634 see below.

This work consists of three distinct parts: (1) 'L'Arithmetiqve,' in two books, the first treating of powers and roots, and particularly of surds, and the second of operations on numerical and algebraic expressions and of the solution of equations; (2) 'Les qvatre premiers Livres d'Algebre de Diophante d'Alexandrie,' translated by Stevin, apparently from Xylander's text; (3) 'La Pratiqve d'Arithmetiqve,' an attempt at a practical textbook, but too scholarly for its purposes. The Pratiqve contains 'La Regle d'Interest avec ses tables,' the 1582 work above mentioned, 'La Disme. Enſeignant facilement expedier par nombres entiers ſans rompuz, tous comptes ſe rencontrans aux affaires des Hommes. Premierement deſcripte en Flameng, & maintenant conuertie en François,' and a 'Traicte des incommensvrables Grandevrs.' The interest centers in 'La Disme,' in which decimal fractions are for the first time treated in any elaborate way (see Fig. 194).

SIMON STEVIN.

Ed. pr. of the arithmetic, 1585. Leyden, 1634.

See above.

Title. 'Les // Œuvres // Mathematiques // de // Simon Stevin, // Augmentées // Par Albert Girard.' (P. 1.)

Page 3 reads: Les // Œuvres // Mathematiques // de Simon Stevin de Bruges.// Ou ſont inſerées les // Memoires Mathematiqves,// Eſquelles s'eſt exercé le Tres-haut & Tres-illuſtre Prince

L'ARITHMETIQVE
DE SIMON STEVIN
DE BRVGES:

Contenant les computations des nombres
Arithmetiques ou vulgaires:
Aussi l'Algebre, auec les equations de cinc quantitez.
Ensemble les quatre premiers liures d'Algebre
de Diophante d'Alexandrie, maintenant pre-
mierement traduicts en François.

Encore vn liure particulier de la Pratique d'Arithmetique,
contenant entre autres, Les Tables d'Interest, La Disme;
Et vn traicté des Incommensurables grandeurs:
Auec l'Explication du Dixiesme Liure d'Euclide.

A LEYDE,
De l'Imprimerie de Christophle Plantin.
cIↃ. IↃ. LXXXV.

FIG. 193. TITLE PAGE OF THE 1585 STEVIN

SECONDE PARTIE DE LA DISME DE L'OPE-RATION.

PROPOSITION I, DE L'ADDITION.

Estant donnez nombres de Diſme à ajouſter: Trouver leur ſomme:

Explication du donné. Il y a trois ordres de nombres de Diſme, deſquels le premier 27⓪8①4②7③, le deuxieſme 37⓪8①7②5③, le troiſieſme 875⓪7①8②2③.

Explication du requis. Il nous faut trouver leur ſomme. *Conſtruction.* On mettra les nombres donnez en ordre comme ci joignant, les aiouſtant ſelon la vulgaire maniere d'aiouſter nombres entiers, en ceſte ſorte:

			⓪	①	②	③
		2	7	8	4	7
		3	7	6	7	5
	8	7	5	7	8	2
	9	4	1	3	0	4

Donne ſomme (par le 1 probleme de l'Arithmetique) 941304, qui ſont (ce que demonſtrent les ſignes deſſus les nombres) 941⓪3①0②4③. Ie di, que les meſmes ſont la ſomme requiſe. *Demonſtration.* Les 27⓪8①4②7③ donnez, ſont (par la 3e definition) $27\frac{8}{10}, \frac{4}{100}, \frac{7}{1000}$, enſemble $27\frac{847}{1000}$, & par meſme raiſon les 37⓪6①7②5③ vallent $37\frac{675}{1000}$, & les 875⓪7①8②4③ feront $875\frac{782}{1000}$, leſquels trois nombres, comme $27\frac{847}{1000}$, $37\frac{675}{1000}$, $875\frac{782}{1000}$, font enſemble (par le 10e probleme de l'Arith.) $941\frac{304}{1000}$, mais autant vaut auſſi la ſomme 941⓪3①0②4③,

c'eſt

FIG. 194. FROM THE 1634 EDITION OF STEVIN

Maurice // de Nassau, Prince d'Aurenge, Gouverneur des Provinces des // Païs-bas unis, General par Mer & par Terre, &c.// Le tout reveu, corrigé, & augmenté // Par Albert Girard Samielois, Mathematicien.// A Leyde // Chez Bonaventure & Abraham Elſevier, Imprimeurs ordinaires // de l'Univerſité, Anno cIɔ Iɔ c xxxiv.'

Description. Fol., 21.7 × 34 cm., printed in double columns, each being 8.1 × 28 cm. 910 pp. (232 on arithmetic, 10 unnumb.), 63 ll. Leyden, 1634.

Editions. See p. 386.

Works of 1585. [497;541] Benedetti, p. 364, 1580; Bungus, p. 382, 1583; Caesar, p. 370, 1582; Clavius, p. 378, 1583; Digges, p. 340, 1572; Gemma, p. 200, 1540; Lonicerus, p. 253, 1551; Lossius, p. 289, 1557; Psellus, p. 168, 1532; Riese, p. 139, 1522; Savonne, p. 314, 1563; Io. Frans. Fulconis, 'Cisterna Fulconica, libro d'abaco in lingua provenzale,' s. l. (Lyons?), 8°; Thilman Ofenlach, 'Rechenbüchlein mit der Ziffer und auf den Linien mit Zahlpfennigen,' Basel, 8°; Johann Schreckenberger, 'Rechenbuchlein auff den Linien und der Federn,' Strasburg, 8°; Samuel Eisenmenger (pseud. Siderocrates), 'Cyclopaedia Paracelsica Christiana,' s. l., with a section on arithmetic.

Works of 1586. Clavius, p. 378, 1583; Gray, p. 353, 1577; Mellema, p. 375, 1582; Ramus, p. 331, 1569; Riese, p. 139, 1522; Schonerus, p. 331 (Ramus, 1569); Stevin, p. 386, 1585; Stifel, p. 226, 1544; Tagliente, p. 115, 1515; Anonymous (Sterner mentions a Rechenbüchlein without title page, Magdeburg); Georg Höflein, 'Rechenbüchlein mit der Ziffer und mit den Zahlpfennigen auf der Linie,' Strasburg, 8°; Paulus Alexandrinus, 'Rudimenta,' Greek and Latin, Wittenberg, 4° (with some notes on Jewish arithmetic).

JOHANNES PADOVANIUS. Ed. pr. 1587. Verona, 1587

GIOVANNI PADOVANI. A Veronese mathematician of the second half of the sixteenth century.

Title. See Fig. 195.

Description. 4°, 14.4 × 20.4 cm., the text being 9.6 × 15.7 cm. 7 pp. unnumb. + 73 numb. = 80 pp., 29–32 ll. Verona, 1587.

Editions. Padovani published a work in Venice in 1565 consisting of six parts, the fourth of which was 'De Arithmetica.'

IOANNIS PADOVANII VERONENSIS,

DE ARITHMETICA OPVS, IN QVO NON ſolum omnis generis numerandi ars tam Latino ſermone, quàm Græco perdiſcitur: uerum etiam quicquid ad quaſcunque ratiocinationes pertinet, facili doctrina aperitur,

Omnibus, & in primis reipublicæ literariæ ſtudioſis perneceſſarium.

VERONAE,

Ex Typographia Sebaſtiani à Donnis. 1587.

De licentia Superiorum.

FIG. 195. TITLE PAGE OF PADOVANIUS

That work is probably the same as this. This seems to be the first and only separate edition, and is very rare.

This is in no sense a practical treatise, but it discusses the subject of arithmetic in a learned way, at first rather on the Greek plan. It then considers the Arabic arithmetic, taking up the four processes, roots, proportion, partnership, and two or three other applications. The style in which it is written is such that it would hardly have appealed to the practical merchant even if it had not been in Latin. For Padovani's other works see Riccardi, I, 1, 251.

CASPAR THIERFELDER.

Ed. pr. 1587. Nürnberg, 1587.

A German Rechenmeister of Steyer, born c. 1525.

Title. See Fig. 196.

Colophon. 'Gedruckt zu Nůrnberg/ // durch Leonhardt // Heufzler.' (P. 369.)

Description. 8°, 9.4 × 15.2 cm., the text being 7 × 13.3 cm. 30 pp. unnumb. + 369 numb. = 399 pp., 25–26 ll. Nürnberg, 1587.

Editions. There was no other edition. Thierfelder also published a book with Ulman at Freiburg in 1564, 8°, entitled 'Neues Kunft-Rechenbuch auf der Linie und Feder; dergleichen weder in lateinifchen noch deutfcher Sprache ausgegangen.'

This is a commercial arithmetic, based on Rudolff and other German writers, and with no particular individuality. It contains eighteen chapters, the last one being upon mathematical recreations, 'Von der Schimpff Rechnung/ vnd Erfindung derfelben Regeln,' a subject that began to come into prominence about this time.

Other works of 1587. Finaeus, p. 160, 1530–32; Riese, p. 139, 1522; Michael Gempelius, 'Arithmetik,' s. l., 8°; Aurelio Marinati, 'Della prima parte della somma di tutte le scienze,' Rome, 4° (with a brief treatment of arithmetic); Conr. Poeppingius, 'Neues Rechenbüchlein auf Linien und Federn,' Braunschweig, 8° (four editions after 1600). 497

Works of 1588. Gemma, p. 208, 1540; Helmreich, p. 306, 1561; Rudolff, p. 152, 1526; John Mellis, 'A brief instruction and maner hovv to keepe bookes of Accompts,' London, 8° (containing a short chapter on arithmetic); Heinrich Strübe, 'Arithmetica oder new künftliches Rechenbüchlein,' Zürich, 8°, with an edition ib., 1599, 12°. 542

Arithmetica Oder Rechenbuch / Auff den Linien vnd Ziffern / mit Vortheyl vnd Behendigkeit / auff allerley gebräuchliche Hauß / vnd Kauffmanns Rechnung / Müntzſchlag / Beſchickung deß Thigels / Kunſtrechnung / gründlich beſchrieben / inn Frag vnd Antwort geſtellet: Durch Caſpar Thierfeldern / Schul vnd Rechenmeyſter zu Steyer.

Gedruckt zu Nürnberg / durch Leonhard Heußler.
M. D. LXXXVII.

FIG. 196. TITLE PAGE OF THIERFELDER

BERNAERT STOCKMANS.

Ed. pr. 1589. Dordrecht, 1609

A French schoolmaster at Dordrecht (Dort), in the second half of the sixteenth century.

Title. 'Een corte ende een-//vuldige Inſtructie/ om lichtelijckē // en by hemſelvē/ ſonder eenige meeſter oft onderwij-//ſer te leeren cijfferen. Seer nut eñ profijtelijcken // alle menſchen/ die in de Conſte van Arith-//metica heel ſtecht/ onervaren // ende eenvuldich zijn.// Geſtelt ende by een vergadert, door Bernaert // Stockmans Ianſz. Françoyſche Schoolmeeſter // inde vermaerde Coop-ſtadt Dordrecht.// Hier zijn ooc bygevoecht de Differentien van de co-//ren-Matẽ der voornaẽſte ſtedẽ in Hollant/ t'Sticht/ // Zeelant/ Brabant/ Vlaenderen/ Gelderlant/ En-//gelant/ Vrancrijck/ eñ Ooſtlant/ tot dienſt // van allen Graen-coopers.// Van nieus overſien ende verbetert door C. P. Boeye.// Item/ noch van nieus bygevoecht een clare onder-// wyſinge om de tafelen van Intereſt te leeren maken/ // alles tot dienſt vanden onervarenen.// Sapiente 11. verſ. 22.// Maer ghy hebbet alles in Mate/ Tellen/ // ende Ghewichte gheordineert.// Tot Dordrecht // By my Pieter Verhagen/ woonende inde Druc // kerije/ teghen over de Wijnbrugge. 1609.' (F. 1, r.)

Colophon. 'Tot Dordrecht.// By Adrien Ianſz Bot.// Anno 1609.' (F. 214, v.)

Description. 8°, 9.1 × 14 cm., the text being 7.4 × 11.9 cm. 4 ff. blank + 211 unnumb. = 215 ff., 31–33 ll. Dordrecht, 1609.

Editions. This book was first printed in 1589, and the dedicatory epistle is dated 'den 20. deſer Maent Julij / 1589.' There was no other edition in the sixteenth century.

This is one of the noteworthy arithmetics of Holland, and it went through a number of editions after 1601 (see p. 394). It is a commercial work, and like Vander Schuere's book it gives an excellent view of the mercantile life of Holland in this period.

Other works of 1589. Jacob, p. 298, 1560; Moya, p. 310, 1562; Fabio Paolini (Paulinus), 'Hebdomades, sive septem de septenario libri,' Venice, 4°, containing a little arithmetic, and ib., 1598 (?).

Works of 1590. Bungus, p. 382, 1583; Digges, p. 340, 1572; Jacob, p. 298, 1560; Lapazzaia, p. 324, 1566; Psellus, p. 170, 1532; Recorde, p. 214, c. 1542; Heizo Buscher (Boscherus), 'Arithmeticæ libri duo,' Helmstadt, sm. 8°, with editions, ib., 1591, 8°; Hamburg, 1592, 8°; ib., 1597, 8°; Frankfort, 1600, 8°, and later; Franciscus Brasser, an arithmetic, Lübeck, with later editions; Cyprian Lucar, 'A treatise named Lucarsolace,' London, 4°; Bartolomé Solorzano, 'Libro de caxa y Manual de cuentas de Mercaderes,' Madrid. 542; 543

BERNAERT STOCKMANS. Ed. pr. 1589. Gouda, 1644.

See p. 392.

Title. 'Aritmetica,// Door Bernardus Stockmã // eertyts Franſoyſche ſchool-//meeſter inde vermaerde coopſtadt // Dorderecht nu van nieu[s] curieus // gecorigert ende verbetert noch is // hier bÿ gevoecht een tafelken om te // rabatteeren op ſulcken tÿt ofte intereſt // men begeert door // Abel. W. Waesenaer // Rekenmeeſter tot Vtrecht // (Portrait of author.) Gedruckt tot Vtrecht by // Eſdras Willemſsen Snellaert.// boeckvercooper // Anno. 1637.' (P. 1.)

Colophon. 'Ter Govde,// Gedruckt by Pieter Rammazeyn, Boeck-// drucker in't vergulde A B C.// Door Eſdras Willemſz Snellaert, Boeck-//vercooper tot Vtrecht by de waert-poort 1644.' (P. 421.)

Description. 8°, 9.3 × 13.9 cm., the text being 7.5 × 12.4 cm. 429 pp. (410 numb.), 32–33 ll. Gouda, 1644.

Editions. See p. 393. The engraved title page of the 1637 edition has been used with this edition, which, as the colophon shows, was printed in 1644.

See p. 393.

FRANCESCO PAGANI. Ed. pr. 1591. Ferrara, 1591.

An Italian arithmetician of the sixteenth century, born at Bagnacavallo.

Title. See Fig. 197.

Description. 4°, 14.2 × 17.9 cm., the text being 10.1 × 15.9 cm. 210 pp. (200 numb.), 38 ll. Ferrara, 1591.

Editions. There was no other edition.

This rare and almost unknown work is based upon the Borghi model (see p. 16) which served for so many of the best Italian writers. It was written, as the dedicatory epistle states, at Bagnacavallo, and is one of the few books on mathematics published in the sixteenth century at Ferrara. It has no merit save as its applied problems give a view of the business life of the time. In its numerical work it is reactionary, making, for example, a strong plea for the galley as opposed to the 'a danda' or modern method of division.

ARITHMETICA

PRATTICA VTILISSIMA,

ARTIFICIOSAMENTE ORDINATA

Da M. FRANCESCO PAGANI da Bagnacauallo,

Nella quale ſi contiene il vero, & facile modo di conteggiare.

Con molti Queſiti importanti, & neceſſarij à Ragionieri, à Mercanti, & ad ogni perſona, in tutti i Paeſi.

AL MOLTO ILLVSTRE SIGNOR RAFFAELE RASPONI, &c.

IN FERRARA,

Appreſſo Vittorio Baldini.

Con licenza de' Superiori. M. D. XCI.

FIG. 197. TITLE PAGE OF PAGANI

RENÉ BUDEL, et al. Ed. pr. 1591. Cologne, 1591.

For biographies, see below.

Title. See Fig. 198.

Description. 4°, 18.2 × 23.6 cm., printed in double columns, each being 6.4 × 18.6 cm. 38 pp. unnumb. + 798 numb. = 836 pp. (PP. 271–350 are not in the volume, and if they were ever bound in any copies they must have constituted a section by themselves.) 46–49 ll. Cologne, 1591.

Editions. There was no other edition.

This rather massive treatise on the history of monetary measures consists of two books by Budel (Budelius), director of the Bavarian mint, and several appended chapters by the following writers: Albertus Brunus (1461–1541), counselor to Louis of France, and ambassador; Johannes Aquila, friend of the astrologer Stöffler who died in 1531; Bilibaldus Pirkheymer (1470–1530), a celebrated humanist; Martinus Garatus Laudensis, who writes a chapter 'De monetis'; Franciscus Curtius, and Joannes Regnaudus of Avignon, who write on the same topic; Carolus Molinæus (1500–1566); Didacus Covarrubias (1512–1577), bishop of Ciudad Rodrigo; Henricus Mameranus, a Belgian printer; Henricus Hornmannus; Franciscus de Oretio (1418–1483), a celebrated lawyer of Arezzo; Nicolas Everardus (1473–1532), a celebrated Dutch lawyer, of Middelburg; Jacobus Menochius (1531–1607), an Italian lawyer, and various others. It is a monumental work, and is helpful in the investigation of the history of monetary tables.

Other works of 1591. Baker, p. 327, 1568; Bungus, p. 382, 1583; Buscher, p. 393, 1590; Gemma, p. 200, 1540; Mariani, p. 181, 1535; Petri, p. 325, 1567; Psellus, p. 168, 1532; Ramus, p. 331, 1569; Vincent de Beauvais, p. 10, 1473; Jordanus Bruno, 'De Monade numero et Figura liber,' Frankfort, 8°; Johann Kaudler, 'Arithmetica oder Rechnung auf der Linien und mit Ziffern,' Regensburg, 8°; Hans Jacob Mewrer, 'Bericht von dem Rechnen mit den Zahlpfennigen oder auf der Linien,' Zürich, 8°. 544

THOMAS HYLLES. Ed. pr. 1592. London, 1600.

An English mathematician of the latter part of the sixteenth century.

Title. See Fig. 199.

Description. 4°, 14.2 × 19.6 cm., the text being 10.8 × 16.9 cm. 15 ff. unnumb. + 270 numb. = 285 ff., with 2 charts at the end; 40 ll. London, 1600.

DE

MONETIS, ET RE NVMA= RIA, LIBRI DVO:

QVORVM PRIMVS ARTEM CVDENDAE MO NETAE: SECVNDVS VERO QVAESTIONVM MONETARIARVM DECISIONES CONTINET.

HIS ACCESSERVNT TRACTATVS VARII ATQVE VTILES, NECNON CONSILIA, SINGVLARESQVE ADDItione tam veterum, quàm Neotericorum Authorum, qui de Monetis, earundemque valore, liga, pondere, poteſtate, mutatione, variatione, falſitate, ac ſimilibus ſcripſerunt.

Quorum omnium Catalogum pagina duodecima indicat.

AVTHORE ET COLLECTORE CLARISS. VIRO RENERO BVDELIO RVREMVNDANO, IC. NECNON REVErendiſs. atque Illuſtriſs. Principis ac D. Domini Erneſti Electoris Colonien Bauariæ Duc. &c. Monetarum, tam Rhenenſium, quàm Vveſtphalicarum Archiepiſcopalium Præfecto.

CVM SVMMARIIS ET INDICE COPIOSO.

COLONIAE AGRIPPINAE,

APVD IOANNEM GYMNICVM,

SVB MONOCEROTE.

ANNO M. D. LXXXXI.

CVM PRIVILEG. CAES. MAIEST. AD SEXENNIVM.

FIG. 198. TITLE PAGE OF BUDEL ET AL.

The Arte of vulgar arith-

meticke, both in Integers and Fractions, *deuided into two Bookes: whereof the first is called* *Nomodidactus Numerorum*, and the second *Portus Proportionum*, with certeine Demonstrations, reduced into so plaine and perfect Method, *as the like hath not hetherto beene published in English.* *Whereunto* is added a third Booke, entituled *Musa Mercatorum*: Comprehending all the most necessarie and profitable Rules *vsed in the trade of Merchandise.*

In all which three Bookes: the Rules, Precepts, and Maxims, are not *onely composed in meeter for the better retaining of them in memorie,* but also the operations, examples, demonstrations, and questions, *are in most easie wise expounded and explaned, in the forme* of a Dialogue, for the Readers more cleere vnderstanding.

A knowledge pleasant for Gentlemen, commendable for Capteines and Soldiers, profitable for Merchants, and generally *necessarie for all estates and degrees.*

Newly collected, digested, and in some part deuised by *a Welwiller to the Mathematicals.*

Ecclesiasticus. cap. 19.

Learning vnto fooles is as fetters on their feete and Manicles vpon their right hand: but to the wise it is a Iewell of golde, and like a Bracelet vpon his right arme.

Boetius libr. 1. Arith. cap. 2.

Omnia quecunque a primaeua natura constructa sunt, Numerorum videntur racione formata, Hoc enim fuit principale in animo conditoris examplar.

Imprinted at London by *Gabriel Simson*, dwelling in Fleete-lane. 1600.

FIG. 199. TITLE PAGE OF THE 1600 HYLLES

Editions. London, 1592; ib., 1600, 4° (here described).

This rather ponderous work of 570 pages is written in the form of a dialogue, following the popular textbooks of Recorde. Hylles introduces his rules and definitions in verse, presumably with the idea that they can be more easily memorized in this form. An illustration of this feature is seen in the following description of the first case of 'Barters or Trucques,' a chapter now obsolete, but one of which we have a reminder in the words 'barter' and 'truck':

> Of Barters or trucques, there are diuers kindes,
> VVhereof the firſt, is when the Trucquers take,
> But ware for ware, by agreement of mindes,
>   No partie grating, greater gaines to make,
> Thequalitie of which exchange of wares,
>   The compound rule aſcending ſole declares.' (F. 255.)

The author seems to take up every rule known to the English arithmeticians of the time, and his book is a good source of information concerning British commerce of the period. It was not a popular work, probably because it elaborated its rules too much to be usable.

JOANNES ANTONIUS MAGINUS.

Ed. pr. 1592. Venice, 1592.

GIOVANNI ANTONIO MAGINI. Born at Padua, June 13, 1555; died at Bologna, February 11, 1617. He was professor of astronomy and mathematics at Bologna, and wrote numerous works on these subjects.

Title. 'Io.// Antonii // Magini Patavini // Mathematicarvm in almo // Bononienſi Gymnaſio profeſſoris.// De Planis Triangvlis // Liber Vnicus.// Eiuſdem // de Dimetiendi ratione // per Quadrantem, & Geometricum Quadratum,// Libri Qvinqve.// Opus valdè vtile Geometris, Aſtronomis, Geographis, Mechanicis, Ar-//chitectis, Militibus, Agrorum menſoribus, & denique // omnibus Mathematicarum profeſſoribus.// Cvm Privilegio.// Veneтiis, apvd Robertvm Meiettvm.// M. D. XCII.' (F. 1, r.)

Colophon. 'Bononiæ,// apud Iannem Baptiſtam Ciottum,// Typis Victorij Benacij,// Anno Domini, M. D. XCII.// Superiorum permiſſu.' (F. 114, v., and 132, v.)

Description. 4°, 16 × 21.6 cm., the text being 11.8 × 18.2 cm. 4 ff. unnumb. + 110 numb. (in the first book) + 124 numb. + 4 unnumb. (in the second book) = 242 ff., 34–38 ll. Venice (but printed in Bologna, as the colophon states), 1592.

Editions. There was no other edition.

Following some other bibliographers, I have included this treatise on plane triangles, because of the 'Expositio, ac vsvs tabvlae tetragonicae, ſeu Quadratorum numerorum cum ſuis radicibus iuxta ſequentes octo Canones' (f. 5, r.) which it contains. The 'Tabvla nvmerorvm quadratorum' begins on f. 41 and extends to f. 64, inclusive. It is the most extensive table of squares and roots that had appeared up to 1592.

THOMAS MASTERSON.

Ed. pr. 1592–95. London, 1592–95.

An English mathematician of the latter part of the sixteenth century.

Title. See Fig. 200.

Description. 4°, 12.8 × 18.1 cm., the text being 10.1 × 15.6 cm. 255 pp. (230 numb.), 34–35 ll. London, 1592–95. Book II (148 pp.) bears the date 1592; Book III (78 pp.) bears the date 1595.

Editions. There was no other edition. There was, however, an addition to the 'First Booke' published at London in 1594.

The author says in his dedicatory epistle, dated 'London this 20 Auguſt Anno. 1592,' that he has 'vnder taken to write and publiſhe ſixe bookes of the Art of Arithmeticke: with this order and methode, that the firſt, third, and fift bookes, ſhalbe as a ſummarie and ground, teaching, the true ingenious, inuentions, and the perfect figuratiue and caractericall operations of the ſame Art . . . Then the ſeconde, fourth, and ſixt bookes, ſhalbe of Arithmeticall queſtions and demaundes, with the application of the definitions, common ſentēces and inſtructions of the firſt, third, and fift bookes.' The plan was not carried out beyond the publication of the three books and the supplement mentioned above.

Book I is on the fundamental operations with integers and fractions, and makes no advance in the ordinary textbooks of the period. Book II is a collection of practical problems representing the mercantile activities of London at the close of the sixteenth century (see Fig. 201). Book III would now be classed as part of algebra (see Fig. 202), since it refers chiefly to irrational numbers.

Other works of 1592. Buscher, p. 393, 1590; Capella, p. 68, 1499; Clavius, p. 375, 1583; Fischer (Piscator), p. 247, 1549; Gemma,

THOMAS MASTERSON HIS FIRST BOOKE OF ARITHMETICKE.

Shewing the ingenious inuentions, and figuratiue operations, by which to calculate the true solution or answeres of Arithmeticall questions: after a more perfect, plaine, briefe, well ordered Arithmeticall way, then any other heretofore published: verie necessarie for all men.

Nothing vvithout labour.
All things vvith reason.

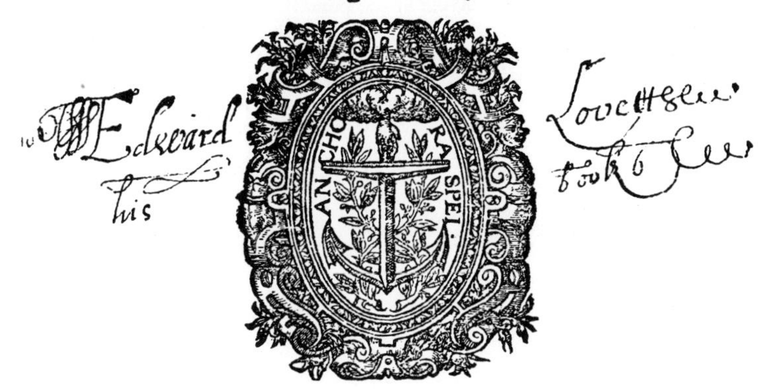

Imprinted at London by Richard Field, dwelling in the Blacke friers neare Ludgate.

1592.

1626
1592
0034

FIG. 200. TITLE PAGE OF MASTERSON

p. 200, 1540; Nonius, p. 315, 1564; Psellus, p. 168, 1532; Ramus, p. 331, 1569; Riese, p. 139, 1522; Tartaglia, p. 279, 1556; Rutilio Cosentino Benicansa, 'Corona di tutte le scientie de Abaco,' Naples,

122 THOMAS MASTERSON HIS

5 . . 200 . . 3 *Facit* 120
200 . . 4 . . 800
100
300 . . 4 . . 1200
150
150 . . 4 . . 600
80 . . 2600 . . 120
Facit 3900.
400 . . 5 . . 2000
1900
$271\frac{3}{7}$
Facit B tooke out $128\frac{4}{7}$ pound.

538 Two marchants made a companie, A put in 300 pound for 2 monethes, and then putteth yet in 100 pound, and 6 monethes after that taketh out 200 poūd, and with the reſt remaineth vntill the yeares end. B put in 100 pound for one moneth, and then putteth yet in 700 pound, and 6 monethes after that taketh out a certaine ſumme of money, and with the reſt remaineth vntill the yeares end. and then finde to haue gained together 400 poūd, whereof B muſt haue 80 pound more then A, the queſtion is how much money B tooke out of the companie, without reckening intereſt vpon intereſt.

300 . . 2 . . 600 — 400
100 — 80
400 . . 6 . . 2400 — 320
200 — 160
200 . . 4 . . 800
160 . . 3800 . . 240
19
5700
100 . . 1 . . 100
700
800 . . 6 . . 4800 *Facit* B tooke out 640 pound.
4900
800
160

If

FIG. 201. FROM MASTERSON

8°; Julius Caesar of Padua, 'Gewiſſe Erinnerung einer allgemeinen arithmetiſchen Practic,' Cologne, 16° (see also p. 368, 1582); Johann Krafft, 'Ein neues vnd wohlgegründtes Rechenbuch,' Ulm.

6 THOMAS MASTERSON HIS

hauing here their number written ouer them.

1. 2. 3. 4. 5. 6. 7. 8. 9. 10. 11. 12.
℞. ℨ. ℭ. ℨℨ. ſß. ℨℭ. bſß. ℨℨℨ. ℭℭ. ℨſß. cſß. ℨℨℭ.
13. 14. 15.
dſß. ℨbſß. ℭſß. &c.

The firſt caracter written thus ℞, doeth ſignifie any number before which it is written, to be the firſt number giuen, taken, or imagined;and is called *radix*, or roote,for that all the other caracters haue their originall or of-ſpring of it. The ſecond written thus ℨ, is called *zenſe* or ſquare, and doth ſignifie any number before which it is written, to be the product of the firſt multiplication of the roote by it ſelfe: that is of the roote two times taken and multiplied. The third written thus ℭ, is called *cube*, and doth ſignifie the number following the ſame to be the product of the ſecond multiplication of the roote,three times taked and multiplied, that is of the ℨ multiplied by the ℞. The fourth is called *zenſezenſe*, and doth ſignifie the number following the ſame,to be the product of the third multiplication,that is the product of the ℞ foure times taken and multiplied. The fift is called *ſurſolide*, and doth ſignifie the number following the ſame, to be the product of the fourth multiplication. The ſixt is called *zenſecube*, and doth ſignifie the number following the ſame,to be the product of the fift multiplication. The ſeuenth is called *bſurſolide*, or ſecond ſurſolide, and doeth ſignifie the number following the ſame,to be the product of the ſixt multiplication. The eight is called *zenſezenſezenſe*, & doth ſignifie the number following the ſame,to be the product of the ſeuenth multiplication. The ninth is called *cubecube*, and doth ſignifie, the number following the ſame,to be the product of the eight multiplication. The tenth is called *zenſeſurſolide*, & doth ſignifie the number following the ſame, to be the product &c. The eleuenth, is called *cſurſolide*,or third ſurſolide. The twelfth is called *zenſezenſecube*. The thirteenth is called *dſurſolide*, or fourth ſurſolide. You may proceede further at your pleaſure, if you marke that the fift caracter is the firſt ſurſolide, the ſeuenth the ſecond ſurſolide, and the next vncompound number following,the next ſurſolide,& ſo infinitely euer the next vncompound

FIG. 202. FROM MASTERSON, SHOWING ALGEBRAIC SYMBOLS

SIGISMUNDUS SUEVUS. Ed. pr. 1593. Breslau, 1593.

A German priest, living at Breslau. Born c. 1550. 544

Title. See Fig. 203.

Colophon. 'Gedruckt zu Brefslaw/ durch // Georgium Bawmann/ J. Jn Mitvorle-//gung Andreæ Wolcken. Jm Iahre :// M. D. XCIII.' (P. 523.)

Description. 4°, 14.8 × 19.5 cm., the text being 11.2 × 15.8 cm. 524 pp. (455 numb.), 28–31 ll. Breslau, 1593.

Editions. There was no other edition.

This work was probably intended to be a practical arithmetic, but the author's theological interests unfitted him for the task of writing such a book. Although the fundamental operations and the common rules of the day are treated in somewhat the usual way, the problems are largely biblical, or mystical. The size of Goliath's armor and Gematria used to foretell the famine of Poland represent the applications.

ALESSANDRO ALAMAGNI.

Ed. pr. 1593. Venice, 1593.

A Venetian arithmetician of the sixteenth century.

Title. See Fig. 204.

Description. 12°, 8.7 × 15.4 cm., the text being 6.3 × 13.2 cm. 1 f. unnumb. + 96 numb. = 97 ff., printed in double columns, 30 ll. Venice, 1593.

Editions. There was no other edition.

The word 'tariffa' in these early books means a table used by merchants to assist in their computations involving the measures and awkward monetary systems of the time. See Paxi, p. 77; Mariani, p. 180.

Other works of 1593. Gemma, p. 200, 1540; Tartaglia, p. 279, 1556; Horatio Galasso, 'Giochi di Carte,' Venice, 8°, with an edition at Verona in 1597, 12°, and a French translation in 1603 (containing several 'giochi d'abbaco').

MIGUEL GERONIMO SANTA CRUZ.

Ed. pr. 1594 Madrid, 1643.

A Spanish merchant and arithmetician of the second half of the sixteenth century, born at Valencia. He lived at Seville.

Title. 'Libro de // Arithmetica // especvlativa, y pra-//tica, intitvlado, el Dorado // Contador, contiene la fineza y reglas de

RITHMETICA
HISTORICA.
Die Löbliche Rechenkunst.
Durch alle Species vnd fürnembste Regeln/
mit schönen gedenckwirdigen Historien vnd Exempeln /
Auch mit Hebraischer / Grichischer/ vnd Römischer Müntze / Ge-
wicht vnd Maß/ deren in Heiliger Schrifft vnd gutten Geschicht-
Büchern gedacht wird / Der lieben Jugend zu gutte erkleret.
Auch denen die nicht rechnen können / wegen vieler
schönen Historien vnd derselbigen bedeutun-
gen lustig vnd lieblich zu lesen.

Aus viel gutten Büchern vnd Schrifften mit fleis
zusammen getragen.
Durch
Sigismundum Sueuum Freystadiensem,
Diener des H. Göttlichen Worts der Kirchen Chri-
sti zu Breslaw / Probst zum H. Geiste / vnd
Pfarrherr zu S. Bernardin in der Newstadt.

Gott hat alles geordnet mit Maß/ Zal
vnd Gewichte. *Sapient. 11.*

Gedruckt zu Breslaw/ durch Georgium Bawman/
Im Jhare M. D. XCiij.
CVM GRATIA ET PRIVILEGIO.

Fig. 203. Title page of Suevus

TARIFFA
NVOVA,

DELLA VALUTA
delli Cecchini, da Lire
dieci, e ſoldi vno,

Fino à £ 12 ß 8. Laqual ſerue
anco per ogni ſorte di
Mercantie;

FATTA DA M.
Aleſſandro Alamagni.
Con Preuilegio

IN VENETIA,

Appreſſo Gio. Ant. Rampazetto. 1593.

FIG. 204. TITLE PAGE OF ALAMAGNI

contar // oro y plata, y los Aneages de // Flandes.// Por moderno y com-//pendiofo eftilo.// Compvesto por Migvel Gero-//nimo de Santa-Cruz, natural de la Ciudad, y // Reino de Valencia, y vezino de // Seuilla.// Al L[do] Don Pero de Baraiz,// Tenience mayor de Corregidor de la infigne Villa de // Madrid, Corte del Rei nueftro Señor.// Año 1643 // Con licencia.// En Madrid, Por Francifco Martinez.// A cofta de Iuan Bautifta Tabàno, Mercader de libros, vendefe al // lado del Colegio de Atocha.' (F. 1, r.)

Description. 4°, 13.8 × 20 cm., the text being 9.9 × 16.8 cm. 6 ff. unnumb. + 238 numb. = 244 ff., 32–34 ll. Madrid, 1643.

Editions. I know of no sixteenth-century edition, but the original privilege is dated May 8, 1594. This edition of 1643 seems by the dedication of Tabàno to have been published after the author's death. There was an edition as late as 1794.

The work follows the general plan of the Spanish and Italian mercantile arithmetics of the sixteenth century. It is not, however, a very practical book, the author having been too much influenced by the theoretical side of works like Tartaglia's.

Other works of 1594. Barlaamo, p. 343, 1572; Masterson, p. 400, 1592; Recorde, p. 217, c. 1542; Reisch, p. 82, 1503; Wenceslaus, p. 421, 1599; Thomas Blundevile, 'His Exercises, containing sixe Treatises,' London, 4° (the first being on arithmetic; there was an edition ib., 1597, 4°, and others **after 1601); Smiraldo Borghetti, 'Opera** d'abbaco,' **Venice, 8°; Jerónimo Cortés, 'Arithmetica practica,' Valencia.** [544]

Works of 1595. Belli, p. 343, 1573; Helmreich, p. 306, 1561; Masterson, p. 400, 1592; Urstisius, p. 361, 1579. [544; 546]

JOANNES BILSTENIUS. Ed. pr. 1596. Basel, 1596.

A sixteenth-century German educator.

Title. Syntagma // Philippo-//rameum // Artium Li-//beralium, Methodo brevi ac per-//fpicua concinnatum // per // Ioan. Bilstenium // Marfbergianum.// In // Gratiam Tyronum partim difficilibus // vocibus Germanica adiecta // est explicatio.// Basileæ,// Typis Conr. Waldkirch.// CIↃ IↃ XCVI.' (F. 1, r.)

Description. 8°, 10 × 16.1 cm., the text being 6.8 × 12 cm. 16 pp. unnumb. + 586 numb. = 602 pp., 26–28 ll. Basel, 1596.

Editions. There was no other edition.

This work consists of twenty parts, treating respectively of the leading branches of knowledge as considered in the sixteenth century. Of these the eleventh is 'De Arithmetica,' and gives in 43 pages a succinct account of algorism. It begins with the catechism form: 'Quid est Arithmetica? Arithmetica, die Rechenkūſt/ est ars benē numerandi. Subjectum Arithmeticæ eſt Numerus.' There is nothing progressive in the treatment, division, for example, being performed by the galley method only, and the applications being confined largely to the 'Aurea Regula' (the 'Golden Rule' of three).

Other works of 1596. Helmreich, p. 303, 1561; Hood (translator; see Urstisius), p. 361, 1579; Petri, p. 325, 1567; Ramus, p. 330, 1569; Recorde, p. 219, c. 1542; Snellius (see Ramus), p. 333, 1569; Tartaglia, p. 278, 1556; Urstisius, p. 361, 1579; Anonymous (William Parley (?), translator), 'The Pathway to Knowledge,' London, 4° (containing 'Thirty days hath September,' see p. 33); Anonymous, 'Arithmetica,' Frankfort, 8°; Sebastian Brandt, 'Plenaria artis Arithmeticae reſolutio,' Frankfort, 8°; C. M. Glysonius, 'Arithmetica practica,' Venice, 4°, with editions as late as 1783; Giacomo Trevisano, 'Memoriale di abbaco,' Venice, 8° (title page 1597, colophon 1596); Bernardo Vila, 'Reglas breves de Arithmetica,' Barcelona, sm. 8°; Antonio Rodriguez, 'Aritmetica pratica y theoretica,' Salamanca, 8°. **497; 545**

ANONYMOUS. Ed. pr. 1597. Luyck, 1597.

Title. 'Cleynen // Catholyken // Catechismvs. // . . . Tot Luyck by Jan Voes. // M.D.XCVII. // Met priuilegie van ſes Iaren.' (F. 1, r.)

Colophon. 'Tot Luyck by Ian Voes, 1597.' (F. 12, v.)

Description. 4°, 9 × 14 cm., the text being 6.2 × 11.8 cm. 12 ff. unnumb., 16–21 ll. Luyck, 1597.

This is one of many similar children's books in Mr. Plimpton's library. It is introduced here merely as a type of those primers that taught some work in number in connection with grammar or the catechism. In this book the child learns the Roman and Arabic numerals to 84. There are also in the library numerous Latin and Greek grammars of the sixteenth century in which numeration is taught, but these have been omitted from this bibliography.

Other works of 1597. Blundevile, p. 407, 1594; Bonocchio, p. 346, 1574; Buscher, p. 393, 1590; Galasso, p. 404, 1593; Gemma, p. 200, 1540; Trevisano, above, 1596; Francis Meres, 'God's Arithmetique,' London, 8°. **545**

HENRY DE SUBERVILLE. Ed. pr. 1598. Paris, 1598.

A French scholar of the close of the sixteenth century. He describes himself as canon of the cathedral at 'Xaintes,' and 'Aduocat en la Cour de Parlement de Bourdeaux,' and signs his name with the birthplace 'Breton-Bearnois.' The dedicatory epistle is dated at 'Kimpercorentin.'

Title. 'L'Henry-metre,// instrvment royal, et // vniversel, avec sa theoriqve,// vsage, et pratiqve demonstree par // les Propoſitions Elementaires d'Euclide, & regles familieres // d'Arithmetique: & auſſi ſans Arithmetique: Lequel prend toutes meſures Geometriques, & Aſtronomiques, qui luy // ſont circulairement oppoſées tant au Ciel, qu'en la Terre,// svr vne sevle station, par vn // ſeul triangle Orthogone, ſans le bouger de ſa place, ny aller // meſurer aucune diſtance de ſtation, ainſi qu'on eſt con-//trainct de faire auec les autres Inſtruments Geometriques.// De l'inuention // D'Henry de Suberuille Breton, Chanoine en l'Egliſe Cathedrale S. Pierre // de Xaintes: & Aduocat en la Cour de Parlement de Bourdeaux.// Item,// Vn petit traicté ſur la Theorique, & Pratique que de l'Extraction des racines quarrées,// pour dreſſer les Scadrons, & Bataillons quarrés.// Dediés au Roy.// Diev a diſpoſé toutes choſes en Poids, Nombre, & // Mesvre. Sap. 11. 21.// A Paris,// Chez, Adrien Perier, ruë ſainct Iaques en la // boutique de Plantin au Compas.// 1598.// Auec Priuilege du Roy.' (F. 1, r.)

Description. 4°, 17 × 22.1 cm., the text being 11.8 × 17.8 cm. 39 pp. unnumb. + 225 numb. = 264 pp., 39–42 ll. Paris, 1598.

Editions. There was no other edition.

Although this is primarily a treatise on mensuration by the use of a trigonometric instrument somewhat resembling a quadrans, it is properly included in this list because it contains several chapters on fractions and denominate numbers. These are of no special merit, and are introduced as a preliminary to the calculations involved in the use of the 'Henry-metre' which the author invented. The book is an interesting effort to perpetuate an inventor's name by a work of no special scholarship describing an instrument of no particular value. There is, however, a value in the general study of all of these early instruments. Many of them are easily constructed, the quadrans for example, and their more extensive use in the teaching of trigonometry would be very helpful.

JOHANN FRIDOLIN LAUTENSCHLAGER.

Ed. pr. 1598. Freiburg in Uchtland, 1598.

A Freiburg Rechenmeister of the latter half of the sixteenth century.

Title. See Fig. 205.

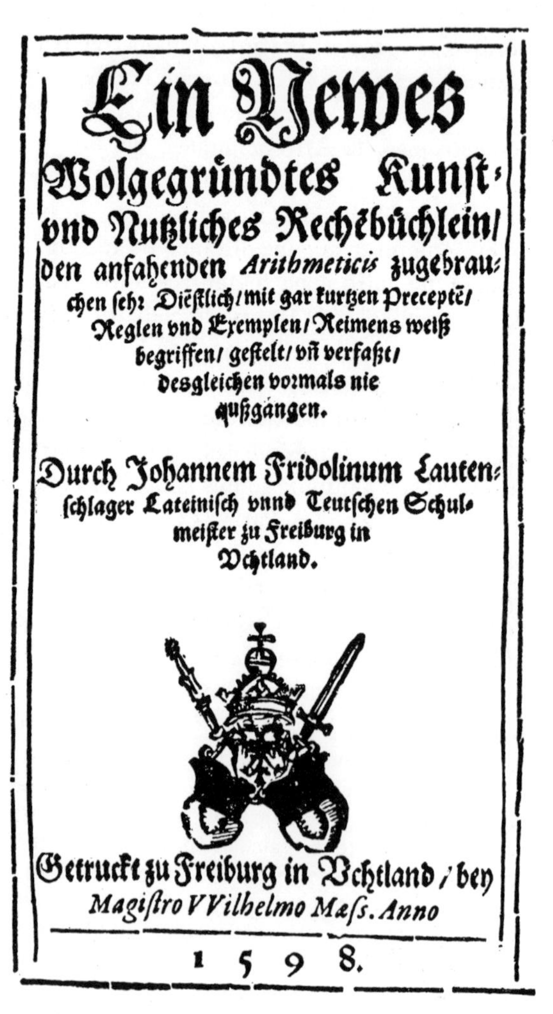

Ein Newes Wolgegründtes Kunst- vnd Nutzliches Rechēbüchlein/ den anfahenden *Arithmeticis* zugebrauchen sehr Diēstlich/ mit gar kurtzen Preceptē/ Reglen vnd Exemplen/ Reimens weiß begriffen/ gestelt/ vñ verfaßt/ desgleichen vormals nie außgangen.

Durch Johannem Fridolinum Lautenschlager Lateinisch vnnd Teutschen Schulmeister zu Freiburg in Vchtland.

Getruckt zu Freiburg in Vchtland/ bey *Magistro VVilhelmo Mæss. Anno*

1598.

FIG. 205. TITLE PAGE OF LAUTENSCHLAGER

Description. 8°, 9.4 × 14.7 cm., the text being 6.9 × 12.3 cm. 4 pp. unnumb. + 58 numb. = 62 pp., 21–30 ll. Freiburg in Uchtland, 1598.

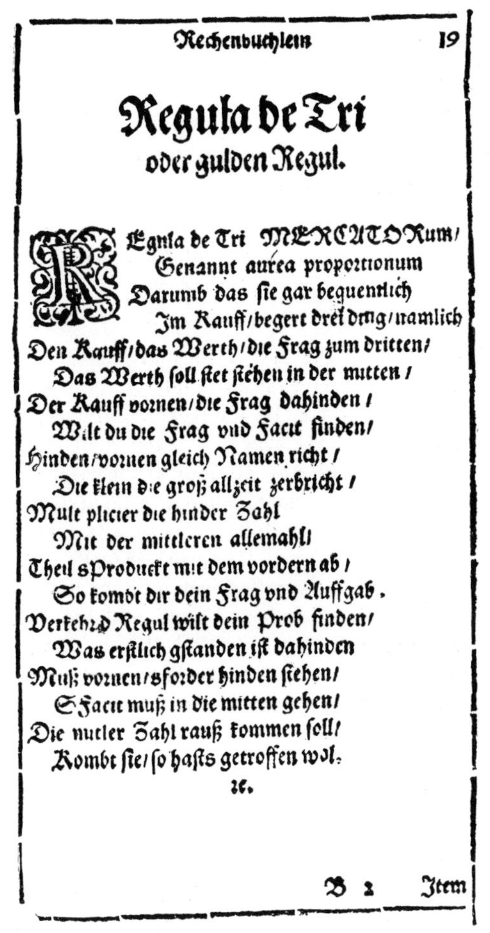

Rechenbuchlein 19

Regula de Tri

oder gulden Regul.

REgula de Tri MERCATORum/
Genannt aurea proportionum
Darumb das sie gar bequemlich
Im Kauff/begert drey ding/namlich
Den Kauff/das Werth/die Frag zum dritten/
Das Werth soll stet stehen in der mitten/
Der Kauff vornen/die Frag dahinden/
Wilt du die Frag vnd Facit finden/
Hinden/vornen gleich Namen richt/
Die klein die groß allzeit zerbricht/
Multplicier die hinder Zahl
Mit der mittleren allemahl/
Theil sProduckt mit dem vordern ab/
So kombt dir dein Frag vnd Auffgab.
Verkehrd Regul wilt dein Prob finden/
Was erstlich gstanden ist dahinden
Muß vornen/sforder hinden stehen/
S Facit muß in die mitten gehen/
Die mitler Zahl rauß kommen soll/
Kombt sie/so hasts getroffen wol.
&c.

B 2 Item

FIG. 206. FROM LAUTENSCHLAGER

Editions. There was no other edition.

This is the first arithmetic that I have seen composed entirely in rhyme. Various others had already contained verses, but this is substantially all in meter. Like all such educational vagaries, it is weak in subject matter, the most difficult example in division being that of 65443 by 6. The illustration (Fig. 206) shows the rule of three in verse.

CASPAR SCHLEUPNER. Ed. pr. 1598 Leipzig, 1598.

A Breslau Rechenmeister, born at Nürnberg c. 1535.

Title. See Fig. 207.

Colophon. 'Gedruckt zu // Leipzig / bey Frantz // Schnellboltz. // Typis hæredum Beyeri. // (Woodcut.) Im Iahr // M. D. XCviij.' (F. 99, v.)

Description. 8°, 9.3 × 14.7 cm., the text being 6.9 × 12.2 cm. 99 ff. unnumb., 22–26 ll. Leipzig, 1598.

Editions. Leipzig, 1598, 8° (here described); Breslau, 1599, 8°.

Schleupner was one of the last serious advocates of the old 'line reckoning' with counters. He was a disciple of the Newdorffer school of Nürnberg Rechenmeisters, as he states in his preface, and had himself taught at 'Neyſe' and 'Breſzlaw' for many years when he decided to write this book. He makes an attempt at an easy method of presenting the fundamental operations, hoping, as he says, to set forth the doctrines of Adam Riese and Johan Seckerwitz (another Breslau Rechenmeister) in simple fashion. The work is made up of a series of impossible questions and answers between a father and his son. The latter always begins with 'Geliebter Vater,' and the father's replies, while always clear, are often very extended, the entire 99 pages covering little more than the four operations, together with a few insignificant problems. As a simple presentation of line reckoning, however, the book has few equals.

JOSEPHUS UNICORNUS. Ed. pr. 1598. Venice, 1598.

See p. 298.

Title. See Fig. 208.

'Parte Seconda,' of the same date, follows f. 204, the two being bound in one, and the pages being numbered consecutively.

Colophon. 'In Venetia, // Appreſſo Franceſco de' Franceſchi Seneſe. // M D XCVIII.' (F. 413, v.)

Description. 4°, 15.5 × 21.6 cm., the text being 10.4 × 17.6 cm. 12 ff. unnumb. + 395 numb. = 407 ff., 29–39 ll. Venice, 1598.

Rechenbüchlein/
Auff der Linien,
Inn
Allen dreyen
Stenden/als im Wehr/Lehr
vnd Nehr stande/ sehr nützlichen zu
gebrauchen.
Daraus ein fleissiger Leser/ wel-
cher alleine sein Ein mal Eins biß auff vier mal
Neune/außwendig kan/ ohne einigen Mündlichen be-
richt/ für sich selbst / die Linien Rechnung lernen kan/
Wil geschweigen der / welcher zuuor schon
seine Species auff der Federn ge-
lernet hat.
Dergleichen hieuor/ in Deutscher
sprach/nicht außgangen.
Gesprech weis gestelt/
Durch Caspar Schleupner/
von Nürmberg / Deudschen Schulmeister zu Breßlaw. Menniglichen / vnd
beuor der lieben Jugend/ zu nutz
verfertiget.
Mit Röm. Key. May. Freyheit.
In vorlegung Andres Wolcken Erben / vnd
Dauid Albrechts. Anno 1598.

FIG. 207. TITLE PAGE OF SCHLEUPNER

DE
L'ARITHMETICA
VNIVERSALE
DEL SIG.
IOSEPPO VNICORNO,
MATHEMATICO
ECCELLENTISSIMO,

Parte Prima:

Nellaquale ſi contiene non ſolo la Theorica di tutti i Numeri,
ma ancora la Prattica appartenente a
tutti i negotij humani.

*Trattata, & amplificata con ſomma eruditione, e con
noui, & iſquiſiti modi di chiarezza.*

CON PRIVILEGIO.

In Venetia, appreſſo Franceſco de' Franceſchi 1598.
Con licenza de Superiori.

FIG. 208. TITLE PAGE OF UNICORNUS

Editions. There was no other edition.

This is one of the most elaborate treatises on arithmetic published in Italy in the sixteenth century. It consists of six books, the first four making up Part I. The first book treats in a detailed fashion of the fundamental operations. Unicornus, for example, gives six methods of multiplication, a treatment that recalls those of Paciuolo and Tartaglia. There is a good discussion of the two general methods of dividing, the downward ('a danda') method having as much attention as the galley plan. Fractions are also treated in Book I. Book II deals with the theory of numbers after the Boethian method. Book III treats of roots, surds, and proportion; Book IV, of the rules of three and false; and Book V, of business arithmetic, including exchange, interest, and alligation.

The work was too theoretical to be popular, but it is an excellent source for the study of the development of elementary mathematics. Unicornus gives a number of interesting historical references.

Other works of 1598. Cassiodorus, p. 211, 1540; Fischer (Piscator), p. 247, 1549; Moya, p. 310, 1562; Paolini, p. 393, 1589; Raymundus Lullius, 'Opera ea quae ad adinventam ab ipso artem universalem . . . pertinent cum diversorum commentariis,' Strasburg, 1598, 8° (see also p. 457); Fabricio Mordente, 'Le propositioni di Mordenti,' Rome, 4° (geometric, but with a little arithmetic).

ANTHON NEWDÖRFFER. Ed. pr. 1599. Nürnberg, 1599.

One of the famous Newdörffer family of Nürnberg Rechenmeisters.

Title. 'Künftliche vnd Ordentliche An//weyfzung der gantzen Practic vff // den Jetzigen fchlag vnd derfelbenn // herlichen gefchwinden Exempel // vffs kürtzt zŭfammen getzogen &c.// Meinen lieben Difcipeln zu fon-//derlichem Nŭtzen geftelt. Durch // mich Anthonium Newdörffer // Rechenmaifter vnd Modift der // Statt Nürnberg // Anno .M.D.IC.' (F. 2, r.) On f. 1, r., is an engraved frontispiece with figures of Euclid and Pythagoras, and a triangular multiplication table. On f. 1, v., are 14 ll. of verse 'Ad Stvdiosos Artis Nvmerandi.'

Colophon. 'Gedruckt zu Nůrnberg/ // durch Paulum Kauffmann.// M. D. XCIX.' (F. 65, r.)

Description. 4°, 15 × 19.3 cm., the text being 11.4 × 15.5 cm. 65 ff. unnumb., 22–32 ll. Nürnberg, 1599.

Editions. There were at least three editions. [547]

This is an excellent illustration of the work of the celebrated Nürnberg Rechenmeisters. It is divided into twelve books, of which 'Das erſte Bůchlein handelt von der Venetianiſchen oder Kauffmenniſchen Practic,' usually called Welsch practice by the German writers. It is composed chiefly of problems, the rules and explanations being left for the master. Directions are given, however, in the case of fractions. Book II relates to denominate numbers and the 'Regula de Tribus.' Book III treats of the weighing of commercial products. Book IV 'Handelt von der Rechnung eines Caſſierers,' including exchange. Book V is entitled 'Iornates, Das iſt/ Rechnung von allerley Handtierung,' and consists of practical problems relating to the purchase of goods. Book VI 'Tractiert von der Regel Converſa vnd Quinque,' inverse and compound proportion. The rest of the twelve books are devoted to practical business questions of the day.

OBERTO CANTONE. Ed. pr. 1599. Naples, 1599.

A Genoese arithmetician of the sixteenth century, residing in Naples.

Title. See Fig. 209.

Description. Fol., 14.5 × 20.1 cm., the text being 10.4 × 15.7 cm. 304 pp. (292 numb.), 34–37 ll. Naples, 1599.

Editions. This is the first edition of this work, three other editions appearing in the seventeenth century.

Arithmetics written by Genoese masters in the sixteenth century are rare, even in manuscript form, and this is probably the only such work printed in Naples. Genoa was a mercantile center, but its dialect was not conducive to the success of a textbook, as is seen in the case of Zucchetti's treatise (p. 425). Naples was too far from the path of international commerce to produce many mercantile works. But although Oberto Cantone was a Genoese, he was a 'professor delle discipline matematiche' in Naples, and he dates his dedicatory epistle 'Di Napoli li 15. Iuglio 1599.' Naples, however, had a commerce of its own, even if not as extensive as that of Venice or Florence, and by the close of the century it was natural to expect works of this kind.

The book is mercantile and is based upon Borghi and similar writers of the North. It gives our present method of multiplication, but makes no use of the Venetian or Florentine names, with the exception of 'per colonna.' Two or three short methods are given, but no such extended treatment of varied forms appears as in Tartaglia, or his great predecessor Paciuolo. Division, except in simple cases, is postponed to p. 142,

L'VSO PRATTICO
DELL'ARITMETICA
DI
OBERTO CANTONE
DA GENOVA,
PROFESSOR DELLE DISCIPLINE
Matematiche.

NEL QVALE CON NVOVA INVENTIONE s'insegna in materia di conti, l'vso tanto della Regia Camera della Sommaria, quanto di Negotianti, Mercadanti, & Artegiani. e come Napoli cambij, e recambij in ciascuna piazza.

CON PRIVILEGIO.

IN NAPOLI,

Appresso TARQVINIO LONGHO. *M. D. IC.*

Si vendono dal medesimo Autore à Banchi nuoui.

FIG. 209. TITLE PAGE OF CANTONE

and is there treated in the modern form, 'a danda.' Most of the applied problems are in exchange, although a few other types are given. The book is poorly constructed, being too prolix in the treatment of the operations, and too narrow in its applications.

JOANNES MARIANA. Ed. pr. 1599. Toledo, 1599.

Born at Talavera de la Reina, in 1536; died at Toledo, February 17, 1624. He was a Spanish Jesuit and a famous historian.

Title. See Fig. 210.

Colophon. 'Toleti, Apud Thomam Guſ-//manium, Anno. 1599.' (P. 206.)

Description. 4°, 14 × 18.6 cm., the text being 9.3 × 15.6 cm. 6 pp. blank + 8 unnumb. + 192 numb. = 206 pp., 26 ll. Toledo, 1599.

Editions. There was no other edition. 497

The author is not the same as the Giovanni Mariani mentioned on p. 180, but a Spanish Jesuit of some fifty years later. The work is on the history of the weights and measures used in Spain in the sixteenth century. It traces these measures from the Roman, Greek, and Hebrew sources, and is valuable for the study of the history of the subject. It closes with a table of comparative measures.

MARTIN WENCESLAUS. Ed. pr. 1599. Middelburg, 1599.

A Dutch arithmetician of the sixteenth century.

Title. See Fig. 211.

Colophon. 'Eynde deſes onſes eerſten Vo-//lumens: Ghedruct int Jaer ons Hee-//ren 1599. In deſe vermaerde Coop-//ſtadt van Middleburgh in // Zeelandt.'// Also in French: 'Fin de ce noſtre premier Volu-//men: Imprimé l'An de noſtre Seig-//neur 1599. A la treſ-renommée // ville Marchande de Middel-//bourg en Zeelande.' (F. 157, r.)

Description. 4°, 15.2 × 20.1 cm., printed in double columns, one in Dutch and the other in French, the Dutch being 6.7 × 15.7 cm., the French 5 × 15 cm. 4 ff. blank + 18 unnumb. + 137 numb. = 159 ff. Dutch: 30–41 ll.; French: 16–26 ll. Middelburg, 1599.

Editions. This is evidently the first edition, since the dedication is dated 'Den 10. dach van Decem. ſtijlo nouo. Anno Chriſti.

IOANNIS

MARIANAE

Hispani,

E SOCIE. IESV,

DE PONDERIBVS ET

mensuris.

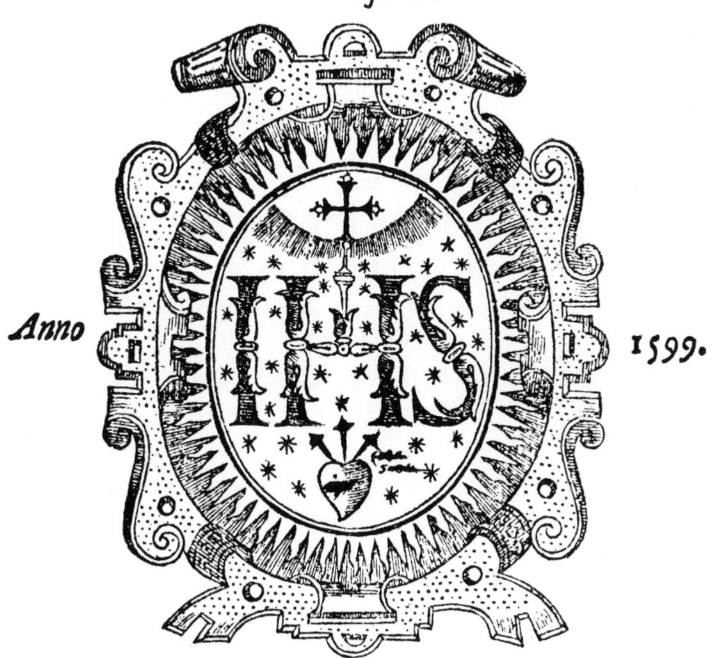

CVM PRIVILEGIO.

Toleti, Apud Thomam Gusmanium.

FIG. 210. TITLE PAGE OF MARIANA

T' FONDAMENT
Van Arithmetica: mette Italiaensche Practijck / midtsgaders d'aller nootwendichste stucken van den Reghel van Interest.
Beydes in Nederduyts ende in Franchois/ met redelicke ouereenstemminghe ofte Concordantien.
Alles
Door MARTINVM VVENCESLAVM, AQVISGRANENSEM.

LE FONDEMENT
DE L'ARITHMETIQVE AVEC LA PRACTICQVE ITALIENNE ENSEMble les pieces, les plus necessaires de la Regle d'Interest.
ENSEMBLE EN BAS ALLEMAN ET EN François, avec raisonnable Harmonie ou Concordance.
LE TOVT
Par *MARTHINVM WENCESLAVM. Aquisgranensis.*

MIDDELBVRGH.
By *Symon Moulert*, woonende inde Druckerije. 1599.

Ende men vintse te coope/ by Adriaen van de Vivere, Boeckvercooper/ woonende by de nieuwe Burse/inden verghulden Bybel.

FIG. 211. TITLE PAGE OF WENCESLAUS

1598,' 'le 10. Iour de Decembre ſtylo nouo. Anno. Christi- 1598.' The long and stupid preface is dated November 30 of the same year. Wenceslaus had already published two works before this one, the 'Proportionale ghesolveerde Tafelen van Interest,' 1594, 8°, and the 'Boukhoudens Instruction,' 1595.

The book is interesting, both because of its arrangement of the Dutch and French texts in parallel columns, thus serving a purpose in language teaching in a bilingual country, and because of the mercantile problems which reveal, as is particularly true of the Dutch books, the contemporary life of the people.

Other works of 1599. St. Augustine (see Reisch), p. 82, 1503; Barlaamo, p. 343, 1572; Benedetti, p. 364, 1580; Boethius (see Reisch), p. 82, 1503; Bungus, p. 384, 1583–84; Capella, p. 66, 1499; Clichtoveus (see Reisch), p. 82, 1503; Faber Stapulensis (see Reisch), p. 82, 1503; Finaeus (see Reisch), p. 82, 1503; Gallucci (see Reisch), p. 82, 1503; Jacob, p. 298, 1560; Jordanus (see Reisch), p. 82, 1503; Ramus, p. 330, 1569; Reisch, p. 82, 1503; Schleupner, p. 412, 1598; Strübe, p. 391, 1588; Johann Heere, 'Rechenbüchlein von allerhand gebraüchlichen Fragen,' Nürnberg, 8°; Andreas Reinhard, 'Drey Regiſter Arithmetiſcher Anfang zur Practik,' Leipzig, 8°, with a second edition in 1600. **547**

JACOB VANDER SCHUERE.

Ed. pr. 1600. Haarlem, 1600.

A Dutch arithmetician of Meenen, c. 1550–1620.

Title. See Fig. 212.

Description. 8°, 9.4 × 14.4 cm., the text being 7.2 × 12.6 cm. 2 ff. unnumb. + 202 numb. = 204 ff., 22–23 ll. Haarlem, 1600.

Editions. That the date of the first edition is 1600 appears in the Voor-Reden of the 1625 edition, where his son, Denys, says that the book was published by the father 'eerst in't Iaer 1600. ende dit is al de vierde mael dat het gedrukt is.' There were various editions in the seventeenth century, including the following: Haarlem, 1611; Rotterdam-Schiedam, 1624, 8° (p. 423); Haarlem, 1625; c. 1630, 8° (p. 423); Gouda, 1634, 8° (p. 424); Amsterdam, 1643, 8° (p. 424); Rotterdam, 1653, 8° (p. 425); Amsterdam, 1675.

ARITHMETICA,

Oft Reken-conſt/

Verchiert met veel ſchoone Exempelen/ſeer nut voor alle Cooplieden/ Facteurs/ Caſſiers/ Ontfanghers/etc. Ghemaeckt/ Door

IAQVES VAN DER SCHVERE

VAN MEENEN.

Nu ter tijdt Francoyſche School-meeſter tot HAERLEM.

TOT HAERLEM,

By Gillis Rooman Boeckdrucker/in de Jacobijne-ſtrate/ in de verguldeParſſe, 1600.

FIG. 212. TITLE PAGE OF VANDER SCHUERE

Vander Schuere's work on bookkeeping is mentioned under the later editions (p. 424), although it was not published in the sixteenth century. No effort has here been made to complete the list of seventeenth-century editions.

This is one of the many practical arithmetics that appeared in Holland about this time. It takes up the fundamental processes, rule of three, fractions, the rule of practice, partnership, commissions, inheritance problems, profit and loss, interest, exchange, barter, alligation, and the various other rules in use at that time. It was one of the most successful Dutch textbooks.

JACOB VANDER SCHUERE.

Ed. pr. 1600. Rotterdam-Schiedam, 1624.

See p. 421.

Title. 'Arithmetica//Oft//Reken-konſt;//Verciert met veel ſchoone// Exempelen/zeer nut voor alle vlijtighe// Oeffenaers ende leer-ghierighe Aenvanghers//deſer Konſt/etc. Ghemaeckt door// Iacob Vander Schuere, Meenenaer,// Nu ter tijdt Franſoyſche. School-meeſter//tot Haerlem.//Ende nu int herdrucken overſien ende ghebetert.// (Woodcut of Vander Schuere with motto: Door siet den Grond.)//Tot Rotterdam,// Voor Pieter van Waeſberghe/ //Anno 1642.' (F. 1, r.)

Colophon. 'Tot Schiedam,//Ghedruckt by Adriaen van Delf.// Anno 1624.' (F. 208, r.)

Description. 8°, 8.7 × 14.7 cm., the text being 7.6 × 12.9 cm. 2 ff. unnumb. + 206 numb. = 208 ff., 22–27 ll. Rotterdam-Schiedam, 1624.

See above.

JACOB VANDER SCHUERE.

Ed. pr. 1600. S. l. a. (?), c. 1630.

See p. 421.

Title. 'Arithmetica oft Reken-konst.'

This edition is without date, the title page being missing, but is probably c. 1630. F. 1 is missing. 9 × 15.6 cm., the text being 7.3 × 12.2 cm. 5 ff. unnumb. + 252 numb. = 257 ff., 35 ll.

JACOB VANDER SCHUERE. Ed. pr. 1600. Gouda, 1634.

See p. 421.

Title. 'Arithmetica//oft//Reken-konſt.//Door Iakob vander Schuere, Meenenaer // Eertÿts Françoÿsche Schoolmeeſter//tot Haerlem,// En in't herdrucken bÿ hem ouerſien // verbetert en vermeerdert, en noch bÿ-//geuoecht een kort onderricht van// 'tItaliaens Boeck-houden // ter Goude // Bÿ Pieter Rammeſeyn, Boeck-vérkooper // inde Korte Groenen-dal, in't Vergult A, B, C. // aº 1634. W Akerſl : fec.' (The title page is elaborately engraved, with a portrait of the author surrounded by the following: 'Doorseit den Grond Iakob Vander Schvere æt 50.') (F. 1, r.)

Description. 8°, 8.8 × 14.2 cm., the text being 7 × 12.7 cm. 8 ff. unnumb. + 208 numb. = 216 ff. Bound with this is 'Kort onderricht // over het // Italiaens // Boek-houden;// Nu int licht ghebracht // Door Iakob vander Schvere,' etc., with 1 f. unnumb. + 37 numb. = 38 ff., making a total of 254 ff. in the book, 28–33 ll. Gouda, 1634.

See p. 423.

JACOB VANDER SCHUERE.
Ed. pr. 1600. Amsterdam, 1643.

See p. 421.

Title. 'Arithmetica//oft // Reken-konſt.// En een kort onderricht van't Italiaens Boeckhoudē // Door Iacob vander Schvere Meenenaer. // Bÿ den Autheur overſien, verbeetert en vermeerdert.//Tot Amsterdam,//voor Michiel de Groot Boeckverkooper // op de nieuwendyk inde Bieſtkens Bybel. 1643.' (Engraving of author with the following wording: 'Iacob vander Schvere Meenenaer. Out 67. Iaer. Doorsiet den Grondt. Anno 1643.') (F. 1, r.)

Description. 8°, 9.5 × 14.7 cm., the text being 7.2 × 12.5 cm. 272 ff. (216 numb + 8 unnumb. in the arithmetic), 26–31 ll. Amsterdam, 1643. Bound with this is 'Kort onder-richt // Over het // Italiaens // Boeck-houden.// In't Licht gebracht // door // Jacob van der Schuere.// t'Amsteldam,// By Michiel de Groot,

Boeckverkooper // op de Nieuwendijck/tuſſchen de twee Haer-// lemmer Sluyſen. A°. 1675.' (F. 1, r., of the bookkeeping.)

See p. 423.

JACOB VANDER SCHUERE.

Ed pr. 1600. Rotterdam, 1653.

See p. 421.

Title. 'Arithmetica // ofte // Reken-konſt, // En een kort onder-richt van't Italiaens // Boeck-houden.// Door // Iacob van der Schuere Meenenaer.// By den Autheur over-ſien, verbetert en // vermeerdert.// Tot Rotterdam,//Gedruckt by Pieter Waeſberge, woo-//nende op't Steyger/ in de gekroonde // Leeuw/ Anno 1653.' (F. 1, r.)

Description. 8°, 8.7 × 14.1 cm., the text being 7 × 12.1 cm. 275 pp. (219 numb. + 8 unnumb. in the arithmetic), 26–33 ll. Rotterdam, 1653. Bound with this is 'Kort onder-richt // over het // Italiaens Boeck-houden. // In't Licht gebracht // door // Jacob van der Schuere.// Tot Rotterdam,// Gedruckt by Pieter van Waeſberge,// Ordinaris Drucker/ woonende op't Steyger // in de gekroonde Leeuw/ Anno 1653.' (F. 1, r., of the bookkeeping.)

See p. 423.

GIOVANNI BATTISTA ZUCHETTA.

Ed. pr. 1600. Brescia, 1600.

Born April 21, 1550. A Genoese arithmetician.

Title. See Fig. 213. The privilege is dated 'In Genoua nel di 30. di Genaro. M. CCCCCC.'

Description. Fol., 24.2 × 33.4 cm., the text being 17.1 × 26 cm. 444 pp. (412 numb.), 33–37 ll. Brescia, 1600.

Editions. There was no other edition. Brunet says this 'Prima Parte' is the only one that appeared. 497;547

The work has several interesting features, not the least one being the apology 'Al generoso lettore,' in which the author speaks of the criticism liable to be directed against a Genoese author on account of his provincial Italian. The 'Prologo' is a curious dissertation on the 'Arti,

PRIMA PARTE DELLA ARIMMETICA DI GIO. BATTISTA ZVCHETTA CITTADINO GENOVESE.

Per la quale con mirabile ordine, & nuoue Regole si risolue con marauigliosa facilità ogni dubbio mercantesco.

Con vn Trattato che risolue qualunque quesito bisognoso à Zecchieri, Orefici, & Argentari.

Copiosa di Postille, & Tauole.

CON PRIVILEGI.

IN BRESCIA

PER VINCENZO SABBIO. M. DC.

Con licenza de' Superiori.

FIG. 213. TITLE PAGE OF ZUCHETTA

Scienze, & altro,' with some ninety-eight arguments to show the need for arithmetic on the part of all classes of humanity. The farmer, the musician, the thief, the cook, the prelate, all are shown to have need of number; and Nature, Intelligence, and even God himself make use of it.

The book presupposes a knowledge of the arithmetic of integers, and opens with a treatment of fractions. The rule of three, in all of its forms, and with most unbusinesslike numbers, is then discussed at great length, and this is followed by various complications of the Regola del Cattaino, 'così detta da gli Arabi inuentori di quello, ch' in lingua nostra ſignifica falſa poſizione.' The latter part of the book treats of such topics as partnership, barter, and alligation. The work was not of a nature to have any influence on Italian arithmetic.

WILHELM SCHEY. Ed. pr. 1600. Basel, 1600.

A German Rechenmeister at Solothurn. Born c. 1560.

Title. See Fig. 214.

Description. 4°, 15.1 × 19.4 cm., the text being 10 × 15 cm. 486 pp. (470 numb.), 36–37 ll. Basel, 1600.

Editions. That this is the first edition appears from the dedication: 'Datum den 12. Septembris. Anno Domini 1600.'

The book is an attempt at a complete commercial arithmetic, but is not well constructed. The author likes to arrange his computations in a bizarre fashion, for effect. He extends his explanations unduly, and for so large a book the mercantile information is not as complete as it should be. There is a curious arrangement of the figures in the proofs by casting out nines, as here shown, although whether this is a typographical matter or a notion of Schey's there is nothing to indicate. In the 'Regula Falsi' there is a rather early use of ÷ for the minus sign. In general the book is reactionary, giving only the galley division (with much attempt at effect), and mentioning, although not treating, 'Duplatio' and 'Mediatio.'

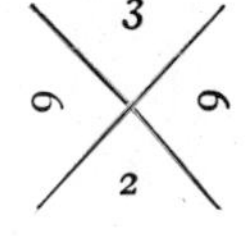

Other works of 1600. Barlaamo, p. 343, 1572; Buscher, p. 393, 1590; Chambers (see Barlaamo), p. 343, 1572; Herodianus, p. 60, 1495; Hylles, p. 396, 1592; Jacob, p. 298, 1560; Lonicerus, p. 253, 1551; Ramus, p. 263, 1555; Reisch, p. 82, 1503; Reinhard, p. 421, 1599; Schulze, p. 383, 1584; M. van den Dycke, 'Chyfer-Boeck' (second edition, I do not know the first), Antwerp, 8°, and 'La vraye reigle d'Arithmétique' (a translation of the other?), ib., 8°; Georg Gleitsmann, 'Künſtliches Rechenbuch ſowohl auf Linien als mit Ziffern

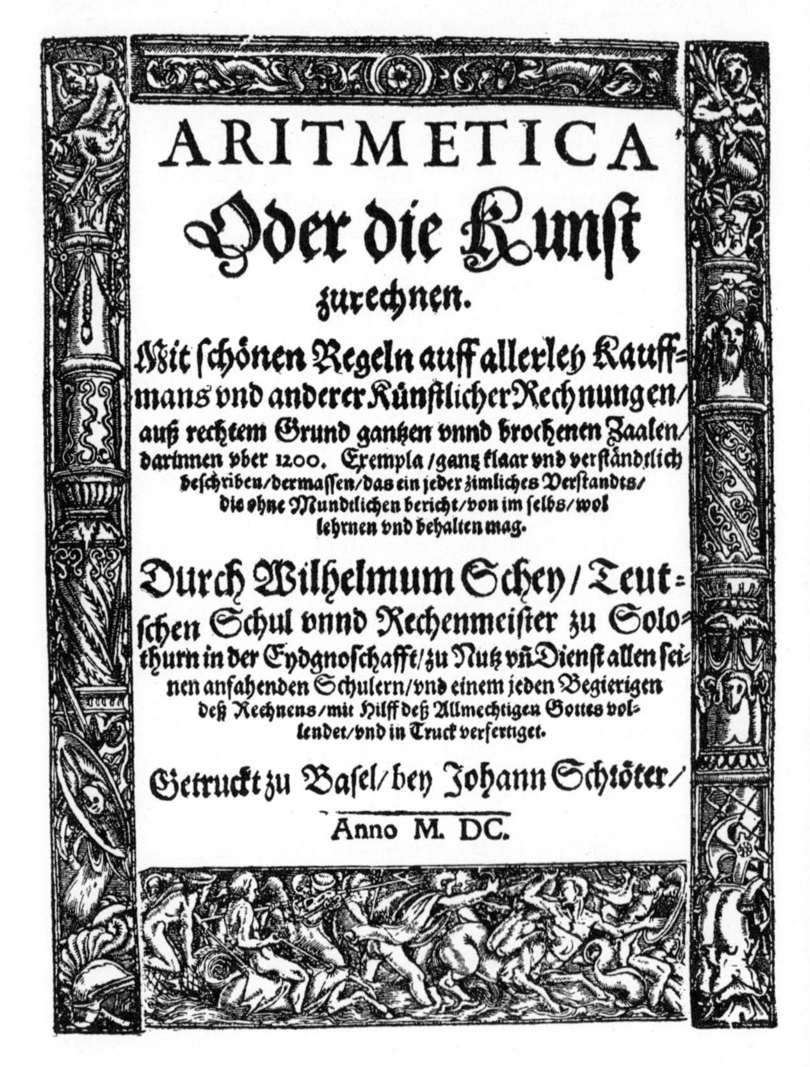

ARITMETICA
Oder die Kunst
zurechnen.

Mit schönen Regeln auff allerley Kauffmans vnd anderer Künstlicher Rechnungen/
auß rechtem Grund gantzen vnnd brochenen Zaalen/
darinnen vber 1200. Exempla / gantz klaar vnd verständtlich
beschriben/dermassen/das ein jeder zimliches Verstandts/
die ohne Mundtlichen bericht/von im selbs/wol
lehrnen vnd behalten mag.

Durch Wilhelmum Schey/Teut-
schen Schul vnnd Rechenmeister zu Solo-
thurn in der Eydgnoschafft/zu Nutz vñ Dienst allen seinen anfahenden Schulern/vnd einem jeden Begierigen
deß Rechnens/mit Hilff deß Allmechtigen Gottes vol-
lendet/vnd in Truck verfertiget.

Getruckt zu Basel/bey Johann Schtöter/

Anno M. DC.

FIG. 214. TITLE PAGE OF SCHEY

nach defs Rami Arithmetica geftellt,' Frankfort, 8°; M. Johann Taf, 'Schönes neues . . . Rechenbuch,' Cologne, 4°. There was also published, s. a., but c. 1600, a work by Vincenzo da Bergamo, 'Arithmetiche instruttioni.' 498; 547; 542

There were many other arithmetics published in the sixteenth century without date, including the following: Anonymous, 'Art et Science de Arismetique,' Paris, 12°, 96 ff. (Boncompagni sale); Anonymous, 'Livre des gects,' s. l., 4° (see p. 130, 1520); Anonymous, 'Abbaco di succincte dimostrazioni,' s. l. (Milan?) (Brunet); Anonymous, '(Q)ui apresso e ināci col nome di dio intēdo di tractare e scriuere alquātimo di e regole sopra larte del numero altrimēti chiamato algurismo,' s. l., 33 ff. (Riccardi); Joachim Ammonius, 'Isagoge Arithmetices . . . cum praefatione P. Melanchthoni,' Wittenberg; Angelus Mutinens, p. 140, 1525; Sarafino da Campora, 'Della ragione dell' Abbaco,' and a work on the calendar (Messina, 1559, and Rome, 1560) (Riccardi); Lauro Quirini, 'De mysterio numerorum;' Matteo Ricci, 'T'ung-wen suanchih' (Practical arithmetic in 11 books, the work of a Jesuit missionary in China, b. 1552, d. 1610; possibly not published during his lifetime).

PART II
MANUSCRIPTS

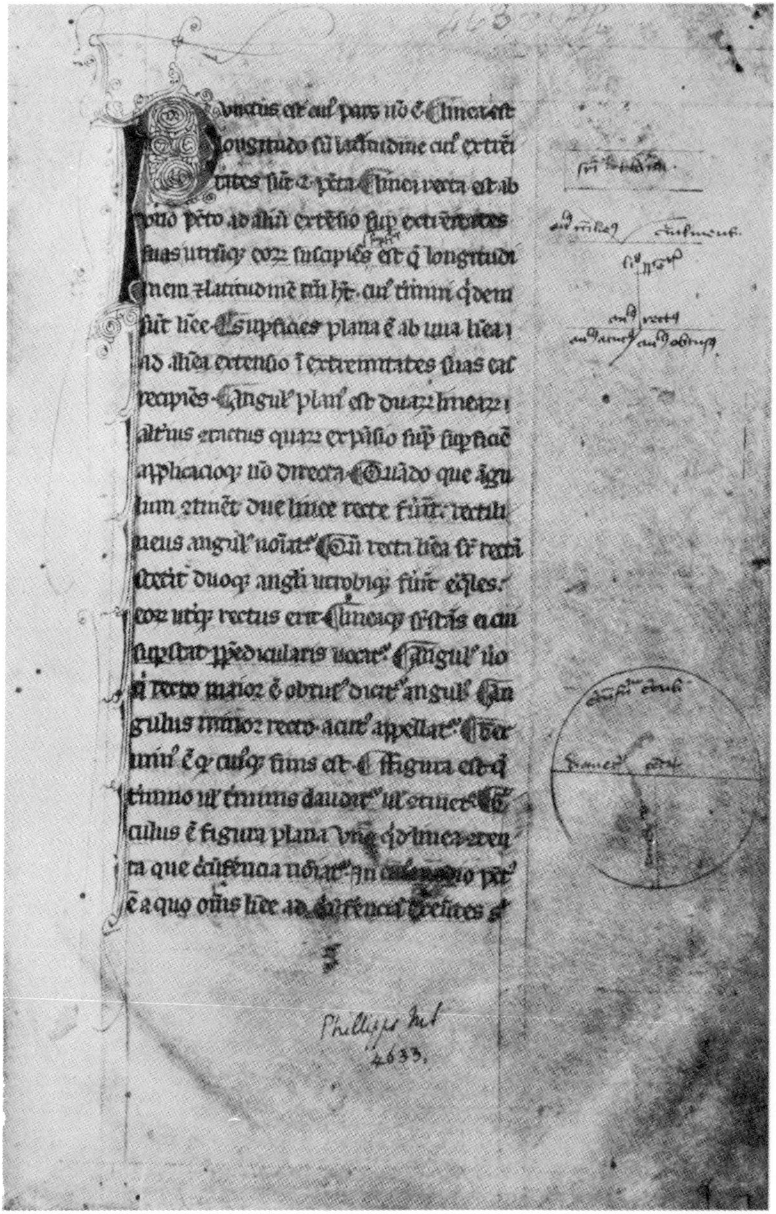

PLATE IV. FROM THE CAMPANUS MANUSCRIPT OF EUCLID, C. 1260

MANUSCRIPTS

EUCLID. Latin MS., c. 1260.

See p. 11.

Title. 'In hoc libro ꝯtinet' geomet'a euclidis // cū ꝯmento magrī campani.' (F. 1, v.)

Colophon. '¶Explicit geometria euclidis cū cōmen//to magiſtri campani.' (F. 165, v.)

Description. Fol., 17.8 × 25.8 cm., the text being 8.5 × 17.2 cm. besides the marginal figures. 165 ff., 23–44 ll. Written on vellum c. 1260.

Editions. This work being primarily a geometry, I have not given a list of the editions. For the arithmetical books, which (except for Book V, on proportion) are not included in this manuscript, see p. 237.

This is a Latin manuscript of Euclid, with the commentary (proofs) of Campanus, written on vellum about 1260. The translation of the theorems is that made by Adelard (Æthelhard) of Bath, c. 1120, but in the early printed editions of the 'Elements' it is generally referred to as that of Campanus.

Of Campanus himself not very much is known. His first name was probably Johannes, and he is known to have prepared a set of planetary tables, and to have been chaplain to Urban IV. (See Boncompagni's *Bulletino*, I, 5, and XIX, 591.)

This manuscript has been studied by Mr. C. S. Peirce (*Science*, N.S., XIII, 809), who believes it to be the copy given by Campanus himself to Jacques Pantaléon when the latter was Patriarch of Jerusalem, hence before August 29, 1261, when he became Pope Urban IV. He bases his belief on a sentence written in a cursive hand just below the colophon, containing the words 'Jacobus Dei gratia Patriarcha Jerusalemitorum.' The complete sentence is: 'In nōīē dn̄ī amē Jacobus dei g^ia^

patriarcha Jerufalemitar omĩbus xpi fidelibus falutem defiderium,' — 'In the name of the Lord, amen; Jacob by the grace of God patriarch of the Jerusalemites to all the faithful of Christ, greeting and love.' While this seems more of a blessing or quasi imprimatur than a mark of ownership, it is equally valuable in serving to fix the date. On f. 1, r., is the inscription in a fourteenth-century hand, 'mgri adolphi di Werda,' and a statement that the manuscript belonged to the Phillips collection, no. 4633. On f. 165, v., is an inscription in an English hand of c. 1400, 'lib^e ifte fuit Di armachani,' — 'This book belonged to Dominus Armachanus.' (For a page of this MS. see Plate IV.)

ANICIUS MANLIUS SEVERINUS BOETHIUS.

Latin MS., c. 1294.

See p. 25.

Title. The manuscript begins: 'Incipit prolog̃ in arithmeticā boetii.' (F. 1, v.)

Colophon. 'Explic^t arifme^ca // boetij ad fymacu⁹ patriciu^c.' (F. 28, r.)

Description. Fol., 19.8 × 27 cm., in double columns, each being 7 × 20.1 cm. 28 ff., 41 ll. Written on vellum c. 1294. Bound with the Euclid described below but in a different hand. It is beautifully written and illuminated. The contemporary pigskin binding has the inscription, 'lib' arifmetice boecij.' The text is practically that followed by the Friedlein edition (Leipzig, 1867). (See Plate I.)

Editions. For printed editions see p. 27.

See p. 27.

EUCLID.

Latin MS., c. 1294.

See p. 11.

Title. The work begins without title: 'Punct⁹ eft cui⁹ ꝑs ñe.' (F. 29, v.)

Colophon. 'Explicit geometri euclidis cum comto campani.' (F. 111, v.)

Description Fol., 19.8 × 27 cm., the text being 17.5 × 19.5 cm. 111 ff. (the first 28 ff. being the arithmetic of Boethius described on p. 434), 44 ll. Written on vellum, c. 1294. It has beautifully executed figures and is a fine specimen of the work of the mediæval scribe. It is written in a different hand from that of the Boethius with which it is bound. The cover, which seems contemporary with the manuscript, has the number cclxxxxiiij, possibly for the date 1294, the M being omitted as is often the case. (See Fig. 215 and Plate V.)

ANICIUS MANLIUS SEVERINUS BOETHIUS.
Latin MS., c. 1300.

See p. 25.

Title. The manuscript begins: 'Incipit libr̄ arismetice art . . .'

Description. Fol., 13 × 18.6 cm., the text being 8.9 × 14.1 cm. 37 ff., 32–36 ll. Beautifully written on vellum. The work is complete, and, like the manuscript described on p. 434, this shows a text very similar to that followed by the Friedlein edition. At the end of the manuscript are two folios of commentary, closing with two almost illegible lines containing the words 'Com . . . campani . . .,' referring to the commentary of Campanus. Roman numerals are used throughout the text, which was not always the case in manuscripts of this date. (See Plate VI.) The commentary, which seems to have been added about a century later, has some Hindu-Arabic numerals.

PAOLO DAGOMARI. Italian MS., c. 1339.

Paolo dell'Abaco, Paolo Astrologo, Paolo Geometra, Paolo Arismetra, Paul of the Abacus. Born in Prato, c. 1281; died at Florence in 1374, or, according to some writers, in 1365. He was a celebrated Florentine arithmetician, 'geometra grandissimo, e peritissimo aritmetico . . . diligentissimo osservatore delle Stelle, e del movimento de' cieli,' as Villani (*Le Vite d'Uomini ilustri Fiorentini*) calls him. 547

Title. 'Trattato d'Abbaco, d'Astrono-//mia, e di segreti naturali // e medicinali.' (F. 1, r.)

Description. Fol., 21.7 × 29.3 cm., the text being 15 × 21.4 cm. 138 ff. (7 blank), 32–35 ll., clearly written on paper, c. 1339 (possibly copied later).

FIG. 215. FROM THE 1294 MANUSCRIPT OF EUCLID

Bernardino Baldi testifies to the esteem in which the author of this work was held, in the following words : ‘ Di patria Fiorentino fu Pauolo ; il quale, per l’ eccellenza ch’ egli hebbe ne le Matematiche, lasciato il proprio cognome, fu chiamato da tutti il Geometra. Come apunto fra gli antichi auenne ad Apollonio Pergeo.’

This manuscript is primarily a treatise on arithmetic. The writer, however, left a number of blank pages at the end, and these have been filled in from time to time by various owners.

[illegible]

Plate V. From a manuscript of Euclid, c. 1294

PLATE VI. FROM A MANUSCRIPT OF BOETHIUS, C. 1300

The arithmetic begins (f. 1, r.): 'El nome sia di Dio et a reverentia //della suo potentia et della santa trinitade. Et dello suo madre // uirgino sempre santa maria Et del beato Sco Giouani batisto // . . . Al chominiciamento del nostro trattato . . .' On f. 1, r., is a table of

FIG. 216. FROM THE DAGOMARI MANUSCRIPT

money. F. 2, r.,—f. 3, v., is a table of contents: 'Questi sono echapitoli del nostro tractato,' and this states that the work includes the 'Regholvzze del Maestro Pagolo Astrologo.' (ff. 121, r., — 131, r.). This part of the volume is an ordinary commercial arithmetic such as the Florentine teachers produced in the fourteenth and fifteenth centuries. The writing and the forms of the numerals indicate this period, and are not unlike that of a fourteenth-century computus (see p. 445) in this library. This may, however, be a fifteenth-century copy. The examples in division are unique, since they follow neither the galley nor the 'a danda' method, as is here shown in the case of 49289 ÷ 23 = 2143. The peculiar position of the remainders should be noted. The o indicates no remainder, and the 2, 2, and 1 are excesses of 7's in the proof.

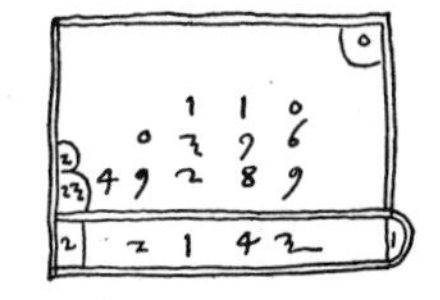

That the author considers this the 'a danda' plan (the forerunner of our present long division) appears in his use of this name immediately after. Another odd feature is the placing of the divisor second in the

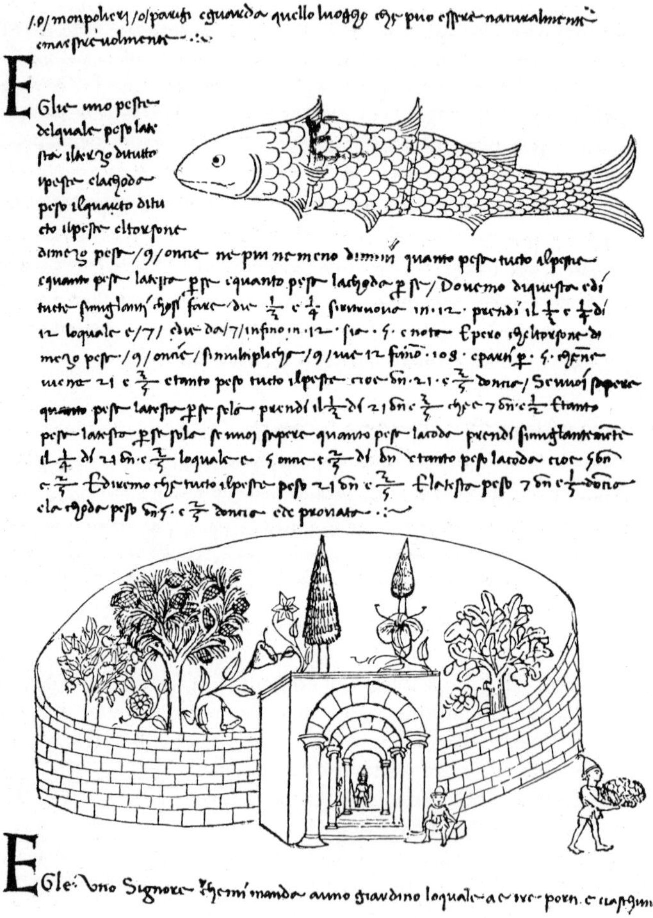

FIG. 217. FROM THE DAGOMARI MANUSCRIPT

division of fractions, it usually appearing first in that period. The problems are of the usual Florentine mercantile character. The first few are followed by a treatise on 'Nvmeri perfetti' (f. 67, r.), and this

(f. 73, r.) by further business problems. The work contains several curious illustrations (Figs. 217, 218).

The most interesting feature of the treatise is, however, the internal evidence as to its date. It is usually possible to determine quite accurately the date of a Florentine manuscript on arithmetic by the examples in equation of payments, a favorite application with the Tuscan arithmeticians, and one requiring the year to be stated. This is the case here, where the dates in these examples are all, save two, 1339. The writer has also used approximately these dates in other examples, in part as follows: 1329–1332 (a problem on the calendar, f. 27, r.); 1310-1404 (an astronomical table, which would naturally extend well into the future, f. 79, r.); 1330 (f. 116, r.), and 1339 frequently. A table of the 19-year cycle (f. 123, v.) begins with 1337, as is easily computed from a marginal note in a different hand bearing the dates 1394 and 1412. This part of the work ends on f. 131, r., and there can be no doubt that it was written about 1339, the date so frequently used in the problems.

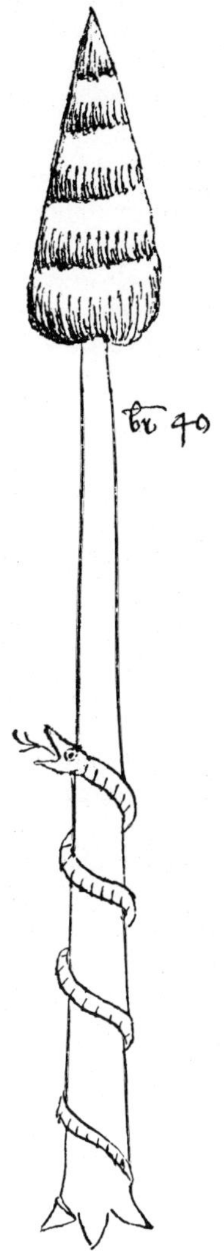

Fig. 218. From the Dagomari manuscript

F. 121, r., begins 'Regholvzze del Maestro Pagolo Astrologo,' Dagomari being referred to in the table of contents as Paul of the Abacus, 'Regholuze del Maeſtro pagholo delabacho . . .' The 'Regholvzze' was first published in Libri's *Histoire*, vol. III, p. 296.

Dagomari is included in the list of Bernardino Baldi's (1589) biographies, published in the Boncompagni *Bulletino*, XIX, 600.

Ff. 131, v., and 132, r., are in a different hand. These and the following leaves were originally blank, and after ff. 132, v., — 133, v., had been written upon, some owner used these two blank pages. He has also left his dates, viz. 1400, 1402, 1406, 1412, 1435, the 1402 being used several times. This was therefore written about 1400–1435. In this occurs the first per-cent sign I have met, other than p. 100, which is in the earlier part of this volume. This unknown writer of about 1425 uses a symbol which, by natural stages, developed into our present %.

Instead of writing 'per 100,' 'p̄ 100,' or 'p̄ cento,' as had commonly been done before him, he wrote 'p̄ $\overset{o}{\frown}$' for 'p̄ $\overset{o}{c}$,' just as the Italians wrote $\overset{o}{1}$, $\overset{o}{2}$, . . . and 1°, 2°, . . . for primo, secundo, etc. In the manuscripts which I have examined the evolution is easily traced, the $\overset{o}{\frown}$ becoming $\frac{0}{0}$ about 1650, the original meaning having even then been

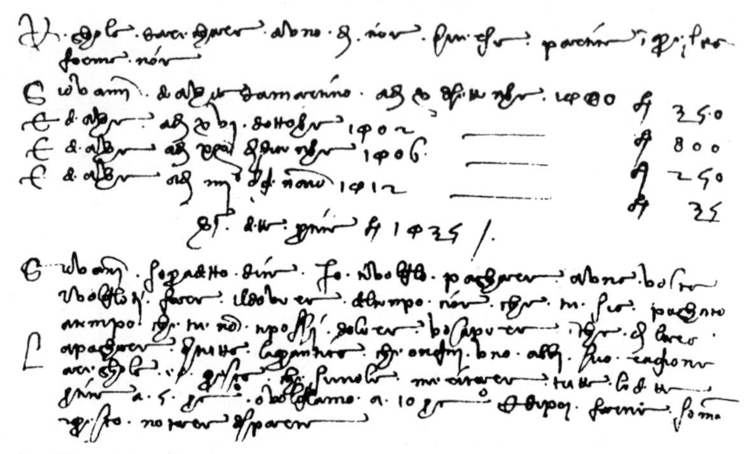

FIG. 219. FROM THE ADDITION (C. 1400–1435) TO DAGOMARI

lost. Of late the 'per' has been dropped, leaving only $\frac{0}{0}$ or %. See Figs. 219, 220.

Ff. 132, v., and 132, r., are in a different hand. They refer to the calendar, and contain the dates 1380 and 1382, each twice. F. 133, v., is in a still different hand, although also on the calendar. The date of this part is fixed by the expression, 'et ī iſto preſēti āno. ſ. 1447 finit circuluſ anni.'

EUCLID. Latin MS., c. 1350.

See p. 11.

Title. None. Fragments of the 'Elements.'

Description. 4°, 16.5 × 21.4 cm., the text being 10.5 × 15.8 cm., with marginal drawings. 6 ff. unnumb., 32–41 ll. A Latin manuscript, written on vellum in a hand of c. 1350.

The manuscript is a fragment of 6 folios, and includes part of Books III and IV, and a list of propositions evidently based upon, but not identical with, Euclid's sequence.

Si Domanda quanto fù comprata la Libbra di questa Mercanzia la quale vendendosi 75 d.e 6 la libbra disse il Mercante guadagnare ...

122	100 — 5 6
20	12
2440	6600
12	20
4880	132000
2440	14880
29280	12
4 . 6 $\frac{1}{10}$	178560
	28880
	292 | 80
	14
	146
	7
	73
	1
	10

FIG. 220. FROM A 1684 MANUSCRIPT, NOT HERE CATALOGUED

ALBERTUS MAGNUS. Latin MS., c. 1350.

ALBERTUS TEUTONICUS, DE COLONIA, or RATISBONENSIS. Born at Lauingen, Swabia, c. 1193; died at Cologne, November 25, 1280. He was Count of Bollstädt, a Dominican priest, and Bishop of Regensburg. He studied at Padua and taught at Bologna, Strasburg, Freiburg, Cologne, and Paris. He was so prominent as a philosopher that he was known as 'Doctor Universalis.'

Title. 'De Cælo et Mundo.' (F. 1, r.)

Description. 22.4 × 32 cm., the text being 14 × 21.3 cm. without the marginal notes, and arranged in double columns, each 6.5 cm. wide. 90 ff., 50 ll.

This is a beautifully written manuscript, in a fourteenth-century hand, with fine initials in red, black, and blue. The first folio contains part of the calendar, and a few random memoranda, including an old price mark of three ducats. The twenty-ninth folio is blank except for some crude circles, and one folio has been cut out. There is a break here in the manuscript, f. 28, v., closing with these words from the fourth tractatus of liber I: 'Rei a' gn̄ate ē ultim̄ ꝛ finis in tp̄r aut actu aut pō. quia si ē corrupta tūc hī actu finē ꝛ si ē adhuc corrupt'.' ('Rei autem generatæ est ultimum et finis in tempore aut actu aut potentia: quia si est corrupta, tunc habet actu finem: et si est adhuc corruptabilis'— the rest of the sentence being 'habet finem potentia.') After the blank folio, f. 31, r., opens with these words from tractatus I of liber II: 'que eſt ſb'a sep̲ata' ('quæ est substantia separata,' the preceding missing words of the sentence being 'quæ tamen non limitant operationem formæ ejus'). Aside from this the manuscript is complete. F. 89 closes in col. 1 with the words 'Explicit liber de celo et mūdo fratris alberti deo agamus gratias.' F. 90, r., is blank, but 90, v., has the zodiac and planets.

There is a brief reference to Pythagorean arithmetic in liber I, tractatus I, caput II.

The best edition of this work is in 'B. Alberti Magni // Ratisbonensis Episcopi, ordinis Prædicatorum,// opera omnia,// . . . cura ac labore // Augusti Borgnet.// Volumen quartum // Parisiis . . . MDCCCXC.' The first edition of the Opera Omnia appeared in Leyden in 1651.

EUCLID. Latin MS., 1375.

See p. 11.

Title. The work begins: 'Punctus eſt cui⁹ p̲s nō ē.'

Description. Fol., 20.9 × 30.1 cm., the text being 11.9 × 19.5 cm. 39 ff. unnumb. + 1 blank = 40 ff., 44–47 ll. A Latin manuscript, written on paper, c. 1375.

This manuscript includes the first five books of Euclid. This is followed by a treatise upon astronomy and mensuration, also in Latin, in a different but contemporary hand. The mensuration includes some work on areas and volumes.

ANONYMOUS. Latin MS., 1384.

Title. None. A Computus manualis.

Description. 4°, 13.3 × 18.8 cm., the text being 9 × 13.3 cm. 33 ff. (3 blank), 40–44 ll., written on paper.

The first written folio has been torn, and the opening lines are missing. The work is that particular kind of mediæval computus (see p. 7) in which a finger mnemonic system is used (see p. 34). It is a copy of an older treatise, the text of which is here written in Gothic characters, the copyist's notes appearing in a smaller hand. As in most computi, the numerals are generally in Roman, both in the text and in the commentary, but the date is twice given, as follows (f. 19, r.): 'an̄o dn̄j 1000.300.80.4'; 'anno dnj̄ 1000.300.80.4' (see Fig. 221). The text is in Latin, except for one page which is in French. A later owner has written the date of his ownership, 1600 (f. 2, r.).

ANONYMOUS. Italian MS., 1393.

Title. '℄Qui chomincia ilpologho d·l conpoto // d·lcorſo d·l folo ꝛ d·lla luno // Prolagho.' (F. 3, r., Fig. 222.)

Description. Fol., 22 × 29.7 cm., written in double columns, each being 7.5 × 20.5 cm. 70 ff. (6 blank), 35 ll.

This is an excellent example of a computus (see p. 7). The verse 'Thirty days hath September,' the only relic of the old computi now familiar to most people, appears in this manuscript in the following Italian form: 'Trenta di a noumbre apile // giugno & ſettenbre di uentotto (. . . . .) vno tutti glialtrj ſono trentuno.' (F. 12, v.) (See also p. 33.)

GIOVANNI, the son of Luca da Firenze. Italian MS., 1422.

The son of Maestro Luca, a celebrated Florentine arithmetician, mentioned on p. 468.

Title. 'Trattato di aritmetica.' (F. 1, r.)

Description. 4°, 14.9 × 21.9 cm., the text being 9.3 × 16.3 cm. 145 ff. 27–30 ll.

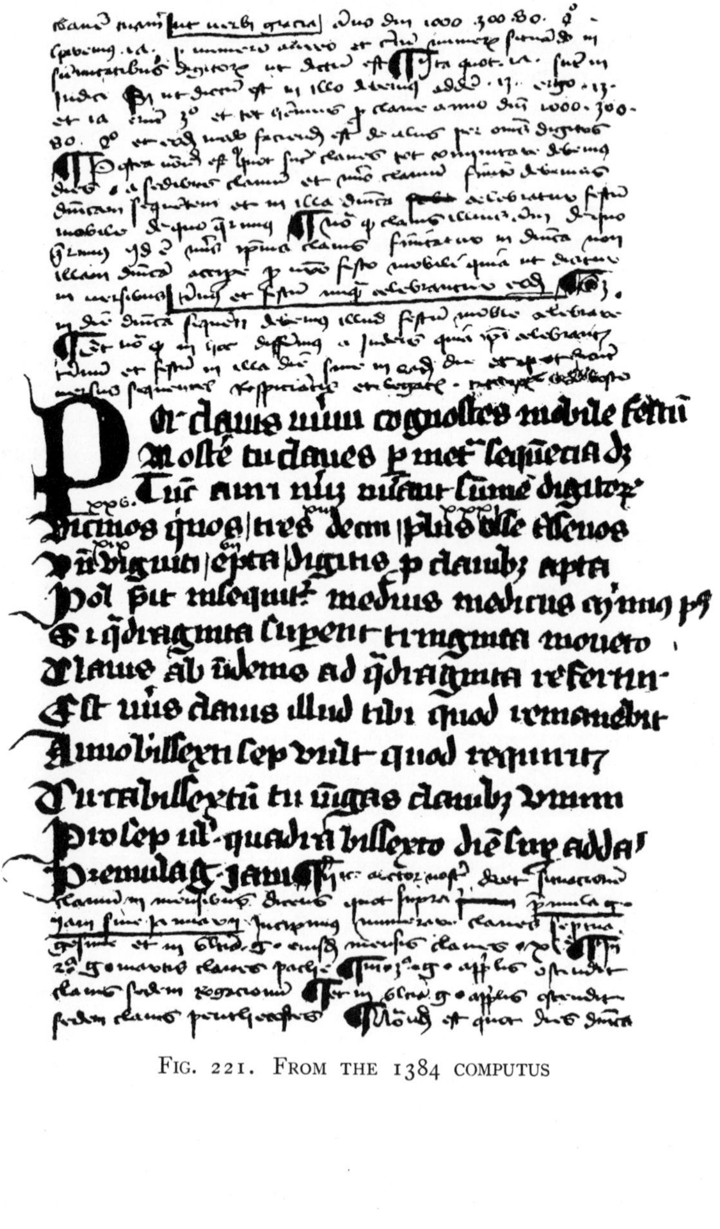

[illegible]

Per clavis minimi cognosces mobile festi
Mostre tu claves p[er] met[ra] sequentia d[i]c[imu]s
Tu[n]c anni nu[m]er[us] ducatur summe digitor[um]
Decimos quos tres deca[m] plus ille ? excessus
V[e]l viginti septa digitus p claub[us] apta
Pol sit insequit[ur] medius medicus q[ue] minimus p[ost]
Si q[ua]dragi[n]ta supe[ra]nt triginta moveto
Clavis ab undenis ad q[ua]dragi[n]ta referu[n]t
Est un[us] clavis illud tibi quod remanebit
Anno bisexti sep[er] vult quod requirit
Tu[n]c a bisexti tu nigras clavib[us] unam
Pro sep[tuagesima] q[ua]drag[esima] bisexto d[ic]e sup[er] adda[s]
Premulag[ue] Janu[arii] [illegible]

[illegible]

FIG. 221. FROM THE 1384 COMPUTUS

This is one of the best examples of an early fifteenth-century commercial arithmetic known. Florence was at this time an important financial center, and the arithmeticians of the city were highly esteemed. Several of the applications found in arithmetics for the next three centuries had their origin here. Subjects like equation of payments and partnership involving time, customs like 'days of grace' in exchange, and forms like time drafts in sets of two or three can be studied to good advantage in the arithmetical manuscripts of this period.

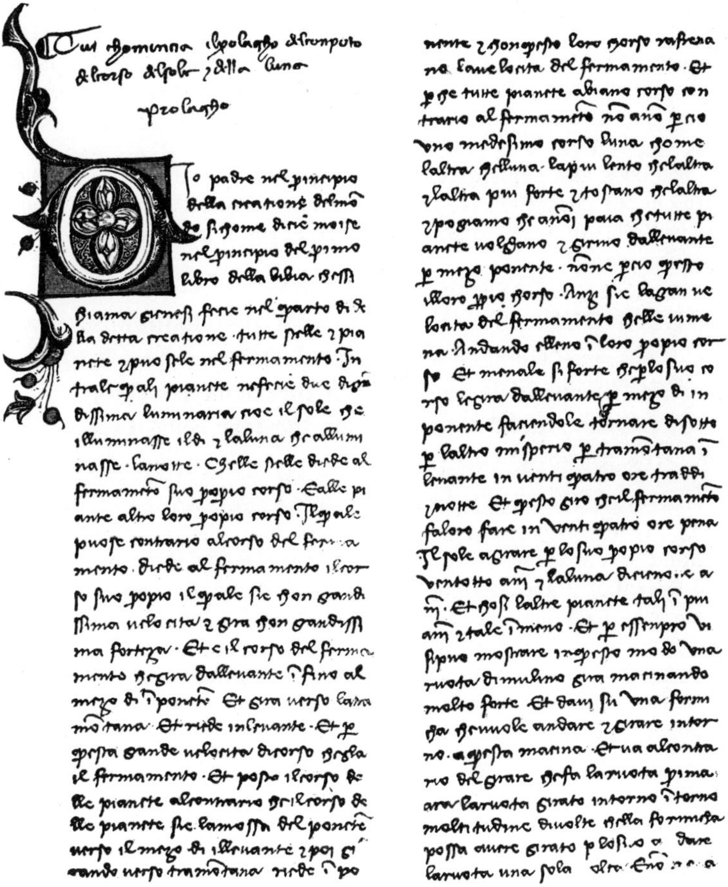

FIG. 222. FIRST PAGE OF THE 1393 COMPUTUS

On f. 1, v., the writer says he proposes to treat of arithmetic or the abacus: 'arifmetricho volgharemente e chiamata abacho.' The word 'abacus' had come at this time, in Italy, to mean simply arithmetic, the original meaning having been lost. Following a custom of Florentine arithmeticians, Giovanni gives an extensive multiplication table, for purposes of reference, and then begins at once with examples in compound numbers and fractions, thus presupposing a knowledge of the fundamental operations. These examples are of a mercantile character and constitute the entire portion devoted to arithmetic. The examples in the equation of payments serve, as usual, to fix the date of the manuscript. They all refer to the years 1418–1426 (ff. 113–122). The date is, however, fixed exactly by the closing lines of a folio near the end: 'quefto libro Ifcriffe Giouannj del maeftro lucho dellabacho e finillo quefto d dottobre 1422 —' (f. 136, v.). In the first page is the date 'ad 28 dottobre 1422.'

The book also contains a section on mensuration and the calendar, with curious illustrations of the months (Pl. VII). The last five folios, originally blank, seem to have been written by a different hand about the same period.

ROLLANDUS. Latin MS., 1424.

A native of Lisbon, canon of Sainte-Chapelle, Paris, c. 1425.

Title. 'Scientia de numero ac virtute numeri.' (F. 3, r.)

Description. Fol., 21 × 29.8 cm., the text being 15.5 × 21.5 cm. 168 ff., 28–31 ll.

This is an exceedingly interesting manuscript, written in the year 1424. It was prepared at the command of John of Lancaster, Duke of Bedford, son of Henry IV of England, at one time Protector of England and Regent of France. To him Rollandus dedicates the treatise: 'Illuftriffimo ac fereniffimo principi metuendiffimo domino domino Iohanni patruo domini noftri regis ffrancie et anglie regenti Regnum ffrancie. duci bethfordie Rollandus fcriptoris veftre celfitudinis phyficus vlexbonenfis fe ipfum ex debito iuramenti.' In 1423 Lancaster issued an ordinance for the restoration of studies in the University of Paris, and it was probably as a result of this that this textbook was written by a Portuguese physician, Rollandus, who was then a canon of Sainte-Chapelle in Paris. The dedication sets forth Lancaster's interest in learning in France and the status of mathematics at that time. Rollandus covers all of theoretical arithmetic as then known, but takes up no practical problems. He also treats of irrational numbers, a topic

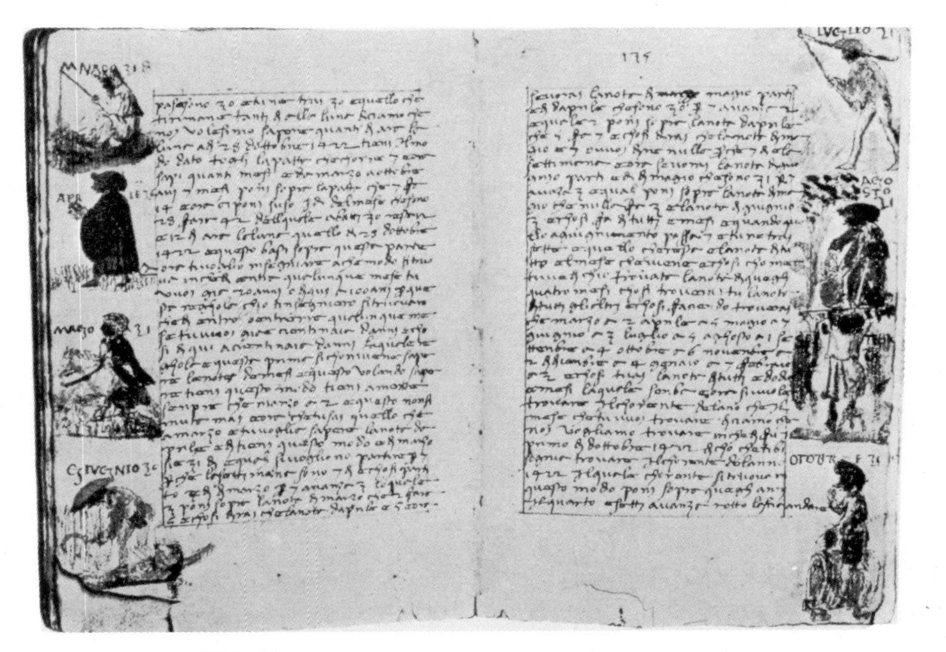

PLATE VII. FROM A MANUSCRIPT OF GIOVANNI DA FIRENZE, 1422

which is now considered part of algebra. It is doubtful if there is a manuscript extant which throws more light upon the nature of French university mathematics at the time this was written. Rollandus also wrote a work on surgery and one on physiognomy. He may possibly be the Rolland who in 1410 was rector of the University. Since this manuscript is evidently a copy, others must have existed, but I have found no reference to them. Fig. 223 shows the forms of the numerals used at this time. It should be noticed that the numerals that have changed materially in form since the twelfth century are 4, 5, and 7. These are shown opposite the letters *b*, *i*, and *p*, respectively, in Fig. 223. The changes in the other forms have been more evidently due to the fashion in handwriting. All of the forms are, however, quite different from the primitive ones found in the cave inscriptions of India.

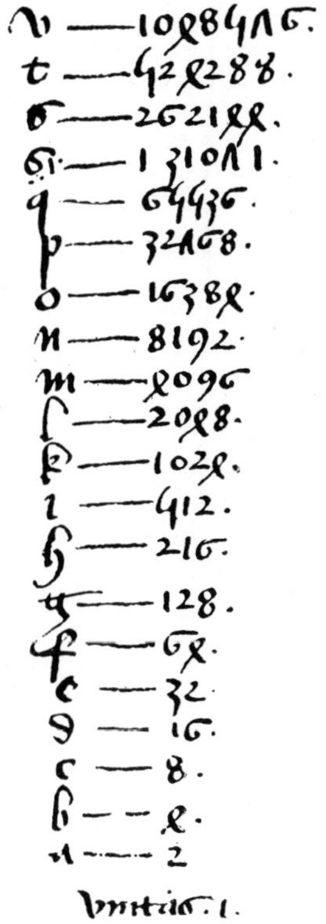

FIG. 223. FROM THE 1424 ROLLANDUS

ANONYMOUS.

Italian and Latin MSS., c. 1430 and 1478.

Title. The first folio is missing.

Description. Fol., 17 × 22.1 cm., the text being 11.2 × 14.4 cm. 183 ff., 25–28 ll.

This volume consists of two Florentine manuscripts, one on commercial arithmetic and the other on the computus. The portion on arithmetic was probably written between 1420 and 1444, these being the extremes of the dates in the examples in equation of payments. It is not unlike the other contemporary Florentine arithmetics described on pp. 443 and 464. The author assumes the student's ability to perform the fundamental operations with integers, although, after numerous multiplication tables, he gives (see Fig. 224) an example under 'Multiplicha p̱ modo de barichocholo' (the Florentine name for our present method), and one under 'Multiplicha p̱ modo de Quadrato' (the 'gelosia' method of the Venetians).

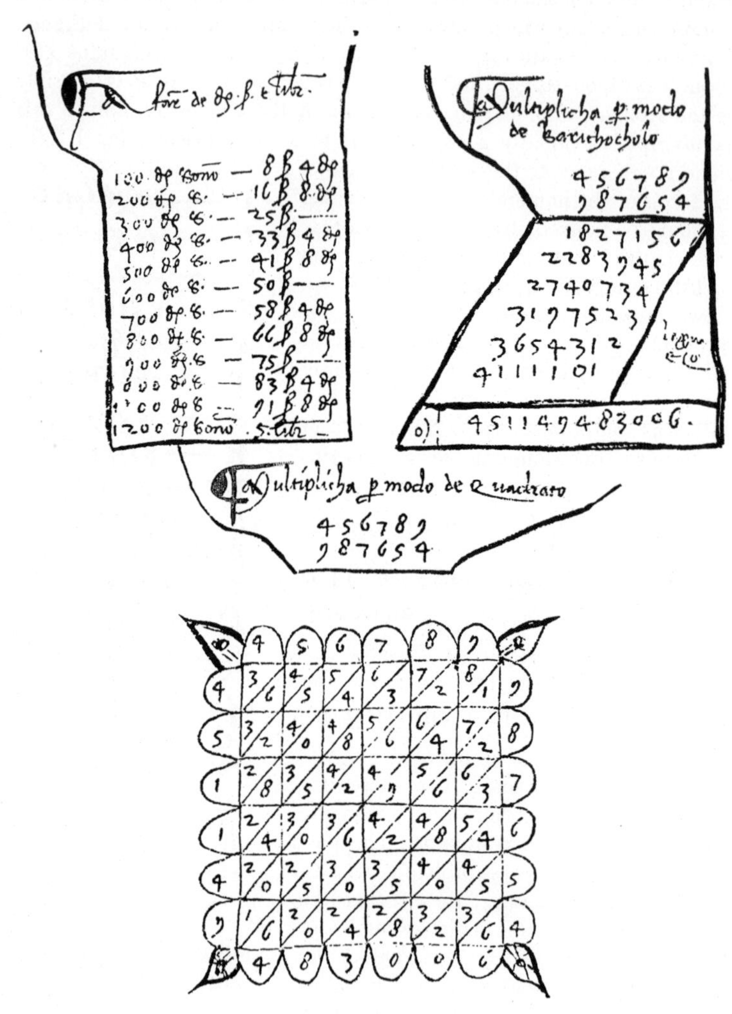

FIG. 224. FROM THE ANONYMOUS MANUSCRIPT OF C. 1430

In the latter part of the work, as if it were a new topic, the author has introduced (f. 126, v.) a chapter entitled 'Partire a danda.' He closes with several folios on mensuration.

The second part (beginning on f. 154, v.) is in a different hand. It is a computus, and from the dates it would seem that it was written in 1475. There is a third part (beginning on f. 172, r.) consisting of a set of religious verses, in a still different hand, bearing the date 1478. The language is Italian, excepting for the computus which is Latin.

ANONYMOUS. Latin MS., c. 1435.

Title. 'De tempore de compositione quadrātis et compositione astrolabii plani,' etc.

Description. Bound with the 'Computus cyrometralis' of 1476. Written on paper, in a German hand. Some dates which it contains indicate that the book was written about 1435. It is related to the history of arithmetic only in the forms of numerals used.

JOANNES DE GMUNDEN. Latin MS., c. 1439.

See p. 117.

Title. None. A treatise on the computus.

Colophon. On f. 17, v., are the words, 'Explicit kalendariū mgrī Joh'is gmünd.'

Description. Fol., 19.1 × 25.7 cm., the written part being about 14.1 × 18.5 cm. (The tables vary and are larger.) 20 ff., 32–38 ll. Latin manuscript. Written on vellum, in red and black, about 1439, as shown by the dates on f. 16.

Although entitled a 'Kalendarium,' it is so much like the mediæval computi as to have a place in this list. It is really a semiarithmetical treatise on the Church calendar.

ANONYMOUS. Latin MS., c. 1441.

Title. None. A computus.

Description. Fol., 21.5 × 28.5 cm. (varies), 10 ff. (1 blank); the number of lines to each folio varies considerably. Latin manuscript, c. 1441.

The first part of this manuscript is a computus, not very extended. It relates rather to the calendar itself than to the computations upon which it is founded. The last folio is in a later hand, and from the tables which it contains it seems to have been written c. 1524.

ANONYMOUS. Latin MS., c. 1442.

Title. None. A treatise on the planets.

Description. Bound with the 'Computus cyrometralis' of 1476. Written on paper, in a German hand. It contains the date 1442 in two places (f. 102). Its value in connection with arithmetic is confined to the forms of the numerals used.

JOHANNES SACROBOSCO. Latin MS., c. 1442.

See p. 31.

Title. An algorismus, beginning with the words, 'Omnia que a primeua rerum origine processerunt.' (F. 3. r.)

Description. 4°, 14.1 × 20 cm., the text being 10.5 × 17.6 cm. 21 ff., 25–39 ll.

The first folio has a picture of an astronomer with a celestial sphere, and the name of the student who copied the MS., 'hainricus muglinchk' (f, 1, r.). On the next page (f. 1, v.) Mugling's name again appears, together with three dates: 'Item hainric9 mugling aſtronim9,' 1442, 1443, 1444. The 'Algorismus' begins on folio 2, r.: 'Omnia que a primeua // rerū orignꝫ,' and ends on f. 10, v.: 'Explicat algoraſmus.' Then follows a picture (Pl. VIII) in colors, representing a master teaching his pupil the Hindu numerals from a kind of large hornbook, with a motto: 'Ich pin algoriſm9 genant // Das . . . (?) . . . hau ich in mein // nes hant.' A table explaining place value is given on f. 11, r. Beginning on f. 11, v., is another treatise on arithmetic, giving the fundamental operations and some work in progression, and ending with a multiplication table (ff. 18, v.; 19, r.). A later hand has added three pages on progressions, rule of three, partnership, and interest. (Ff. 19, v.; 20.)

Sacrobosco's algorismus was the first arithmetic, based on the new numerals, written by an English scholar. It consists of eleven chapters, viz. Numeratio, Additio, Subtractio, Mediatio, Duplatio, Multiplicatio, Divisio, Progressio, Perambulum ad radicum extractionem, Extractio radicum in cubicis. For the various editions of this work see p. 32. The title of the Paris edition of 1510 is 'Opvscvlvm de praxi numerorum quod algoriſmum vocant,' and the work consists merely of four folios containing the chapter 'De arte numerandi.' There was also published at Antwerp in 1547 (with later editions, Paris 1550, Venice 1564,

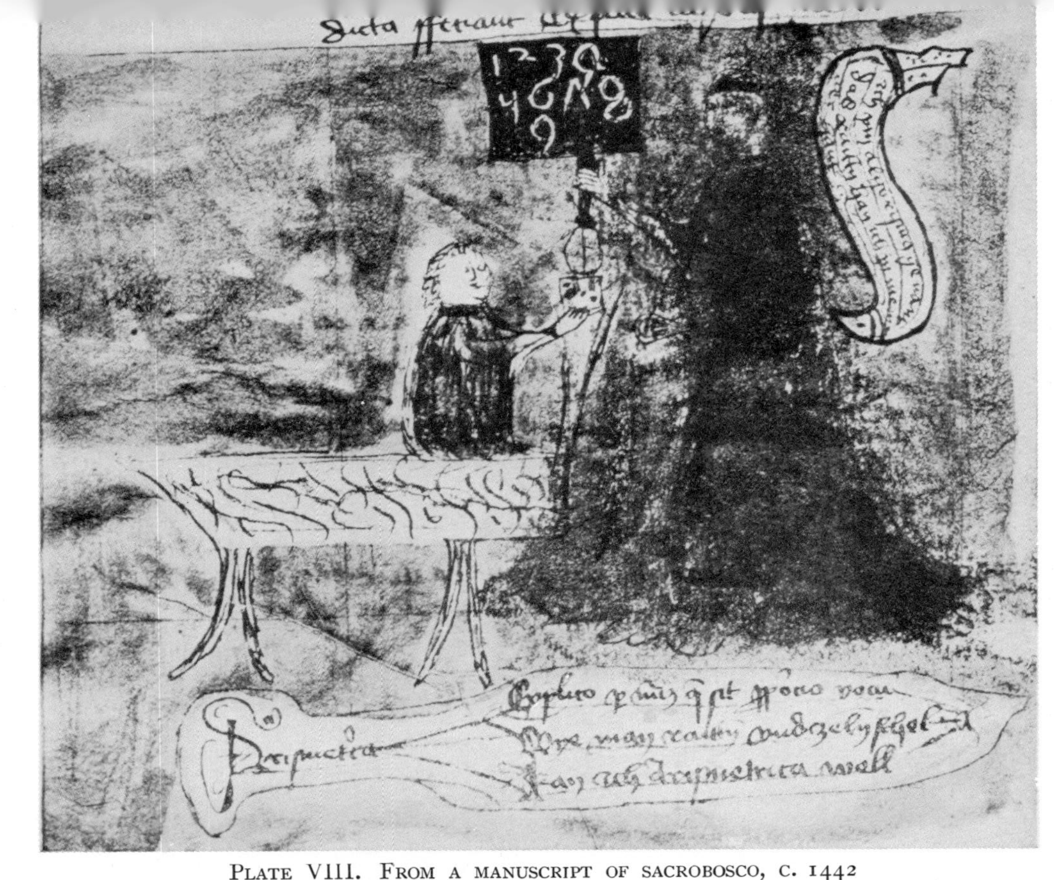

PLATE VIII. FROM A MANUSCRIPT OF SACROBOSCO, C. 1442

Wittenberg 1578), Sacrobosco's 'Libellvs, de anni ratione: seu vt vocatur vulgo, Computus ecclesiasticus.' His work on the Sphere was published in 1488 at Venice. For a discussion of the authenticity of the Algorismus, see De Morgan, p. 13.

As stated on p. 31, the date of the death of Sacrobosco is uncertain. It is either 1244 or 1256, according as we interpret certain lines on his tomb:

> 'M. Christi bis C quarto deno quater anno
> De Sacro Bosco discrevit tempora ramus,
> Gratia cui dederat nomen divina Johannes.'

MARO ANTONIO ROZINO. Latin MS., 1447.

Title. 'Qoñes Marci Antonij rozoni artiū // doctoris sacre theologie magrī et papię p̄hiam legentis.' (F. 1, r.)

Description. Fol., 20.3 × 28.1 cm., the text being 13.7 × 24 cm. 99 ff., 32–33 ll.

This is not an arithmetic, and it has been included in this list only because it is semi-mathematical and is bound with Bradwardin's treatise on the theory of proportion. It is a Latin treatise on the theory of perspective. It is written in a clear Italian hand, and was part of a volume numbered 493 in the Boncompagni sale, containing four manuscripts. Some dealer has removed the first of these manuscripts, the 'Perspectiva communis' of John Peckham, archbishop of Canterbury. A memorandum in the Boncompagni catalogue, probably from a leaf removed with the first treatise, shows that the manuscript was copied in 1447: 'scripte per me antonium confaronesium ut (vocatur?) de lavilata Anno dñi M"ccccxlvij.' The other manuscripts bound with this are described below and on p. 452.

THOMAS BRADWARDIN. Latin MS., 1447.

See p. 61.

Title. None. A treatise on proportion.

Colophon. 'Explicint propor^nes thomi brardi // scripte p̱ me.' (F. 115, r.)

Description. Fol., 20.3 × 28.1 cm., the text being 13.7 × 22.8 cm. 15 ff., 33 ll.

This forms part of the volume last mentioned, and is written in the same hand. Bradwardin's treatise on proportion was published in Paris, 1495 (p. 61).

ANONYMOUS. Latin MS., 1447.

Title. A treatise on lenses.

Description. Fol., 20.3 × 28.1 cm., written in two columns, each being 6.6 × 22.9 cm. 7 ff., 24–33 ll.

This rather early work on optics forms part of the volume mentioned on p. 451, and is written in the same hand.

ST. BERNARD OF SIENA. Latin MS., c. 1450.

Born at Massa, Tuscany, in 1383; died at Aquila in 1444. He was a zealous founder of monasteries, and wrote various religious treatises.

Title. This book of sermons begins as follows: 'Smo .33.9 .de reſtitutiōe.// Doñi .q̂. in quadrageſſīa ordo dicendoꝝ ꝑ ſeptimanā ſequēteꝫ ī li° de x'ana re//ligione .a. R.do p. S. B.no de ſeniſ ordoniſ minoꝝ edito.' (F. 1, r.)

Colophon. 'Explicit tra//ctatuſ de uſꝫiſ ꝛ ꝯctibꝫ ſꝫuꝫ Bntuꝫ// Bnardiniꝫ (?) de ſeniſ. ordīſ minorꝝ.' (F. 129, r.)

Description. 4°, 14.8 × 19.8 cm., the text being 10 × 13.8 cm. 129 ff., part vellum and part paper. After f. 60 the pages are arranged in double columns, each being 5 × 14.5 cm., 44–46 ll.

This beautifully written manuscript of c. 1450 is included in this list because it contains several sermons bearing upon the mercantile customs of the time, including 'De usuris,' 'De cambiis,' 'De contractibus,' and 'De mercatoribus.'

Part of this work was translated into Italian in the fifteenth century, as appears from a codex in the Biblioteca Comunale at Siena. One of the sermons was published by Riccomanni, in the *Scelta di Curiosità Letterarie inedite o rare* (no. 13) of Romagnoli, Bologna, 1862. This sermon, 'Sulle Soccite di Bestiami,' contains considerable information as to the business problems of monastic institutions of the fifteenth century. A copy of several of these sermons, made by one Eustachio da Feltre in 1469, is mentioned on p. 466.

ANONYMOUS. Latin MS., c. 1450.

Title. None. On the Quadrivium.

Description. 4°, 14.7 × 21 cm., the text being 14.5 × 17 cm., 47 pp., 18–40 ll.

This is a Latin manuscript on the Quadrivium. It is written on paper, in a German hand of about 1450. It includes a brief treatment of arithmetic (5 ff. + 1 blank, the text occupying ff. 18 – 22, v.), and a brief treatment of the calendar and the zodiac (ff. 15, r. – 17, v.). The geometry begins with modifications of Euclid's definitions : 'Pvnct9 eſt c9 pars nō eſt/Linea ē longi^{do} ſn̄ latitu^{ne} ꝛ ꝓfūditate.' It closes : '¶Explicat prim9 liber euclidis cū ꝯmeto cāpani.'

JOHANNES SACROBOSCO. Latin MS., c. 1450.

See p. 31.

Title. 'Spera mundi secundum Joha//nnem de ſacroboſco.' (F. 1, r.)

Colophon. 'Explicit tractat9 de ſꝑa ẜm // Jōhem de ſacroboſco ām.' (F. 35, v.)

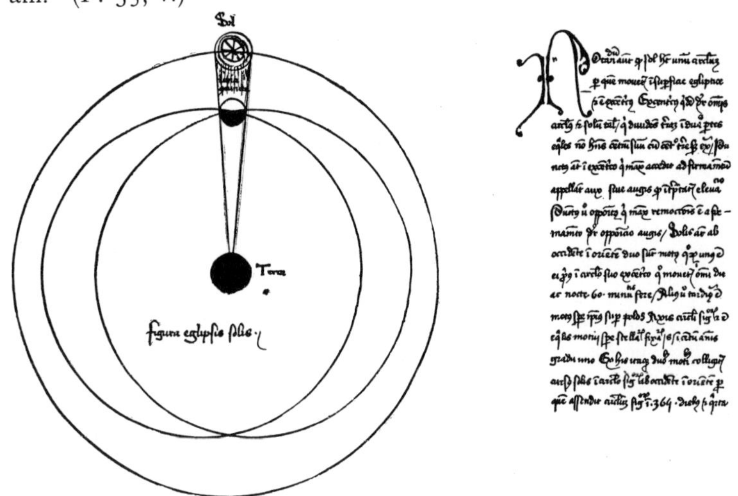

FIG. 225. FROM SACROBOSCO'S *Spera*, C. 1450

Description. Fol., 35 ff., written on vellum.

Printed works of this kind have not been included in this catalogue. This manuscript, however, shows the state of the numerals about 1450, and therefore is mentioned.

The treatise begins (f. 1, r.) : 'De ſpera in quatuor capitula diſtingui9 // dicentes p'mo quid ſit ſpera/quid eius // centrum/ quid axis ſpere / et quid ſit po//lus.'

The first four folios have a marginal commentary, closely written in a later hand. The figures are carefully drawn throughout (see Fig. 225).

There is nothing to show the exact date of the MS., but the handwriting and numerals indicate the last half of the fifteenth century.

Sacrobosco's 'Sphere' was published in 1488, and often thereafter. It was the great mediæval work on astronomy.

ANONYMOUS. Latin MS., c. 1450.

Title. 'Incipiūt floreſ Ariſmetrice.' (F. 1, r.)

Description. 4°, 15 × 21.8 cm., the text being 8.4 × 15.5 cm. 70 ff. unnumb. + 2 blank = 72 ff., 45 ll. Latin MS., in a German hand, c. 1450.

This is a theoretical treatise on arithmetic and algebra, written respectively on the Boethian and Al-Khowarazmian models. Only a little elementary treatment of the fundamental operations (chiefly multiplication) is given, the writer devoting most of his attention in the first part of the work to subjects like progressions, ratios, and proportions. The latter part of the book is algebraic and may prove to be a copy of some mediæval work of importance. It resembles in some places the work of al-Khowarazmi, a manuscript of which follows this one in the same volume.

MOHAMMED IBN MUSA AL-KHOWARAZMI. Latin MS., 1456.

ABU 'ABD ALLAH MOHAMMED IBN MUSA AL-KHUWARIZMI. Born in the province of Khwarazm (whence his name), died c. 831. The most celebrated algebraist of his time, and the first to write a book bearing the title *Algebra*. From his name comes the word *algorism* (see p. 7).

Title. 'Liber mahometi de Algebra et almuchabila × compāriſ et oppōſ.' (See Fig. 226.)

Colophon. There is none, but f. numb. 85 bears the date 1456, and the forms of the numerals and letters are of that period.

Description. 4°, 15 × 21.8 cm., the text being 12 × 17.2 cm. 23 ff. unnumb. + 1 blank = 24 ff., 44–48 ll. Latin MS. in German hand, 1456.

This interesting manuscript of the first book bearing the name algebra is more complete than the one found by Libri in the Bibliothèque Nationale (*Histoire des sciences mathématiques*, I, note XII). It more nearly resembles the one which I found in the Columbia University Library in 1904, and showed to be in the handwriting of Scheubel. The two deserve to be edited and compared with the Rosen translation

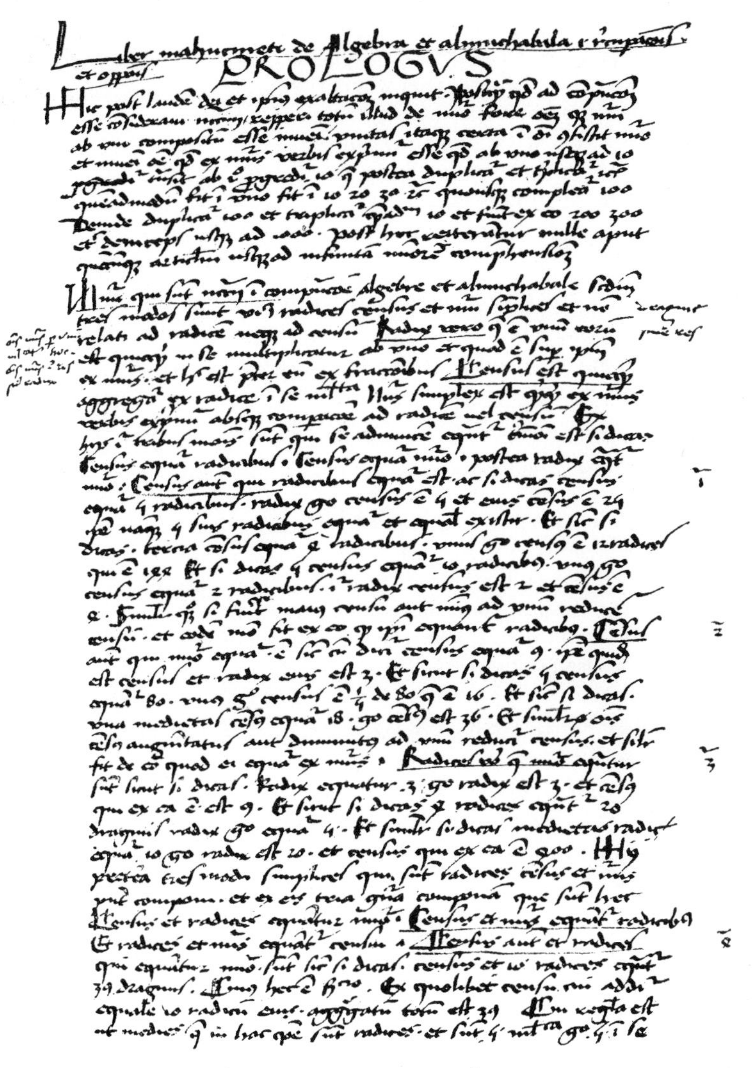

FIG. 226. FIRST PAGE OF THE 1456 AL-KHOWARAZMI

(London, 1831). This manuscript is particularly valuable because, unlike the one published by Libri, it has the Arabic numerals and the mediæval algebraic symbols. The *et* used for plus so closely resembles the + as to leave little doubt that the latter was derived from this Latin word. Like Regiomontanus, the writer uses $\overline{\imath 9}$ for minus.

This is followed (ff. numb. 97) by a brief treatise on rhetoric, and by three manuscripts on mathematics of 1501, c. 1475, and c. 1550, described on pp. 480, 468, 486.

ANONYMOUS. Italian MS., c. 1450.

Title. A treatise on mensuration, roots, and algorism.

Description. Fol., 21.6 × 30.3 cm., the text being 10.5 × 15.8 cm. 177 pp. (2 blank), 20–25 ll. Written on paper, in a Florentine hand, c. 1450.

This is a general treatise on mathematics, with divisions as follows: A 'praticha de Geometria' (ff. 2–11); fractions (ff. 13–16); square root (ff. 17–19, r.); cube root (ff. 19, v., – 21); ratios (ff. 22–26); algebra (ff. 26–30); mensuration (ff. 31–46); algorism (ff. 47–177).

The algorism is evidently the work of some Florentine teacher, and the handwriting is that of the middle of the fifteenth century. It includes the usual applications of the period, such as profit and loss, partnership, and interest, and it makes prominent the rule of three and the rule of false position. 497

PETER PAUL VERGERIUS. Latin MS., c. 1450.

Born at Capo d'Istria, Venice, c. 1350; died in Hungary c. 1420, or possibly in 1444. Bishop of Capo d'Istria.

Title. 'De Ingeniis Moribus et liberalibus Studiis. Franciscvs//ſenior auus tuus cui⁹ ut extant//plurime res magnifice geſte ita//et multa paſſim ab eo ſapiēter i//dicta memorantur. . . .' (F. 1, r.)

Colophon. 'Petrj pauli uergerij de uiſtinopoli ad uber-//tinuȝ Carárienſem de ingenijs moribs ꝛ//liberalibs ſtudijs adoleſcentie liber felici-//ter. Explicit.//Amen.' (F. 23, v.)

Description. 4°, 14 × 20.3 cm., the text being 8.8 × 13.6 cm. 23 ff. (10 vellum, 13 paper), 26–36 ll. The first 10 (vellum) folios are in Gothic script; the last 14 are plainly written in cursive characters.

The treatise refers to 'arifmeticha' and 'geumetria' (f. 15, r.) among the liberal arts. It is bound with several other manuscripts.

RAYMUNDUS LULLIUS. Latin MS., c. 1450.

RAYMOND LULLY, RAMON LULL. Born in Palma, Majorca, c. 1234; died in 1315. He was a Catalan alchemist, philosopher, and missionary, and was known as 'Doctor illuminatus.'

Title. 'Ars Brevis.' A note by a somewhat later hand, written on f. 1, r., reads: 'Ars breuis Raymundi Luli quam scripsit Pisis // in monasterio Sancti Dominici anno 1307. à. c.23.' The *Ars brevis* begins on f. 5, r.: 'Deus cū tua grā fapiā et amore Incipit ars breuis que eft // ymago artis gn̄alis. Nā ifta fata ab intelectu fb'tili et fun//dato ipē pōt fcire gnalē artē.'

Colophon. 'Finiuit Raymundus artē breuen pifis in monaft'io fāc//ti dominici Anno ab incarnatione dn̄i. 1307.// Explicit feliciter.' (F. 23.)

Description. 4°, 15.4 × 20.8 cm., the text being 9.9 × 13.2 cm. 24 ff., 28 ll.

This is the second part of a manuscript of 59 folios, of which the first is the *Sensuale* of Lullius. This and several other manuscripts are bound with the Vergerius already described (p. 456).

The *Ars brevis* was originally written in 1307; this copy was made about 1450. Although not an arithmetic, the work contains several mathematical definitions.

JOHANNES ROS. Latin MS., 1450.

A Valencian priest.

Title. 'Artificium artis arithmeticæ.' (F. 1, r.) '[D]Eus qui es unus . . . Incip̄ artifim? ātis aīfmet'ce De alphabeto.' (F. 24, r.)

Colophon. 'Ad laudem // oīpotentis dei et uirginis marie . . . fi//niuit frat' Johānis ros de Valencia ꝑuincie aragonii hoc // artificiū arifmetice pad . . . ī loco fratrum minoꝝ de ofpitali // 1450 die .5. Januarij ī uigilia epiphanie domini.' (F. 14 = 37, r.)

Description. 4°, 15.1 × 19 cm., the text being 9.2 × 13.7 cm. 36 ff., 28 ll. Bound with the Lullius and Vergerius described above and on p. 456.

This rather early Spanish monastic treatise on algorism is the third of three manuscripts in the same hand, written c. 1450, of which the first two are mentioned on pp. 456, 457. It begins on f. 24, r., and ends on f. 37, r. After some definitions, tables, and computus figures, the author takes up the 'nine subjects': 'Nouē ſ'biecta ponūt ī ariſmetica' (f. 3 = 26, v.). These are treated very briefly. This is followed by a religious work, in the same hand (to f. 57).

ANONYMOUS. Latin MS., 1442.

Title. 'De ymagine mundi.' (F. 2, r.)

Colophon. '℄deo grat. añj. (amen ?) Año. dñj. 1442 . . . ℄Explicit lib.9 de ymagine mundi deo grāt . . .'

Description. 8°, 14.5 × 21.5 cm., the text being 11.5 × 19 cm. 158 ff. in the entire manuscript, 11 ff. in this portion. The other portions are described on pp. 477, 478.

This is the first manuscript in a collection of 156 folios, and begins with the lines 'I. H. S.//℄de ymagine Mundi' (f. 2, r.). It is written in Latin and treats of physical and descriptive geography. It has some interest in the history of arithmetic through the Roman numerals which are generally employed except in the case of large numbers. This manuscript and the others bound with it are described in Narducci's catalogue of the Boncompagni manuscripts (Rome, 1862, no. 81). As there stated, at one time it belonged to Alessandro Padovani, a celebrated collector of the sixteenth century.

ANONYMOUS. Italian MS., c. 1456.

Title. The title is missing.

Description. 4°, 14.6 × 21 cm., the text being 11 × 15.6 cm. 111 ff., (12 with drawings), 30 ll. No. 168 in the Boncompagni sale catalogue.

This is a business arithmetic, written in northern Italy, and completed, as appears from a note, July 15, 1456. As in most of the business arithmetics of the period, the first pages contain a set of tables, these being followed by a discussion of the fundamental operations with denominate numbers. Although written in 1456 it is probably a copy of an earlier work of about 1420, for the examples in the equation of payments involve dates from 1418 to 1425 (ff. 78–84). The work is also interesting because it contains the early form of the sign % (see Fig. 227) already mentioned on p. 439. The column tables used by merchants in their multiplication 'per colonna,' and common in the Italian manuscripts of this nature, are shown in Fig. 228.

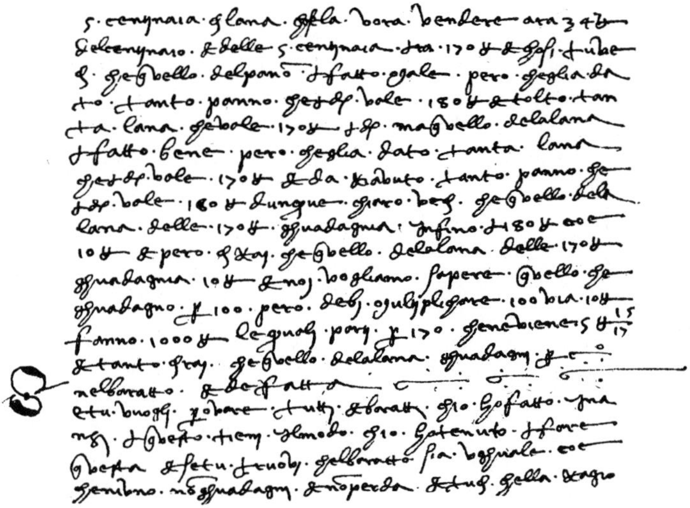

FIG. 227. FROM AN ITALIAN ARITHMETIC OF C. 1456

ANONYMOUS. Italian MS., c. 1460.

Possibly by Raffaele Canacci, a Florentine mathematician.

Title. There is none given, but the work is a general treatise on mathematics.

Description. Fol., 28.2 × 39 cm., the text being 16.5 × 28 cm. 2 blank + 322 numbered ff. = 324 ff., 51 ll. Italian manuscript, c. 1460.

This is an Italian manuscript, beautifully written on vellum, with finely executed initials in colors and gold at the beginning of each of its sixteen books. It belonged at one time to Libri, and later to Boncompagni. Narducci describes it in the catalogue of the latter's manuscripts (no. 14). The author begins (f. 1) with a description of the work: 'Come e in che modo eldetto trattato e diuiſo/ cioe cio che lopa cōtiene.' The successive chapters are as follows:

2	41	82
3	41	123
4	41	164
5	41	205
6	41	246
7	41	287
8	41	328
9	41	369
1	41	410

41	100	4100
41	90	3690
41	80	3280
41	70	2870
41	60	2460
41	50	2050
41	40	1640
41	30	1230
41	20	820

2	43	86
3	43	129
4	43	172
5	43	215
6	43	258
7	43	301
8	43	344
9	43	387
10	43	430

43	100	4300
43	90	3870
43	80	3440
43	70	3010
43	60	2580
43	50	2150
43	40	1720
43	30	1290
43	20	860

2	47	94
3	47	141
4	47	188
5	47	235
6	47	282
7	47	329
8	47	376
9	47	423
10	47	470

47	100	4700
47	90	4230
47	80	3760
47	70	3290
47	60	2820
47	50	2350
47	40	1880
47	30	1410
47	20	940

FIG. 228. FROM AN ITALIAN ARITHMETIC OF C. 1456

I. ‘Qui chomincia el ꝑimo libro del detto trattato, & ꝑima pone la diuiſione del detto primo libro/ laquale e achapitoli/ cioe e diuiſo i .4. capitolj.’ (F. 1, r.) The four ‘capitoli’ are as follows:

1. ‘El primo capitolo del ꝑimo libro/ doue ſimoſtra lordine e modo del numerare le fighure cheſuſano aſcriuere enumeri.’ (F. 1, v.) In this are explained the Hindu-Arabic notation, the nine ‘figure ſignifichatiue,’ and the 0, ‘che i arabia ſidice çero.’

2. The title of chapter 2 is wanting, f. 3 having for some reason been left blank. It related to the addition of integers and compound numbers.

3. ‘Qui chomincia el terzço capitolo del ꝑimo libro/ doue ſitratta del modo & hordine del trarre el numero minore del numero maggiore.’ (F. 5, r.) The method is that of borrowing and repaying.

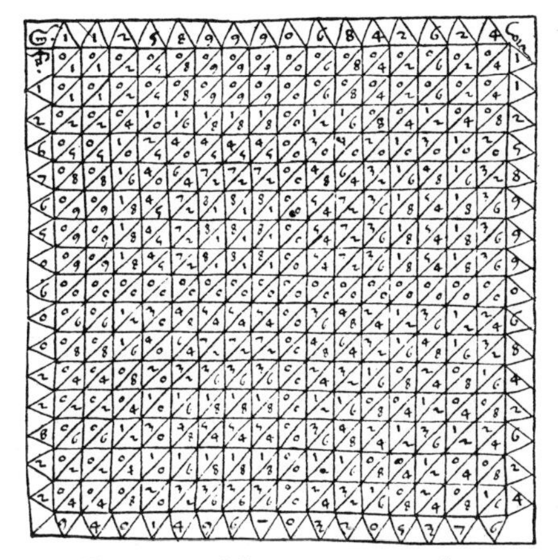
FIG. 229. MULTIPLICATION ‘PER QUADRATO,’ C. 1460

4. ‘El quarto capitolo del primo libro di queſto trattato/ Doue ſitratta del modo & hordine del ⓜchare e numerj.’ (F. 6, v.) The column form of the multiplication table is first given, and is followed by various methods of multiplying. The names ‘El berichuocholo’ (f. 10, v.) and ‘ꝑ quadrato’ (f. 11, r.) show the work to be Florentine rather than Venetian, and the handwriting and the numerous references to Florence confirm this fact. The method ‘per quadrato’ is shown in Fig. 229. Five pages of multiplication tables are given, such an elaborate treatment having been rather common in the arithmetics of that city. Division is not treated in Book I.

II. ‘El ſecondo libro del detto trattato. Nelquale ſi contiene la natura & proprieta De numerj. & prima come e diuiſo & aprj lontelletto.’ (F. 17, r.) This is divided into two chapters:

1. ‘El ꝑimo capitolo del 2o libro/ Nelqual ſitratta la natura che a īſe el nūō con diffinitionj ſapute.’ (F. 17, r.) This relates to such ancient classifications of number as odd and even, prime and composite.

2. ‘El ſecondo capitolo del 2o libro/ Doue ſitratta de numerj nominati per nomj apropiati alle fighure geometre.’ (Figurate numbers.)

III. 'El terzço libro della praticha dariſmetricha. Nelquale ſitratta el modo di partire ꝑ nūj. & ꝓima ī che forma e modo el detto libro e diuiſo.' (F. 23, r.) The subject of division is now taken up in three chapters:

1. By the use of the table, 'el modo di partire per gli numerj ſcripti ī ſulle librettinj.' (F. 23, v.)

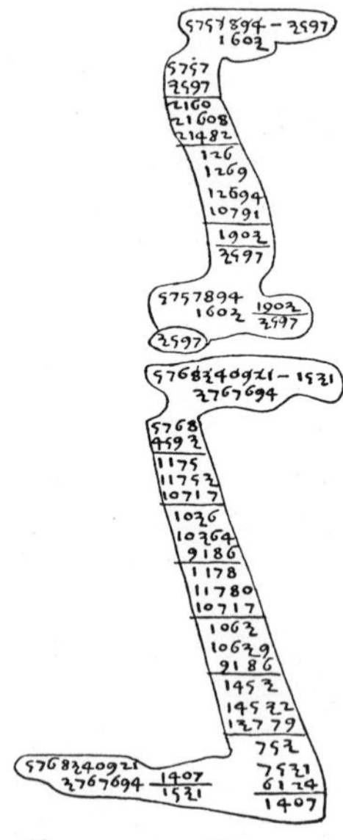

FIG. 230. FROM AN ITALIAN MANUSCRIPT OF C. 1460

2. The 'ripieghi' method (f. 28, v.), by the successive factors of the divisor.

3. Long division. (F. 32, r.) The author gives not only the galley method, but an early description of the method 'a danda,' substantially our present plan (see Fig. 230).

IV. This treats of common fractions. (F. 33, v.)

V. Ratios, 'quantita proportionalj.' (F. 46, v.)

VI. Mercantile arithmetic (f. 58, v.), with numerous references to the customs of Florentine merchants. Sixty-six large folios (132 pp.) are given to this book.

VII. The rule of false position: 'El ſeptimo libro diqueſto trattato nelquale ſitratta del modo delaſoluere de chaſi ꝑ loſemplice modo delchatain che ꝑ moltj ſidicono principij del chatain.' (F. 124, v.)

VIII. Simple and compound interest (f. 134, v.), with 2 pages (4 incomplete) of tables giving the interest on £100 for 1–21 yrs. at rates varying from '5 ꝑ 100. laño' (f. 140, v.) to '40 ꝑ 100. laño' (f. 150, v.). Equation of payments is also presented in this book (f. 152, v.), the dates ranging from 1458 to 1464.

IX. False position as treated by Leonardo of Pisa. (F. 170, v.) Leonardo is mentioned in Pl. IX.

X. Miscellaneous problems. (F. 176, r.) These are largely traditional examples, and include the hare and hound, finding numbers satisfying given conditions, problems about eating, the jealous husbands, and the testament complication.

XI. Proportion, based on Euclid V. (F. 225, r.)

XII. Algebra, 'della regola della algebra.' (F. 233, v.) Unfortunately few of the figures for this book were drawn. The treatment is rhetorical, practically no symbolism being used.

xm. .CCLxxbmj°.

Incomincia il ij° libro di q° trattato nel qual si contiene come e che modo s'asso
luono e casi p la regola de Algebra amucabale. Cp° la divisione del detto libro
Rendiamo gratie al altissimo. Cosi comincia el testo de lagha
bar arabico nella regola del geber laquale noi diciamo al
gebra. laquale regola dalgebra secondo ghuglielmo de luna
traslatore i posta i q° 7 nomi cioe. Geber. Elmelchel. El
chal. Elchelif. Elfatiar. Di ssarelbuzam. Eltermem. E quali
nomi secondo el detto guglielmo sono cosi i terpetrati. geber.
E quanto a dire recuperatione. i po che come p lo seguente si coprendera
nella recuperatione di 2 pte iguali s'asolue il caso. Elmelchel. E quato
a dire exemplo ouero asomigliamento i po che la absolutione de casi si trou
oua p absomigliare la quantita posta al caso dato. Elchal. E quato a dire
opositione p che di 2 quantita trouate luna e oposta alaltra. E q'ado no so
no 2 quantita oposte il caso e insolubile. Elchelif. E quato a dire dispo
sitione i po che benche lle 2 quantita oposte sieno e no abbino dispositione
auso delle regole lo caso sarebbe fuori delle regole e po abisogna le q'ati
ta disposte. Elfatiar. E detto differentia che differentia e i [illegible] nomi della
detta regola che no [illegible] differentia la regola sarebbe una e sarebbe
contro allo inpossibile. Di ssarelbuzam. E detto ragione i po che co rag
ione tutto si mostra e ragioneuole sono e casi p la regola absoluti. El
termem. E detto finitione i po che l fine ragioneuole della regola e
trouato q'ando p oppositione delle parti abiamo trouato [illegible] Elionardo
pisano nella terza pte del 15 cap° dice la regola della algebra amucha
bale e detta regola d'oppositione e di ristoratione cioe di ristoramento
Come chiaro si vedra pe casi [illegible] cacio che questa regola sia bene co presa
diuidero el detto libro i 3 cap°. Nel p° fia la dimostratione chiara p figu
re delle dette 3 regole. Nel s° fia come e nomi ouero propieta della de
tta regola abisogneuoli si co[illegible]hano. Nel terzo e ultimo certe regole sopa
le 6 passate e le 6 ancora dando 2 casi exemplari a ciaschuna regola.

El primo capitolo del ij° libro nel quale si chiarifica p figure geometrice
le 3 regole composte de lalgebra.

Al computamento della regola de lalgebra e amucabale 3 propieta
che sono i ciascuno nu° si co sidera. E la p° e quando si co sidera p radice
e chiamiallo cosa. El s° e q'ado si co sidera p quadrato e chiamasi censo. E la
terza considerandolo p suo esse se plice e allora lo diciamo drame. E il luo
go delle q'atita la drama e uilissima quasi come il punto nella linea.

La cosa adunque e ra° del censo onde lo censo e quadrato della cosa e p che
ogni quadrato e fatto della sua ra° i se [illegible]hata lo censo nasce della cosa
i se [illegible]hata i npo che e sua ra°.

Le drame sono dette nu° semplice i po che lle si co pongono solamente du
nitadi ma lla cosa e il censo sono conposte i loro essere e po le drame
no ano rispetto alle cose ne alli censi cioe a dire del nu° sono sapute lu
nitadi ma della cosa ne del censo no sono sapute lunitadi e po diciamo dra
ma essere nu° semplice e lla cosa e ra° cioe quello che i se [illegible]hato fa
quel nu° e il censo e quello che nasce della cosa cioe della ra° i se [illegible]ha
ta lo nu° semplice cioe drame e q'atita che co parole si dice i tal modo che
no a saputa proportione alle ra° ne llo censo.

Quando della ra° cioe cosa o del censo si uoglono sapere lunitadi e di bisog
nio che i fra lloro s'aguaglino lo quale aguagliamento i 6 modi si puo fa
re de quali modi e 3 si chiamano semplici e gli altri 3 conposti. Onde p°
ma diremo degli aguagliamenti semplici.

PLATE IX. FROM AN ANONYMOUS MANUSCRIPT, c. 1460

XIII. Algebra continued, 'la regola de Algebra amucabale.' (F. 279, r.) This is a very interesting treatment of the subject 'ſecondo ghuglielmo de lunis' and 'Lionardo piſano,' and it throws some light upon algebra as studied in the fifteenth century. (See Plate IX.)

XIV. Algebra continued (f. 295, v.), according to Master Biagio ('certj caſi che ſcriue m.º biagio nel ſuo trattato di pratica') of 1340, Master Gratia de Castellani ('ſecōdo che ſcriue m.º gratia de caſtellani), and Leonardo of Pisa.

XV. Algebra continued (f. 312, r.), according to certain 'maeſtrj antichi,' viz.: 'Maeſtro paolo,' 'm.º Antonio,' 'm.º giouañj,' 'leonardo piſano,' 'm.º biagio che circha al. 1340. añj morj,' 'm.º paolo fiorj che circha al. 1360. duro,' 'm.º michele padre di m.º mariano,' 'm.º lucha,' and 'un altro m.º biagio.'

XVI. This is missing, the manuscript ending with Book XV not quite completed. On f. 1, r., there is mentioned the 'ſedecimo e ultimo' chapter, and this might have contained the name of the author had the work been finished.

EUCLID. Latin MS., c. 1460.

See p. 11.

Title. None. The first book of the Elements.

Colophon. '¶Explicat prim⁹ liber euclidis cū ꝯmēto cāpani.' (F. 14, v.)

Description. 4°, 14.7 × 21.2 cm., the text being 9.9 × 15.8 cm. 22 ff., 18–40 ll.

This manuscript of the first book of Euclid, with the commentary by Campanus, has been included on account of the forms of the numerals used. It is written in a German hand of c. 1460, and is bound (ff. 1–14) with the two manuscripts next described.

ANONYMOUS. Latin MS., c. 1460.

Title. None. A treatise on the calendar.

Description. See the preceding manuscript. This is ff. 15–17 of the volume.

This manuscript, bound with the Euclid just described, was probably written by the same hand. It is a brief treatise on the calendar, and was intended, as usual, for the Church schools. In it occurs the date 1460.

ANONYMOUS. Latin MS., c. 1460.

Title. None. On the Quadrivium.

Description. See p. 463. This is ff. 18–22 of the volume.

This is a general treatise on the quadrivium, and therefore contains a chapter on arithmetic.

BENEDETTO DA FIRENZE. Italian MS., c. 1460.

A Florentine arithmetician of the first half of the fifteenth century.

Title. 'Inchomincia el trattato darifme//tricha efpelialmēte quella p̃te // che e fotto pofta alla mercatātia // e comminciando alnome didio.' (F. 11, r.)

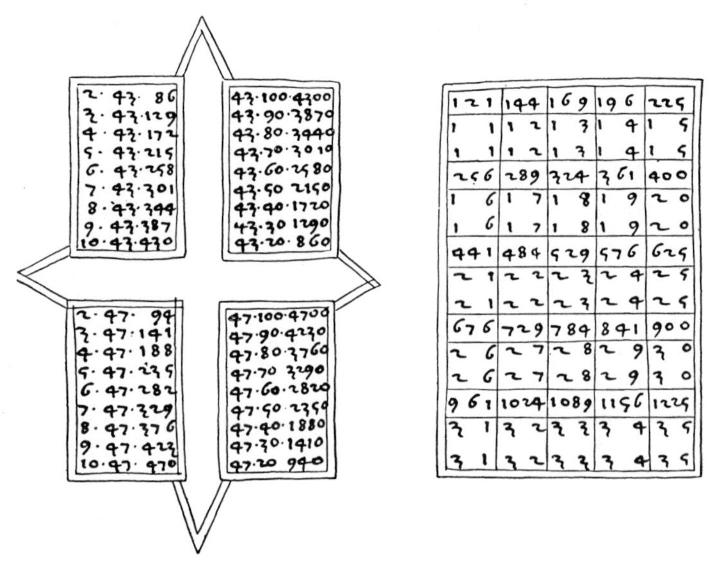

FIG. 231. TABLES FROM BENEDETTO DA FIRENZE

Description. 8°, 8.3 × 11.7 cm., the text measurement varying. 2 ff. missing + 348 ff. unnumb. = 350 ff., 20 ll. Manuscript on parchment.

This is a parchment codex written about the middle of the fifteenth century. It lacks the first two folios and possibly also a few at the end. It is one of the best examples of the mercantile arithmetics in Italy preceding the printed works. The author begins, as usual, with several pages of tables (see Fig. 231), the multiplication table including the prime numbers below fifty. There is also a table of squares and one for the multiplication of compound numbers. Benedetto presupposes that the reader is able to perform the fundamental operations with integers, and he begins at once with operations on compound numbers and fractions. The applications include exchange, partnership, and equation of

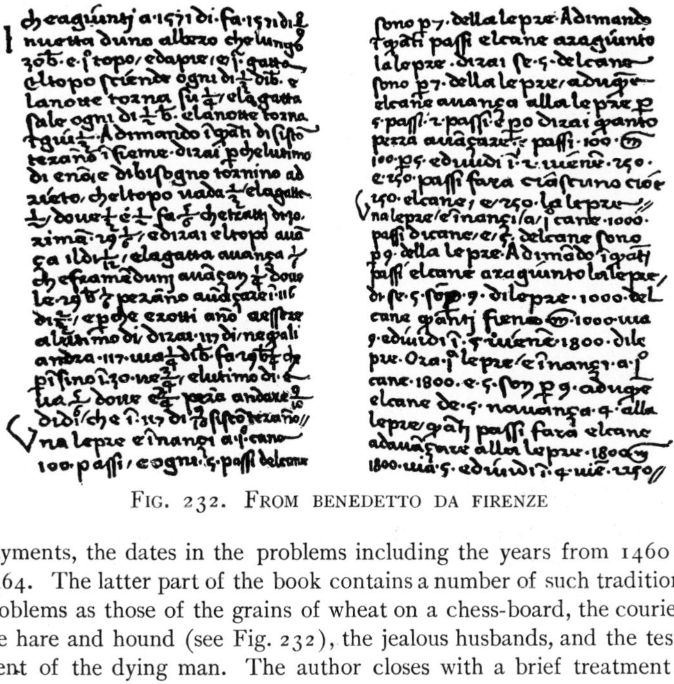

FIG. 232. FROM BENEDETTO DA FIRENZE

payments, the dates in the problems including the years from 1460 to 1464. The latter part of the book contains a number of such traditional problems as those of the grains of wheat on a chess-board, the couriers, the hare and hound (see Fig. 232), the jealous husbands, and the testament of the dying man. The author closes with a brief treatment of mensuration.

ALBERT OF SAXONY. Latin MS., 1462.

See p. 9.

Title. 'Tractatus proportionum. Incipiūt ꝑportioēſ cōpote a dn̄o albertutio.// [ ] Roportio coītr dā ē duoꝝ ꝯperitoꝝ in aco t'rio unico // ad inuicē hītudo.' (F. 1, v.)

Colophon. 'Explicāt ꝑportioēſ ꝯpote // ꝑ reuēdo mr̄o alberto deſiſona // Finis //' (F. 12, v.)

Description. 4°, 14.5 × 21.8 cm., the text being 11.8 × 13.8 cm. (varies). 19 ff., 23 ll. Bound with the Vergerius and Lullius mentioned on pp. 456, 457.

This is the treatise on proportion that was printed without date, probably at Venice c. 1478 (p. 9). It is followed by another treatise in the same hand, 'De latitudinibus formarum,' which bears the date 1462.

ST. BERNARD OF SIENA. Latin MS., 1469.

See p. 452.

Title. None. Sermons of St. Bernard (Bernardinus) of Siena.

Description. 4°, 15 × 20.6 cm., the text being 11.5 × 14 cm. 130 ff. (12 blank), 28 ll.

Editions. The works of St. Bernard were first printed in Venice in 1591, 4°, and again in Paris in 1636, 2 vol., fol. Some of his sermons were, however, printed in Florence, and some in Venice, in 1495.

This manuscript, containing several of the sermons of St. Bernard, was copied by one Eustachio de Feltre in 1469. Two dates are given showing the completion of parts of the work, September 22, 1469, and October 6, 1469. For the reasons for including this manuscript, see p. 452.

ANONYMOUS. Italian MS., 1473.

Title. 'Trattato della Arithmetica.' (F. 1, r.)

Description. Fol., 16.7 × 23.7 cm., the text being 10 × 16.2 cm. 190 ff., 35 ll.

This treatise was composed, as the first folio states (Fig. 233), in the year 1473. In the examples in partial payments the dates given are about 1490, and this particular manuscript may have been copied about that time. There is, however, one example with the date 1392, which would seem to indicate that at least part of the book was copied from some earlier writer. There is also (f. 183) a brief treatment of the calendar with two dates 1443. On f. 180 there are also the dates 1452 and 1453. In the margins some sixteenth-century dates have been added by a later owner.

The work is beautifully written on vellum. The first part consists entirely of arithmetic, treated from the mercantile standpoint. That it is a copy of an earlier work also appears from the fact that a few pages

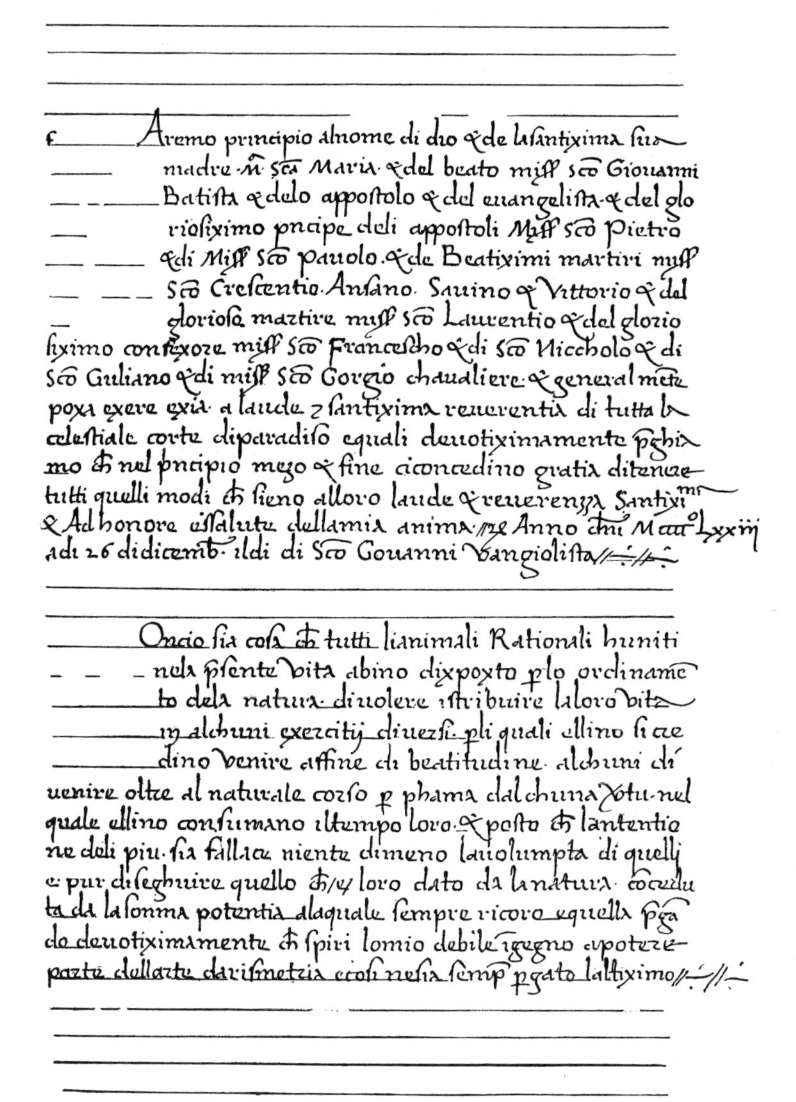

Aremo principio alnome di dio & de lasantixima sua
madre · M · Sca Maria · & del beato miss sco Giouanni
Batista & delo appostolo & del euangelista · & del glo
riosiximo pncipe deli appostoli Miss sco Pietro
& di Miss sco pauolo · & de Beatiximi martiri miss
Sco Crescentio · Ansano · Sauino & Vittorio & del
gloriosa martire miss sco Laurentio & del glorio
siximo confexore miss sco Francescho & di sco Nicchola & di
Sco Guliano & di miss sco Gorgio chaualiere · & generalmete
poxa exere exia · a laude z santixima reuerentia di tutta la
celestiale corte diparadiso equali deuotiximamente pghia
mo ch nel pncipio mezo & fine ciconcedino gratia ditenere
tutti quelli modi ch sieno alloro laude & reuerenzza Santiximi
& Ad honore essalute dellamia anima · Anno dni Mcccc Lxxiij
adi 26 didicemb · ildi di Sco Gouanni Vangiolista

Oncio sia cosa ch tutti lianimali Rationali huniti
nela psente vita abino dixpoxto plo ordiname
to dela natura diuolere istribuire laloro vita
in alchuni exercitij diuersi pli quali ellino si cre
dino venire affine di beatitudine · alchuni di
uenire oltre al naturale corso p phama dalchuna vertu · nel
quale ellino consumano iltempo loro · & posto ch lantentio
ne deli piu · sia fallace niente dimeno lauolumpta di quelli
e pur diseghuire quello ch/e loro dato da la natura · conceedu
ta da la somma potentia alaquale sempre ricoro equella pga
do deuotiximamente ch spiri lomio debile igegno apotere
parte dellarte darismetria ecosi nesia semp pgato laltiximo

FIG. 233. FIRST PAGE OF THE 1473 *Trattato*

are left blank for subsequent insertions, the scribe having written : 'Hic dificiunt quatuō // chartæ q̄ non ſunt // ī exemplario' (f. 65, v.) The second part of the work consists of practical geometry and mensuration (ff. 103 to 128). The third part (f. 129) relates to irrational numbers. On f. 135 begins the 'Regole de la Arcibra,' a chapter on algebra. There is nothing to indicate the name of the author or the copyist.

ANONYMOUS. Latin MS., c. 1475.

Title. 'De duplirj Arti Viſorir.' (F. 1, r.)

Description. 4°, 15 × 21.8 cm., the text being 10 × 17 cm. (varies). 11 ff. + 1 blank = 12 ff., 40–45 ll. Written in a German hand of c. 1450–1475. Bound with the manuscript of Mohammed ibn Musa (p. 454).

This manuscript gives some account of gauging, and closes with a brief treatment of trigonometry. The mediæval numeral forms are used throughout. A copy also appears in the 1501 manuscript described on p. 480.

LUCA DA FIRENZE. Italian MS., c. 1475.

Maestro Luca da Firenze lived in the fifteenth century, and was the son of the celebrated Florentine arithmetician Matteo, who was born in the fourteenth century.

Title. 'Inprencipio darte dabaco.' (F. 1, r.) F. 2 begins: 'Inprencipio darte dabacho ſecondo loſtile dinſegniare del ma-// eſtro luca di Matteo da ffirençe.' (Fig. 234.)

Description. 4°, 16.8 × 23.3 cm., the text being 11.7 × 13.8 cm. 46 ff., 29 ll.

Although Fabbroni's *Storia dell' Università di Pisa* (I, 97) says that Luca's son Giovanni went with Lorenzo dei Medici to the University of Pisa in 1515, thus putting his birth about 1495 and Luca's birth perhaps about 1450, I feel that either this is incorrect, or it is to some other Giovanni and Luca that he refers. One of the best evidences of the date of an arithmetic is found in the dates given in its problems. Authors usually mention years that are not remote from the time when they write, and in the examples in equation of payments (f. 29, v.) Luca uses dates from 1410 to 1441. I therefore think that either he copied a problem from his father (Matteo), or, what is more probable, he himself wrote about that time, say c. 1425, and that in either case this Giovanni was a descendant but not a son.

The arithmetic resembles numerous others written in Florence about this time, such as those described on pp. 443 and 464. The author begins with the fundamental operations and follows these by a treatment

FIG. 234. FIRST PAGE OF LUCA DA FIRENZE

of fractions and denominate numbers, closing with a series of applications to the business problems of the time. Subjects like equation of payments, partnership, and exchange are given the most attention. The handwriting indicates that the manuscript is a copy made about

1475–1500. The symbols in Fig. 235 may also throw some light on the disputed origin of our symbol for dollars, \$. No one seriously considers such fanciful theories as the combination of U and S, or the Spanish banner about the Pillars of Hercules. The symbol first appears in print in *The American Accomptant*, by Chauncey Lee (Lansingburgh, 1797), but in a very different form from that now used. It was common among merchants for some time before it was cast in type form, for a note in one of the early American arithmetics says that the symbol was in use, but that there was no type for it. The third edition of Pike's arithmetic (1798) uses m., c., d., D., and E., for mills, cents, dimes, dollars, and eagles, but Daboll's *Schoolmaſter's Aſſiſtant* (4th ed., 1799, p. 20) gives the symbol \$ very nearly in its present form. Now whence come symbols like this? If they are invented *de novo* it is usually easy to find their first appearance, as in the case of symbols like π, e, and i (for $\sqrt{-1}$). But mercantile symbols usually develop slowly, like £ from *libra*, /- from the old *s* (ſ) from *soldi* or *solidi*, and *d* from *denarii*. So it is probable that \$ was simply developed from some earlier symbol of value, such as that for pounds or reales. Now the symbol for pounds (libra, lire) has various forms, appearing in England as £ or lb., but generally in Italy as L, or lb. with two bars across. The former is seen in Fig. 243, from a manuscript of 1545, and the latter in Fig. 216 from Dagomari's work. The latter form, in the fifteenth and sixteenth century manuscripts, appears as practically our dollar sign, as shown in Figs. 235, 237, 246, and it is not improbable that our early American merchants used it for the new unit of value, the dollar, just as the /- for shilling is still used in many parts of our country for $12\frac{1}{2}$ c., although the original meaning is entirely lost. The symbol came into general use in printed books between 1800 and 1825.

ANONYMOUS. Latin MS., 1476.

Title. 'Computus cyrometralis.'

Description. 4°, 15.5 × 20.9 cm. A volume of 180 ff., consisting of four treatises, the other three being elsewhere described.

The first of the four manuscripts is a computus of the usual fifteenth-century type. (See Fig. 236.) In it the Latin original of the verses beginning 'Thirty days hath September' appears (f. 6) as follows:

'Ja. mar
ma. iul. aug. oc. de. deca-
trib9. et. vno. alij. trigēta.
ſʒ. februus. octo. viginte.'

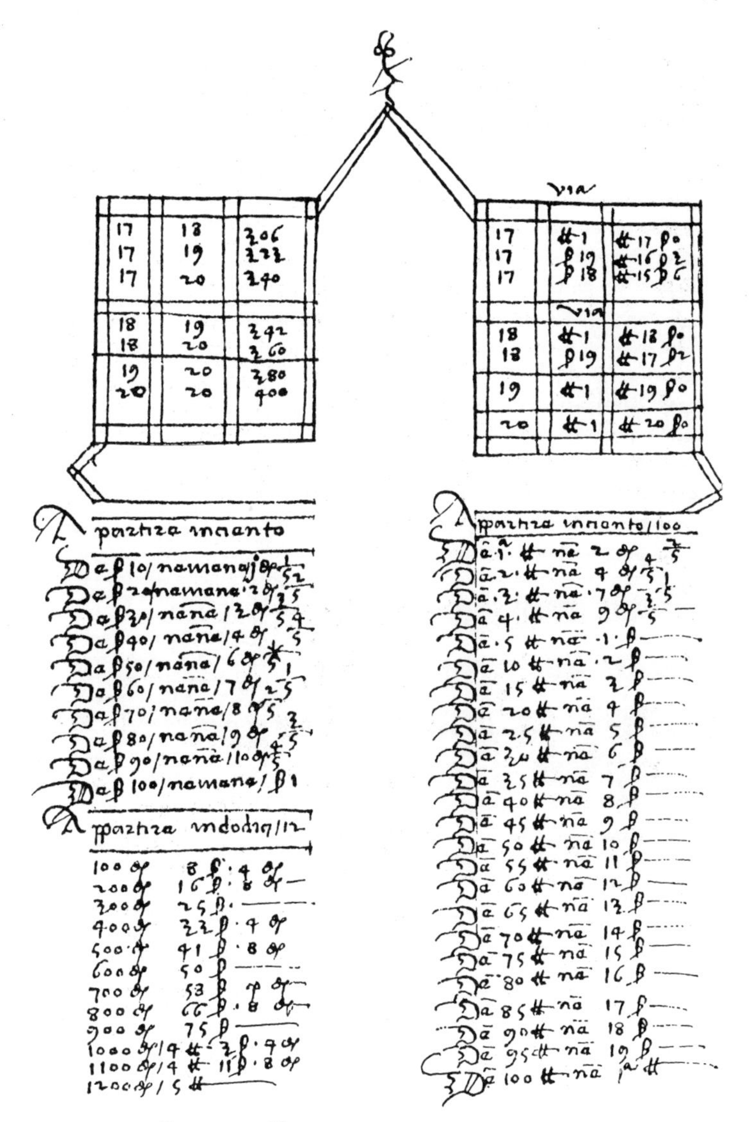

FIG. 235. TABLES FROM LUCA DA FIRENZE

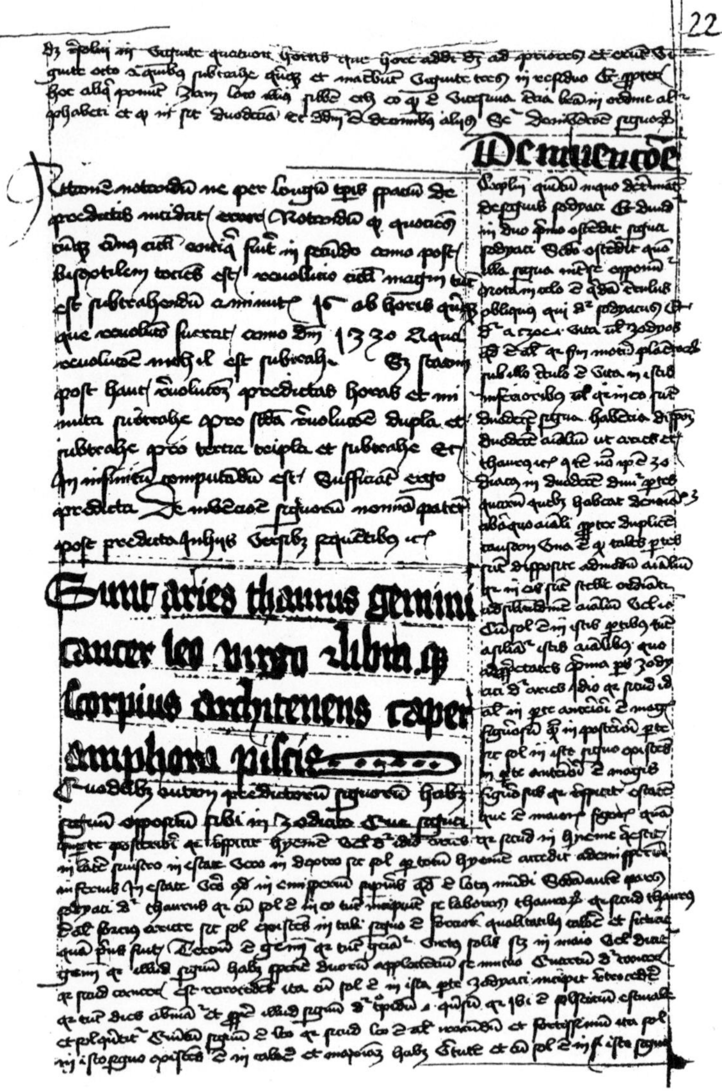
22

Sunt aries thaurus gemini
cancer leo virgo libra
scorpius architenens caper
amphora pisces

FIG. 236. FROM THE 1476 *Computus cyrometralis*

(See also p. 33.) It does not start here, however, but goes back at least as far as Sacrobosco (13th cent., see p. 31), in whose Computus (see p. 451) it appears as follows:

'Sep. No. Iun. Ap. triginta dato, reliquis magis uno.
Ni ſit Biſſextus, Februus minor eſto duobus.'

(1545 edn., f. K 6.)

JACOBUS OBERNHEYM. Latin MS., 1476.

A Nürnberg computist of c. 1431.

Title. 'Computus norembergensis.'

Colophon. 'Anno dnī 1476 completo.'

Description. See the preceding manuscript, with which it is bound.

This is a German computus written by the same hand as the preceding. It is evidently a copy of a work written in 1431, for the following statement appears on f. 65: 'Anno dnī m°cccc°xxxi° quarta feria p⁹ judica finit⁹ ē liber iſte per me Jacobū Obernheym.'

ANONYMOUS. Latin MS., 1477.

Title. None. A calendarium. The first folio begins 'Albēto belo caro decagolo ca⁹ nobilis.'

Description. 4°, 15.4 × 20.4 cm., the text being 10.5 × 16.4 cm. 14 ff., 17 ll. Bound with the Vergerius and Lullius described on pp. 456, 457.

This is a manuscript of 14 folios on the calendar, and has two dates, M̊.cccc°.lviij° and M̊.cccc°.lxxvij, on the first folio. It also has some lunar tables for 1364–1381, the treatise having probably been copied about 1458–1477 from another one of a century earlier.

NICOLÒ DE ORBELLI. Latin MS., 1478.

Title. 'Incipit op⁹ fratris (?) dorbelli ſup . . .' (F. 1, r.). 'Incipit. Matematica' (F. 132, r.).

Colophon. On f. 126, v.: 'Explicunt Scriptā frs̄ nicolai de orbellis // doctoris eximini ſup . . . // Deo gratias et xpō ihū Amen // M°. 478.'

Description. 6°, 10.5 × 15.6 cm., the text being 9.2 × 11.9 cm. 270 ff. (6 blank), 41 ll. Latin MS. on paper, 1478.

The first part of the book is devoted to dialectics and logic. The mathematics begins on f. 132, r., and consists of two folios on the theory of numbers and three on geometry. On f. 140, r., begins a treatise on philosophy, containing (f. 184, v.) a drawing of 'Johs ſcotus docts ſubtilis.' On f. 193, r., begins a treatise on astronomy: 'Hic // Incipit liber de // celo & mundo.' The rest of the work relates to science in general. (See p. 23, 1485.)

LEONARDUS MAYNARDUS. Latin MS., c. 1488.

A fifteenth-century mathematician of Cremona, Italy.

Title. 'Leonardi Cremonenſis artis metri//ce pratice compilatio. Primus tractatus.' (F. 1, r.)

Description. 4°, 17.2 × 24.4 cm., the text being 7 × 10.8 cm. 24 ff., 25 ll., figures on the margins.

This is a treatise on trigonometry, and has been included in this catalogue because of the arithmetical nature of some of the problems.

Favaro's careful investigation of the time when Maynard lived still leaves the matter in doubt. He may have lived in the latter part of the fifteenth century, or he may possibly have been the 'Leonardus de Antoniis de Cremona, ordinis minorum, bacalarius' who lived early in the fifteenth century.

In the catalogue of the Boncompagni sale another Latin manuscript of Leonardo is given as 'in pelle di 33 carte membranacie del secolo XIV.' If this is correct, which is doubtful, Leonardo must have lived before 1400. See also Eneström in *Bibliotheca Mathematica* IV (3), p. 290, and Favaro in the same journal IV (3), p. 334. The latter gives a bibliography, with some quotations.

In 1902 M. Curtze translated and edited one of the three manuscripts of this work known to him. This is in the Venetian dialect, and belongs to the University library at Göttingen. He also consulted the two Latin manuscripts formerly belonging to the Boncompagni library, of which this is the later by a few years. This manuscript belonged to the advocate Cav. Carlo Morbio, in Milan, before Boncompagni secured it. A note on f. 15, r., shows that in 1655 it belonged to Bonifacio or Joseffo Aliprandi. Another note, on f. 1, refers to a passage in a work by Franciscus Arisius, printed at Parma in 1702. This passage is as follows: 'LXXXVIII (i.e. 1488) Leonardus Maynardus Insignis Astronomus, Physicus et Mathematicus, cuius opusculum M. S. Mediolani servatur, mihi indicatum ab eruditissimo Viro Lazaro Augstino Cotta I. C. amico meo nequaquam satis laudato, cui est initium: (Here follow the first few lines of this work, so nearly identical with the manuscript as to show that Arisius probably had this very one at hand.) ... Fuit ante Blasium *Leonardus Maynardus*, qui suo tempore non tantum inter nos, sed etiam inter omnes in iis studiis tenuit principatum.' A similar passage appears in the manuscript *Biografia Cremonese* of Vincenzo Lancetti, now in the civic library at Cremona, from which we may infer that he lived before Battista Piasio, of a noble Cremonese family, a philosopher, physician, and astrologer, who flourished about 1500. This manuscript of Maynardus is described (no. 254) in Narducci's catalogue of the Boncompagni manuscripts (Rome, 1862).

ANONYMOUS. Italian MS., c. 1490.

Title. None. A manuscript on elementary mathematics. 'Choncio ſia choſa che ſono noue fighure nellaba//cho.' (F. 1, r.)

Description. 4°, 14.5 × 21.5 cm., the text being 9 × 15.2 cm. 158 ff. (5 blank), 26 ll.

The arithmetic is of the ordinary commercial type, and includes the fundamental operations together with a considerable range of business applications. It follows the general style of the Florentine arithmetics and uses the Tuscan *berichuocholo* instead of the Venetian *scachero* to designate our present multiplication: 'Volendo multiprichare .2. nūmeri p berichuocolo' (f. 17). The dates in the problems in equation of payments indicate that it was written about 1490 (ff. 83–86–129). The per-cent symbol % here appears as p c̊, p c°, p c°, as well as p 100 (ff. 84, 86, 134, v. See Fig. 237).

The second part of the work (f. 91) relates to algebra: 'Qui aprſſo ſcriuerio lareghola dellarabre (dell' algebra). m°ochabiln° (e muqabala).' The work is rhetorical, there being practically no symbolism employed. This part of the work closes with the words, 'Voglia hora fare fine enondire piu ſopa a que//ſta reghola delagebrem°ghabile ... (f. 109, r.).

The geometry begins on f. 109, r., and is confined to simple mensuration. The work closes (ff. 123–153) with a series of miscellaneous problems.

ANDREA DI GIOVANNI BATTISTA LANFREDUCCI.
Italian MS., c. 1490.

An officer of the Republic of Pisa in 1505.

Title. None. A treatise on arithmetic. 'Choncio ſia choſa che ſono noue figure nellabacho.' (F. 3, r.)

ildi de partire ilghuadagnio delmese.in.30
equello chenevíene tanto ghuadagnia ildi
equando ai veduto quello che ghuadagnia
ilsuo capitale.lano eilmese eildi vedi quanto
tempo entra nelmerito della detta partita
equello tempo chenescie. pollo sopra allaprima
partita equello chessa chosi viene arechato. la
sopradetta partita avendie. hora seghuendo. la
detta ragione dobbiamo vedere quanto tempo
e dalla prima partita alla 2ª coe. da 25 diluglo
nello 89 insino adi 4. disettenbre nello 89
cui uno mese e 9 di a.10.p c°. coe.a. 2 d p lb ilme
se merita 461 lb 8ß 4 d ghuadagniano in j°
mese e.9 di 5 lb 0ß10d. ora dei vedere quanto
tempo e dalla prima partita alla terça coe dadi
25 diluglio nello 89 insino adi 28 di dicenbre
nel 90. cui. 5.mesi.4 di.ora dei meritare .602 lb
16ß 8d in 5 mesi 3 di a 10 p c° coe a.2 d p lb
ilmese ghuadagniano. 25 lb 12ß 5d.hora dei
vedere quanto tempo edallaprima partita.alla
$\frac{1}{4}$ coe dadi. 25.diluglo.nello.89.insino adi p°
dimarço nello 90 cui. 7 mesi e.6.di. hora dei
meritare. 943 lb 0ß 4 d in.7 mesi e. 6 di a.2d
p lb ilmese ghuadagniano 56 lb 11ß 7d. hora
dei vedere quanto e dalla prima partita alla $\frac{1}{5}$
coe dadi.25 diluglio nello 89. p insino adi 5 di
luglio nel. 90. cui. 11 mesi. 10 di hora dei meri

Fig. 237. From the anonymous manuscript of c. 1490

Colophon. 'Qvefto Libro. afcritto. difua. ꝓpia ajano. andrea. Lanfreduccj.' (F. 96, v.)

Description. 4°, 16.8 × 23.4 cm., the text being 12.1 × 16.5 cm. 96 ff. (15 blank), 29 ll.

This book is an Italian commercial arithmetic of the Florentine type written about 1490. After a brief treatment of notation the author takes up the 'Librettine minore,' or the smaller multiplication table to 10 × 10, following this by several examples. He then gives (f. 9, r.) 'le libretine maggiore' or larger table, with multiplication 'ꝑ quadrato' and 'ꝑ berichuochulo,' these names showing the work to be Tuscan. Division is followed by a large number of problems, per cent (il c̃° and ꝑ 100) playing a large part. The problems in equation of payments show the book to have been written between 1489 and 1491 (ff. 72–74). 548

JOHANN NEWDÖRFFER. German MS., 1492.

A Nürnberg Rechenmeister, c. 1450–1500.

Title. 'Hanns Dimpfel//(Multiplication table)//Johann Newdörffer Rechen-//maifter vnd Modift zu N.// 1492.' (F. 1, r.)

Description. 8°, 10.5 × 13.9 cm. 11 ff., 20 ll. Written on vellum.

This is a beautifully written primer apparently done under the guidance of one of the celebrated Rechenmeisters of the Newdörffer family of Nürnberg. It was written in 1492 and is evidently the work of a beginner in commercial arithmetic. It opens with the addition and multiplication of compound numbers, and this is followed by the rule of three, with some applied problems.

The name Hanns Dimpfel is doubtless that of the pupil whose work was done under the direction of Newdörffer. As to the latter, it is not improbable that he was the father of the celebrated Schreib- and Rechenmeister Johann Newdörffer, the founder of the German calligraphy, who was born at Nürnberg in 1497, and died there November 12, 1563, and whose son Johann (b. February 22, 1543, d. October 28, 1581) was also a well-known Rechenmeister.

ANONYMOUS. Latin MS., c. 1500.

Title. 'De com̄utata ꝓporn̄e.' (F. 14.)

Description. Fol., 14.5 × 21.5 cm., the text being 11.5 × 19 cm. This is the second part of the 1442 manuscript, ff. 14–32 (p. 458).

This is a Latin treatise on arithmetic, apparently written about 1450–1500. It relates almost exclusively to Boethian ratios and figurate numbers. It closes with a few pages on the circle, including the following: 'Rem nouā mirabileȝ. quadra//tura; circuli velut iſcritabileȝ.// apud doctoſ ppli' olim. ſ. ſabilē.// pure cernūt oculi vere demrā//bilem nūc ī fine ſeculi' (f. 30, v.); '❡de quadraturā circuli' (f. 32, r.). The table in Fig. 238 shows some work in series such as is common in arith-

1	.2.	.3.	.4.	.5.	.6.	.7.	.8.	.9.	.10.	11
4	.8.	.12.	.16.	.20.	24.	28.	.32.	.36.	.40.	44.
3	.6.	.9.	.12.	.15.	.18.	.21.	.24.	.27.	30	33

FIG. 238. FROM *De cōmutata pporne*, c. 1500

metics of this period. Fig. 239 shows the multiplication table as commonly seen in the Boethian arithmetics (compare p. 26). It also shows (on line 1) the use of the Roman numerals in connection with the Arabic, and (on lines 9 and 10) the absurdly long Latin names for ratios.

ANONYMOUS. c. 1500.

Description. Fol., 14.5 × 21.5 cm. It is bound with the preceding manuscript, but is in a later hand. It consists of ff. 33–38 of the volume.

The manuscript consists of an interesting set of drawings, including one of an astrolabe (f. 33) and several horoscopes (f. 35). The latter serve to fix the date of this portion (c. 1500) and the country (Hungary) in which it was written. They include (f. 35, r.) a horoscope cast at the birth (1456) of Ladislas or Uladislas VII, possibly by Peurbach (see p. 53), and one at his coronation as king of Hungary on September 18, 1490, mentioning his coronation as king of Bohemia on August 25, 1471, and his election as king of Hungary on July 15, 1490.

ANONYMOUS. Latin MS., c. 1500.

Description. Fol., 14.5 × 21.5 cm. It is bound with the preceding manuscript, is possibly in the same hand as the horoscopes, and is certainly on the same paper. It occupies ff. 39–44 of the volume.

This is a treatise on the mensuration of the circle and dates from c. 1500. It has some bearing upon the metrical computations of the time.

Vel dup^x sesquiqrtptis quintis ut xiiii ad v ut septimas ut cxviii ad vii et sic de
variis et multis huius nūi spēb; que variato venit ut ex pte multiplicitatis ut suptcla
ritatis ut vtriusq. Inveniūt aūt et hos nūos aliqui pmi ut minu qui radices dicūr
alioz i quib; ptes sup multiplicitatē excrescentes unitatis sunt. Alii vo sexti
quoz ptes sup multiplicitatē habundantes nūi sūt. que diligens inquisitor q̃ of mo
potit p ea que dicta sunt agnoscere Similr et qui sunt nūi minoris in
equalitatis sub dictis otenti nūis et quali vocari debeat. q2 sic illi qui
dr maioris inequalitatis, nisi q addigi debet eis hec ppositio sub. ut ex
utraq pte ut saltem ex alia ut dicat aliquis nūs sub dup^x sbiptiis ut dup^x
subsbiptis ut sub dup^x subsiptis et sic de aliis Proprietates hoz nūoz
alique tacte sunt et multe alie tangi possēt qui illas scret et tn hic inmorari
vellet. Similr et hij nūi quali pcreet. Ut tn aliqd de hoc appeat et redu
cant ad memoria omnis tacte quique spēs maioris et minoris inequalitatis
poneda est hic quedā valde nobilis nūoz desciptio qua dat nob boe aris
sue libro pmo In qua multa pnt speculari ut q multiplicatis cetis inequali
tatis spēs nalr antecedat. Similr et spēz eius ordo et pcreatio necno et
aliaz ut suppti viii supptictui multiplicui sypti viii et multiplicui supticeui.
Adhuc pot ibi videri qualitas tetragonoz et longilateroz nūoz et alia
multa. Est aut desciptio illa talis.

Expo porte desciptois.

ordo sm lati^ne i primus — Secūda unitas

1	2	3	4	5	6	7	8	9	10
2	4	6	8	10	12	14	16	18	20
3	6	9	12	15	18	21	24	27	30
4	8	12	16	20	24	28	32	36	40
5	10	15	20	25	30	35	40	45	50
6	12	18	24	30	36	42	48	54	60
7	14	21	28	35	42	49	56	63	70
8	16	24	32	40	48	56	64	72	80
9	18	27	36	45	54	63	72	81	90
10	20	30	40	50	60	70	80	90	100

primus ordo sm longi^ne decimus — ordo sm longi^ne decim^9 — prima

Secūda unitas ordo sm lati^ne decim^9 — prima

Intacta de scripta duo notandi sunt. nūoz ordines unus sm longi^ne qui a sursum i deorsum descendit Alius sm lati^ne qui lateralr a sinistris i extrem tedit. Et illoz quilibet x i nūos cotinet ordines Primus ordo sive attēdat sm longi^ne sive sm lati^ne ab

unitate icip et i denariu; nūi tiatur. Sunt etm i illo pmo ordine nūi sm dispoi

FIG. 239. FROM *De comutata pporñe*, c. 1500

The fifth portion of the volume (ff. 45–52) consists of a set of Latin, and the sixth (ff. 53–58) of a set of Italian verses. The seventh (ff. 59–60) contains a brief reference to astrology. The rest of the book is of a literary or astrological character. There are, in all, thirteen different manuscripts in the volume. For details concerning the non-mathematical portions, see Narducci, l. c., p, 31.

ANONYMOUS. Italian MS., c. 1500.

Title. 'Regole per far Orologi da sole // con le sue Figure.'

Description. 4°, 16.8 × 22.3 cm. 22 ff. Bound with the Vergerius manuscript described on p. 456, but written in a later hand.

This is an Italian treatise on dialing, written in a fine hand of the sixteenth century, with well-executed figures. Several other manuscripts, not of a mathematical character, are also bound with it.

ANONYMOUS. Latin MS., 1501.

Title. None. A treatise on mensuration.

Colophon. There is none, but one of the folios bears the date '1501 adj ſeptembr In Nürnberg.'

Description. 4°, 15 × 21.8 cm., the text being 9.8 × 17.1 cm. 89 ff. unnumb. + 7 blank = 96 ff., 27–30 ll. Latin MS., written in a German hand. Bound with the manuscript of Mohammed ibn Musa (p. 454).

This is a treatise on mensuration, of no particular merit save as it shows the nature of the work at the opening of the sixteenth century. It contains a copy of the c. 1475 manuscript described on p. 468. 497

STEPHANO DI BAPTISTA DELLI STEPHANI DA MERCATELLO. Italian MS., Mercatello, c. 1522.

An Italian teacher, born at Mercatello, and living there in 1522. He was a pupil of Paciuolo.

Title. 'Svmme // Arismetice.' (F. 1, r.) 'Stephano.D.B.Dellistpha // ni.damercatello.atvtti.qve // ili.liqvali.in arte.mercan // tile.exercitare.sidilectano.' (F. 1, v.)

Description. 8°, 13.5 × 20.7 cm., the written part being 9.1 × 14.3 cm. 153 ff. numbered (5 blank), 29 ll. Italian MS. on paper, except f. 1, which is on vellum. Written, as the problems show, at Mercatello, c. 1522.

This is a manuscript on commercial arithmetic, unusually complete in its applications, and also unusually well written. It is of the general Florentine type, but, as appears from a date on f. 101, r., was written at Mercatello, a town south of Ferrara and east of Florence. The name of the author appears not only in the dedicatory epistle, but also at the end of a bill of exchange dated '1522 A di./25/maggio in M^llo' (Mercatello), in the latter case as 'Stefano di Bap^ta Stefanj fʒ.' The examples in equation of payments are dated 1371–72, showing that these were copied from some earlier work, and in fact they were taken from Paciuolo ('Diſtinctio nona, tractatus quintus'), who in turn borrowed them from some predecessor.

Stephano states in his dedicatory epistle that he was a pupil of Paciuolo, and that he is chiefly indebted to him for his material: 'et maxime dal mio R^do. et ex^te affme & p̄ceptor'. M°. Luca dal Borgo.' This is quite evident on comparing certain passages; for, while Stephano does not usually copy his master verbatim, there is often a great similarity between them, and sometimes (as in the chapter 'De le. 2. falſe poſitioni') there is evident plagiarism. Stephano, however, omits most of the theory of numbers to be found in Paciuolo, and confines himself to mercantile applications. These cover barter, partnership, various forms of discount and exchange, and other similar topics. There are also given a number of mediæval puzzles, including the testament problem, the sale of the eggs, the hound and hare, and the guessing of numbers. A little work in mensuration and the calendar is given at the end of the book.

BARTOLOMEO ZAMBERTO, editor. Latin MS., c. 1525.

A Venetian scholar of c. 1500. He was born c. 1473.

Title. 'Evclidis // Megarenſis græci philoſo-//phi ex Theone græco com-//mētare Interprete Zāber-//to veneto triplex prīci-piorū // genus prímū diffinitiones: // Signv̄//est cuius pars nulla: // Linea vero, lon-//gitudo illa tabilis.// Lineæ autem limites, ſunt ſigna.// Recta linea, est quæ ex æqua-//li ſua interiacet ſigna. Superfi-//cies, est quæ longitudinem latitu//dinemqʒ tantum habet. Super-//ficiei extrema, ſunt lineae. Pla-//na ſuperficies, est quæ ex æqua-//li, ſuas interiacet lineas.' (F. 2, r.)

Description. 12°, 8 × 11.7 cm. 152 ff. (2 blank), 16–20 ll. Written on parchment.

This beautifully written manuscript has been included in this catalogue of arithmetics without much justification, since it does not contain even Book V. It is, however, valuable in showing the influence of printing upon written numerals. It is a copy of a translation of the first three books of Euclid made by Bartolomeo Zamberti of Venice in 1513. The complete translation was first printed in Venice in 1513, the statement of some bibliographers that it appeared in Paris in 1505 being unsupported by any evidence. It also appeared in Venice in 1517, in Paris in 1516, and in Basel in 1537, 1546, and 1558. See Riccardi, II, 1, 644; Weissenborn, *Die Uebersetzungen des Euklid durch Campano und Zamberti*, Halle, 1882. This manuscript was evidently written about 1525 for some noble family, for it has (f. 2, r.) three illuminated coats of arms.

ANONYMOUS. Latin MS., 1533.

Title. 'Declaratio Calendarii et // Almanach huius Cifte.' (F. 1, r.) 'Ars supputandi cum Denariis.' (F. 66, r.)

Description. 4°, 16.6 × 23.3 cm., the text being 11.8 × 16.3 cm. 81 ff. unnumb., 18 ll. Latin MS., written on vellum, in 1533.

This beautifully written Latin manuscript consists of two distinct works. Of these the first is a computus in twenty-three chapters, written apparently in Salisbury cathedral in 1533. The second part is a treatise on counter reckoning, and consisted originally of six chapters, 'De Numeratione, Additione, Subftractione, Multiplicatione, Diuifione, Fractione minutiis.' The last of these chapters is missing. The manuscript is particularly interesting because it gives the counter reckoning as it was used in England early in the sixteenth century, the numbers all being written in Roman when they are not represented 'on the line.' It is illustrated by numerous diagrams representing the line abacus. The manuscript closes with five pages 'De proportione vel regula Detri,' and 'De Proba regule Detri.'

ANONYMOUS. Italian MS., c. 1535.

Title. 'Trattato d'Aritmetica, e del Misure.' (F. 1, r.)

Description. Fol., 18 × 23.8 cm., the text being 14.5 × 21 cm. 148 ff., 23–27 ll.

This is an Italian manuscript, written, as the dates on folios 67, 68 show, about 1535. It is a commercial treatise, beginning, as was frequently the case, with the fundamental operations with compound numbers. The method of division 'a danda' is preferred to that 'per galea,' although both are given: 'Il partir a galea e molto legiadro et ſpeditiuo, ma non tanto ſicuro per un principiante quanto il partir a danda' (f. 19, v.). The applied problems are generally of a practical type useful to merchants' apprentices in the north of Italy. (Fig. 240.)

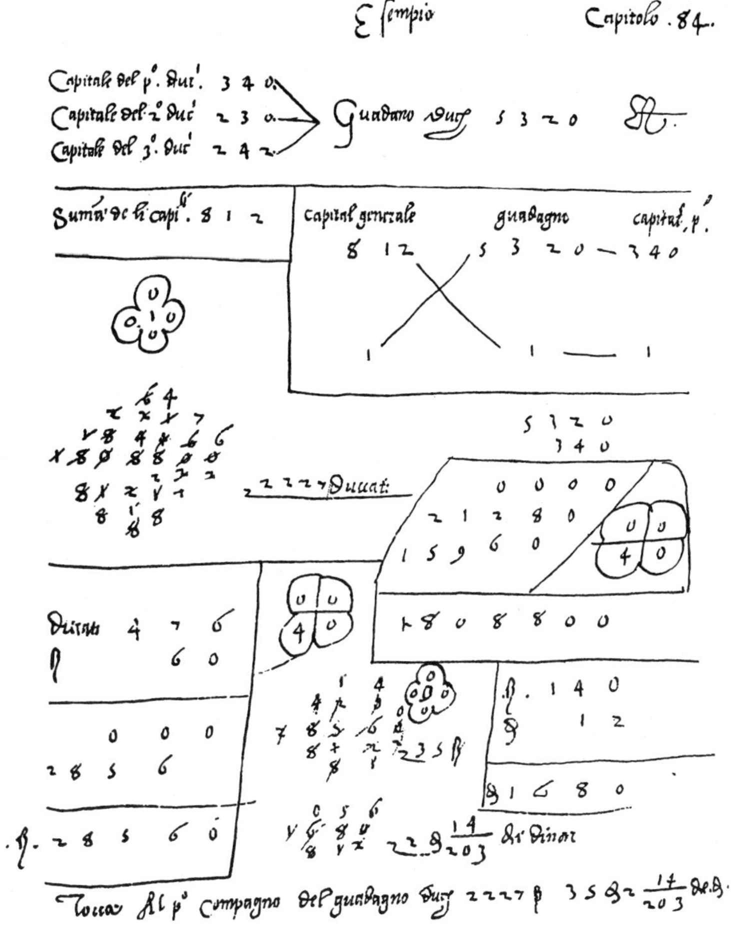

FIG. 240. FROM THE ANONYMOUS *Trattato* OF C. 1535

Much of the latter part of the treatise relates to mensuration. A set of tables and some notes are given at the end. Among the notes is the celebrated problem of the horseshoe. (Fig. 241.)

Un Cavallo che ha quatro ferri soto li piedi et per ogni
ferro ha sei chiodi che sono chiodi 24 Io dimando a un
quatrino l'uno quanti quatrini valera quelli chiodi 24
cominciando dal primo sino al ultimo sempre radopiando
de mano in mano che seguita li numeri sino al 24 come
qui sotto ci vede descritto

1 — 1
2 — 2
4 — 3
8 — 4
16 — 5
32 — 6
64 — 7
128 — 8
256 — 9
512 — 10
1024 — 11
2048 — 12
4096 — 13
8192 — 14
16384 — 15
32768 — 16
65536 — 17
131072 — 18
262144 — 19
524288 — 20
1048576 — 21
2097152 — 22
4194304 — 23
8388608 — 24

16777215:

Si che ci vede che 24 chiodi, sempre
radopiando li numeri, come al incontro
ci vede l'esempio, fano quatrini cento sesa-
nta sette, milioni, e, setanta sette milia, e
due cento, e, quindeci, dico 16777215
Che fanno gazette, a quatrini sei per
gazeta dico 2796202: e quatrini 3
Che sono berlingotti: 279620: et gazette
due, e, quatrini 3
Che fanno scudi trenta nove milia
nove cento, e, quaranta cinque et
berlingotti cinque, e gazette, due, e
quatrini tre dico 39945:5:2:3
dico a quatrini sei per gazetta
et a berlingotti 7 per scudo
Ma volendo che la gazetta valia
quatrini 7 per gazetta farano
se non scudi dico, trenta quatro
milia due cento, e trenta nove et
un berlingotto dico 34239:1:—
ci che da li sei quatrini alli sette
per gazetta vi he di diferentia dico
scudi cinque milia sette cento e sei
berlingoti quatro, gazete due e quatrini 3
dico ———— 5706:4:2:3

FIG. 241. NOTES IN THE ANONYMOUS MANUSCRIPT OF C. 1535

The manuscript is no. 23 in Narducci's catalogue of the Boncompagni manuscripts (Rome, 1862, p. 16).

LUDOVICO ALT DI SALISPURGO. Italian MS., 1545.

A sixteenth-century student.

Title. '.M .D. XLV.// Ludouicho alt de Salifpurga.' (F. 1, r.)

Description. 8°, 12.6 × 17.8 cm. 92 ff.

This is a business arithmetic written by some student in 1545. The author first treats of the fundamental operations with denominate numbers, following this by a treatment of fractions. (Fig. 242.) The last part

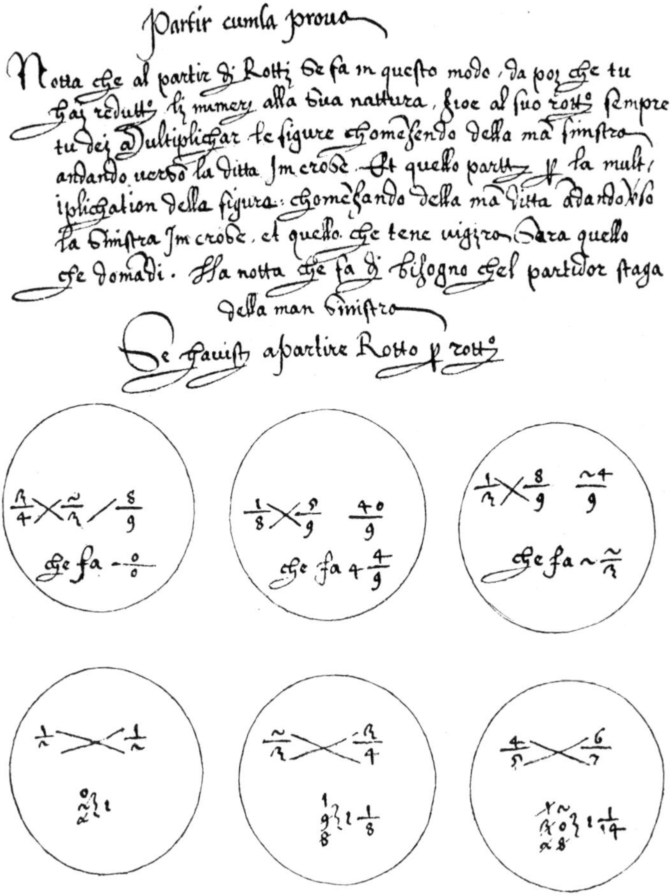

Fig. 242. From the Ludovico Alt

of the work is a 'praticha' (f. 48), and considers the ordinary business arithmetic of the time. The author uses almost the same form of the per-cent symbol as the writer of 1456 referred to on p. 458, viz. p $\overset{o}{\frown}$ (see Fig. 243). The manuscript is described in Narducci's catalogue, p. 152.

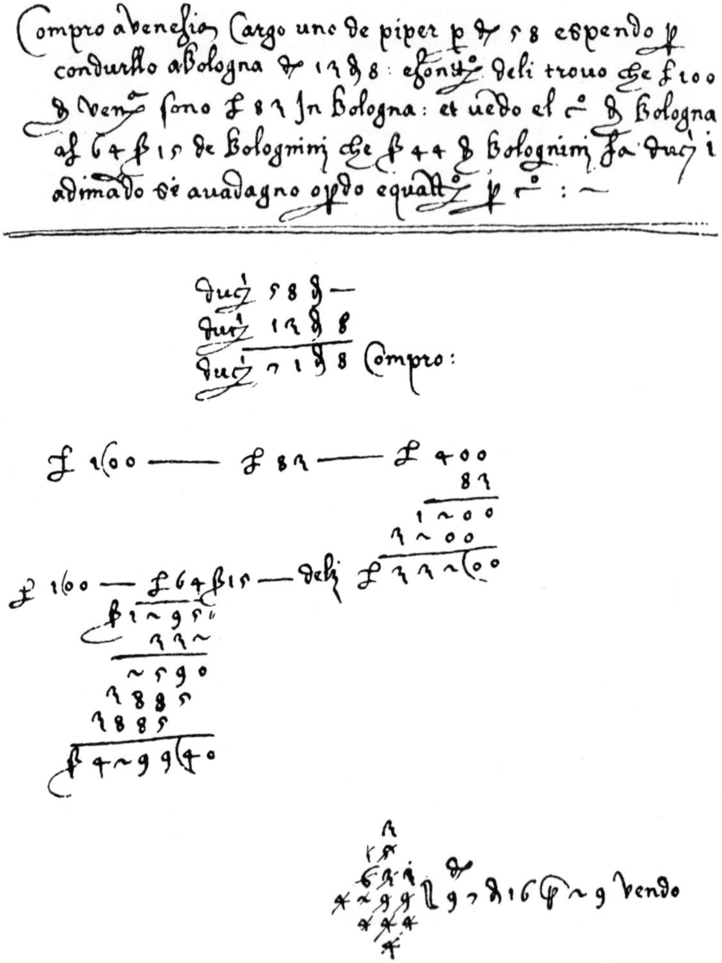

FIG. 243. FROM THE LUDOVICO ALT

ANONYMOUS. Latin MS., c. 1550.

Title. 'Introducoriū breue ſup̄ elementa Euclidis.'

Description. 4°, 15 × 21.8 cm., the text being 14.2 × 19 cm. 23 ff. unnumb. + 2 blank = 25 ff., 28–33 ll. Written in a German hand, c. 1550.

This manuscript is the last one in the volume containing the algebra of Mohammed ibn Musa (p. 454). It consists of introductions, book by book, to books I–XV of Euclid. The symbols $+$, $\div$ (for $-$), $\sqrt{\ }$, and $W\overline{\ }$ (for $\sqrt[3]{\ }$) are used. 498

HONORATUS. Latin MS., c. 1550–1600.

A Venetian monk of the sixteenth century.

Title. 'Opus Arithmeticā D. Honorati vene-//ti monachj coenobij S. Laurētij.' (F. 1.)

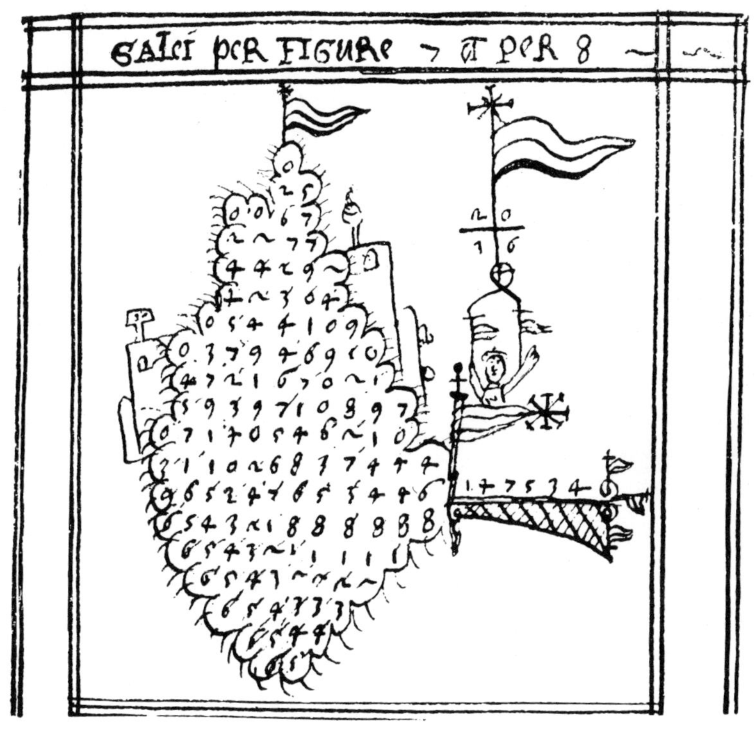

FIG. 244. FROM HONORATUS

Description. 8°, 10.9 × 16 cm., the text being about 9.5 × 15.8 cm. 111 ff.

This manuscript was written by a pupil of a Venetian monk named Honoratus evidently about 1550–1600. It is a practical arithmetic, the author taking up rather fully the fundamental operations with integers and denominate numbers. He also treats of such common applications as partnership and barter. The illustrations are those which a pupil would be likely to make, and prove the manuscript to be the work of an immature hand. The common galley method of division is given, with the ship in full form (Fig. 244), and the sign % appears as per $\overset{\circ}{\sigma}$ (see p. 439).

ANONYMOUS. Italian MS., c. 1560.

Title. 'Dichiarazione per intelligentia de Principiantj // del vso, che merchantilmente tiene la Citta dj Firenze // sopra le monete, pesi, e Misure.' (F. 2, r.) On f. 1, r., the coat of arms of the Angelotti (?) family is painted on parchment.

Description. Fol., 18.7 × 25.5 cm., the text being 12.5 × 20 cm. 166 ff. (3 blank), 8–42 ll.

This is a very clearly written Florentine manuscript on commercial arithmetic. The 8's are all made like S on its side, thus (ᔕ), a feature not infrequently seen in the second half of the sixteenth and even in the seventeenth century. As was often the case, the author presupposed a knowledge of the fundamental operations, reviewing them only briefly with compound numbers. The problems are of the common mercantile type of the period. The manuscript seems to have been written about 1560. (See Fig. 245.)

ANONYMOUS. Latin MS., c. 1565.

Title. 'Trattato d'aritmetica mercantile.' (Cover.)

Description. Sm. 4°, 11.9 × 16.8 cm., 279 ff., 20–25 ll.

It appears by the problems that this mercantile arithmetic was written at Bologna about 1565. Like so many similar treatises, it opens with a set of column multiplication tables. The first operation is 'partiri piccoli,' or short division, this being necessary for the simple reductions in the addition of denominate numbers. This is followed by the addition of pounds, shillings, and pence ('Sommare di lire e ſs e de'). The next operation is multiplication, several methods being given. The first is cross multiplication ('modo di mcare p̧ +'); the second, from left to right ('modo di multrip^{re}, per la dirieto,' multiply always appearing as 'multriply,' 'multripich,' in this manuscript); the third, by reference to the tables ('p̧ colonna'); the fourth, our common form, the 'bericuocolo' of the Florentines, but here called 'modo per bilicuocolo'

Vno hà riceuto d'interessj ₤ 961-16-9 però à ragione di ₤ 7 $\frac{3}{4}$ per 100-sj domanda dungue quanto fù il fondo.

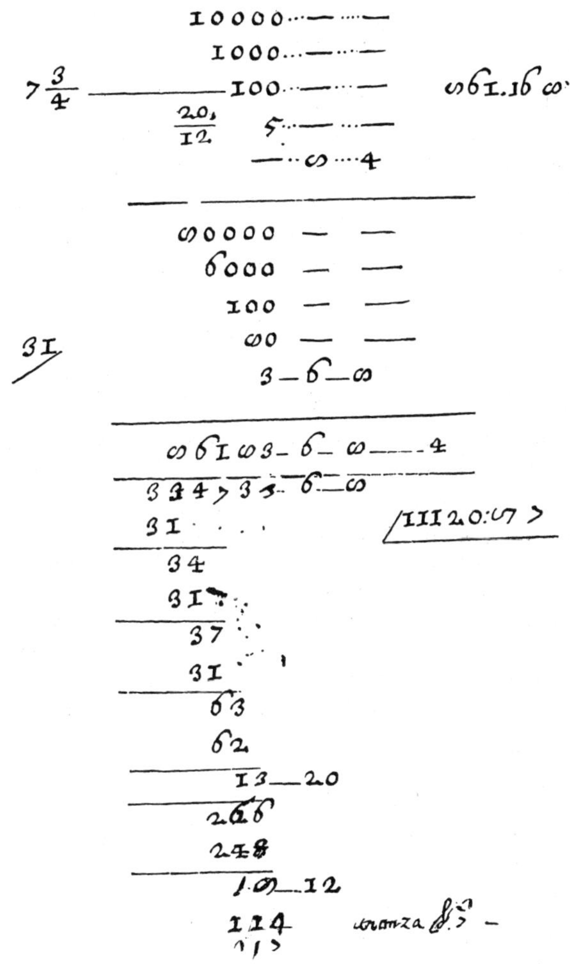

FIG. 245. FROM AN ITALIAN MANUSCRIPT OF C. 1560

(also 'bilicocolo'). This is followed by the subtraction of denominate numbers, and this by a more extended treatment of multiplication. Division follows, at first 'per ripiegho,' and then 'per danda' (Fig. 246), the galley method not being given. This is followed by a treatment of fractions, percentage ('Commicano e centi dalcuna merchantia'), the rule of three ('Commica lareghola del tre'), profit and loss, equation of payments, exchange, and the other mercantile rules of the period. Altogether it is one of the best of the sixteenth-century manuscripts on the commercial arithmetic of Northern Italy.

GAUDIOSO FRASCADA. Italian MS., 1568.

A schoolmaster of Brescia, about the middle of the sixteenth century.

Title. 'Libro di Arithmetica // et Geometria.' (F. 1, r.)

Colophon. 'Il p̄nte libro cioue opera di abaco et geometria scritto ad // instantia de - - - - - - - figliolo di m Bertholameo sachetto // habitante nella terra de' l'orala algise' scritto per mi // Camillo sachetto l'anno 1568 a honor de dio et della // uergine maria laus deo' (f. 43, v.). But f. 5, v., has the following: 'Questo libbro e ſtato ſcritto p̱ Gaudioso fraſcada cittadino di // Bressa a di 4 Nouembrio lanno 1555 laus dei.'

Description. Fol., 20.2 × 28.6 cm., the text being 14.7 × 24.2 cm. 46 ff. (3 blank), 17–29 ll.

This is a copy, made in 1568, of a treatise composed by Frascada, a schoolmaster of Brescia, in 1555. The author uses both the Florentine and the Venetian forms and names for the operations, as in the expression 'Multiplicar per ſchacchiere ſeu baricocolo.' Several forms of multiplication are given, as in the work of Paciuolo, but in division only the galley method appears. The examples are generally of the ordinary business nature, and the rules include 'Raggioni p̱ lo cattayno, cioe position false.' The last few folios refer to practical geometry.

ANONYMOUS. German MS., c. 1575.

Title. 'Von Küſtlicher Abmeſſung aller groſſe,// ebene, oder nidere, in die lenge, höhe, breite vnd // tiefte als gräben Ciſternen, vnd Brunnen.' (F. 1, r.)

Description. 12°, 10.2 × 15 cm., the text being 8.6 × 13 cm. 30 ff. unnumb. + 4 blank = 34 ff., 24 ll. Written in a German hand and in the German language, c. 1575.

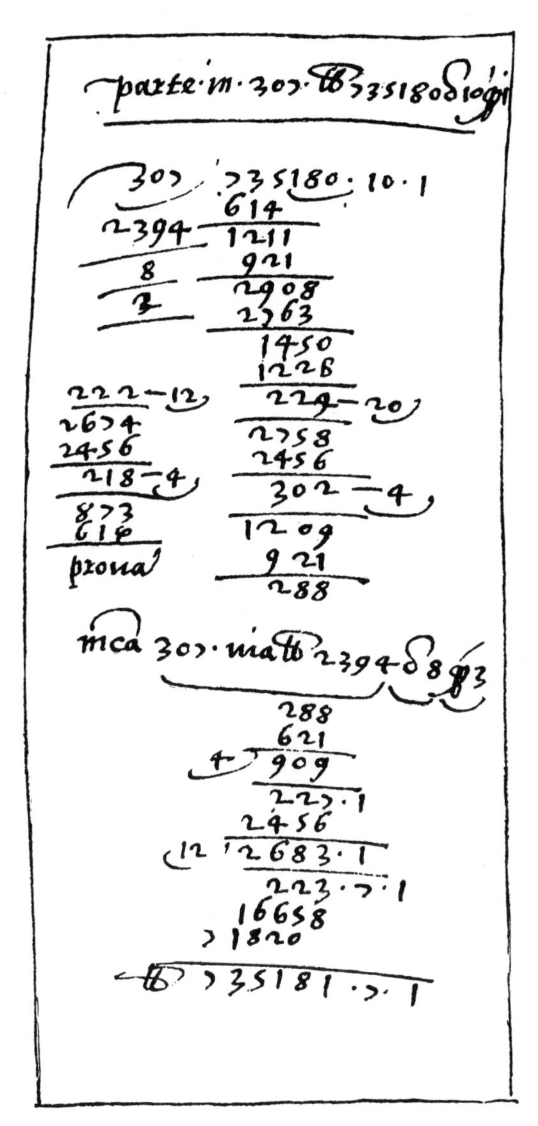

FIG. 246. FROM AN ITALIAN MANUSCRIPT OF C. 1565

This manuscript contains a few numerical problems in connection with mensuration. The work is of no merit save as it throws light upon the mensuration of the sixteenth century in Germany. Like most books of this kind, written at that time, it has little or no explanation of the rules used.

DOMENICHO DA BIEN DE VALSUGANA.

Italian MS., 1579.

A Venetian student of 1579.

Title. None. An elementary algorism.

Colophon. On p. 21 there is this statement: 'L'anno. 1579. a di .9. del // mese di marzo in liedolo // Per Domenicho da Bien de Valsugana.'

Description. 12°, 9.5 × 14.5 cm., the text being about 7.5 × 12.5 cm. 48 pp., varying number of lines. Probably written at or near Venice. 1579.

This is a student's manuscript on commercial arithmetic. It opens with the fundamental operations, the treatment of 'Multipticare Per scachiero' (p. 12) showing the Venetian instead of the Florentine influence. Division is performed only 'per galia, ouero per battello' ('p Galera' Domenicho calls it elsewhere), or 'per Collona,' the modern form not appearing. The examples are wholly mercantile, most of the applications being in the rule of three. The work also contains some simple examples in mensuration.

FRANCESCO GIUNTINI. Italian MS., Lyons, 1579.

An Italian student or teacher of astronomy. I judge from his horoscope that he was born November 14, 1522, at Florence.

Title. None.

Colophon. 'Di Lione adi 13. di Maggio 1579: Francesco Giuntinj.'

Description. 4°, 15.5 × 21.8 cm., 32 ff. (2 blank), 26–30 ll. Written on paper.

This manuscript is clearly written in Italian, and relates to astrology, in particular to Giuntini's own horoscope. Its interest in the history of arithmetic lies wholly in such symbols as that for degrees, and in the forms of numerals.

JOHANNES KLUMPIUS. Latin MS., 1598–99.

Title. 'Varij Tractatus Mathematices // à Joanne Klumpio philiæ // ſtudioſo excpt Ingol-//ſtadij Anno // 1598 et ſequentj.' (P. 1.)

Colophon. At the end of the 'Arithmetica practica' are the words 'Abſoluim9 15 Januarij // Anno 1599.'

Description. 4°, 15.3 × 20 cm., the text being 10.5 × 16.5 cm. (varying). 70 pp. blank + 442 pp. written = 512 pp., 19–21 ll. Latin MS., written in a German hand, Ingolstadt, 1598–99.

This is a set of lecture notes on general mathematics, including geometry, trigonometry, arithmetic, and astronomy. The arithmetic consists of two distinct parts, probably the result of two courses of lectures. The first treats of the four fundamental operations with integers and common fractions; the second of practical arithmetic, as Klumpius calls it, although it simply gives the fundamental operations, progressions, roots, and a few rules like the rule of three, partnership, alligation, and the rule of false.

B. ROTH. German MS., 1599.

A German student of c. 1600.

Title. 'Das Fünfft Capitel.// Item Im Funfften Capitel iſt von dem Algoriſmo // oder Coſs.' (F. 4, r.)

Colophon. Not entirely legible. It contains the date, August 4, 1599.

Description. 4°, 18.8 × 24.6 cm., the text being 13.5 × 19.8 cm. 321 ff. (13 blank), the lines varying.

This is a German manuscript containing the solutions of the problems in Stifel's edition of Rudolff's Coss, beginning with Chapter 5. (See p. 258.) This manuscript was written, as the colophon shows, in 1599. The name of the writer is not entirely legible, but it seems to be Britenus Roth. The solutions are written in a very clear hand, and furnish an excellent example of the symbolism of that period.

ANONYMOUS. Dutch MS., Louvain, c. 1600.

Title. None. A treatise on mensuration.

Description. Sm. 8°, 10.5 × 16 cm., the text being about 8 × 13 cm. 80 pp., 33–46 ll. Written on vellum.

This is a Dutch manuscript on gauging and general mensuration. The gauger's tables are as clearly written as in the best Florentine manuscripts.

ANONYMOUS. Latin MS., c. 1600.

Title. None. A treatise on arithmetic.

Description. Sm. 4°, 10.5 × 14.1 cm., the text being about 6 × 12 cm. 197 ff. numb. (there have been added 8 ff. in the back, written on vellum, making a total of 205 ff.), 22 ll. (varies). Latin MS., written in a German hand.

This manuscript begins with the fundamental operations after the manner of algorism, and then takes up the theory of numbers according to the Boethian system. Figurate numbers and proportions (ratios) are treated at great length, as in the arithmetic of Boethius, although this is not a copy of that work. It is not common to find manuscripts written as late as this that go so fully into the ancient theory of pure arithmetic. At the end of the book the author has placed eight folios of 'Arithmetica tabulaīs et formularis,' clearly written on vellum, containing multiplication (or 'area') tables of primes, of 'oddly even' and 'evenly odd' numbers, of 'solid numbers,' and the like.

ANONYMOUS. German MS., c. 1600.

Title. 'Nützlicher Gebraüch // Der Weldt Kügel.' (P. 3.)

Description. Fol., 20.5 × 33 cm., the text being 19.5 × 27 cm. (varies), 160 pp. (several blank), 24–29 ll. (varies). Written on paper, in German, c. 1600.

Although the first part of this manuscript is on the terrestrial sphere, the second part, beginning on p. 83, is on arithmetic as needed by cosmographers. This includes the fundamental operations, including square and cube roots. The galley method of division is used exclusively.

ADDENDA

Since the completion of the manuscript for the *édition de luxe* of this work, two years ago, numerous additions have been made to Mr. Plimpton's collection of early arithmetics. Excluding a number of early Arabic manuscripts, these acquisitions, with references to the pages on which they would naturally appear, are as follows:

PAGE 16. The 1561 edition of Borghi has been acquired.

PAGE 23. The 1485 edition of Nicolò de Orbelli has been acquired. See also page 473.

PAGE 36. There has been acquired a work on the calendar by Regiomontanus: 'Almanach magiſtri Johãnis // de monteregio ad ãnos. xviij.// acuratiſſime calculata.' *Colophon:* ... 'Erhardi Ratdolt Auguſ-teñ // Vindelico ... M.cccc.lxxxviij.' This is not an arithmetic, but it is interesting on account of its mathematical treatment of the calendar.

PAGE 62. There has been acquired a Cracow edition of a work by Faber Stapulensis, containing a little arithmetic: 'Jacobi Fa-//bri Stapv-//leñ ĩ Artiũ diuiſione ĩtroductio ...' *Colophon:* 'Cracouię ... 1534.' 4°, 13.8 × 18.7 cm., the text being 10.2 × 16 cm.

PAGE 82. The first (1503) edition of the Margarita Philosophica has been acquired. The colophon is as follows: 'Chalchographatum primiciali hac // peſſura/ Friburgi ꝑ Ioannẽ Scho//ttũ Argeñ. citra feſtũ Margarethę // anno gratiæ M:CCCCC.III.'

PAGE 114. An edition of Tagliente's 'Libro // de abaco,' 'Vinegia ... M.D.XLIIII' has been acquired. Also the Venice edition of 1570, and an edition s. a.

PAGE 115. The 1570 Milan edition of Tagliente has been acquired: 'Libro // de Abbaco che inse-//gna a fare ogni ragione // mercadantile ... Milano 1570.'

PAGE 139. The 1525 edition of Riese's first book has been acquired: 'Rechnung auff der linihen // gemacht durch Adam Rieſen vonn Staffel-//ſteyn/ in maſſen man es pflegt tzu lern in allen // rechenſchulen gruntlich begriffen anno 1518.// vleyſigklich vberleſen/ vnd zum andern mall // in trugk vorfertiget. // ℭ Getruckt zu Erffordt zcum // Schwartzen Horn.// 1525.' *Colophon:* 'ℭ Gedruckt tzu Erffordt/ durch // Mathes Maler. M.//CCCCC.xxv. Jar.' 8°, 8.8 × 14.6 cm., the text being 7.1 × 10.7 cm. 44 ff., 17–20 ll.

PAGE 139. Another Erfurt edition of Riese's second work has been acquired: 'Rechnung auff // der Lynihen vñ Federn/ // Auff allerley handthirung/ gemacht durch // Adam Ryſen.// Zum andern mal vberſehen/ // vnd gemehrt.// Anno M. D. XXvij.' *Colophon:* 'Gedruckt zu Erffurdt zum Schwartzen Horn.'

PAGE 148. The 1570 Venice edition of Feliciano has been acquired. 4 + 79 ff.

PAGE 167. 'Ulrich Kern von Freysing Eyn new Kunſtlichs wolge-// gründts Viſierbuch/ gar grviß vnnd behend // auſz rechter art der Geometria/ . . . M.D.XXXI.' Fol., 18 × 27.8 cm., the text being 12.7 × 23 cm. 57 ff., 46 ll. Strasburg, 1531. A work on guaging.

PAGE 181. Two more editions of Mariani's 'Tariffa' have been acquired, Venice 1564 and 1572.

PAGE 223. There is mentioned an arithmetic by Medlerus, 1543, with a second edition in 1550. This work, recently acquired, may be all that there is of the 1550 edition: 'Facili-//ma et exactis-//sima ratio extra//hendi radicem Quadratam // & Cubicam, a Doctore // Nicolao Medlero in // gratiam ſtudioſæ // iuuentutis // ædita. // Anno Domini. // M.D.L.' *Colophon:* 'Impressvm VVitem- // bergae, per Vi-//tum Creutzer.' 8°, 9 × 14.8 cm., the text being 7 × 11.2 cm 7 ff., 27 ll.

PAGE 322. The 1575 edition of Lapazzaia has been acquired. It is not the same, however, as the work described on page 322, either in title or in contents. 'Opera // terza // de aritmeti-//ca et geo-//metria. // Dell' Abbate Georgio Lapazaia // da Monopoli. // Intitolata il Ramaglietto.// In Napoli // Appreſſo Mattio Cancer. M.D.LXXV.' 14.2 × 20 cm., the text being 11 × 16 cm. 4 + 169 pp., 29 ll.

PAGE 322. The 1566 edition of Lapazzaia has been acquired: 'Familiarita // d'arithmetica, e geometria // con l'vsitata prattica Napo-// litana, Compoſta & ordinata per Abbate Geor-//gio Lapizzaya Canonico Monopolitano // Nuouamente con ſomma diligentia // Riſtampata, e corretta.// . . . In Napoli // Appreſſo Horatio Saluiani // MDLXVI.' 4°, 14 × 22 cm., the text being 11.2 × 16.2 cm. 2 + 62 ff., 28 ll.

PAGE 340. 'L'Arithmetiqve//militaire d'Alexandre//Vandenbussche Flandrois//departie en deux liures.//... A Paris...' (s. a.). 8°, 14.8 × 20.8 cm., the text being 9.8 × 16 cm. 35 ff., 34 ll. Paris 1571. The Proeme is dated "De la groſſe tour de Bourges le douzieſme d'o-// ctobre. 1571." An arithmetic for instruction in military circles.

PAGE 347. A 1590 Antwerp edition of Menher has been acquired: 'Livre // d'Arithme-//tique contenant pluſieurs belles que-//ſtions & demandes, bien propres & // vtiles à tous Marchans, // Par M. Valentin Mennher // de Kempten. // Reueu par M. Melchior van Elſtaer.//...'

PAGE 375. The 1584 Cologne edition of Clavius has been acquired.

PAGE 383. 'Arithmetica // oder // Rechenbuch // .... durch // Anthonium Schultzen/... Zur Liegnitz... 1600.' Fol., 13.6 × 18.5 cm., the text being 10.2 × 13.5 cm. 4 + 259 ff., besides an appendix on 'Buchhalten,' 30 ll. The first edition appeared in 1583. A book of no particular merit.

PAGE 389. In 1585 Fr. Barocio published a Cosmographia, at Venice. It contains 5 pp. on arithmetic. [541]

PAGE 391. The work of Aurelio Marinati has been acquired: 'La prima parte // della // somma di tvtte // le scienze // nella qvale si tratta delle // sette arti liberali //... Roma... 1587.' 4°, 16 × 21.5 cm., the text being 11 × 17.7 cm. 8 + 156 pp. The chapter 'Dell' aritmetica' begins on p. 99 and ends on p. 114.

PAGE 408. Vila's arithmetic has been acquired: 'Reglas // brevs de Arith-//metica... per Bernat // Vila,... Barcelona... Any. M.D.-LXXXXVI.' 8°, 9.8 × 14.3 cm., the text being 7.5 × 12 cm. 8 + 136 ff.

PAGE 418. The 1611 edition of Mariana's work has been acquired. 'Ioannis // Marianæ // Hiſpani //, e socie. Iesv, // De Ponderibvs // et Mensvris. // Typis Wechelianis. // Anno M.DC.XI.'

PAGE 425. A manuscript of Zuchetta's work, copied in 1692, has been acquired.

PAGE 440. There has been acquired one folio of manuscript on parchment, in a fourteenth-century hand, containing part of the Etymologies of Isidorus. 31 × 45.7 cm. See also p. 8.

PAGE 456. There has been acquired an anonymous manuscript on astrology, with some directions for arithmetical computations. It bears no date but was written c. 1450. 5 ff., on paper, in Latin, in a German hand.

PAGE 480. A manuscript of Bede, 'De Scientia computandi,' written in Latin, c. 1520, has been acquired. Fol., 22.7 × 33 cm., the text being 13 × 22.5 cm. In the same volume and written by the same hand is 'Franconis De Quadratura Circuli lib. V.'

PAGE 487. A manuscript of Boethius and Gerbert, on geometry, written in Latin, c. 1550, has been acquired. The manuscript of Gerbert begins 'Incipit liber geometrię artis æditvs ā Dnō // Gerberto Papa et Philosopho. Qvi et Silvester // secvndvs est nominatvs.' Each manuscript contains some interesting number work. Fol. on paper, 22 × 33.5 cm., the text being 15.5 × 23.5 cm.

As stated in the preface, there will naturally be found from time to time numerous additions to the bibliography contained in this work. The following have recently come to the attention of the author:

PAGE 70. Portius. There was also an edition, Rome, 1524, 4°.

PAGE 81. Boethius, Clichtoveus, and Stapulensis. There was also an edition, Paris, 1514, fol.

PAGE 97. There is an anonymous 'Algorithmus linealis,' 1513 at Göttingen, — probably Licht.

PAGE 106. Köbel. There was an edition of the 'Vysierbuch,' Oppenheim, 1519.

PAGE 123. Grammateus. There was also an edition, Frankfort, 1554.

PAGE 139. Riese. There was also an edition, 1559, and a Frankfort edition, 1563, of No. 2.

PAGE 180. Albert. There was also an edition, Magdeburg, 1588.

PAGE 195. Noviomagus. There was also an edition, Cologne, 1539.

PAGE 214. Recorde. There was also an edition, London, 1658.

PAGE 257. Gülfferich. There was also an edition, Frankfort, 1561.

PAGE 286. There was published a small Tariffa by Marcello in 1566.

PAGE 359. There may have been a Rechenbuch by Junge published in 1577.

PAGE 396. Lindebergius published at Rostock, in 1591, a work 'De præcipuorum . . .,' containing a little arithmetic.

PAGE 429. There is also a work by Lachar, 'Algorithmus mercatorum,' s. l. a.

The statement in the preface, page ix, relating to the De Morgan library should be modified. While some of the books were sold (for example, see page 122), most of them were purchased by Lord Overstone and presented to the University of London, where they may now be seen.

ADDENDA TO
RARA ARITHMETICA

WHICH DESCRIBED IN 1908 SUCH EUROPEAN ARITHMETICS PRINTED BEFORE 1601 AS WERE THEN IN THE LIBRARY OF THE LATE GEORGE ARTHUR PLIMPTON, BY DAVID EUGENE SMITH

PRINTED BOOKS

ADDENDA TO RARA ARITHMETICA

A reference like **Page 7** in heavy type at the beginning of a paragraph refers to the *Rara Arithmetica*.

Page 7. RODERICUS ZAMORA.

Ed. pr. [1471]. s. l. a. [1471/2].

Title. 'Speculum Humanorum.'

Description. It has a brief section on arithmetic, but neither this nor the Magni book below should be ranked with the Treviso textbook of 1478 mentioned in the *Rara*.

Page 7. JACOBUS MAGNI. Ed. pr.? Strasburg, [c. 1475].

Title. 'Sophologium.'

Mentioned by Hain-Copinger, *10471, and Proctor, 240. It was printed by Adolf Rush at Strasburg. It contains a section 'De aresmetrica' (Book I, Tract II, chapters 6 and 7).

Page 9

Editions. There were other editions of the 'tractatus' of Albert of Saxony, Padua, 1477; ib., 1487; Paris, 1485; Rouen, 1493; ib., 1494 and 1500; Venice, 1476; ib., 1494; Trent, 1475, — including Italian and German translations.

Page 16

To the list of Borghi's 'Qui comenza' editions add a sixth edition, Venice, 1509, mentioned on p. 16. It contains 6 pp. of additions.

There are two editions in 1561. In one of them the last two signatures, M and N, have been reset. In 1560 one was set, and in 1561 this was used, the reset ones being used to complete the book. See *Rara*, p. 495.

Page 23

Klebs gives three editions of the 'Ars Numerandi,' Cologne, 1482; Paris, 1490; Florence, 1491. See also pp. 473, 495, of *Rara*.

Page 27

In the list of Boethius-Stapulensis editions here and on p. 62 of *Rara*, include 'Arithme//tica specvlativa Boe-//tij per Iacobum Fabrum Stapulen-//sem in compendium redacta.//' Colophon: 'Basileae . . . Mense Martio,// Anno // M.D.XXXVI.'

Page 32. ANIANUS. Ed. pr. 1488. Paris, 1527.

Title. 'Compot³ nouissi//me ampliatus ac//emendat³ & fami// liari elucidatus cōmentario: vna // cū manib³ / figurisqȝ suis in locis // decēter adiectis M.d viii. . . . Compendium deniqȝ su//p additū est N. Bonespei trecēsis //. . . . Au Pellican.'

Colophon. 'Impressus pictauis in domo honesti vi//ri. Enguilberti de Marnef Bibliopola // dicte Vniuersitatis ad insigniuȝ // Pellicani . . . Millesimo quingentesimo // Vigesimo septimo.'

Description. 9 × 13 cm., 68 leaves, numbered fol. lxvij and 1 unnumbered. This is the 1508 edition mentioned on p. 32 of *Rara*, but reprinted in Paris 1527. It has several illustrations, one of which is given in Fig. 1.

Pages 33–35

There were at least 70 editions of Anianus before 1600 (34 being before 1501) with or without the added name of Sacrobosco, besides 8 of the 'Cuer de Philosophie' which contains a so-called translation of the Compotus. Among the editions in the sixteenth century some are in verse, — 'Compotus cū commēta familiari.

una cum figuris et manibus necessariis . . . Lugduni, 1509.' In the preceding year (1508) there was one 'Compot³ nouissi//me ampliatus ac//emendat³ & fami//liari elucidatus cōmentario: vna//

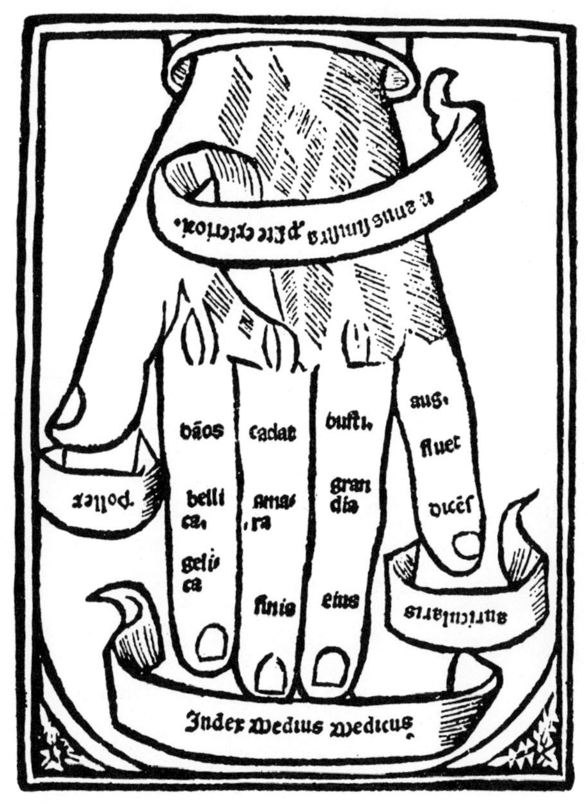

FIG. 1. FROM THE ANIANUS *Compotus* OF 1527

cū manib³ figurisq3 suis in locis // decēter adiectis M.d viii,' mentioned above. This contains numerous illustrations (see Fig. 1), verses, church calendars, and the length of days. In 'Le Comput Manuel de Magister Anianus,' by D. E. Smith, Paris, 1928, pp. 80–103, these additional editions are listed: Lyons, 1488–9, 2 editions; ib., 1490–3, 6 editions; 1494–5, 4 editions; Paris,

1483; ib., 1490, 4 editions; ib., 1492, 2 editions. This work was so popular that many other editions appeared before 1601, and later.

Page 36

See p. 495 of *Rara.*

Page 40

From variations in spelling and other evidence there were two editions in Augsburg, 1526, possibly due to changing type in certain pages while in the process of printing.

Page 46

Add to l. 3 of editions: Hain, 125–6, gives two editions s. l. a.

Page 53

Add 'Vienna, 1498; ib., 1500.' In first title, for 'Practis' read 'Fractis.' † The section 'De Numeris Fractis' seems to be the work of some anonymous writer.

Page 53

The arithmetic of Peurbach mentioned as 'Opus Algorithmi' is now in the Plimpton Library. There are 8 leaves, 15.3 × 21.4 cm. The title page reads: 'Opus Algorithmi iucūdissimū // Magistri Georgii Puerbachij wiēnēsis (preceptoris singula-//ris Magistri Joannis de Mōteregio) . . .' The colophon reads: 'Et tantum de hoc opere Algoristico Anno // Christi Jiesu Millesimo quīgentesimo septīo // ꝑ Baccalariū Martinū Herpipolen̄ īpresso.'

Page 56

Paciuolo wrote another book, the 'Libro dà Abacho,' Venice, Bern. de Bindoni, 1540; 100 numbered leaves.

Page 60

There were three Leipzig editions of 'Algorismus linealis' in 1495 and one in 1500. Probably printed by Landsberg, c. 1495, 33 ll., 11 woodcuts, possibly written by Widman or Licht.

Page 61

Add to Editions (of Bradwardin): Paris, c. 1515. Also 'Regula falsi apud philosophantes augmenti et documenti appelata,' Leipzig, Martin Landsberg, between 1490 and 1495, 20 leaves. The first edition of any work on false position. There was also an algorism, 'Algorithmus minutiarum physicarum,' Leipzig, Martin Landsberg, 1490–95, treating of sexagesimal fractions, 6 leaves, possibly the work of Widman. From the same publisher, another 'Algorithmus minutiarum vulgarium.' The preface says that it goes back to Campanus, thirteenth century. There is a 'Computus novus,' s. l. a., sometimes quoted as Leipzig, 1495, but it is probably ib., 1499. (See Hain 5600.)

Page 62

See p. 495 of *Rara.*

Page 70

Add to Portius editions: Rome, 1534.

Page 81

Add to Editions, l. 3, after 1510, 'and also an edition in 1514.'

Page 82

This copy in the Plimpton Library includes all the arithmetic, but not the folding map of the world. See p. 495 of *Rara.*

Page 83. CONRADUS NORICUS.

Ed. pr. 1503. Leipzig, 1503.

The author was a member of the faculty of Liberal Arts in the University of Leipzig.

Title. 'Commentatio Arithmetice commu//nis a Conrado Norico artium libera//lium: academie Lipsensis Magistro . . .'

Colophon. '. . . Vale ex Liptzck Sole in tercia parte aquarij Anno salutis. 1503.'

Description. 20.8 × 30.4 cm. It is the 'Textus arithmeticæ' mentioned on p. 83 of the *Rara*, acquired by Mr. Plimpton after the *Rara* was printed. Fol., 25 leaves. Bound with two other works, the first being the 'Mathemalogiū pri-//me p[r]tis Andree alex//andri Ratisbonēsis // mathematici sup[r] no//uam et veterem loy-//gam Aristotelis,' devoted to the Aristotelian and the [then] modern logic. The second part (38 leaves) treats of perspective and is profusely illustrated with geometric figures. The third part is the work above mentioned and seems to have been published by the same printer as the others — Melchior Lotter of Leipzig, the type, style, and paper being the same. It is divided into two sections, one being related to the theory of numbers and the other to computations, explained by numerous marginal notes in a sixteenth-century hand.

Page 83. STROMER, HENRICUS.

Ed. pr. 1504. Leipzig, 1510.

Known also as STROMER AUERBACHENSIS.

Title. 'Algorithmus līealis // numerationem Additionem Subtra//ctionem Duplationem Mediationē // Multiplicationē Diuisionē et progressionē vna cum re//gula de Tri perstringens.' This is given on a separate preliminary leaf. The text begins on Aii.

Colophon. '.·. Impressum Lyptzk per Jacobum // Thanner Herbipolensem Anno domi-//ni Millesimoquingentesimodecimo.' Followed by the device of Thanner.

Description. 4°, 16 × 21.3 cm., 7 leaves. The text includes diagrams of computation by counters.

To the Stromer editions given in *Rara* on page 83, add 'Cracow, 1536.' †

Page 87. HIERONIMUS DE HANGEST.

Ed. pr. 1508. Paris, 1508.

Title. 'Liber proportio//nū magistri hie//ronimi de han-// gest. . . . Iehan Petit' (with emblem).

Colophon. 'Liber proportionum a Magistro hieronymo de hangest editus // Parrhisijs nouissime Impressus opera Johannis Barbier . . . feliciter finit. Anno ab orbe redempto Millesimo quingente-//simooctauo. xxij. Junij.'

Description. Fol., 20 × 28 cm. 29 fol. The colophon is on fol. d. iiii, v. Then begins a supplement of two pages, 'Antidotum memorie atqȝ ītelle-//ct.'

The book is concerned with the theory of numbers, including proportion, and the relation of numbers to a geometric type of algebra and to the theory of music. It is divided into five books, the 'Liber quintus' being 'De proportionibus harmonicis.' In its field it is one of the best works of the period.

Page 91

To *Other works of 1510* add: Ambrosius Lachar, 'Algorithmus mercatorū magistri // Ambrosij Lacher de Mersspurg Mathematici de integro & fracto // numero . . .' *Colophon:* 'Impressum Lyptzigk pr Bac-//calaureū Martinū Herbipolē// sem Anno dñi 1510.'

In 1511 there was printed by Cornelis of Zyrickzee, at Cologne, an 'Arithmeticae Lilium triplicis practice,' which is verbatim like Huswirt's 'Enchiridion' of 1501 illustrated on p. 75 of the *Rara.*

There was also another edition of 'Arithmetice Lillium triplicis practice que pulcherrime ut puta . . .,' Cologne, 1511.

Page 93

After the description of Ortega's 'Tratado' insert: There was a translation of the Spanish work, Lyon [1515], 'Euure [Oeuvre] tressubtille ∴ profitable de lart ∴ science de Aristmeticques : ∴ Geometrie translate nouuellement despaignol en frācoys [par moy Frere Glaude Plantin . . .'].

Colophon. 'Impr. a Lyon . . . par Estienne Balaud' . . . [1515]. It differs somewhat from the first Spanish edition and may be considered a separate work. The name of the author was L'Ortie or l'Ortega, each meaning 'nettle,' and appears in French as Juan de l'Ortega and Juan de Ortega.

Page 97. LANDSHUT, KARL VON (Johann Lanzut, and Lanczut). Ed. pr. 1513. Cracow, 1517.

Known in this edition as JOANNES DE LANCZUT Magister.

Title. 'Algorismus linealis // cum pulchris conditiōibus duarū // regula .·. De trivna de integris: // altera vero de fractis: regl'isq .·. // socialibus: .·. semper exem-//plis ydoneis adunctis. // In florētissimo studio // Cracouiensi editus // nō minus litteris // erudit .·. qȝ merca//toribus vtilis.// .·. maxime // incipientibus.' See Fig. 2.

Colophon. 'Impressum Cracouie Opera .·. impensis // prouidi viri domini Joannis Haller // ciuis Cracouieñ. Anno Cristi.// 17. supra millesimum // quingētesimū.'

Description. 4°, 10.4 × 15.8 cm., the text being 9.9 × 14.8 cm. Type Gothic. The date, 'Data Cracouie vi. Kalendas.// martias. M. D. Xv.' is evidently that of the completion of the manuscript copy. The text includes eight diagrams of the abacus and counters.

Editions. Hain, no. 828, mentions a possible incunabulum, s. l. a. This may be the *Algorithmus integrorum* mentioned in the *Rara Arithmetica* as of 1504 (p. 83). Other editions of 1515 and 1519 (*Rara*, p. 97). Delete the duplicate list under *Other works of 1513*.

There was another edition of Landshut (Lanzut), Cracow, 1538. See also p. 498 of *Rara*. Joannes Martinus is also called Sciliceus.

Page 99

In the editions include one by Budaeus and Alciatus, Venice, 1532, and a French edition, 'Summaire ou Epitome du livre de Asse,' Paris, 1522.

Algorismus linealis
cum pulchris conditiōibus duarū
regularū De tri vna de integris:
altera vero de fractis: regulisq;
socialibus: z semper exemplis ydoneis adiunctis.
In florētissimo studio
Cracouiensi editus
nō minus litteris
eruditis q̄ merca
toribus vtilis.
z maxime
incipientibus.

¶ Ad Lectorem Octosticon.
Quisqs Arithmeticen optas cōtingere sacram
Quam Samius nimium concelebrauit auus
Hunc (moneo) legito studiosa mente libellum
Lectaq; sub memori rite repone sinu
Hæc nāq; ad reliquas iter amplum dirigit artes
Discutit errores, iurgia dira secat
Quam bene qui norit res cunctas nosse putat
Constat em̄ numeris quicqd in orbe viget.

FIG. 2. TITLE PAGE OF LANDSHUT'S *Algorismus* OF 1517

Tractatus Arithmeti
ce Practice qui dicitur Algorismus cum additioni
bus vtiliter adiunctis

Veneunt in edibus Joannis Gourmontij e
regione Collegij Coqueretici

FIG. 3. TITLE PAGE OF THE 1514 EDITION OF THE *Tractatus*

Page 106

In l. 2, after '(p. 113)' insert 'also 1519 at Oppenheim or [Frankfort].' In l. 2, before 'Frankfort,' insert 'ib., 1519;' To *Other works of 1514* add Martinez Guijeno's 'Arithmetica in theoriam et praxim scissa,' Paris, 1519 and frequently reprinted.

In the *Other works of 1514* the 'Algorithmus linealis Proiectiliū . . .' listed as due to Buschius is due to Joannes Cusarrus (Cusa), as shown in the colophon '. . . diligenti opera per Magistrum // Joannem Cusanum . . .'

Page 106

Among the *Other works of 1514* include a 'Tractatus Arithmeti//ce Practice qui dicitur Algorismus cum additioni//bus vtiliter adiunctis,' s. l. The book is in the Plimpton Library and the title page is reproduced in Fig. 3. It has 8 leaves in signature A8 and 6 in B6. In this copy the binder has misplaced certain leaves. It is possibly the work referred to in some catalogues under the title 'Arithmeticae practicae Tractatus,' s. l. a., anonymous, but probably by J. Martinez Siliceo. (See page 106 above for change of name.)

Page 114

The 1550 edition of Tagliente mentioned in the last line has now been acquired for the Plimpton Library. A facsimile of page 2 is shown in Fig. 4. See also *Rara*, p. 495.

Page 115

There was also another edition, Venice, 1570, 'per Francesco de Leon.' One edition in Gothic type s. l. a. See p. 495 of *Rara*.

Page 122. VOLVMNIVS RODVLPHUS SPOLETANUS.

Ed. pr. 1516. Rome, 1516.

Title. 'Volvmnii Rodvlphi Spoletani de pro//portione proportionvm // dispvtatio.'

FIG. 4. PAGE FROM THE 1550 TAGLIENTE

Colophon. 'Impressum Romæ per Iacobum Mazochium. Anno dñi.// M.CCCCC.XVI. Die XXVII. Septembris.'

Description. 12.6 × 19.4 cm. 28 leaves. One of several books of the sixteenth century treating of proportion from the geometric standpoint while at the same time making use of arithmetic. The text is preceded by two Latin poems, and the book is dedicated to 'Amplissimo in Xp̄o patri / & Dño. D. Laurentio Puccio.// S.R: E. . . .'

Editions. There were no other editions.

Page 123

To the editions of Grammateus add 'Erfurt, n. d., but post July, 1518.' Also in line 2 of Editions read 's. l. a. (Frankfort, 1544).' † Also see p. 498 of *Rara.*

Page 126

In l. 11, for 145 read 146. †

Page 130

The colophon of the Algorismus in l. 2 reads: 'In officina Sigismundi Grim̄ Medici ac // Marci Vuirsung, Augustae Vinde-//licorum. Die. 23. Jañ. An. 1520.'

After l. 8 insert: 'This was probably the first arithmetic printed in French,' and add: 'J. Furst, Compendiosa ars numerandi, Vienna, 1520.' See *Rara*, p. 93, for another work by Furst.

Page 130

In *Other works of 1520* include a small book printed in Paris [1520], entitled 'L Art & sci//ence de arismetique // moult vtille et prffitable a toutes // gēs & facille . . . imprime a Paris'; see Fig. 5 on page 514. On fol. 6, r, the anonymous author gives 'les cinq rei//gles principalles qui sont Nume-//ration. Adition. Sustraction. mul//tiplication. et partition.//' As a compendium of the operations needed in business it is exceptionally good.

Page 132. FRANCESCO GHALIGAI.

Ed. pr. 1521. Florence, 1548.

Title. 'Pratica // d'Arithmetica,// di Francesco Ghaligai // Fiorentino.// Rivista & ristampata // con diligentia.// In Firenze,// MDXLVIII.' See Fig. 6.

Colophon. 'In Firenze // appresso Bernardo Giunta.'

Editions. Add the edition of 1560, Bologna, by Pelegrino Bonardo, which is now in the Plimpton Library. Also insert: There was a Nuova Edizione of the Pratica, Florence, 1552, printed by the Giunti as in the 1548 ed., 'con somma Diligenza ristampata.'

L'Art & sci
ence de arismetique
moult vtille et pffitable a toutes
ges & facille a entendre par la plu
me & par le gect subtil/pour ceulx
qui ne scauent lyre ne escripre noù
uellement imprime a Paris.

On les vend a paris en la rue neu/
ue nostre dame a lescu de France.

FIG. 5. TITLE PAGE OF *L'Art & Science de arismetique*, PARIS, 1520

Page 138

Add to l. 4 that Berlet in his biography of Riese gives the date of his birth as 1492.

Page 139

To the editions No. 2, add 'Frankfort, 1533, 1534, 1559, and 1563. There was also a revision by Sebastian Kurz (Curtius), Nürnberg, 1610 (the first of his revisions).' See also *Rara*, pp. 495, 496.

PRATICA
D'ARITHMETICA,
DI FRANCESCO GHALIGAI
FIORENTINO.

RIVISTA & RISTAMPATA
CON DILIGENTIA.

IN FIRENZE,
MDXLVIII.

FIG. 6. TITLE PAGE OF GHALIGAI'S *Pratica* OF 1548

Page 140. CONRAD FEMEN. Erfurt, 1523.

Title. 'Eyn gut new rechen buch-//leyn ist erdacht.' . . . See Fig. 7.

Colophon. 'Gedruckt zu Erffordt durch Matthes // Maler

Eyn gut new rechen buch-
leyn iſt erdacht vnd iſt nach der ſchle-
chtē zall gemacht Als hye geſchri-
ben ſteth Als man hundert
nach eynander zelt.
∴

Gedruckt zu Erffordt durch
Matthes Maler:.

FIG. 7. TITLE PAGE OF FEMEN'S *Rechen buchleyn*

nach Christi geburdt. M.// CCCCC. vnnd ym Drey//vndzwentzigisten // Jar.'

Description. 8°, 8.8 × 14 cm. 11 leaves. It is written in rhyme throughout. The author's name appears at the end: 'Gedicht vn volbracht / durch Conradum Femen gemacht.'

Page 142

There was also an edition of the Riese-Helm book printed in 1570.

Page 146. FRANCESCO FELICIANO da Lazesio.

Ed. pr. 1517 [1518]. Venice, 1524.

Title. See Fig. 8.

Colophon. 'Stampato nella inclita citta di Vineggia // per Francesco Bindoni, & Mapheo Pa-//sini compagni, Nel anno. 1524.// Del mese di Decembre.'

Editions. This is the third edition of the work mentioned on p. 146 of the *Rara* together with other editions. The 1526 edition, pictured on p. 147 of the *Rara*, was probably begun in 1526 and completed in 1527.

Page 148

See p. 496 of *Rara*.

Page 151

The ed. pr. of Christoff Rudolff's 'Kunstliche Rechnung' was printed in Vienna in 1526. A copy is now in the Plimpton Library. See Fig. 9. Rudolff also published the 'Behend vnnd Hübsch Rechnung durch die Kunstreichen regeln Algebre, so gemeincklich die coss genēt werden,' Strasburg, 1525.

Page 160

In l. 7, note an edition of 1533. Under Finaeus, the French spelling Fine is generally preferred to Finé. †

Page 164

Under Alciatus, Editions, l. 5 b., note another edition with Budaeus and Melanchthon, Venice, 1532.

Libro de Abacho nouamente
composto per magiſtro Fráceſcho da Lazeſio ve
roneſe : ilquale inſegna a fare molte ragione
mercantile : & come riſpondeno li Pre/
cii : & Monete nouamente ſtampato.

FIG. 8. TITLE PAGE OF FELICIANO'S *Libro de Abacho*

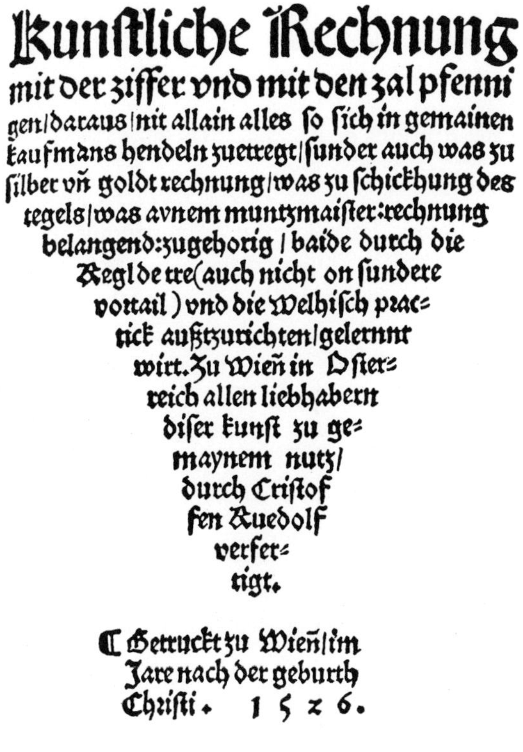

Kunstliche Rechnung
mit der ziffer vnd mit den zal pfenni
gen/daraus/nit allain alles so sich in gemainen
kaufmans hendeln zuetregt/sunder auch was zu
silber vñ goldt rechnung/was zu schickhung des
tegels/was avnem muntzmaister:rechnung
belangend:zugehorig / baide durch die
Regl de tre(auch nicht on sundere
vortail) vnd die Welhisch prac=
tick außzurichten/gelernnt
wirt. Zu Wieñ in Oster=
reich allen liebhabern
diser kunst zu ge=
maynem nutz/
durch Cristof
fen Ruedolf
verfer=
tigt.

Getruckt zu Wieñ/im
Jare nach der geburth
Christi. 1526.

Mit F.D.gnad vnd Priuilegien.

FIG. 9. TITLE PAGE OF RUDOLFF'S *Kunstliche Rechnung*

Page 165

Read *van* Ringelbergh. †

Page 167

To the list of works of 1531 add a 'Visierbuch' by Kern, Strasburg, 1531. See *Rara*, p. 496.

Page 168

Add to the Editions of Psellus: 'Tournon, 1592.'

Page 178

Under Editions: Possibly the 'Algorithmus Demonstratus' was written by 'Magister Gernardus.'

The name Albert appears in the Ed. pr. as the German Albrecht and in the 1561 edition, according to Professor Q. Vetter, as Albert, — 'Johannovi Albertovi autoru aritmeticky revněž z r. 1534.'

Page 180

To the Albert editions add Wittenberg, 1534, 1541, 1553, 1556; Frankfort, 1541 (colophon 1542).

Page 180

Under *Other works of 1534* mention is made of Rabbi Elias Misrachi's arithmetic. The commentary of Schreckenfuchs runs parallel to the Hebrew text, as shown in Fig. 10, where the numbers show the use of the zero in connection with the alphabetic numerals. It is also interesting to observe that the page is numbered at the top in Hindu-Arabic numerals. The book is now in the Plimpton Library.

Page 181

See *Rara*, p. 496. Also add 'Ein New Behendes und ganz grundtlichs Rechenbüchlin Auff der linien und Federn,' Nicholaus Felner, Erfurt.

המלאכת

832

כי אין שם המכול' ודרך אל ידיעתו
לכן תירה מן המוזכר לנו להקן ל
לוח כולל כל מיני הכמות הפרטים
עם הפרטים איזה הפרטים שיהיו
עד שירגיל כי האדם עצמו.

הנה הלוח הכולל הוא זה

ט	ח	ז	ו	ה	ד	ג	ב	א
חא	וא	דא	בא	אס	ח	ו	ד	ב
זב	דב	אב	חא	הא	בא	ט		ג
וג	בג	חב	דב	בס	וא			ד
הד	דס	הג	גס	הב				ה
דה	חד	בד	וג					ו
גו	וה	טד						ז
בז	דו							ח
אח								ט

Tabula multiplicationis digitorum inter se in qua situs literarum haud aliter quàm in cifris est aduertendus, ut בס *idem est quod* 20. *&* וג 36.

ואחר שכבר ידעת זה על פה הנה
בכר תוכל להכנס בדרך ידיעה זאת
המין ונאמר שכבר בארנו שההכא
הוא קבוץ מספרים רבים כלתי מ
מונחים מקובלים רק האחר כי לולא
זה לא נדע איזה מין מספרים הם
הנקבצים ולכן יהויב עלינו שנכתוב
בטור

Multiplicatio, inquit, est additio multorum numerorum, qui tamē expressi non scribuntur per summas suas, sed semel dūtaxat signamur, & alius multiplicans ducitur in multiplicandum.

Exem-

Fig. 10. Page from Misrachi's arithmetic

Page 182. SIR RICHARD BENESE.

Ed. pr. 1536. London, [1562].

Title. See Fig. 11.

Description. Sm. 8°. 8.5 × 14 cm. Black letter. 56 fol.

Editions. As stated in the *Rara*, p. 182, there were six or seven editions of which this is the fifth. There seems also to have been an edition of 1556 in addition to those in *Rara*, p. 182.

The work is essentially one of land measure, the necessary numerical operations being presupposed. There are numerical tables to avoid computation. The most interesting part is 'The Preface of Thomas Paynell, Chanon of Marton to the gentell Reader,' giving a crude history of the subject. He speaks of 'Pytagoras the fyrst Phylosopher,' 'the noble Archytas,' 'Archimenides,' and other ancients. The book is divided into two sections: (1) on 'The maner of measurynge of Lande,' including 'Of dyuersytie of lynes and angles'; and (2) 'To measure Tymber or Stone.' Because of the interest in this sixteenth-century English arithmetic, three facsimiles have been given (Figs. 11, 12, 13).

Page 182

After 'Editions,' l. 2, add 'Wittenberg, 1546.' Add 'J. Schlichtung, Rechnung mit Federen und uff die Reinströmisch und Schilling in Gold . . .' Strasburg, 1536. After *Other works of 1536*, add: There was an anonymous 'Algo-rithmus Linea,' 28 pp., Cracow. The works of Boethius and Morsianus were sections of a work entitled 'Arithme//tica specvlativa Boe-//tij per Iacobum Fabrum Stapulen-//sem in compendium redacta. // Aritmetica Practica // Christierni Morssiani in quinq̄ // partes digesta.// Prima de numeris integris.// Secvnda de fractionibus uul//garibus & physicis.// Tertia de regulis quibusdam. // Qvarta de progresione & ra-//dicum extractione.// Qvinta de proportionibus.// Basileae excvdebat // Henricus Petrvs.' 1536.

Page 183

Change 'Ed. pr.' of Vander Hoecke to 1514. † An examination of the editions in the British Museum shows that the date 1514 is correct.

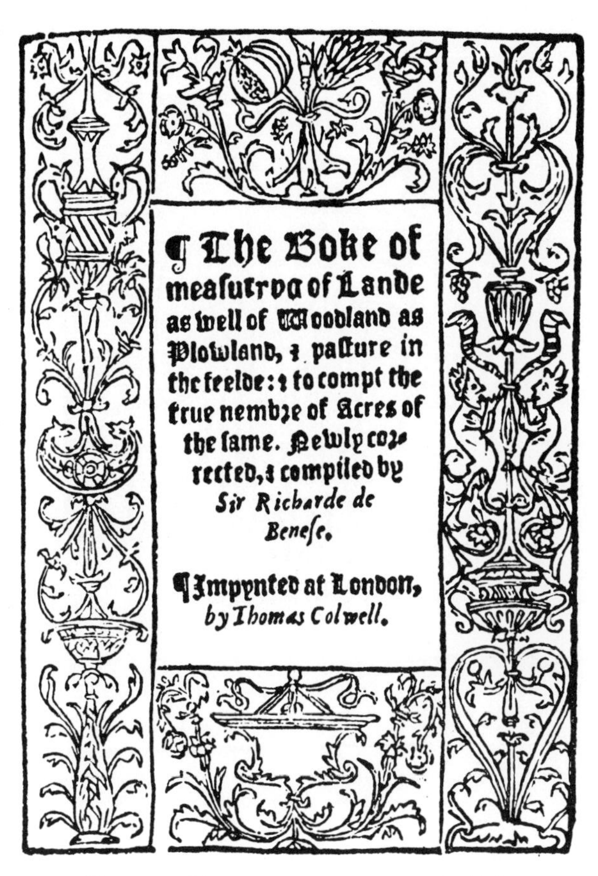

¶ The Boke of
measurynge of Lande
as well of Woodland as
Plowland, & pasture in
the feelde: & to compt the
true nombre of Acres of
the same. Newly cor-
rected, & compiled by
Sir Richarde de
Benese.

¶ Imprynted at London,
by Thomas Colwell.

FIG. 11. TITLE PAGE OF BENESE'S *Boke*

¶ Here foloweth the Booke to measure all Landes.

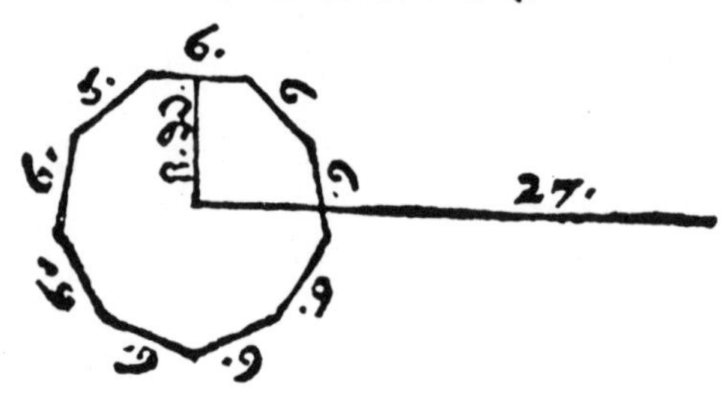

Fig. 12. From benese's *Boke*

¶ The maner of measu-
rynge of Lande, and coump-
tynge the nombre of acres
of the same.

y cause in measu
ringe of Lande, meny
men somtyme the sel-
lers, sel more measure
than ryght, somtyme
the byers by lesse measure thā right
be greately deceiued, by ye meaters
therof, the which be not experte and
connynge, both in true measurynge
of Lande, and also in true coūptinge
and summynge the numbre of acres
of the same. ¶ Therfore in this lytle
booke, ye shall reede certayne rules
much necessary for ye perfecte knowledge bothe of true measurynge of
Lande, and also of true comptynge
and summynge the numbre of acres
of the same.

¶ Of

FIG. 13. FROM BENESE'S *Boke*

Page 192

Loritus Glareanus also published 'Liber de Asse et partibus eius, cum Rerum et Vocum . . .,' Basel, 1551, which deals with weights as used in current problems.

Page 192

There was also another edition of Glareanus, Cracow, 1554.

Page 195

In Editions, after Cologne insert '1539; ib.' To *Other works of 1539* add 'Balthasar Wreedt, Ein kurz vnd // behend Rechenbüchlin vff // Linien vnd Ziffern . . .', Collen (Cologne), 1539.

Page 200

There were also editions of Gemma Frisius: Wittenberg, 1587; Leipzig, 1595; Cologne, 1578. The 1543 edition is now in the Plimpton Library. It is 'Paris, 1543,' 8°. See Fig. 14.

Page 214

To the list of editions of Recorde's 'Ground of Arts,' carried into the seventeenth century because of its importance to English schools, the following have come to the writer's attention: 1540, 1561, 1570, 1575, 1583, 1600, 1607, 1610, 1648, 1652, 1654, 1658, 1662, and ten more in the seventeenth century.

Page 216

There was another work in 1542, by Ch. Stiltz, 'Eyn neu Rechenbuch,' Strasburg.

Page 223

There was another work in 1543, H. Hoff, 'Ein kunstlich Rechenbiechlin mit der Ziffer und Zallpfenningen, Freiburg im Breisgau.' See p. 496.

ARITHMETICAE
PRACTICAE METHODVS FA
CILIS PER GEMMAM FRISIVM
Medicum ac Mathematicum.

PARISIIS,
Apud Ioannem Lodoicum Tiletanum, ex
aduerso Collegij Remensis.
1543

FIG. 14. TITLE PAGE OF GEMMA FRISIUS, 1543 EDITION

Page 231. LEONHARD GRUEBER.

Ed. pr. 1544. Strasburg, 1544.

Title. 'Rechenbuechel der // Linien vnnd Zyffer / . . . Durch Leonhardū Grueber // Burgerssune zu Oting / Camrer // vnnd Lateinischer Schül-//maister im Closter zu // Sewn gepra-// cticiert.'

Description. 38 fol., sm. 8°. A small work for merchants.

Page 233

There was a revision of Scheubel's 'Compendium,' Basel, 1560.

Page 236

There was also an edition of Sacrobosco at Antwerp, 1559. Other editions, Wittenberg, 1534, 1538, 1545, 1556, 1574, 1588, are in the Columbia University Library.

Page 238

Under *Other works of 1545* add Cardan, 'De proportionibus.' Change 'Pedro Espinosa' to 'Petro a Spinosa,' Salamanca, 1545. †

Page 244

After *Works of 1548* add 'J. Sekgerwitz, Rechenbüchlein auff allerley Hanthierung. Auffs new gebessert.' Leipzig, 1548.

Page 247

To the editions of Fischer's 'Arithmeticae Compendium' add s. l. 1562, and Leipzig, 1577. Change Ed. pr. to 1545, and 'Wittenberg, 1592,' to 'Wittenberg, 1545; ib., 1592.' †

Page 249

In *Other works of 1549,* add: Casper Wueyglerus (Weigler), 'Prima arithmetices practicae rudimenta,' Breslau.

The name Juan de Jciar, Vejar, Yciar is variously spelled. There was another book by him, 'Arte subtilissimo,' Saragossa,

1550, which contains in the second and third editions (ib., 1553, 1559) a brief treatment of arithmetic.

Page 250

On this page it is stated that Valentin Menher wrote three or four arithmetics, these being there listed, and on pages 249 and 346 two of them are described with analyses of their contents. The Plimpton Library has a fourth revision of the 'Livre // d'Arithme-//tique . . . Reueu par M. Melchior van Elstaer,' published at Antwerp in 1590, sm. 8°, A_1 to H_7, 9×14.4 cm. Printed forty years after his 'Practiqve,' it is naturally different from that in detail but not in purpose.

Page 252. GIOVANNE ROCHA. Ed. pr. 1550. Venice, 1550.

Title. 'Libro // . . . abacho il quale insegna fare ogni // ragione mercantile: & pertegare le // terre con larte della geometria & al-//tre nobilissime ragione straordina-//rie con la Tariffa . . . citta di Vinegia Il qua//le Lib. se chiama Thesoro vniuersale.'

Colophon. 'Stampata in Vinegia per Giouanni // Padouano. Nell' anno del // Signore.// M.D.L.'

Description. Sm. 8°. 80 unnumbered fol. Numerous illustrations. On fol. 79, v, is an address to the reader: 'Giovanne Rocha // allo Lettore.'

A mercantile arithmetic with applications to home, commerce, foreign exchange, and metrical geometry.

Page 252

For Scheubel's Ed. pr. 1551, read 1550. † This appeared as an introduction to Euclid's 'Elements.' Note this under 'Editions.' In l. 23 change 223 to 233. † Under Buckley's work note an edition c. 1570, and two editions of Seton's Logic, London, 1599, and Cambridge, 1631.

Page 253

To *Other works of 1551* add G. Stichel, 'Arithmetica // Wolgegründete Rech-//nung mit schöner forteiliger Anwei//sung zusamen bracht / Durch Georgium Sticheln // zu Leipzig. M.D.L.I.,' the only arithmetic printed that year in Leipzig.

Page 257

To *Other works of 1552* add 1561 to the Frankfort editions; also at the end, 'An Introduction for to learne to recken wyth the Pen or with counters,' London, 1552; ib., 1562; M. Stötter, 'Ein schön nutzlich Rechenbüchlin auff allerlei Kauffmans Rechnung,' Tübingen, 1552; D. Schedel, 'Ein nutzbar new Rechen und Exempel Büchlein . . .,' Nürnberg, 1552. To the Gülfferich list add 'Frankfort, 1561.'

Page 260

To *Other works of 1553* add 'Epitome in Arith. Boëthii, Stapulensis,' Basel, 1553; D. Manzoni, 'La Brieve Risolutione di Arithmetica Universale,' Venice, [1553].

Page 262

De Boissière also published 'Art Poetique redvict et abrege en singulier ordre et souveraine methode,' 1554; 'L'Art d'Arythmetique. Le tres excellent et ancien jeu Pythagorique dit Rithmomachie. . . . pour obtenir vsage et prompte habitude en tout nombre et proportion, nouvellement illustre par maistre Claude de Boissière . . . Paris, imp. de A. Brière,' 1554.

Page 263

Camerarius also wrote 'De graecis latinisque Numerorum Notis, et praeterea Sarracenicis seu Indicis . . .,' Leipzig, 1569. Mention may also be made of one of the editions after 1600, Frankfort, 1627.

Page 268

Omit the '(?)' after Cathalan's name.† It appears in an edition of 1599, which date should also be added to the 'Editions.'†

Page 286 CASPARUS PEUCERUS BUDISSINUS.

Ed. pr. 1556. Wittenberg, 1556.

CASPAR (KASPER) PEUCER (1525–1602) was a professor of medicine and mathematics (not at that time an unusual combination).

Title. 'Logisti-//ce Astronomi-//ca Hexacontadon et //scrvpv-lorvm sexagesimo-//rum . . . Item :// Logistice Regvlae // Arithmeticæ, Qvam Cossam // & Algebram quadratam uocant, . . . Avtore Casparo // Pevcero Bvdissino // Vitebergæ . . . M.D.LVI.' See Fig. 15.

Description. 8°, sign. A–V given to the theory of arithmetic and an introduction to algebra. The copy in the Plimpton Library is bound in parchment with a work by Joachim Camerarius, 'Elementa Rhetoricae, Lipsiae, M.D.LXIIII,' of 402 pages and an index. On Camerarius, see *Rara*, pp. 262, 186, 211.

Page 271

The Ed. pr. of Spänlin should be given as 1556 without the mark '(?).'†

Page 278

There was an edition of the 'Quesiti et Inventioni' by N. de Basearini . . . Tartaglia autore . . . 1554, but I have not seen it. It would make the publication extend over seven instead of five years.

Page 284

The title page reads: 'L'Arithmeticque // De P. Forcadel // de Beziers.// En laqvelle sont traictées // quatre reigles brefves. . . .' In l. 3, read 'Royal.'†

LOGISTI-
CE ASTRONOMI-
CA HEXACONTADΩN ET
SCRVPVLORVM SEXAGESIMO-
rum, quam Algorythmum minutiarum
Physicalium uocant, Regulis ex-
plicata & demonstra-
tionibus.

ITEM:
LOGISTICE REGVLAE
ARITHMETICÆ, QVAM COSSAM
& Algebram quadratam uocant, compendio
tractata & illustrata Exemplis, ut
Scholarum usui sit accom-
modata.

AVTORE CASPARO
PEVCERO BVDISSINO.

VITEBERGÆ EXCVDEBANT HÆ-
REDES GEORGII RHAVV,
ANNO M. D. LVI.

FIG. 15. TITLE PAGE OF THE *Logistice* OF PEUCERUS

Page 286

To *Other works of 1556* add: Ercole Marcello, 'Degli otto modi di Tariffa sopra l'oro e l'argento,' Venice, 8°. (Sometimes given as 1566.)

Page 290. GIOVANNI FRANCESCO PEVERONE.

Ed. pr. 1558. Lyons, 1558.

Title. 'Due breui e facili trattati,// Il primo d'Arithmetica: l'altro di Geometria:// ne i quali si contengono alcune cose nuoue // piaceuoli e utili, si à gentilhuomini come ar-//tegiani. Del Sig. Gio. Francesco // Peverone di Cvneo.// In Lione // per Gio. di Tovrnes.// M.D.LVIII.'

Description. This is the Ed. pr. described in *Rara*, p. 290, where the portrait of the author appears in the second edition of the book.

In line 1 add to the editions of Frankfort an der Oder, 1563, 1589; s. l., 1564; and Leipzig, 1595. Also add to the Cressfelt list an edition s. l., 1563.

Page 298

Among the works of 1560 there may be mentioned one of 48 pp. by Oliviero de Fonduli, 'Pratiche di Fioretti Merchantili,' a kind of commercial handbook of problems involving the weights and measures in various parts of Italy. Also to *Other works of 1560* add: Christofor Wiltvogel, 'Ein New // Künstlich Re-// chenbuch / Auff den Linien // vnd mit der Federn . . .' [1559–60]. Also, under Editions (1) add 1559. † The 1600 edition of Simon Jacob's work has the date 1552 in the Preface.

Page 298. GASPARO RIZZO Venetiano.

Ed. pr. 1560. Venice, 1560.

Title. 'Abbaco Nvovo,// molto copioso et // artificiosamente ordi-//nato . . . Con molte belle // proposte . . . Opera vtilissima et vni-//uersalmente in tutte le Citta & Prouincie // molto necessaria. . . .'

Colophon. . . . 'Opera de Gasparo Rizzo Venetiano.// Stampato in Venetia l'anno del Signore 1560.'

Description. 120 leaves, A–O. 10 × 16 cm. Unusual arrangement. The book begins with directions for the use of tables and explanations of the numerical operations. Most of the book is devoted to mercantile problems at the top of each page, the rest of the page being left for computations relating to the problems. The pupil therefore combines instruction with solutions.

Editions. There seem to be no others. This one is referred to in the *Rara*, p. 298, but is mentioned here because of its unusual arrangement and the fact that it has lately been secured for the Plimpton Library.

Page 310. JUAN PEREZ DE MOYA.

Ed. pr. 1573. Alcalá, 1573.

Title. 'Tratado de // Mathematicas en // Qve se contienen cosas de Arithme-//tica, Geometria, . . . y Philosophia natural.'

This is mentioned, with other works of the same author, in the *Rara.* The title page has a large woodcut of a royal seal and the words 'Con licencia, y priuilegio Real de Castilla y Aragon.// En Alcala de Henares.// Por Iuan Gracian. Año de 1573.' It is in folio form, 19.5 × 29 cm., 752 pp., divided into ten books, with various tables of contents. It contains a considerable amount of historical material and is one of the most interesting works of its kind that appeared in the sixteenth century. This copy was acquired after the *Rara* appeared.

Page 310

There was also an edition of the Arithmetica Practica y Especulativa of Juan Perez de Moya, Madrid, 1609; and another, Alcalá, 1569.

Page 311

To *Other works of 1562* add: Jiřík Mikuláš Brněnský wrote an arithmetic in Bohemian, published at Praha (Prague) in

1562. Also, in l. 6, note editions as late as 1776 and 1784. † Under Strigelius, for the date of printing see *Rara*, p. 311, l. 7 b.

Page 314. WOLFFGANG HOBEL.

Ed. pr. (1563). Nürnberg, s. a. (1563).

Title. 'Ein nützlich Re-//chenbüchlein / Mit vil schö-//nen Regeln und Fragstucken . . .' See Fig. 16.

Description. 8°, 9.8 × 15 cm. The dedication is dated 'Nürnberg / den 29 // Nouemb. im 1563. Jar. // . . .,' but no imprint date is given.

Colophon. 'Gedruckt zu Nürmberg // Bey Valentin // Newber. //Durch Wolff Hobel Rechenmeyster // in der Bindergassen woh-//nende / verleget.'

Editions. (1563), 1565, 1577.

To *Other works of 1563* add: A. Lonicer, 'Arithmetices brevis introductio ab autore de integro recognita,' Frankfort, 1563. In 1563 there was published at Vitebergae (Wittenberg) a 'Logistica Astronomica,' written by Sebastianus Theodoricus Winshemius, containing a little arithmetic.

Page 315

In 1564 Johann Sthen published s. l. an 'Arithmetices // Evclideae // Liber // Primus,' but on the theory of numbers only. As to Euclid, his 'Elementa' includes some of the theory of number, but is not an arithmetic in the present American use of the term. Such are not in general listed. To the last line add 'the algebra not being included.'

Page 319

To *Other works of 1565* add: 'Matthäus Nefe, Arithmetica, zwei newe Rechenbücher, das Erste auff der Linien und Federn. . . . Breslau, 1565.' Add also: H. Henning, 'Gerechnet Rechenbüchlein, darinnen alle yetztgebreuchiche Kauffmans, . . . Nürnberg, 1565.' Ib., J. Kaltenbrunneh, 'Ein newgestellt künstlich Rechenbüchlein. 1565.'

Ein nützlich Re-
chenbüchlein / Mit vil schö-
nen Regeln vnd Fragstucken / Alle new erfun-
den / auff allerley Kauffschlag: deren vor
keine an tag kommen. Durch
Wolff Hobel Rechenmeyster / der hoch
löblichen Reichsstat Nürnberg zu ehrn / allen liebhabern der Arithmetica zur fürderung
inn Truck beschriben.

Mit Röm. Key. May. Freyheit in 10 Jarn nit nach [illegible] / Bey
peen 10 mr. lötigs golts / vnd verlier [illegible] Bücher.

FIG. 16. TITLE PAGE OF HOBEL'S *Rechenbüchlein*

In l. 2 b. insert 'Graevius, Ratio computandi per digitos, Paris, 1565.'

Page 320

Among Trenchant editions after 1606 add Lyons, 1617, 1618. The first edition should be given as 1563, two copies being in Dublin and Oxford, and probably others not in lists examined.

Page 322

The edition of 1575 was printed in Naples. Among the books of 1566 may be noted three by W. Frantz: (1) Jar rechnung, (2) Treidt vnnd Zentner rechnung, (3) Win Rechnung, although they are chiefly composed of numerical tables. See *Rara*, p. 496.

Page 325

To the editions of Petri's work add 's. l. 1598.'

Page 327

Under Humphrey Baker, change Ed. pr. to 1562, and do the same on pp. 328, 329. † To Editions (London) add 1564 and 1598. There are several editions after 1600, the one of Henry Phillippes, greatly enlarged, appearing in 1670. In the Plimpton Library there is an edition 'Now newly perused, augmented, // & amended in all the three parts. At London, // Printed by Tho: Purfoot . . . 1631.' 6 leaves unnumbered + 198 numbered + 12 unnumbered. Sm. 8°. 7.8 × 12.7 cm.

Page 330

There was also a reprint of the Ramus work, Frankfort, 1627.

Page 333

Besides Rudolphus Snellius, who revised the Arithmetica of Ramus, already mentioned, there was his son Willebrordus

Snellius. He wrote a later work, 'De // Re Nummaria // Liber singularis,' Amsterdam, 1635, 72 pp., and 'P. Rami // Arithmeticæ // Libri dvo,// cum commentariis // Wilebrordi Snellii R. F.// Ex Officina Plantiniana // Raphelengii, // M.D.CXIII.' This consists of two 'libri' of 72 and 106 pp., 10.2 × 17 cm.

Page 335

The Ramus work also appeared with the Schonerus edition of the geometry at Frankfort in 1627. A Snellius edition also appeared at Antwerp, Plantin Press, in 1613.

Page 338

Add to the works of 1569: N. Werner, 'Rechen Buch von allerley Kaufmannschlag auff sonderlichen Vorthel der Regel Detri . . .,' Frankfort am Main, 1569.

Page 340

See *Rara*, p. 497.

Page 346. PETER BEAUSARDUS.

Ed. pr. (?). Louvain, 1573.

Title. 'Arithmetices // praxis, ad qvam // vetervm permvlta ex-//empla revocata expli-//cantur, & illustrantur,// Petro Beausardo, Doct. Med. & Ma-//thematum Professore Re-//gio, autore . . . Lovanii // . . . M.D.LXXIII. . . .'

Description. 8°, 9.7 × 15 cm. 30 pages unnumbered + 119 numbered folios. It is divided into four parts, the first being 'De integrorvm svp-//putatione' (fol. 1, r.), the second 'De fragmentorvm // svppvtatione' (fol. 40, r.), the third 'De praxeos arith-//metices regvlis' (fol. 53, v.), and the fourth 'De rationibvs' (fol. 106, r.). Each part is divided into chapters. For the time a fairly satisfactory textbook.

Page 347

There may have been a 1547 edition of Sekgerwitz, or the 1547 may have been printed by mistake for 1574. (H. Grosse, *Historische Rechenbücher des 16. und 17. Jahrhunderts*, Leipzig, 1901, p. 67, gives 1547.)

Page 353

In l. 5, the German name for Giřjka . . . is Georg Gehrl. See *Rara*, p. 359, l. 17.

Page 353. BERNARDUS SALIGNACUS.

Ed. pr. 1575. Frankfort, 1580.

Title. 'Bernardi // Salignaci Bvr-//degalensis Arithmeticæ // Libri duo,// et // Algebræ totidem,// cum demonstrationibus.// Francofurti // Apud Andream Wechelum,// MDLXXX.' See Fig. 17.

See *Rara*, p. 359, for editions and another spelling of Bernard.

Page 359

There were no other editions of Chauvet Champenois before 1601, but after 1600 there were several, including 1615, 1619, 1640, and 1645. The 'Addenda' does not usually mention editions after 1600.

Page 361

In l. 5 add 'Kandler, Arithmetic, Regensburg.' Under the Urstisius editions add 'Basel, 1586.' In l. 3 b., after 1569, 4°, insert 'ib., 1586.'

Page 367

In l. 4, after 1577 add 'Frankfort, 1580.'

BERNARDI SALIGNACI BVR-DEGALENSIS ARITHMETICÆ LIBRI DUO,

ET

ALGEBRÆ TOTIDEM,
cum demonſtrationibus.

FRANCOFURTI
Apud Andream Wechelum,
M D L X X X.

FIG. 17. TITLE PAGE OF THE *Arithmetica* OF BERNARDUS SALIGNACUS

Page 375

There was an edition of the 'Epitome Arithmeticæ Practicæ' of Clavius at Cologne, 1607. To the other editions add 'Mogant (Mainz), 1614.' See *Rara*, p. 497.

Page 383. ANTON SCHULTZE.

Ed. pr. 1584. Frankfort an der Oder, 1584.

Title. 'Rechenbuch auff // Muentz vnd gewicht in Schle=// sien / sampt vergleichung derer / in. . . . Lei=//pzig vnd Franckfurt an der Oder . . . Durch // Antonium Schultzen Arithmeticum // vnd Deutschen Schulhalter in // Lignitz . . . Gedruckt zu Franckfurt an der Oder/ bey // Andream Eichorn / Anno 1584.'

Description. 8°, 168 leaves, 9 × 15.5 cm. One of the best commercial arithmetics of its time.

Colophon. 'Beschlusz an den Leser,' followed by ten lines in verse, beginning with

Hie endt ich nun in Gottes Nam.

See also *Rara*, pp. 383 and 497.

Page 386

Change Ed. pr. to 1582. † There were only four copies known to the bibliographer Zeper. There are many differences between this edition and the later ones.

Page 389

As stated in *Rara*, p. 497, in the Cosmographia of Franciscus Barocius (Barocio, Barozzi), Venice, 1585; Pavia, 1560 (2 ed.), there are five pages on arithmetic.

ANONYMOUS. Ed. pr. c. 1585. Paris, [1585].

Title. 'L'Arithme-//tiqve et maniere // d'apprendre à chiffrer & compter // par la plume, & par les gects . . . Auec la maniere de tailler // la plume.// . . . A Paris,// Chez Pierre Ménier . . .'

Description. 6.7 × 11.8 cm. 16°, 79 fol. numbered + 5 fol. unnumbered. The last folio, verso, has the 'Wheel of Pythagoras,' with the name 'Rout de Pythagore.' There is another one from the same woodcut, on fol. 80, v. After this, beginning with fol. 81, r., there are 7 pages of rules and problems in verse. One tells how to make and repair quill pens and how to use them to advantage, and how to choose paper, ink, and parchment. It is evidently a book for merchant apprentices. It contains the usual rules for computation and a number of simple puzzle problems.

Page 391. Add Albert, Magdeburg, 1588.

Page 393

To works bearing the date 1590 add: Giacomo Campolini, 'Positioni Arithmetiche & Operationi de Numeri Rotti . . .,' Venice, 1590, but the date should be 1600. To the last line add: Eberhardus Popping von Münster, 'Ein Newes Rechen Büchlein auf Linien vn Federn,' [Helmstadt,] 1590.

Page 393. EBERHARD POPPING. Braunschweig, 1590.

Title. 'Ein Newes // Rechen Büch-//lein auff Linien un Federn // jetzundt von newem wieder vberse-//hen / und mit mehren denn zuuor nützli-//chen / vnd nötigen Exempeln / auch mit // der Regul von Schiffs Parten // vnd andern sonderlichen // behenden vorteilen // gebessert.// Der gemeinen Jugendt und an-//fahenden der Arithmetic Kunst // zu gute verfertiget // Durch // Eberhardum Popping von // Münster / verordenten Deudschen // Schreib: oñ Rechenmeister der weit be-//rhümpten und löblichen Stad // Braunschweig.// . . . uliusfriedenstedt. 1590.' See Fig. 18.

Colophon. M.D.CX.

The preface is dated 1587, the title 1590, and the colophon 1610, showing the time required to produce the book.

Ein Newes
Rechen Büch-
lein auff Linien vñ Federn/
jetzundt von newem wieder vberse-
hen/ vnd mit mehren denn zuuor nütz-
chen vnd nötigen Exempeln/ auch mit
der Regul von Schiffs Parten
vnd andern sonderlichen
behenden vorteilen
gebessert.

Der gemeinen Jugendt vnd anfahenden der Arithmetic Kunst
zu gute verfertiget/

Durch

Eberhardum Popping von
Münster/ verordenten Deudschen
Schreib: vñ Rechenmeister der weitbe-
rhümpten vnd löblichen Stad
Braunschweig.

Ulwsfriedenstedt. 1590.

FIG. 18. TITLE PAGE OF POPPING'S *Rechen Büchlein*

Page 396

Change Aretio (l. 20) to Oretio. † Under *Other works of 1591* add: Lindebergius, 'De precipuorum,' Rostock, 1591, containing a few pages on arithmetic. For Lindebergius see *Rara*, p. 498.

Page 404

In l. 2 note that Suevus was born July 26, 1526, and died May 15, 1596.

Page 407

To the publications of 1594 add 'Anonymous, Lyons, 1594,' with title 'Aritmetiqve // facile a appren-//dre a chiffrer et // compter par la // plume & par les gects :// tres-vtile à tou-//tes gens.// Ensemble plusieurs excellentes sentences moral-//les, faictes par quatrains, & ordre alpha-//betique: Auec la maniere de // tailler la plume.// Nouuellement reueuë & corrigee.// A Lyon,// Par Benoist Rigavd.// M.D.XCIIII.' See Fig. 19.

Description. 12mo., 7.3 × 11.3 cm. 79 pp. Following the 'Aritmetique' are collections of moral sentences and adages in meter, arranged alphabetically. It seems to be the only book of its kind to appear before the seventeenth century. It is rare and was only recently obtained for the Plimpton Library.

Page 407. BARTHOLD WIESACK.

Ed. pr. 1595. Stargart, 1595.

Title. 'Ein Neuw // Rechenbüchlein // auff der Linien und Federn, / auff // allerley Kauffhandelunge / dem einfelti-//gen gemeinen Man / vnd anheben-//den der Arithmetica Liebha-// bern zu gut.// Durch // Barthold / Wiesack / Rech-//enmeister zu Stargardt / auff der // Ihna / Auffs fleissigest zusam-//men gedragen.// Gedruckt zu Alten Stet-//tyn / bey Andreas Kell-// ner / S. Erben.// Anno 1595.' See Fig. 20.

Page 408. GIACOMO TRIVISANO. Ed. pr. 1597. 1597.

Title. 'Memoriale // di Abbaco,// ritrovato vn nvovo modo // & ordine di tariffa . . . // ad ogni sorte di persone // E in ogni sorte di negotio // . . . In Venetia M.D.XCVII.'

ARITMETIQVE
FACILE A APPREN-
DRE A CHIFFRER ET
COMPTER PAR LA
plume & par les gects:
tres-vtile à tou-
tes gens.
Ensemble plusieurs excellentes sentences moralles, faictes par quatrains, & ordre alphabetique: Auec la maniere de tailler la plume.
Nouuellement reueuë & corrigee.

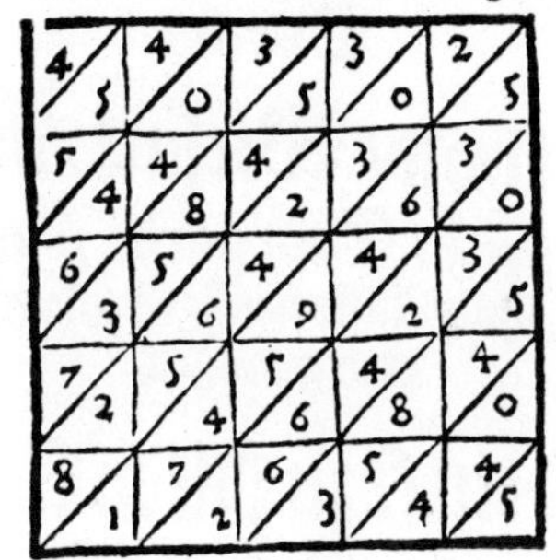

A LYON,
PAR BENOIST RIGAVD.
M. D. XCIIII.

FIG. 19. TITLE PAGE OF THE *Aritmetique* OF 1594

Description. This is the book mentioned among *Other works of 1596* on p. 408, and is now in the Plimpton Library. 10.5 × 17 cm. [40] + 377 numbered pages. The 'Dechiaratione della prima parte' of 36 pages, unnumbered, is signed 'Primo Auttor Giacomo Triuisan // di questa Tariffa.' The book is not an arithmetic in the ordinary sense. It consists of 377 pages of tables for easy computation, beginning with products of 2 × 2 and extending to 995 × 50, then (with breaks) to 3000000 × 50. The rest of the tables consists of products involving monetary fractions, corresponding to pounds, shillings, and pence.

Ein Neuw Rechenbüchlein /
auff der Linien vnd Federn / auff
allerley Kauffhandelunge/ dem einfelti-
gen gemeinen Man/ vnd anhebenden der Arithmetica Liebhabern zu gut

Durch

Bartholde Wiesack / Rech-
enmeister zu Stargardt / auff der
Ihna/ Auffs fleissigest zusammen gedragen.

Gedruckt zu Alten Stet-
tyn/ bey Andreas Kell-
ner/ S. Erben.
Anno 1595.

FIG. 20. TITLE PAGE OF WIESACK'S *Rechenbüchlein* OF 1595

Page 408

See *Rara*, p. 497.

Page 412

In l. 20, for Nürnberg read Breslau. †

Pages 415–416

In connection with Newdörffer mention should be made of an edition as late as 1634, 'Kuenst- vnd ordentliche // Anweisung // in die Arithmetic / als // eine Mutter vieler Kuensten; . . . In XIII. Buechlein verfasset . . . Editio V.// Nuernberg / . . . Im Jahr 1634.' 14.6 × 8.8 cm. Among the various topics is an extensive treatment of 'Die Regel Helcataim oder Positionum.'

Page 416

After 'Editions' (l. 1) read, 'There were at least three editions. One dated 1613 says that it was the third edition.' Cancel 'There was no other edition.' †

Page 418

See *Rara*, p. 497.

Page 421

To *Other works of 1599* add: Jean Coutereel, 'Cyffer-Boeck,' Middleburg, 1599. Coutereel was a Belgian and his 'Konstig cyfferboek' was translated into French by H. Cole.

Page 425

There is a manuscript of Zuchetta's work in the Plimpton Library, written in 1692. *Rara*, p. 497.

Page 429

Add to *Other works of 1600*: Lachar, 'Algorithmus Mercatorum,' s. l. a., but c. 1600. *Rara*, p. 498.

Page 435

Late authorities place the death of Dagomari in 1373 or 1374.

Page 477. ANONYMOUS. Italian MS., 1490.

Title. 'Questo sie uno libro di abacho . . .' (F. 1, r.). The text begins: '9. 8. 7. 6. 5. 4. 3. 2. 1.// La decima figura sie questa o laqual uien chiamata zero . . .'

Description. 4°, 15 × 20 cm., the text being 10.5 × 13 cm. Ff. 1–68, r., 12–23 ll. Ink black and red, the latter being faded. A master's or pupil's textbook on arithmetic designed especially for mercantile purposes. The problems in division make use of the 'scratch method.' Numerous numerical tables of multiplication and division somewhat like those illustrated in the *Rara* on pp. 464 and 471. Bound with two other treatises: Addresses and formulæ of letters (ff. 68, v., to 84, r.) and Problems in practical geometry (ff. 84, v., to 92).

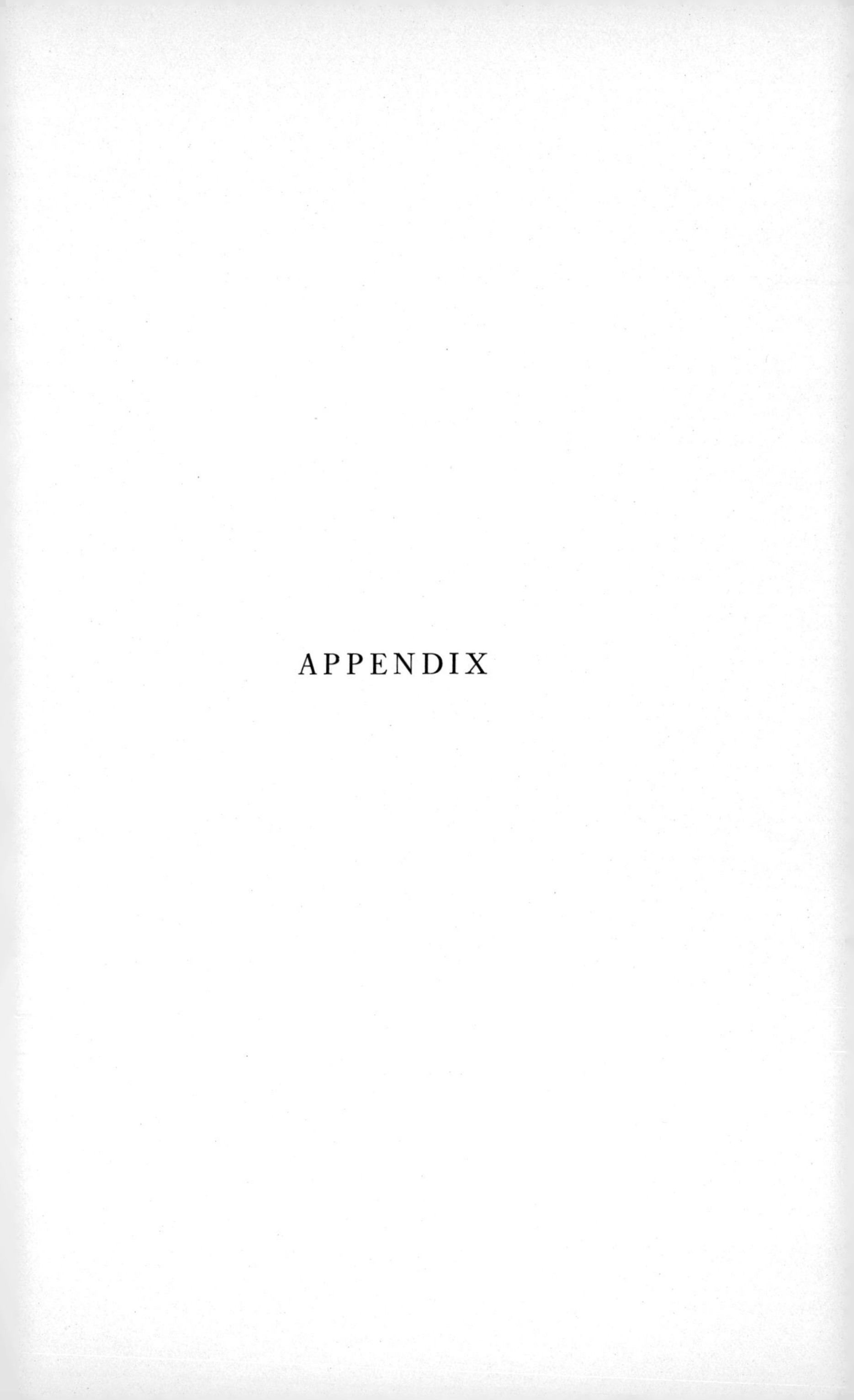

APPENDIX

ARITHMETICAL BOOKS

FROM

THE INVENTION OF PRINTING TO THE PRESENT TIME

BEING

BRIEF NOTICES OF A LARGE NUMBER OF WORKS

DRAWN UP FROM ACTUAL INSPECTION

BY

AUGUSTUS DE MORGAN

OF TRINITY COLLEGE, CAMBRIDGE
SECRETARY OF THE ROYAL ASTRONOMICAL SOCIETY: FELLOW OF THE CAMBRIDGE PHILOSOPHICAL SOCIETY
AND PROFESSOR OF MATHEMATICS IN UNIVERSITY COLLEGE, LONDON.

"Much surprised, no doubt, would the worthy man have been, had any one told him that two hundred years after his death, when no man alive would think his ideas on the nature of mathematics worth a look, the absence of better materials would make his list of" arithmeticians "not only valuable, but absolutely the only authority on several points."—DUBLIN REVIEW, No. XLI.

LONDON

TAYLOR AND WALTON

BOOKSELLERS AND PUBLISHERS TO UNIVERSITY COLLEGE

28 UPPER GOWER STREET

1847

TO THE

VERY REVEREND GEORGE PEACOCK, D.D.

DEAN OF ELY, LOWNDEAN PROFESSOR,

&c. &c. &c.

MY DEAR SIR,

IT never entered into my head till now to adorn the front of any book of mine with an eminent name: and the reason I take to be, that I have hitherto never chanced to write a separate work* upon any subject with which the name of one individual was especially associated in the minds of those who study it. But you are the only Englishman now living who is known, by the proof of publication, to have investigated both the scientific and bibliographical history of Arithmetic: and this compliment, be the same worth more or less, is your due, and would have been, though my knowledge of you had been confined to your writings. And it is the more cordially paid from the remembrance of nearly a quarter of a century of personal acquaintance, and of many acts of friendship on your part.

I have rather grown than made this catalogue. It never occurred to me to publish on the subject, till I found, on a casual review of what I had collected, that I could furnish from my own books a more extensive list than Murhard, Scheibel, Heilbronner, or any mathematical bibliographer of my acquaintance, has described *from his own inspection.* Knowing, from sufficient experience, the general inaccuracy and incompleteness of scientific lists, I therefore determined to do what I could towards the correction of both, by describing as many works as I could manage to see. From

* Had the regulations of the work in which it appeared permitted, it would have been most peculiarly appropriate to have inscribed my treatise on the *Calculus of Functions* to my friend Mr. Babbage.

the Royal Society's library, the stock of Mr. Maynard the mathematical bookseller, and my own collections, with a few from the British Museum and the libraries of private friends, including three or four of great rarity from yourself, I have accordingly compiled the present catalogue. I have also given in the Index, in addition to the names of the authors whom I have examined, those of all whom I could find recorded as having written on the subject of Arithmetic, whether as teachers, historians, or compilers of special tables for aid in the main operations, independently of logarithms and trigonometry.

A great number of persons are employed in teaching arithmetic in the United Kingdom. In publishing this work, I have the hope of placing before many of them more materials for the prevention of inaccurate knowledge of the literature of their science than they have hitherto been able to command, without both expense and research. Your History, unfortunately for them, is locked up in the valuable, but bulky and costly, Encyclopædia for which it was written. I may have the gratification of knowing that some, at least, of the class to which I belong, have been led by my catalogue to make that comparison of the minds of different ages which is one of the most valuable of disciplines, and without which the man of *science* is not the man of *knowledge.*

The most worthless book of a bygone day is a record worthy of preservation. Like a telescopic star, its obscurity may render it unavailable for most purposes; but it serves, in hands which know how to use it, to determine the places of more important bodies.

The effect of this work which would please me best, would be, that the professed bibliographer should find it too arithmetical, and that the student of the history of the science should find it too bibliographical. I might certainly have entered more into the methods of the several works, with advantage to the reader. But I could not attempt to write the complete annals of arithmetic; this would require still more books: neither could I, without losing sight of my plan altogether, combine the information here presented with that derived from other sources. That plan has been, to attempt some rectification of the numerous inaccuracies of existing catalogues, by recording only what I have seen myself. And it is a sufficient justification of the course I have taken, that I have produced in this way a *larger* catalogue than yet exists among those devoted, in whole or in part, to this particular subject.

None but those who have confronted the existing lists with the works they profess to describe, know how inaccurate the former are; and none but those who have tried to make a catalogue know how difficult it is to attain common correctness. There is now a prospect of this country possessing in time *such a record of books as can be safely consulted in aid of the history of literature,*—I refer to the intended catalogue of the library in the British Museum. I, for one, can only hope that the chance will not be lost by any attempt to expedite its formation, in deference to the opinion of those who either are not aware how bad existing lists are, or are willing to take more than a chance of having nothing better. If, through negligence or fear on the part of those who have really compared book-lists and books, the expression of public feeling which any *primâ facie* case against public officers so easily obtains, should succeed in hurrying the execution of this national undertaking, the result will be one more of those magazines from which non-existing books take their origin, and existing ones are consigned to oblivion by incorrect description. However extensive the demand for spoiled paper may be, it should be remembered that the supply is immense, and that there is no need to insist on Parliament, at least, furnishing a larger contingent than it is obliged to do already.

I make no apology for troubling you to read this diatribe: it is your affair, and mine, and that of every one who believes accuracy to be an essential characteristic of useful knowledge.

I remain,

My dear Sir,

Yours sincerely,

A. DE MORGAN.

UNIVERSITY COLLEGE, LONDON,
April 29, 1847.

INTRODUCTION.

At the end of this work will be found an Index containing, besides the names of the authors mentioned in the Catalogue, who may be known by the paginal figures opposite, every name which I have found *any where* as belonging to a writer on arithmetic before 1800. Since that period I have not swelled the list from mere *catalogues*, but have contented myself with what I had either seen, or learnt from some other source. For the existence of such writers, or of their works, of course I do not vouch, except in those cases in which the work is found in the body of the Catalogue. Of those who stand unpaged in the index, all I can say is, that they are taken from many sources, and that there is in each case plenty of reason to make inquiry for their writings, and strong presumption that their works are still to be found.

A word with an asterisk before it is the name of a writer who is referred to by Dr. Peacock. The great apparent preponderance of German names arises partly from the number of works written on the subject in Germany being very considerable, partly from the bibliographical catalogues of that country being more full than those of any other, and partly from the names which seem to be German, really including Danes, Dutch, Belgians, and some Swiss. I am far from thinking that this list contains even the third part of the names of those who have really written on the subject. I have not been particular in searching for any thing after 1750, though I have not refused what came in my way. I might have very much increased the list of recent Germans, from Rogg's *Bibliotheca Selecta* (Tubingen, 1830). Looking at the various countries which enjoyed the art of printing from 1500 downwards, I have an impression, from all that I have gathered, which would lead me to suppose that the number of works on arithmetic published in Latin, French, German, Dutch, Italian,

Spanish, and English, up to the middle of last century, cannot be less than three thousand, which gives to each language less than an average of one a year. Few of these would seem to fall within the province of the historian. Dr. Peacock refers to about a hundred and fifty. Unfortunately, history must of necessity be written mostly upon those works which, by being in advance of their age, have therefore become well known. It ought to be otherwise, but it cannot be, without better preservation and classification of the minor works which people actually use, and from which the great mass of those who study take their habits and opinions. The *Principia* of Newton is, if we believe the title-page, a work of the seventeenth century: but the account of the effect which it produced on science belongs to the eighteenth. It was not till many years after the publication of the *Principia,* that its predecessor in doctrine, the great work of Copernicus, produced its full effect upon general thought and habit. Nor have we any reason to suppose that it could have been otherwise. The great exceptions will always bear, perhaps, as large a ratio to the average power as ever they did: it is as likely as not, that if the intelligence of the sixteenth century had been sufficient to verify and receive the opinion of Copernicus at once, some predecessor of his might have been Copernicus, and he or another of his day might have been Newton.

It is then essential to true history, that the minor and secondary phenomena of the progress of mind should be more carefully examined than they have been. We must distinguish between the progress of possibilities and that of actual occurrences. Our written annals shew us too much what might have been, and too little what was: they give some words to the slow reception of an improvement, and more sentences to the account of the one man who was able to make it before the world at large could appreciate it.

The public is beginning to demand that civil history shall contain something more than an account of how great generals fought, great orators spoke, and great kings rewarded both for serving their turn. The progress of nations might well be described, for the most part, with much less mention of any of the three: but the parallel does not hold of knowledge. Copernicus and Newton would fill a large space, though the history were written down to that of every individual who ever opened a book: but it seems to me that they, and their peers, are made to fill *all the space.* Nor will it be otherwise until the historian has at his command a readier access to second and third rate works in large numbers; so that he may write upon effects as well as causes.

This list contains upwards of fifteen hundred names, of which a few may be duplicates of others, arising from the wrong spelling of my authorities.* But these must be much more than counterbalanced by the number of names which belong to two or more authors, but which only appear once in my list. Thus there are two Digges's, three Riese's, two named Le Gendre, Walker, Newton, Wallis, &c., and several Taylors, Butlers, &c., though each name is only mentioned once. There are several cases in which I have not ventured to strike out one of two names, though there is every reason to suppose one is a mistake for the other; as Caraldus and Cataldus, Cappaus and Cuppaus, Döhren and Dühren. But I have often been deceived in this way; and have more than once been obliged to re-insert names which I had struck out, supposing them to be only different spellings of names already in the list. There are several whom I have seen in more places than one, who are clearly Germans with names metamorphosed by wrong reading of the black letter, or use of the genitive for a nominative, or both. Thus Petzoldt has become Pekoldts, Seckgerwitz has been made Sectgerwik, Schultze has been made Schulken, and so on. These obvious mistakes of course have not been admitted. Moreover, several persons are, I suppose, down in the list under two names by which they are known. Thus, Fortius may be Ringelbergius (whose real name was Sterk), Blasius may be Pelacanus. But as I cannot undertake to assert that there is no Fortius except Ringelbergius, &c., I have let both names stand in the list. Again, next to Buckley comes Budæus. Many foreign writers, Heilbronner among the rest, have turned *Buclæus* into *Budæus*, so that in all probability the second of these names is a mistake for the first. And yet there may be also some Budæus who has written on arithmetic; though, having excluded writers on weights and measures only, I have not put down the author of the work *De Asse*.

I have had, in one or two instances, to throw away German *authors*, for a very obvious reason. The reader will not find the works of *Anleitung*, or *Grundriss*, or *Rechenbuch* in my list, which is more than can be said of every one which has preceded it.

I have not attempted to translate the names of those who

* It is necessary, for instance, to keep close watch upon a writer who introduces among his English authors *Gul. de Cavendy dux de Xeucathle.*

wrote in Latin, at a time when that language was the universal medium of communication. In every such case I consider that the Latin name is that which the author has left to posterity; and that the practice of retaining it is convenient, as marking, to a certain extent, the epoch of his writings, and as being the appellation by which his contemporaries and successors cite him. It is well to know that Copernicus, Dasypodius, Xylander, Regiomontanus, and Clavius were Zepernik, Rauchfuss, Holtzmann, Muller, and Schlüssel. But as the butchers' bills of these eminent men are all lost, and their writings only remain, it is best to designate them by the name which they bear on the latter, rather than the former. In some cases, as in that of Regiomontanus, both names are frequently used: in others, where the Latin consists in a termination only, as Tonstallus for Tonstall, or Paciolus for Pacioli, it matters nothing which is used. It may happen that errors are introduced by returning to the vernacular in a wrong way. I should like to know how it is shewn that Orontius Fineus was Oronce Finée: or in what respect this reading is more finically correct than Horonce Phine, which has great antiquity in its favour? In the case of Vieta, Viète is certainly wrong: he was called by his contemporaries Viet (which I suspect to be derived from the Latin form) and de Viette, never Viète.

It may also be asked, why the unlatinizing process should, for confusion's sake, be practised by the learned only; when it is pretty certain that the world at large will never reconvert Melancthon into Schwartzerd, or Confucius into Kaen-foo-tzee. Neither will they restore to the Popes and other priests of the Roman Catholic Church the names under which they were born and educated.

In many cases it would be impossible to recover these last. For myself, I am well content with the name under which an author was known in literature to his contemporaries, and has been handed down to us, his successors. I know of no canon under which it is imperative to speak of a writer rather as his personal acquaintance than as his reader: and, so far as feeling of congruity is concerned, I think *Alexander ab Alexandro* looks better at the head of a Latin preface than *Saunders Saunderson.* Those who really wish to catch the tone of the middle ages, and shew themselves quite at home, should dwell on the *Christian* name, and make the surname a secondary distinction; should learn to think of *Nicolas* (who happened to be called Copernicus, or Zepernik, it should matter little which) and *Christopher,* to whom the calendar was entrusted.

That this list must be very imperfect I am well aware, for I have been able to add many names to it which I never found in any catalogue. But it will not be useless. It may furnish a reason for preserving any work whatsoever which comes in the way of the reader of this book. If the name be in the list, a book should not be destroyed which has been somewhere catalogued and recognised as a portion of the existing materials for the history of science. But if the name be not in the list, it is obvious that there is some curiosity about a writer whose name is not in the most numerous catalogue of arithmetical authors that has ever been collected in any one place. And it is undeniable that every name must be in one or the other predicament.

I now come to the Catalogue which forms the body of this work. The defects, which every one who has examined the lists that are extant, knows to prevail, arise, in great measure, from the titles and descriptions of books being copied by those who had never seen the books themselves. This is not the worst; for a true copy of a true copy is a true copy: many of these accounts have finally become formal representations of informal titles, which, though not originally intended for more than sufficient indication of the book mentioned, have been used as if they had been full and accurate descriptions. There is a large number of works, not much distinguished in the history of science, each of which has, nevertheless, done its part in its day. The minor points of the same history depend much upon these books, which, being neither of typographical curiosity, nor of literary fame, are gradually finding their way either to the waste-paper warehouse, or the public library. These two depositories are almost equally unfavourable to works of no note assuming their place in the annals of the knowledge to the progress of which they have contributed. Take the library of the British Museum, for instance, valuable and useful and accessible as it is: what chance has a work of being known to be there, merely because it is there? If it be wanted, it can be asked for; but to be wanted, it must be known. Nobody can *rummage* the library, except those officially employed there, who will only now and then have leisure to turn their opportunities to account in any independent literary undertaking. And it would perhaps be difficult to make any regulation, under which persons not belonging to the institution might have access to see what is there. Nor will the publication of the catalogue do much towards supplying the defect. Titles of books are but vague indicators of their contents; and a catalogue* of half a million of entries, even if its contents could be guessed at by the titles of the books, is not made to be read through.

It would be something towards a complete collection of mathematical bibliography, if those who have occasion to examine old works, and take pleasure in doing it, would add each his quotum, in the shape of description of such works as he has actually seen, without any attempt to appear more learned than his opportunities have made him. This is what I have done in the descriptive catalogue of works on arithmetic, without selection, or other arrangement than order of date. The only reason for a work being in this list, is that it has come in my way: the only reason for one being out of it is, that it has not come in my way, at least at the time of compilation. Whenever any statement is made which is taken from any other writer, it will be put in brackets [ ], except when the statement itself cites an authority: though I have sometimes put the brackets even in the latter case. The only other mode of proceeding would be, to collect lists from authority, naming the sources of information. But, having found so many errors in these sources, when my opportunities have enabled me to bring them to the test, I did not feel inclined to be the tenth transmitter of inaccurate copies. My mistakes shall be of my own making, and it would not be easy to invent one which should want high precedent for its species.

The description of the books in the Catalogue is uniformly as follows. There are given:

1. The place at which the work was printed, in Italics, and generally in English.

2. The date of the title-page, or colophon, or, when both are wanting, of the preface, in words; but not quite at length. Thus instead of fifteen-hundred-and-eighty-eight will be found fifteen-eighty-eight. I am sure that dates will never be given correctly until this plan is universally adopted: for two reasons. First, the chance of error in printing is very much diminished: particularly the risk of a transposition of figures *at the press.* Secondly, the writer, who is, on the whole, and

* When the great catalogue of the Museum is published, those who can give house-room to forty or fifty volumes, and time enough to their examination, may have some of the advantage which they would derive from actual access to the books themselves. And those known to be engaged in research will derive a still larger portion of the same advantage from the readiness with which the officers of the Museum will go out of the usual routine of duty to help them.

one date with another, more to be feared than the printer, has time to be accurate. A glance at four numerals and four strokes with the pen, is too rapid a process for certainty: and those who think they can rely upon themselves to stop *every time,* and look at what they have done, will frequently find reason to wish they had been less confident. But *Incidit in Scyllam,* &c.: I had very nearly announced an edition of Cocker as being unmistakeably the seventeen-hundred-and-twentieth (see page 56).

3. The author's name. When an initial only is given, it is because the author has left no more, either in title or preface.

4. As much of the title as will certainly identify the book. In spelling, initial capitals, &c., I have generally followed the author closely. When there is any defect in this respect, I suppose it will be in those works which I had not any opportunity of re-examining while the sheets were in proof. Imitation of type I have not attempted. To have given the full titles would have swelled my book too much: at certain periods the authors of elementary works were much given to write out descriptive chapters in their title-pages; scores of them would each have filled two or three pages of even the smaller type used in the Catalogue.

5. The form in which the work is printed; a matter which will require some explanation.

A folio, quarto, octavo, duodecimo, or smaller work, is now generally known by its size, though not always. In the folio the sheet of paper makes two leaves or four pages, in the quarto four leaves, in the octavo eight, in the duodecimo twelve, and so on. But even a publisher thinks more of size than of the folding of the sheet when he talks about octavo or quarto; and accordingly, when he folds a sheet of paper into *six* leaves, making what ought to be a *sexto* book, he calls it a duodecimo printed in half sheets, because such printing is always done with half-sized paper, or with half-sheets, so as to give a duodecimo size. From a very early period it has been universal to distinguish the sheets by different letters called *signatures.* In the book now before the reader, which is a half-duodecimo (or what I call a *duodecimo in threes*), the first sheet which follows the prefatory matter, B, has B on the first leaf, and B 2 on the third; which is enough for the folder's purpose. But in former times the signatures were generally carried on through half the sheet, and sometimes through the whole. Again, in modern times, no sheet ever goes into and forms part of another; that is, no leaf of any one sheet ever lies between two leaves of another. But in the sixteenth century, and even later in Italy, it was common enough to print in *quire-fashion*. Imagine a common copybook, written through straightforward, and the string then cut: and suppose it then separates into four double leaves besides the cover. It would then have sixteen pages, the separate double leaves containing severally pages 1, 2, 15, 16; 3, 4, 13, 14; 5, 6, 11, 12; 7, 8, 9, 10. If a book were printed in this way, it would certainly be a folio, if the four double leaves of any one quire or gathering were each a separate sheet: and if the sheet were the usual size, it would give the common folio size. But if each gathering had the same letter on all its sheets, if the above for instance were marked A_1 on page 1, A_2 on page 3, A_3 on page 5, and A_4 on page 7; the book, when made up, would have all the appearance of a more recent octavo in its signatures. In order to give the size of a book, and at the same time to give the means of identifying the edition by its signatures, I have adopted the plan which gives the following rules:

(a) The words *folio, quarto, octavo, duodecimo, decimo-octavo*, refer entirely to size, as completely as in a modern sale-catalogue, the maker of which never looks at the inside of a book to tell its form. All the very modern distinctions of *imperial, royal, crown, atlas, demy*, &c. &c. &c. I have relinquished to paper-makers and publishers, who alone are able to understand them. But in old books, the reader must expect to see the several sizes, each one of them, smaller than in the modern books. When the work is decidedly small of its name, I have noted it by the word 'small.'

(b) When the single word occurs, without any thing more, the signatures are as in the genuine meaning of the word. Thus, as to signatures, *folio* has two leaves to one letter of signature, *quarto* four leaves, *octavo* eight, *duodecimo* twelve; and of course double the number of pages.

(c) When the modern word occurs with the addition of *in twos*, or *in threes*, &c. the addition expresses the number of *double leaves* which belong to one letter of signature: and which I believe would be found, if the books were taken to pieces, to be in each quire or gathering. Thus, *folio in ones*, or *quarto in twos*, or *octavo in fours*, or *duodecimo in sixes*, would in each case be unnecessary repetition; for the first word, when alone, is intended to express the gathering in the third. But *folio in twos* would mean the folio size with two double leaves in one quire, *folio in fours* with four double leaves. Thus, a book of the octavo size, with the quarto signatures, is *octavo in twos:* had it been larger, I should have called it *quarto.*

By this means there is something as to size, and something as to signatures, in every description. But whether any book which I call *octavo in twos,* for instance, really was printed on whole sheets, or on half sheets, that is, really was a small quarto, or a divided octavo, is more than I can in any case undertake to say. All I know is, that with these rules, the reader has two indications in every case, to guide him in determining whether the book he has in his hand be the one I describe or not.

This is no unnecessary excess of description. For so frequently, in the sixteenth and seventeenth centuries, were there issues of the same impression under different titles, that all I have done will in many cases only give a presumption as to whether a book in hand is or is not the edition I have described. Were I to begin this work again, I would in every instance make a reference to some battered letter, or defect of lineation, or something which would be pretty certain not to recur in any real *reprint.* Ordinary errata would not be conclusive: for these might be reprinted for want of perceiving the error.

Rules are given for determining the form of printing by the waterlines of the paper, and by the catchwords. It is supposed that the latter are always at the end of the sheet, and also that the waterlines are perpendicular in folio, octavo, and decimo-octavo books, and horizontal in quarto and duodecimo. But in the first place, a great many old books have catchwords at the bottom of every page, many have none at all; and as to the rule of waterlines, I have found exceptions to every case of it. Pacioli's Euclid, *Venice,* fifteen-nine, *folio in fours,* has horizontal waterlines: the Hypomnemata of Stevinus, an undoubted folio, has thick waterlines *both ways.*

6. In the smaller type are entered such remarks as suggested themselves on the manner or matter of the work, or on any point arising out of it. In some cases these have extended themselves to short dissertations, such as, on the geometrical foot, page 5,—on Sacrobosco's knowledge of the Arabic numerals, page 14,—on the invention of + and —, page 19,—on the age of Diophantus, page 47,—on the genuineness of Cocker's Arithmetic, page 56. But for the most part I have contrived to keep within a very moderate compass as to what is said under each work.

The principal point on which I have distinguished one work

from another, is as to the use of the *old* or *new* method of performing division, which, more than any other single point, decides the character of a work. The letters O, N, or ON, will tell whether the work uses the old, the new, or both. Not that the verbal distinction is here very correct; for neither method is older than the other; and both appear in Pacioli. A description of the now disused method is given by Dr. Peacock, p. 433.

With regard to the works themselves, I have made no selection, as before noticed. No book that I *have* seen during the compilation has been held too bad to appear; no book that I *have not* seen too good to be left out. I have had but one discretion to exercise, namely, to determine the extent to which algebra should be considered as arithmetic. In the earlier day, the distinction was slight: I doubt whether I have not overrated it; but it is not an easy line to draw.

The history of Arithmetic, as the simple art of computation, has found little notice from the historians of mathematics in general. They shew themselves deficient in the knowledge of its progress, and of the connexion of that progress with the rest of their subject. The writers whom I can name as having attempted—some more and some less—to supply this defect, are Wallis, Dechâles, Heilbronner, Scheibel, Kästner, Leslie, Delambre, Peacock, and Libri. I speak of the progress of arithmetical writings as works on science, independently of bibliography properly so called, and biography.

Wallis (p. 44 and Additions) was one of those writers whose works remain the standard of the erudition of their day. His *algebra*, so called, is rather the history, theory, and practice of both arithmetic and algebra. Its miscellaneous and badly indexed form prevents any from knowing what is in it, except those who make a study of it, which none of our day will do, unless they intend to go rather deeply into the history of the exact sciences. But many and many a page by which the writer intended to gain the credit of research, will be found to be a transcript from Wallis. As a connected history, however, it is nothing; and as a bibliography, less. For example, that Regiomontanus used decimal *fractions*—a very common story—is the consequence of Wallis's confused method of stating that he introduced the *decimal* instead of the *sexagesimal* radius into trigonometry; the confusion arising from his not having a clear knowledge of what had been published of Regiomontanus. And again, we find him, though the editor of an edition of Oughtred, balancing as to which was written first,

Oughtred's *Clavis*, or Harriot's *Praxis* (which were published in the same year), apparently ignorant that the great method as to which the two authors were chiefly to be compared, did not appear in Oughtred's first edition at all. But for all this, if Wallis be cautiously watched as to books and dates, his works are most valuable magazines of historical suggestion. From them might be collected a much better *scientific* history of arithmetic than existed in his time, or, indeed, in any time preceding the publication of Dr. Peacock's article on the subject.

Claude Francis Milliet Dechâles was a Jesuit, who published, in four very large folios, a complete course of mathematics, including architecture, carpentery, fireworks, and all that was then held to belong to the exact sciences (see page 53). The first volume opens with about a hundred pages (large folio, double column) *de progressu matheseos,* consisting entirely of description of books, in order of date. The part relating to arithmetic fills nine of these pages. The whole is done with much care, and is, for the mode of describing books current at the time, very accurate; and the opinions given on the books shew that Dechâles had read them. But he is strongly addicted to the very common mistake of judging the books according to what ought to have been said of them, if they had been published in his own time. For example, he finds that there is not sufficient demonstration in Tonstall; which is true, absolutely speaking; but Tonstall is a very Euclid by the side of his contemporaries.

Heilbronner's Historia, &c. (page 69), though it professes only to give writers up to the beginning of the sixteenth century, makes a particular exception in favour of arithmetic. Up to the year 1740, about 170 authors are recorded, a great many of whom he had not seen. There is also historical dissertation on points of arithmetic. This is a work of great value to the inquirer: he must not rest upon its statements; but he will find more than usual materials for further research.

Scheibel (Additions) may be considered as partly repetition, partly extension, of Heilbronner. He is one of those bibliographers who collect from various sources the names and dates of more editions than those who know catalogues will readily believe in.

Kästner (Additions) falls under the following censure from Dr. Peacock: "The meagre sketch which Kästner has given of some insulated works on the subject, generally contrives to omit almost every particular which is essentially connected with the history of the progress of the science." This is well merited, inasmuch as the author chose to call his work a history, instead of a *bibliography;* and as the former, nothing can be more incomplete. I have almost as good a right to call my work a history, as Kästner his. But, as a bibliography, it may be urged in defence, that he gave fuller descriptions of books than his predecessors. Scheibel, Heilbronner, and Dechâles sink into title-writers (and bad ones) before Kästner.

The late Professor *Leslie* (page 89) was one of those men, the strength and asperity of whose opinions would make it fair to deal with them as they dealt with others. In his *Philosophy of Arithmetic* he has entered incidentally into much of its history. He was, by taste, a searcher of old books; and various dates, &c. occur, which shew that he had more knowledge of books than can be got from catalogues. A few words will sometimes shew this to a person who has compared the books with the accounts of them. But, writing in a popular manner, he does not give references to his authorities, which is a serious diminution of the value of his work. Of Leslie, as an historian on controverted points, one principal thing to be cautious of is, his almost monomaniac antipathy to every thing Hindoo—a most unfortunate turn for an arithmetical investigator. Those who inquire into this subject will see what he is in Colebrooke's hands; those who do not, may compare his description of the *Lilivati,* "a very poor performance, containing merely a few scanty precepts," with the summary of the contents of that work in the *Penny Cyclopædia,* article *Viga Ganita.* Leslie also generalises most fearfully every now and then. He informs us that it was the *practice throughout Europe* to reduce the rules of arithmetic to memorial verses, and that Buckley's *Arithmetica Memorativa* appears at one period to have *gained possession of the schools and colleges of England.* Now the truth is, that the *verses* attributed to Sacrobosco had never even been printed when Leslie wrote; and Buckley, so far as is known, was printed only once alone, and two or three times as an appendix to a work on logic. Dr. Peacock expresses the truth in saying that, *before the invention of printing,* the practice of *writing* memorial verses was common, as appears by manuscript libraries. It is needless to say that, had the practice of *using them* been common, the presses of the fifteenth and sixteenth centuries would have given them forth in great numbers. But I cannot learn that any metrical work was printed in the fifteenth century, except the *Compotus* of Anianus, and that only once.

Delambre (page 84) wrote on the history of Greek Arithmetic in such a manner as to set that part of the subject fairly going. I doubt if any one since the time of Wallis, at least of their order of note, has exhibited so much of the union of the *scholar* and the *computer*. Of the obligations under which the history of astronomy lies to him, it is unnecessary to speak here, or any where : no one man was ever so closely connected with that science, with its past, its present, and its future, by history, observations, and tables.

Of my opinion of Dr. *Peacock's* work (page 91) I need hardly say any more, after the pains which I have taken to give reasons on every point in which I differ from its author, and to correct every little error which I have found. This I judge to be the most undoubted compliment which can be paid to any work. Having looked carefully over it, with a great many of the works mentioned in it in my hands, it would be a strong evidence in its favour, were it needed, that I have found no more to set right than is there noted, in matters of dates and circumstances. It is much to be wished that this treatise should be published separately :* those who can obtain it will find that it gives life and spirit to the catalogue of books which forms the main part of this work. Up to this moment it is the only work which can be called a history of arithmetic.

M. *Libri's* history of science in Italy (page 95) is the work of a man who, to the character of a mathematician, adds that of a man who is well versed in literature, and a successful collector of the rarest works in that of his own country. There is much which is interesting in its early history of Italian Arithmetic. Unfortunately, it is not yet finished ; and, I am told, is not likely to be speedily resumed.

The history of most of the sciences resembles a river which sinks underground at a certain part of its course, and emerges again at a distant spot, swelled by certain tributaries, which have joined it in the tunnel. Of arithmetic in parti-

* It was once intended to publish these treatises separately. Nine years ago, the proprietors of the *Encyclopædia Metropolitana* so fully intended to publish separately, that they considered themselves aggrieved because I, who had written the *mathematical* article on probabilities, wrote a *popular* work with that word in the title-page, which they alleged, through their agent, was in effect a republication of the former work, &c. Not being able to get them either to litigation or arbitration, I was obliged to write a pamphlet to prove that the charge was frivolous. The pamphlet is unanswered, and all the treatises unpublished (separately) to this day. The latter I regret, for the sake of science. It is a great pity that Sir John Herschel's treatises on light and sound, Dr. Peacock's arithmetic, Mr. Airy's tides, &c. are thus locked up.

cular, we note the disappearance in the seventh century of all that was good in the Greek system; and we see the rise, at the invention of printing, of the most trivial part of it, combined with the additions which are now well known to have been received from Eastern sources.

The manuscript literature of the middle ages will prove no very productive source of information, to judge by all that has been made of it hitherto. But then it is to be remembered how seldom, if ever, it has happened, that the investigator of it has united the character of a sufficient mathematician with that of an industrious and well-trained palæographist. Neither the one nor the other can proceed alone: the former has not only to learn how to read, but what to read; for the actual habitat of the manuscripts, and how to get preliminary knowledge of their existence, is a study of itself. The latter is apt to know no difference between what is sound, and what is worthless.

When Euclid, Ptolemy, &c. are first seen to reappear, they come in as Arabic writers. They may be called Greek, but the first translations are from the Arabic; and their effect upon the literature of Europe is, in the first instance, just what it might have been if the authors had been Persians or Saracens. How the communication took place has been considered as a point of small curiosity compared with the importation into Europe of the Arabic numerals.

This subject, both generally, and with reference to the several countries, has been long on the anvil: not the authorship of the letters of Junius has tasked research and ingenuity more than the introduction of the nine digits and the cipher. I suppose nobody would listen to the hypothesis that the former wrote themselves: but I am much inclined to suspect that the latter introduced themselves. The *endosmose* constantly going on between nations connected by war and commerce would not merely explain so easy a matter, but would render it very difficult to explain how it did not happen, if it had not happened. That the Venetian merchants should not know the system of accounts of those with whom they traded, is incredible: that, at the beginning of the thirteenth century, many priests and some soldiers, returned from their crusades, should not bring back with them the account of so very elementary a difference between themselves and those with whom they had treated, and whom they had in many instances served as slaves taken in war, is unlikely. That Leonard of Pisa, as he is called, was the first who wrote on or in the new system, is pretty generally affirmed by the Italians of the fifteenth and sixteenth centuries: and there can be no doubt of it. But because we trace the first formal user or expounder to Italy, it by no means follows that the other parts of Europe owe the system to that country in the first instance, however certain it may be that they owe much assistance in the course of the general establishment. It was so common in England in the thirteenth century, that Roger Bacon recommends it (page 14) as a study, of course as a practicable study, for the clergy. Had it not been commonly known, at least as to what it *was*, he would have done more than give the name of the science, and the names of its rules; he would have added some description. Those who have searched into the matter, merely with a view to Arabic *numerals*, and without any collateral thoughts about *arithmetic* in their heads, may have passed over much valuable evidence. They did not know, perhaps, that the organised rules of computation always went with the Arabic system, and never with the Roman or Boethian: that if ever they came to the connected mention of addition, subtraction, multiplication, and division, it ought to have been a sign that they were reading on *algorism* as distinguished from *arithmetic*. This passage of Roger Bacon has been neglected; though the mere occurrence of the word *algorism*, in a most interpolable clause of one sentence, occurring in Matthew Paris, has received notice and discussion from Mr. Hallam, from a learned writer in the Archæologia, and probably from others.

The rejection of the work attributed to Sacrobosco may be accounted for from the general ignorance of its having been printed under his name at an early period. Though this does not prove the genuineness of the work, it very much adds to the evidence for it. As early as 1523, the learned world was invited to dissent from the assertion that the treatise in question was written by Sacrobosco, and did not do it. It was often cited afterwards, and never, as far as I have seen, with any doubt. But the only writer of this or the last century that I can find, who describes the *printed* edition, is the Italian editor of Fabricius, *Bibliotheca Latina*.

Come how they might, however, we find ourselves, at the invention of printing, in possession of the knowledge that two distinct systems of arithmetic were current. The streams which had united in one bed had not mingled their waters; and for nearly a century two distinct classes of writings present themselves. Their only common point is the use of the Arabic or Indian numerals and method of notation.

The first, or *algoristic* system, as we may call it, proceeded by systematic rules to the performance of questions of computation. It embraced mercantile arithmetic, and what was then called algebra: and laborious and prolix efforts were made to combine the two; that is, to represent in such a form as we should now call algebraic, as distinguished from arithmetical, the solutions of questions of commerce, or of what might become such, if the connexion were diligently fostered. I cannot suppose that the multifarious problems of exchange of different kinds which abound in Pacioli and his immediate followers, were actually useful to the merchant: but there is an obvious leaning to the idea that they ought to be so. Geometry was also an application of this arithmetic; and in a manner which strongly marks its eastern character, as will appear to those who compare the work of Pacioli with the Indian books.

The second, or *Boethian* system, as I shall call it, because the work of Boethius was its great text-book, did not give any rules of calculation, nor apply itself to any application. The study was the properties of numbers, and particularly of their ratios. There was no art about it; and we have no means of telling whether the philosophers of this school reckoned on their fingers, or used an abacus, or put pen to paper for the performance of some organised method of computation. To judge by the smallness of the numbers used in the instances adduced, we must suppose that the writers left it to their readers to do as they liked best.

For some specimens of the laborious manner by which the Pythagorean Greeks, in the first instance, and afterwards Boethius in Latin, had endeavoured to systematise the expression of numerical ratios, I may refer the reader to the article "Numbers, old appellations of," in the Supplement to the *Penny Cyclopædia*. If I were to give any account of the whole system, on a scale commensurate with the magnitude of the works written on it, the reader's patience would not be *subquatuordecupla subsuperbipartiens septimas* — or, as we should now say, *seven per cent*—of what he would find wanted for the occasion. Not that the books I am speaking of get quite so far as this. I am hardly prepared to say exactly what number under fifty ought to be named as the terminus of the Boethian store of numbers; but certainly we rarely find them choose their instances from numbers above it. The system broke down under its own phraseology, as did that of the Yancos mentioned by Dr. Peacock, who could not get further than *three,* because they could not express this idea by any thing more simple than *Poettarrarorincoaroac.*

The Italian school of *algorists,* with Pacioli at their head, found followers in Germany, England, France, and Spain; and in all but England, the Boethian school also. I cannot discover a single English work which pays any detailed attention to the latter class of arithmeticians. Tonstall, and still more Recorde, for the former appears to have been little known, were the preservers; and before the end of the sixteenth century the ordinary style of commercial arithmetic, which has prevailed among us ever since, was in course of establishment.

This gradual formation of the English school of commercial works will be apparent enough in my list. From the time of Recorde, we always were conspicuous in numerical skill as applied to money. The questions of the English books are harder, involve more figures in the data, and are more skilfully solved. It is possible that I might, if my French, German, and Italian list were more complete, produce exceptions to this rule. But nothing, I think, could arise to alter my conviction that the efforts which were made in this country towards the completion of the logarithmic tables in the seventeenth century, and the instantaneous appreciation of the value of the discovery of logarithms, were the result of that superiority in calculation which I assert to have been formed in the sixteenth. And yet this last-named century produced one man in France, one in Germany, and one in Italy, with either of whom no one English calculator could compare in extent of operations: I allude to Vieta, Rheticus, and Cataldi. There was no opportunity to compete with these men; for the subjects on which they worked were not introduced here. It was only towards the *end* of the sixteenth century that what were then the higher parts of the mathematical sciences began to be disseminated with effect in Britain.

To the commercial school of arithmeticians above noted we owe the destruction of demonstrative arithmetic in this country, or rather the prevention of its growth. It never was much the habit of arithmeticians to prove their rules: and the very word *proof,* in that science, never came to mean more* than a test of the correctness of a particular operation, by reversing the process, casting out the nines, or the like. As

* At first I had written "degenerated into nothing more;" but this is incorrect. The original meaning of the word *proof,* in our language, is testing by trial.

soon as attention was fairly diverted to arithmetic for commercial purposes alone, such rational explanation as had been handed down from the writers of the sixteenth century began to disappear, and was finally extinct in the work of Cocker, or Hawkins, as I think I have shewn reason for supposing it should be called. From this time began the finished school of teachers, whose pupils ask, when a question is given, what rule it is in, and run away, when they grow up, from any numerical statement, with the declaration that any thing may be proved by figures—as it may, to them. Any thing may be unanswerably propounded, by means of figures, to those who cannot think upon number. Towards the end of the last century, we see a succession of works, arising one after the other, all complaining of the state into which arithmetic had fallen, all professing to give rational explanation, and hardly one making a single step in advance of its predecessors.

It may very well be doubted whether the earlier arithmeticians could have given general demonstrations of their processes. It is an unquestionable fact of observation, that the application of elementary principles to their apparently most natural deductions, without drawing upon subsequent, or what ought to be subsequent, combinations, seldom takes place at the commencement of any branch of science. It is the work of advanced thought. But the earlier arithmetician and algebraists had another difficulty to contend with: their fear of their own half-understood conclusions, and the caution with which it obliged them to proceed in extending their half-formed language. It was not merely by oversight, I suspect, that Oughtred so often calculates $ab+ac$ by two multiplications, instead of using $a(b+c)$: but rather from that same general fear of abbreviation, and suspicion that error may lurk in it, which possesses men of business who dare not multiply by 10 by the annexation of a cipher, but proceed with each figure, and carriage, as they would do if the multiplier were 8 or 7. I have seen this often enough; and things nearly as strange times without number. But what shall we say to the following; a most sufficient recommendation of the study of old works to the teacher, as shewing that the difficulties which it is now (I speak to the *teacher* not the *rule-driller*) his business to make smooth to the youngest learners, are precisely those which formerly stood in the way of the greatest minds, and sometimes effectually stopped their progress. Perhaps no man of his day had so much power over mathematical language as Wallis. But the following extract of a letter from him to Collins (Macclesfield Collection, vol. ii. p. 579), in 1673, shews that he once had doubts whether he might dare write down the square root of 12 as being twice the square root of 3, however certain he might be that it is so; because no one had so written it. Speaking of the square root of a negative quantity, he says, "Only, though I had from the first a good mind to it, I durst not without a precedent, when I was so young an algebraist as in the history my late letter reports, take upon me to introduce a new way of notation, which I did not know of any to have used before me. And it was not without some diffidence that I ventured on $2\surd 3$, instead of $\surd 12$, not having then met with any example of a number so prefixed to a surd root; but I found it so expedient, not only for the discovering the root of a binomial, whether quadratic, cubic, or others, but for the adding and subducting of commensurable surds, that I resolved to use it for my own occasions, before I knew whether others would approve of it or no; especially having found in the first edition of Oughtred's Clavis (for I had not then seen the second) one instance or two for it, to justify myself if it should be questioned. But since that time it is grown more common, and I perhaps have somewhat contributed thereunto."

If I were to take such things from old and unknown writers, in which they abound, I should be thought to waste the time of my readers by giving an undue importance to the difficulties of very inferior minds. But those who do not want such a confession as the above from a Wallis, to help to persuade them that the difficulties of progress are of the same character in all ages, will find in old works plentiful and useful supplies of thought for a modern teacher.

Another instance of the halting progress of language is the mode of introduction of the decimal point. Stevinus, the undoubted introducer of the decimal fraction (though others may have seen very clearly the great use of 10, 100, 1000, &c. as divisors), had no such thing, but only a distinct mode of denoting the meaning of each decimal place. Dr. Peacock mentions Napier as being the person to whom the introduction is unquestionably due: a position which I must dispute, upon additional evidence.

The inventor of the single decimal distinction, be it point or line, as in 123·456 or 123|456, is the person who first made this distinction a permanent language: not using it *merely as a rest in a process*, to be useful in pointing out afterwards how another process is to come on, or language is to be applied; but making it his final and permanent indication as well of the way of pointing out where the integers end and the fractions begin, as of the manner in which that distinction modifies operations. Now, first, I submit that Napier did not do this; secondly, that if he did do it, Richard Witt did it before him (page 34).

I have not seen Wright's summary of logarithms, of 1616, to which Dr. Peacock refers for some indications of the decimal point. But I take with confidence his assertion that the first distinct notice of the thing in question, by or by means of Napier, is contained in the *Rabdologia* (page 35).

In this famous tract there are but two instances to the purpose: a scanty number, be it noticed, for a person who had seized the idea of completing the decimal scale. The first is an instance of division which Dr. Peacock cites. Here the quotient is, as we should now say, 1993·273. Napier uses a comma in his quotient, as a rest, and writes 1993,273, and then presents his answer in the form 19932′7″3‴, as Stevinus would have done. Dr. Peacock states this, and notes it as a "partial conformity to the practice of Stevinus." Unfortunately Napier has not given us instances enough to tell us whether he had a *practice* of his own: and in the only other case, which Dr. Peacock also cites (I have verified both citations), he adheres still more closely to the method of Stevinus. For though, in a question of addition, he has drawn a line down the interval where our decimal points would be, which line would be as distinctive in his total as in his *addenda*, he places the exponents of Stevinus in this total, which thus has both distinctions, and stands 1994|9′1″6‴0⁗.

I cannot trace the decimal point in this: but if required to do so, I can see it more distinctly in Witt (page 33), who published four years before Napier. But I can hardly admit him to have arrived at the notation of the decimal point. For though his tables are most distinctly stated to contain only numerators, the denominator of which is always unity followed by ciphers; and though he has arrived at a complete and permanent command of the decimal *separator* (which with him is a vertical line) in every operation, as is proved by many scores of instances; and though he never thinks of multiplying or dividing by a power of 10 in any other way than by altering the place of this decimal separator; yet I cannot see any reason to suppose that he gave a meaning to the quantity with its separator inserted. I apprehend that if asked what his 123|456 was, he would have answered: It *gives* $123\frac{456}{1000}$, not it *is* $123\frac{456}{1000}$. This is a wire-drawn distinction: but what mathematician is there who does not know the great difference which so slight a change of idea has often led to? The person who first distinctly saw that the answer -7 always implies that the problem requires 7 things of the kind diametrically opposed to those which were assumed in the reasoning, made a great step in algebra. But some other stepped over his head, who first proposed to let -7 *stand for* 7 such diametrically opposite things.

Who, then, was the real inventor of the decimal separator, to give both interpretation and operative form? This is a question I will not answer positively: nor attempt to answer till I have pointed out how the case stands with several possible claimants. And first, as to Briggs. He does not indeed arrive at the simple decimal *point* (which is a strong presumption against such a thing being suggested by Napier; for who would have learnt it from him, if not Briggs?)—but he omits the denominator, and draws a line under the decimals. And it is a further presumption against any such idea of abbreviation being common, that Briggs explains his notation as quite peculiar to himself. In the preface, or *Ad Lectorem,* of the *Arithmetica Logarithmica, London,* sixteen-twenty-four, *folio in twos,* he commences thus:

"Ut obscuritatis crimen, quantum in me situm fuerit effugiam, paucula hæc humanissime Lector, te admonendum censui. 1. Si numerus occurrat cui linea subscribitur, notæ illæ quæ supra lineam sunt descriptæ, Numeratorem partium constituunt, quarum Denominator semper intelligitur unitas, cum tot cyphris, quot sunt notæ superius positæ. ut $\underline{75}$ designant $\frac{75}{100}$ vel $\frac{3}{4}$. sic 5 $\underline{9321}$ designant $5\frac{9321}{10000}$."

Oughtred adopted both the vertical and sub-horizontal separators, thus shutting up the numerator in a semi-rectangular outline. In the mean time, Gunter, after adopting Briggs's notation in the first instance, gradually dropt it, and substituted the decimal point. The history of this symbol in his hands is rather curious, on account of its having been so completely overlooked, though all the circumstances were to be seen in widely circulated works.

Gunter published in 1620 his *Canon Logarithmorum,* sines and tangents on Briggs's system. To this table there is no explanation. But an explanation was written by him, and, though I cannot make out that it was ever published separately, yet it appears in the collection of his works. I have what is called the second edition of this collection, *London,* sixteen-thirty-six, *quarto*, which, for various reasons, I suspect to be the first, thinking that the announcement of second edition merely refers to the fact of its having been previously published during Gunter's life. After the *finis*, with new paging and signatures, comes 'The Generall use of the Canon and Table of Logarithmes.' This I suspect to have then made its first appearance: but from the very manner in which the decimal fractions are treated, it is clear that it was written before or with the work on the sector, &c., which first* came out in sixteen-twenty-three; and this, though it makes reference to that work, or at least to its matter. At the beginning of this 'Generall use, &c.' common fractions are used, even when the denominators are decimal. At page 13, Briggs's notation appears, without explanation: and $116\ \underline{04}$ is the third proportional to 100 and 108; this continues through page 14. In page 15, a dot is added to Briggs's notation in one instance: 100*l.* in 20 years at 8 per cent becomes $466.\underline{095}$*l.* At the bottom of this page Briggs's notation disappears thus, "It appeareth before, that 100*l.* due at the yeares end is worth but $92\ \underline{592}$ in ready money: If it be due at the end of 2 yeares, the present worth is 85*l.* $\underline{733}$: then adding these two together, wee have 178*l.* 326 for the present worth of 100. pound annuity for 2 yeeres and so forward." After this change, thus made without warning† in the middle of a sentence, Briggs' notation occurs no more in the part which relates to numbers. But in the following chapter, which is trigonometrical (and which may have been written first, for Gunter's *own* logarithms are only trigonometrical), it reappears, sometimes with and sometimes without the dot. In the previous work on the sector, &c., the simple point is always *used:* but in explanation the fraction is not thus written, but described as parts. Thus, 32.81 feet in operation is 32 feet 81 parts in the description of the question or answer.

It was long before the simple decimal point was fully recognised in all its uses, in England at least: and on the continent the writers were rather behind ours in this matter. As long as Oughtred was widely used, that is, till the end of the seventeenth century, there must always have been a large school of those who were trained to the notation $123|\underline{456}$. To

* A printed title-page has 1623: an additional engraved and ornamented title-page has 1624.

† As far as I can find, the words *cosine* and *cotangent* (which Gunter introduced) appear in the same manner, without warning or explanation.

the first quarter of the eighteenth century, then, we must refer, not only the complete and final victory of the decimal point, but also that of the now universal method of performing the operation of division and extraction of the square root.

This evident slowness in the admission of such important improvements will not appear singular to those who observe what is now taking place with reference to Horner's method for the solution of equations. It will be the business of some successor of mine, a century or two hence, to distinguish in his list, by some appropriate mark, the works of our period which adopt this method from those which do not: just as I have done with the two methods of division. I hope, if I live to publish a second edition, that I may be able to add to my list the name of the *first* work, written for students in the University of Cambridge, which shall contain this most fundamental addition to the processes of pure arithmetic: which, though it has found its way into the public examinations, is not yet in the books which prepare students for them.

This has been a long digression, on what is perhaps the most interesting part of the subject, its language. We owe every thing, almost, to the simplicity of certain modes of expression. Nothing is more clear than that the Greek geometers, with all the acuteness and perseverance necessary to carry results of arithmetical application to a high pitch, and all the taste for numbers which would turn their thoughts that way, were stopped by their insufficient system of numeration, and the tediousness of their processes. The history of language, then, is of the highest order of interest, as well as utility: its suggestions are the best lessons for the future which a reflecting mind can have.

In looking over the list of books, it will be observed, that during the eighteenth and nineteenth centuries there is a great preponderance of English writers. The law I imposed on myself, of entering no books which I could not see, necessarily brought this about. During the last two centuries, the elementary works of the different countries in Europe have not gained great general circulation. They have ceased to be written by the greatest names, for the most part; and they have rarely been cited by the historians of science. Copies which have found their way to this country have probably soon been destroyed. As it is my intention to endeavour to extend this list, whether I publish the extensions or not, and in any case to provide for the preservation of the materials I collect, it will be worth the while of any one who is able, to furnish me with information on works which I have not seen. I think it probable that any one who has had the curiosity to rescue three books of arithmetic from a stall, will find that one of them is not in this list. From any such person I shall thankfully receive the full title of the neglected work, with the form in which it is printed. Nor need it by any means be presumed that because a book is wholly unknown, it proves nothing in the history of the science. A book so thoroughly lost as that of Witt, contains a nearer approach to the decimal point than was made by Napier. Horner found a close approximation to his own method, for the case of the cube root, in an obscure compendium of arithmetic. I might also instance Dary's discovery mentioned in page 48; and other things of the same kind. The history of hints given before the time at which they were (perhaps could be) made to bear fruit, would be a very curious one: and the progress of science will never be well understood until some little account can, in each case, be given of the reason why a notion should be so productive at a particular period, which was so barren at a previous one.

On looking at the fourth edition of Gunter's work (sixteen-sixty-two), I find that some liberties have been taken with the original text: among others, Briggs's notation is restored in several places, though not entirely. It should also be noticed that the point which generally occurs after the first figure of a logarithm is not a fractional separator, but only divides one integer from the rest.

*** In consequence of a little change in the plan of this work, made after the printing was commenced, there may be one or two places in which the reader must consult the *Introduction*, where he is told to consult the *Index;* and the *Additions*, where he is told to consult the *Introduction.*

A LIST

OF

WORKS ON ARITHMETIC,

THEORETICAL AND COMMERCIAL, IN CHRONOLOGICAL ORDER.

*** The letter P. followed by a number, refers to the page of Dr. Peacock's treatise in which the work or its author is mentioned. The brackets [ ] enclose statements which I have taken from others, and cannot therefore affirm from personal inspection. For O and N see those letters in the Index.

Florence, fourteen-ninety-one. **Philip Calandri.** 'Philippi Calandri ad nobilem et studiosum Julianum Laurentii Medicem de Arimethrica opusculum.' *Octavo* (small).

This book, which very few have mentioned at all, and fewer still from inspection, is a part of the rich bequest of the late Mr. Grenville to the nation. It begins with a picture of Pythagoras teaching, headed 'Pictagoras Arithmetrice introductor.' In the preface is the following on Leonard of Pisa: 'Vero e che il modo del notare e numeri con decte figure dice Lionardo pisano haver nel Mcc. incirca rechato dindia in Italia: et decti carateri: o vero figure essere indiane: et appresso deglindi havere imparato la copulatione desse.' He explains the leading rules, except division, for integers and for lire, soldi, and denari. Division he puts down in examples, and appears to have mistaken the mode of working, or to have had an incorrect printer. His notion of a divisor is curious. When he divides by 8, he calls his divisor 7; demanding, as it were, that quotient which, with *seven* more like itself, will make the dividend. He also describes the rules for fractions, and gives some geometrical and other applications. The book is in black letter, but the numerals are in a thin and small-bodied type. At the end of the book is, 'Impresso nella excelsa cipta de Firenze per Lorenzo de Morgiani et Giovanni Thedesco da Maganza finito a di primo di Bennaio 1491.'

Venice, M.cccc.lxliiii., fourteen-ninety-four. **Lucas Pacioli,** de Burgo Sancti Sepulcri. 'Summa de Arithmetica Geometria Proportioni et Proportionalita.' *Folio in fours.*

Tusculano, fifteen-twenty-three. **Lucas Pacioli.** 'Summa de Arithmetica geometria. Proportioni: et proportionalita: Novamente impressa In Toscolano su la riva dil Benacense et unico carpionista Laco: Amenissimo Sito: de li antique et evidenti ruine di la nobil cita Benaco ditta illustrato: Cum numerosita de Imperatorii epithaphii di antique et perfette littere sculpiti dotato: et continens* finissimi et mirabil colone marmorei: inumeri fragmenti di alabastro porphidi et serpentini. Cose certo lettor mio diletto oculata fide miratu digne sotterra se ritrovano.' *Folio in fours* (ON).

These are the full titles of the two editions of this celebrated work. Both editions are by one printer, Paganino of Brescia, and are beautifully printed: the type of the second being a good instance of the black letter in its state of approach to what is now called Roman letter. The work itself has been described by Hutton, Montucla, Peacock, Libri, &c.; but it would yet require a volume of description to do it justice. It is sometimes called the first work on arithmetic printed; but Calandri, Peter Borgo, three already mentioned in the Introduction, and perhaps more, take precedence. But it is certainly the first printed on algebra, and probably the first on book-keeping. P. 414, 424, 429 &c., 432 &c., 451, 460, 462; Hutton, *Tracts*, vol. ii. p. 201.

On comparing my copy of the first edition with that in the British Museum, I found one of those phenomena which so frequently occur in very old printed books. The first leaves of the two copies, to the number of about thirty, and the first leaf of the geometry, are not from the same setting up. The endings of the pages, and the ornaments of the capital letters, are different in the two. Nor does either of them agree with the second edition. A part of the first impression may have been lost, so that a second setting of types was required to replace that part. Several mathematical bibliographers of note enter Pacioli as Lucas di Borgo, and even as Borgodi, Lucas, and some do not mention the first edition.

The Latin *Paciolus* is usually spelt in Italian Paccioli, but Libri spells it Pacioli. I believe that the various assertions that Pacioli wrote Euclid in *Italian*, arise from the geometry of the work above described being Italian: but it is not Euclid, though of course founded on him.

There is another old book of which I have reserved the mention till now, because its author has been confounded with Pacioli. Brunet gives it as 'La Nobel opera di Arithmetica . . . compilata per B. [should be P.] Borgi,' *Venice*, fourteen-eighty-four, *quarto*: and Hain has 'Borgo (Pietro) Venet. Aritmetica,' *Venice*, fourteen-

* Cns in the title.

eighty-two, *quarto*. This edition Hain had not seen: but he had seen the next (the one mentioned by Brunet), which begins 'Qui comenza la nobel opera di arithmetica . . . per Piero Borgi,' *Venice*, fourteen-eighty-four, *quarto*. And he gives two other editions, *Venice*, fourteen-eighty-eight and ninety-one, *quarto*. Accordingly, Mattaire sets down the first edition of *Lucas di Borgo* as of 1484, and Dr. Peacock adopts this date. It is quite certain that Dr. Peacock must have had one of the editions of Pacioli before him when he wrote; and we are therefore to suppose that, in stating the date of the first edition, he followed Mattaire, and presumed an edition earlier than any he had seen. Tartaglia refers to Peter Borgo. P. 458.

No place, no date. **John Muris.** 'Arithmetices Compendium ex Boetii Libris per Johannem Muris excellentis ingenii virum Accurate congestum.' *Quarto*.

This is a different edition from the one presently mentioned. John Muris seems to have escaped the notice of bibliographers as an arithmetician. I think I remember that there is in Hawkins's History of Music some discussion of his ideas on the musical scale, [on which was published his work, *Leipsic*, fourteen-ninety-six, *folio*]. Muris lived before the invention of printing, but when I cannot ascertain. [His astronomical tables are preserved in manuscript.] The present tract has twelve leaves unpaged, and is certainly of the very earliest part of the sixteenth century, if not of the fifteenth.

Venice, fifteen-one. **Geo. Valla.** 'De expetendis et fugiendis rebus opus.' *Folio in fours*.

The first part of this is 'De Arithmetica libri iii, ubi quædam a Boetio pretermissa tractantur' (O). Certainly he supplies some of the omissions of Boethius; the four rules for instance. The work was published by the son, Joh. Pet. Valla. I have seen an earlier edition mentioned, but I cannot find any trustworthy account of such a thing.

Cologne, fifteen-one. **John Huswirt.** 'Enchiridion Novus Algorismi summopere visus De integris. Minutiis vulgaribus Projectilibus Et regulis mercatorum sine figurarum (more ytalorum) deletione percommode tractans,' &c. *Quarto, twos and threes* (O).

A short treatise, apparently one of the earliest printed in Germany on the Arabic system. The *projectilia* are counters. The rules are verified by casting out the nines. The words 'sine figurarum deletione' are not made good in the rule of division. On this work see the *Companion to the Almanac* for 1844, p. 3.

Paris, fifteen-three. **Joh. Faber Stapulensis, Jodocus Chlichtoveus,** and **Carolus Bovillus.** 'In hoc libro contenta Epitome compendiosaque introductio in libros Arithmeticos divi Severini Boethii,' &c. &c. *Folio in fours.*

This book contains an epitome of Boethius by Faber of Étaples; with a commentary and an arithmetical collection of rules by Chlichtoveus, his pupil; a compendium of geometry, a book on the quadrature of the circle, and *cubication* of the sphere, and a book on perspective, by Charles Bovillus; with an astronomical compendium by Faber already mentioned. This book is one of the earliest printed by Henry Stephens and his partner Wolfgang Hopilius. It is the first edition, I have no doubt, of Faber's epitome of Boethius, though Heilbronner and Murhard assert the contrary, perhaps misled by Faber's edition of Jordanus. As to the contents, the Arithmetic of Boethius was the classical work of the middle ages. It consists of statements of the commonest properties of numbers, under a great many classifications, to each of which a name is given. The second work, of arithmetical rules, shews the very low state of the art. It takes pages upon pages to explain the simple rules, though no examples are ventured on which have more than three figures. Montucla says that the quadrature of Bovillus was only saved from the laughter of geometers by its obscurity. But a work printed by Henry Stephens, and containing Boetius and Faber, must have been very far from obscure in its day. The historians of mathematics confine themselves to works the reputation of which has lasted: but they ought not to make the state of their own minds, with respect to the rest, a criterion of that of the contemporary readers.

Basle, fifteen-eight. **Gregorius Reisch.** 'Margarita Philosophica, cum additionibus novis: ab auctore suo studiosissima revisione tertio superadditis.' *Quarto in fours.*

According to Kloss (from his own copies) the first edition of this curious book is *Friberg* fifteen-three, the second and third both *Strasburg* fifteen-four (one printed by Gruninger, which I have seen, and one by Schott), and he calls the one before me the fourth. But Hain marks as the first edition one which appears to be printed at *Heidelberg,* fourteen-ninety-six, *quarto.* The one before me is not the third; it is evident that a former title-page has been reprinted · for though this edition is printed (as appears from the colophon) by Furter and Scotus of Basle, the title-page bears 'Jo. Schottus Argent. lectori S.' Which agrees with Kloss's statement that the previous edition to this one was printed by Schott of Strasburg. The Arithmetic (O), which is a part of the system of philosophy here laid down, has a frontispiece representing Boethius at one table with Arabic numerals before him; and Pythagoras at another with counters. Pythagoras among the Greeks, Apuleius and Boethius among the Romans, were often made the inventors of arithmetic. The arithmetic is divided into speculative and practical. The former is a summary of Boethius, often in the words of John de Muris. The latter is a short treatise on *Algorithm,* as it was called, or the rules of computation by the Arabic numerals. There is also computation by counters, fractions common and sexagesimal, and the rule of three. Many works of fifty years later do no more difficult questions. That Boethius was the author of the Arabic numerals was a common notion at the time, revived in our own day. I have seen another edition of *Strasburg*, fifteen-twelve; and there is said to be another edition by Orontius Fineus, *Paris* (?) fifteen-twenty. There is also an Italian translation, *Venice*, fifteen-ninety-nine, *quarto*, with the additions of Orontius, translated by Giovan Paolo Galucci.

If the number were sufficient of those who wish to take their notions of liberal education in Europe at the time immediately preceding the Reformation from original sources, and not from the reports of others, a reprint of the *Margarita Philosophica* would be made. The diversity of the matters which it treats, and the largeness of its circulation, stamp it as the best book for such a purpose.

The *Margarita Philosophica* is the earliest work I have found in which mention is made of that peculiar system of measures which was current among the mathematicians of the sixteenth century, and which has caused no little confusion among writers on metrology. I have already given some account of this ill-understood system, and shall here endeavour to present the whole case with some additional evidence.

The Roman foot (of 11·62 inches English) was of course established throughout the empire: and with it the Roman pace* of five feet, or 58·1 inches or 4·84 feet English. The natural pace of a man in our day is as nearly as possible five English feet: Paucton's experiments on the walking paces of individuals gave him 59·7 inches English, or 4·98 feet. In the British army the step, both what is called the ordinary step and the quick step, is, by regulation, thirty inches: making a pace of five feet. The Roman pace, by which distances were actually measured, was that of a soldier on the march: and, as might be expected, the weight of his arms and other equipage seems to have shortened his pace a little. But so near were these measures, as actually used by the Romans, to the natural ones from which they derived their names, that it was customary, not only to recur to legal standards, but, in the absence of ready access to them, to make use of the natural foot and pace. And we also know from Roman writers, that a somewhat fanciful relation, though not very far from the truth, was established between the breadth of the hand across the middle of the fingers, the *palm*, and the length of the *foot*. It was taken that four palms made a foot: and a palm was made of course to consist of four (average) finger-*breadths* or digits. This division into palms and digits was the most recognised division of the Roman foot: that into inches, or *unciæ*, is well known not to belong to the foot merely, but to any thing else. Whatever magnitude was called unity, the *uncia* was its twelfth part. Accordingly, *all* the Roman foot-rules which have been found in ruins or excava-

* The pace is derived from the double step, being the distance from the extremity of the heel at the place from which it is removed in walking to the same at the place in which it is set down again.

tions have the digital division, to which *some* (most, I believe) have the uncial division superadded. All this I take to be too well established to require the citation of any authorities.

It is quite out of my power, or, as far as I know, of that of any one else, to trace the gradual alteration of the foot in different countries. It does not appear that any means were taken to institute comparisons of various measures, to be preserved as public records. Such means could have been found: that which was then the common church of Christendom might have easily regulated the weights and measures of Europe, even without appearing to do so. But in all probability, the extent of the variations was not well known until it was too late. We know, however, as a fact, that the geometers were successful in establishing a measure among themselves, and communicating it through Europe on paper. This measure, I have no doubt, they believed to be the true Roman foot: for they divide it in Roman denominations, make use of it in their quotations from Roman authors, and never hint at their having any other notion of a Roman foot. And moreover, the writers who, in the sixteenth century, recovered the true Roman foot, never mention any peculiar geometrical foot in use among mathematicians, or in any way distinguish the latter from the *wrong* Roman foot which they were correcting. That the geometers believed the Roman foot, that is, their own foot, to be the *human* foot, might be easily proved. And, with such belief, they would make their so-called Roman foot too short. From a hundred measures of the feet of adult men, furnished to me by a boot-maker, and taken as they came in his books, I find the average length of the Englishman's foot to be 10·26 inches, in our day: or an inch and a third shorter than the Roman lineal measure.

This *geometrical foot* of the mathematicians is, I make no doubt, the geometrical foot to which writers of the seventeenth century refer, or mean to refer. But, not long after the true restoration of the ancient measures, there arose a disposition among those who inquired into the subject to seek a mystical origin of weights and measures, on the supposition of some body of exact science once existing, but now only seen in its vestiges: a disposition which is not yet entirely extinct. Some speculated on the pyramids of Egypt, and tried to establish that the intention of building those great masses was that a record of measures founded on the most exact principles might exist for ever. But more turned their attention to the measurement of the earth, and, by assuming nothing more difficult than that a degree of the meridian a thousand times more accurate than that of Eratosthenes was in existence hundreds, if not thousands, of years before him, it was easy enough to make out that the whole system of Greek, Roman, Asiatic, Egyptian, &c. measures was a tradition from, or a corruption of, this venerable piece of lost geodesy. There runs through all these national systems a certain resemblance in the measures of length: if a bundle of faggots were made of foot-rules, one from every nation ancient and modern, there would not be any very unreasonable difference in the lengths of the sticks.

The metrologists who treat this subject handle it according to their several theories. Those who have none in particular either neglect it altogether, or speak of its length as uncertain, or define, with Dr. Bernard, the geometrical pace as being five feet of its own kind, without saying what this kind is. Those who have the notion of the old measure of the meridian accommodate it to their supposed ancient measure; but at the same time, those of most research and note make it *less* than their Roman foot. Thus Paucton makes it nine-tenths of the Roman foot, which, with his version of that measure, is 10·9 inches English, and with the true one, 10·5 of the same. Similarly, Romé de L'Isle makes it more than half an inch (French) less than *his* Roman foot. As they do not refer to the geometers of the middle ages, I cannot guess whence they get their notion, otherwise than from their theory.

I now proceed to demonstrate the existence of this geometrical foot, which I believe to have been the effort of mathematicians to perpetuate and make common what they took to be the Roman foot, on the supposition that it was nothing but the average length of the *human foot.*

Passing over the general expressions of writers who refer to the use of the parts of the body in measurement, and who sometimes distinctly state that the determination of the human foot is necessarily that of the Roman measure, I take first the statement of Clavius, whose term of active life was the latter half of the sixteenth century and who says* very distinctly that the mathematicians, to avoid the diversity of national measures, had laid down a system for themselves. The table of measures which he gives (and dozens of other writers before him) is as follows:

1 breadth of a barleycorn.
4 = 1 digit.
16 = 4 = 1 palm (across the middle of the fingers).
64 = 16 = 4 = 1 foot.
96 = 24 = 6 = $1\frac{1}{2}$ = 1 cubit.
160 = 40 = 10 = $2\frac{1}{2}$ = $1\frac{2}{3}$ = 1 step (gressus).
320 = 80 = 20 = 5 = $3\frac{1}{3}$ = 2 = 1 pace.
640 = 160 = 40 = 10 = $6\frac{2}{3}$ = 4 = 2 = 1 perch (pertica).

125 paces 1 Italic stadium.
8 stadia 1 Italic mile.
4 Italic miles a German mile.
5 Italic miles a Swiss mile.

The constant reference to this barleycorn measure (which is seldom, if ever, omitted) induced me to try what it would really make. There is some difference between the breadths of barleycorns. A certain statement of Thevenot (cited in the *History of*

* 'Enumerandæ sunt mensuræ quibus mathematici, maxime geometræ, utuntur. Mathematici enim, ne confusio oriretur ob diversitatem mensurarum in variis regionibus (quælibet namque regio proprias habet propemodum mensuras), utiliter excogitarunt quasdam mensuras, quæ certæ ac ratæ apud omnes nationes haberentur.' —*Comm. in Sacroboscum.*

Astronomy, Lib. Usef. Kn.) makes the breadths of 144 grains of oriental barley give 1½ French feet; at which rate 64 would only give 8·53 inches English. A sample from a London shop gave me (when the largest grains were picked out) 33 to just more than five inches. Some other samples, procured from two different parts of England as the finest which could be got, gave 33 to 5 inches, 33 to 5·1 inches, and 33 to 5·1 inches. The average of these is 9·8 inches to 64 grain-breadths: a result which coincides more nearly than could have been expected with the following determinations.

There is a chain of writers who have studied to perpetuate their geometrical foot by causing a line to be laid down on the page, representing the digit, the palm, or the foot. Sometimes the palm or foot is divided into digits: and of course I rely most on those whose subdivisions are the best. As paper is apt to shrink as it becomes old, the foot deduced from these will be somewhat too small, and it may be afterwards discussed how much it should be lengthened. Leaving this for the present, I give the measurements from different authors.

Margarita Philosophica, above described. In the Strasburg edition of fifteen-four, the length of the geometrical palm is less than 2½ inches by from half to three fourths of the 24th of an inch. Taking it half way between these, the four-palm foot is 9·9 inches English. In the Basle edition of fifteen-eight, in which the woodcuts are of much rougher execution, the palm is 2·64 inches, giving a foot of 10·56 inches. The palm only is given in both cases.

Oppenheim, fifteen-twenty-four. Stöffler. 'Elucidatio Fabricæ Ususque Astrolabii.' *Folio in threes* (quarto size).

The digit, palm, and foot, are separately given; the foot is divided into palms, and all agree excellently well with one another. The foot is exactly 9·75 inches English.

Paris, fifteen-twenty-six. John Fernel. 'Monalosphærium. *Folio in threes.*

Paris, fifteen-twenty-eight. John Fernel. 'Cosmotheoria.' *Folio in threes.*

The historical mistake arising out of these works is the most remarkable circumstance attending the loss of the geometrical foot. While Fernel was publishing the first work, he was meditating (or perhaps executing) his famous measurement of a degree of the meridian. In this first work he lays his geometrical foot down the page, with great care, as he says (*omni molimine*). In two copies of this work which I have examined, the length of the foot is within a sixtieth of an inch of nine inches and two-thirds, giving 9·65 inches. In the second work, in which he announces his measure of the degree, he states that five of his paces and those of men of ordinary stature make six *geometrical* paces; which, he adds, is agreeable to the opinion of Campanus and others (at least a century and a half before), who made the mile of 1000 common paces to be 1200 geometrical paces. Allowing 60 inches (English) to a common pace, which is rather over than under the truth, this gives a geometrical pace of 50 inches, and a foot of *ten* inches. This is enough to shew that Fernel was, in the second instance, speaking in rough terms of the foot which he printed *omni molimine* in the first. The geometrical pace being forgotten, and the *monalosphærium* also, the modern historians have assumed that Fernel used the Paris foot: by which he is made to appear to come very near the real degree, whereas he is fifteen miles wrong.

Paris, fifteen-fifty-two. Jac. Koebelius. 'Astrolabii Declaratio.' *Octavo.*

This worthy astrologer, after referring to the perfect notoriety of the system of measures, gives a digit and a palm. The digit is nine-sixteenths of an inch (English), giving a foot of nine inches. The palm is 2½ inches and one-sixteenth, giving 10 inches and a quarter to the foot. The book is small, and the palm incorrectly subdivided.

Frankfort, sixteen-twenty-one. Peter Ryff. 'Questiones Geometricæ.' *Quarto.*

Four very accordant palms are given, indifferently subdivided into digits. Each palm is 2¼ inches and three-sixteenths, giving a foot of 9·75 English inches.

From these different sources, good and bad, we have for the geometrical foot 9·8, 9·9, 10·56, 9·75, 9·65, 10, 9, 10·25, 9·75 inches: the mean is 9·85. But much the best authorities are Fernel and Stöffler, because they are the greatest names, have given the whole foot, and have taken the greatest pains with the subdivisions. Their results are 9·75 and 9·65, with a mean of 9·7.

Taking this as the foot on paper, it remains to ask how much it must be lengthened to allow for the shrinking of the paper. At first, relying on the plate in Dr. Bernard's work on ancient weights and measures, in which the English foot appears to have shrunk by its 42nd part, I was disposed to lengthen the above in the ratio of 41 to 42. But observing that an older English foot, figured in the 'Pathway to Knowledge,' 1596, has shrunk only by its sixtieth part, I am rather inclined to consider the shrinking of Bernard's as an extreme case. And moreover, the two copies of the Monalosphærium give the same foot within one-hundredth of an inch certainly, and less: and it is very unlikely that if the paper had shrunk much, it should have shrunk so equally in two different copies. But, taking one-fiftieth as the outside, it follows that the geometrical foot is any thing the reader pleases between 9·7 and 9·9 English inches. The result from modern barley* gives 9·8, as above shewn.

It is remarkable how completely the English writers are in ignorance of the existence and use of the geometrical foot among their continental neighbours. Blundeville takes Stöffler, in the work above mentioned, to be speaking of a *German* foot, which, says he, Stöffler makes to be 2½ inches less than the English foot.

* Those who have tried to make the *lengths* of three barleycorns into an inch will probably think little of this mode of judging. But I observed that in samples of barley of very different apparent fineness, the difference was in the length of the corns, the breadths hardly varying at all.

Paris, fifteen-fourteen. **Boethius, Jordanus Nemorarius, Shirewode** (?), **J. Faber** (Stapulensis). 'In hoc opere contenta Arithmetica . . . Musica . . . epitome . . . Boethii : rithmimachie ludus qui et pugna numerorum appellatur.' *Folio in fours.* Second edition.

The first arithmetic is by Jordanus, the second by Boethius, both edited and commented by Faber. The music (a tract on which was in those days little but a tract on fractions under musical names) is also by Faber. The *Rithmimachia* I suppose to be the work of Shirewode's mentioned in the Introduction, but no author is named. It is a short triple dialogue on the properties of numbers, with some sort of numerical game. Libri attributes it to Faber, probably from its appearance in this edition without an author's name, and headed by an address from Faber, the editor. This second edition was printed by H. Stephens [the first, *Paris,* fourteen-ninety-six, *folio,* was printed by Hopilius, whose partner H. Stephens afterwards was, under the superintendence of David *Lauxius,** of Edinburgh].

Paris, fifteen-fourteen. **Nicolas Cusa.** 'Hæc accurata recognitio trium voluminum operum Clariss. P. Nicolai Cusæ Card. &c.' In three volumes, *folio in fours.*

Cardinal Cusa is put down in several arithmetical lists because one of his *opuscula* is entitled *de Arithmeticis Complementis.* But it is not a work on arithmetic, nor does it even proceed by arithmetic. Or it may be that John Cusa, next mentioned, may have been confounded with Nicolas. For Cusa, see the *Companion to the Almanac* for 1846, p. 14, and *Penny Cyclopædia,* 'Motion of the Earth.'

Vienna, fifteen-fourteen. **Joh. Cusanus.** 'Algorithmus linearis projectilium, de integris perpulchris arithmetrice artis regulis,' &c. *Quarto size.*

A tract of seven pages on counters.

Augsburg, fifteen-fourteen. **Jacob Kobel** (printer). 'Ain Nerv geordnet Rechen biechlin auf den linien mit Rechen pfeningen :' &c. *Quarto, twos and threes* (duodecimo size).

Computation by counters and Roman numerals : the Arabic numerals are explained, but not used. In the frontispiece is a cut representing the mistress settling accounts with her maid-servant by an abacus with counters. This book is said by Kloss to have been also printed by Kobel himself at Oppenheym in the same year.

* *Capiat qui capere potest* is the only general principle on which names like this can be read back into the vernacular. I once found James Hume, well known as a mathematician at Paris in the beginning of the seventeenth century, described as *Scotus Theagrius.* What part of the land of cakes this might be, would probably have eluded all the books in existence : but a Scottish friend, to whom I mentioned my difficulty, solved it at once, by telling me that he must have been one of the Humes of *Godscroft,* a place in which it seems certain Humes are, or were, lords of the soil.

Oppenheym, fifteen-fifteen. **Jacob Kobel.** 'Eyn New geordnet Vysirbüch.' *Quarto in threes.*

A work on gauging, in which the Arabic numerals are used.

Vienna, fifteen-fifteen. **Boethius, J. de Muris, Thos. Bradwardin, Nic. Horem, Peurbach, de Gmunden.** 'Contenta in hoc libello. Arithmetica communis, Proportiones breves, de latitudinibus formarum, Algorithmus D. Georgii Peurbachii in integris, Algorithmus Magistri Joh. de Gmunden de minuciis physicis.' *Quarto.*

The name of George Tannstetter as editor of this collection is a fair guarantee for the genuineness of the several works.

The first is an abridgment of Boethius, by John de Muris, of whom elsewhere.

The second is on proportion by Thomas Bradwardin (of the time of Edward III.), who is Bradowardinus, Bragvardinus (as in the work before me), Bragadinus (a confusion perhaps between him and Bragadini), &c. according to the fancy of the speller: Tanner says the book was printed at Paris in 1495. Bradwardine is better known among theologians for his book against Pelagius, than among arithmeticians. He begins thus: 'Proportion is either that which is commonly so called, or properly so called. Proportion commonly so called is the mutual habitude of two things. Proportion properly so called, is the mutual habitude of two things *of the same kind.*' Perhaps the authority of an Archbishop of Canterbury (for he was or was to be nothing less) may induce those who teach the rule of three to remodel their plan according to the archiepiscopal dictum, which is also that of common sense. [Other mathematical works of Bradwardine were printed.]

The third book is on the areas of figures (though nothing to any purpose is done), with the leading idea that the difficulty arises from variation of breadth. Peurbach's work is a summary of operations like that of Sacrobosco; but though written to explain the Arabic notation, it obviously takes for granted that that notation is generally known. John de Gmunden's work is on sexagesimal fractions (*minuciæ physicæ*).

Paris, fifteen-fifteen. **Gaspar Lax.** 'Arithmetica speculativa magistri Gasparis Lax Aragonensis de sarinyena duodecim libris demonstrata.' Also, same date and place, 'Proportiones magistri Gasparis,' &c. *Folio in threes.*

This is a very diffuse and extensive work, in small black letter. It treats only of the simple properties of numbers; though it is evident that to a knowledge of Boethius, Lax added that of the arithmetical books of Euclid. The thing which appears most surprising in this and other works of the same kind, is the apparent difficulty of dealing with numbers. Here are upwards of 250 small black-letter pages in folio, filled with propositions on the simplest properties of numbers, and the reader looks in vain for any number so high as 100 used by way of exemplification. For any thing that appears the author could not count as far as 100. Montucla says that Lax was afterwards Pope; but I am assured by a learned Catholic bishop that there never was any Pope of the name.

Lyons, fifteen-fifteen. **John de Lortse** (Dominican). 'Oeuvre tressubtile et profitable de lart et science de aristmeticque: et geometrie translate nouvellement despaignol en francoys Auquel est demonstre par figure evidemment: tant le nombre entier: nombre rompu: regle de compaignies: soubde fin: que toutes aultres choses qui par geometrie et aristmeticque peuvent estre comprises: comme appert par la table cy apres mise.' *Quarto* (O) (166 folios).

This is perhaps the first book in French on commercial arithmetic. I have not found mention of it in any catalogue. It is a good Italian importation, through Spain: and must have made the readers of the Boethian school stare to see what arithmetic really could do. We have thus two Spanish books published at Paris in one year: another (see Siliceus presently mentioned) had been published there the year before.

Venice, fifteen-twenty. **R. Suiseth.** 'Calculator. Subtilissimi Ricardi Suiseth Anglici Calculationes noviter emendate atque revise.' *Folio in threes.*

This man is Richard at the beginning of the book, Raymund at the end, while Tanner [and Gesner] call him Roger. His name seems to have been Swinshead, latinised into Swincetus, Suicetus, Suineshevedus, &c. The title of his book, *Calculator*, has given rise to the fiction that he was a great arithmetician, and even algebraist; which, as he lived in the fourteenth century, would have made him a very remarkable man. But nothing can be further from the truth: the book is full of philosophy about intension and remission, and siccity, and calidity and frigidity, and other *quisquiliæ Suiceticæ*, as some afterwards called them. That number and magnitude are used occasionally, is all that can be said.

Brucker, according to Enfield, Hallam, &c., says this book is as scarce as a white raven. But probably this refers to one of the previous editions. Hain gives three of the fifteenth century, marked 'Padue (no date),' 'Papiæ, 1488,' and 'Papie, 1498.' All these Suissets are *Richards*. (See Tanner *in verb.* 'Swincet;' Wallis *Op.* t. iii. pp. 673, 675, 680, 685.)

Lyons, fifteen-twenty. **Stephen de la Roche Villefranche.** 'Larismethique nouvellement composee divisee en deux parties dont la premiere tracte des proprietes parfections et regles de la dicte Science,' &c. *Folio in fours* (O).

A large treatise, very full on commercial arithmetic, and containing also geometry. I suppose we should find it largely indebted to the Spanish books above named.

Paris, fifteen-twenty-one. **Boethius.** 'Divi Severini Boetii Arithmetica duobus discreta libris: adjecto commentario, mysticam numerorum applicationem perstringente, declarata.' *Folio in threes.*

The commentary is by Girardus Ruffus, and for absurdity and dulness, it ought to rank high among the mystic commentaries: even to the justification of Cornelius Agrippa, when he asserts that arithmetic is not less superstitious than vain, adding, like a philosopher of his day, that it is only valuable to merchants for the low and mean benefit of keeping their accounts.

London, fifteen-twenty-two. **Tonstall** (Bp. of London). 'De Arte supputandi libri quatuor Cutheberti Tonstalli.' *Quarto.* (Printed by Pynson.)

Paris, fifteen-twenty-nine. The same title. *Quarto in fours.* (Printed by Rob. Stephens.)

Strasburg, fifteen-forty-four. 'De Arte . . . . . Tonstalli, hactenus in Germania nusquam ita impressi.' *Octavo.*

[There is said to have been an earlier Strasburg edition, but I have not seen it: there are several other editions, but no English reprint.] For the life of Tonstall see Anth. Wood, *Ath. Oxon. in verb.* This book was a farewell to the sciences on the author's appointment to the see of London (see the preface): it was published (that is, the colophon is dated) on the 14th of October, and on the 19th the consecration took place. This book is decidedly the most classical which ever was written on the subject in Latin, both in purity of style and goodness of matter. The author had read every thing on the subject, in every language which he knew, as he avers in his dedicatory letter to Thomas More, and had spent much time, he says, *ad ursi exemplum,* in licking what he found into shape. The wonder is, that after this book had been reproduced in other countries, and had become generally known throughout Europe, the trifling speculations of the Boethian school should have excited any further attention. For plain common sense, well expressed, and learning most visible in the habits it had formed, Tonstall's book has been rarely surpassed, and never in the subject of which it treats. It seems to have been very little known to succeeding English writers. P. 419, 426, 427, 439.

Venice, fifteen-twenty-three. **Sacrobosco.** 'Algorismus domini Joh. de Sacro Busco noviter impressum.' *Quarto.*

The edition of this work mentioned by Watt, as by Cirvellus, fourteen-ninety-eight, is a mistake: [it was the work on the sphere which Cirvellus edited in that year] (Hain). Dr. Peacock (p. 416) thinks that this work is attributed to Sacrobosco without sufficient reason, and mentions it only as a manuscript. Mr. Halliwell reprinted it in the *Rara Mathematica*, evidently under the impression that it had never been printed. If it be the work of Sacrobosco, it establishes beyond doubt that he had the Arabic numerals and the method of local value: being nothing but a summary of rules for that arithmetic. The words *noviter impressum* are ambiguous: they may either imply a first impression or any succeeding one, though I have commonly found them used in the latter sense.

I should pause before I rejected as spurious a work which is attributed to Sacrobosco in many manuscript copies extant in different countries, which was printed under his name as early as 1523, and is often cited as his. Dr. Peacock lays stress upon there being no mention of the Arabic system in his other works, those on the sphere and on the calendar. But against the presumption drawn from this circumstance, it may be urged that it was not likely he would introduce mention of a new system of arithmetic in works intended for common use, though he might write a separate work to explain that system. And he would have no motive for alluding to a method which his readers were not acquainted with. There is great probability that Sacrobosco was acquainted with the modern system, which all other evidence goes to shew was introduced into Italy before his time.

It is of course very possible, or, looking at the progress of other things, most probable, that isolated individuals had obtained the Arabic notation from the East, or from Italy, practised it, and written in it, long before it obtained the smallest general currency.

But there is one circumstance which seems to me to lend more than presumption to a still wider supposition, namely, that Arabic notation and rules were known beyond the bounds of Italy, in the time of Sacrobosco, to *more* than a few isolated individuals. Except by importations from the East, it is impossible to say whence the philosophers of the thirteenth century could have got any thing like a set of rules. They certainly had not got any thing Greek, except from the Arabs: the system of Boethius does not give rules of computation. Moreover, the name *algorithm* was in later times so invariably connected with Arabic arithmetic, that the presumption is strong it was so from the beginning. Now Roger Bacon, the contemporary of Sacrobosco, not only used the adjective *algoristicus* several times, but recapitulates the names of a set of rules. The theologian, he says (Jebb, p. 138), should abound in the power of numbering, that he may know all the algoristic modes, not only for integers, but for fractions, to numerate, to add, to subtract, to *mediate* (divide by two), to multiply, to divide, and to extract roots. Now this is precisely the set of rules given in the treatise attributed to Sacrobosco, with one omission: this last treatise distinguishes *duplation* (multiplication by two) from other cases of multiplication, which Bacon does not; and *progression*, which is, however, only a mixture of other rules.

There is a sentence of Bacon's contemporary, Matthew Paris, quoted by Mr. Hallam, in which he speaks of what can be done with Greek notation, as being what cannot be done either in Latin, or in algorithm, *vel in algorismo.* These words might easily be interpolated; so instead of using them, as Mr. Hallam does, to give ground of presumption that Bacon had the Arabic notation, I should rather use what I have drawn from Bacon to strengthen the genuineness of the words of Paris; and I think Mr. Hallam will be ready to do the same.

In 'Rara Mathematica,' *London*, eighteen-forty-one, *octavo*, second edition, Mr. Halliwell has inserted this treatise of Sacrobosco, together with the poem *de algorismo* attributed to him in some of the manuscripts, but probably enough without foundation. The seven rules above mentioned, are thus given :

Septem sunt partes, non plures, istius artis;
Addere, subtrahere, duplareque dimidiare
Sextaque dividere est, sed quinta est multiplicare
Radicem extrahere pars septima dicitur esse :

on which Mr. Halliwell quotes a manuscript of perhaps nearly as early a date as Sacrobosco, which describes the seven rules and the necessity of the *letters called figures*, as follows:

En argorismo devon prendre
Vii especes . . . .
Adision Subtracion
Doubloison mediacion
Monteploie et division
Et de radix enstracion
A chez vii especes savoir
Doit chascvn en memoire avoir
Letres qui figures sont dites
Et qui excellens sont ecrites

In the same 'Rara Mathematica' are to be found an English manuscript of the 14th century 'On the numeration of Algorism,' and John Norfolk 'In artem progressionis summula,' dated fourteen-forty-five at the end.

Paris, fifteen-twenty-six. **John Martin Siliceus.** 'Arithmetica Joh. Martini Silicei theoricen praxinque luculenter complexa' &c. *Folio in threes.*

This is a third edition by Thomas Rhætus; Heilbronner says that the first was *Paris*, fifteen-fourteen, and that the author, who was a professor at Salamanca, named *Gujieno* (Silex), died at Toledo in fifteen-fifty-seven. Montucla says he was Archbishop of Toledo. This is a work of considerable extent of subject, and seems to have no great fault except the usual one of prolixity. It gives a treatise on theoretical arithmetic (Boethian in character), one on the rules of computation, one on the mode of calculating by pebbles or the abacus, and one on fractions. The second edition, fifteen-nineteen, by Orontius Finæus, is in the Royal Society's Library.

Paris, fifteen-twenty-eight. **John Fernel.** 'De proportionibus Libri duo.' *Folio in threes.*

A book on proportion of this date is in great part filled with Boethian arithmetic. This book deserves attention : its author had a much better grasp of Euclid than most of his contemporaries.

Leyden, fifteen-thirty-one. **Joach. Fortius Ringelbergius.** 'De Ratione Studii.' *Octavo* (duodecimo size).

At the end, same place, year, and form, with another title-page, apparently another work, but of this I am not sure, is J. F. R. 'Compendium de scribendis versibus,' to which is attached 'Arithmetica' (O), the most Lilliputian treatise I know of, giving in 17 pages an epitome of Boethian terms, algorithmic rules, and counter calculation. [It is said that the author's name was Sterk.]

Venice, ——. **Giovanni Sfortunati.** 'Nuovo Lume, Libro di Arithmetica . . . . Composto per lo acutissimo prescrutatore delle Archimediane, et Euclidiane dottrine Giovanni Sfortunati da Siena.' *Quarto in fours.*

The end of the work torn out, so I cannot give the date. He mentions L. de Burgo, Calandri, and P. Borgio, and Tartaglia mentions him: which gives certain limits. By the manner in which the author is described, this must be a reprint: it is not likely a work would be originally printed at Venice, one cause of the production of which is stated to be the barbarism of the Venetian dialect. I can find no mention of Sfortunati in catalogues. Cardan, writing in fifteen-thirty-seven, mentions the work of Fortunatus, the same as the above, no doubt. P. 458, 460.

Nuremberg, fifteen-thirty-four. 'Algorithmus demonstratus.' *Quarto.*

John Schoner, the editor, attributes this to Regiomontanus. It is a series of demonstrated propositions of numbers connected with the Arabic notation, involving not only the rules of computation, but such as that the number of figures in the cube cannot exceed three times that in the root, &c. J. Schoner, who edited several writings of Regiomontanus, is likely enough to be right on this one, which is moreover not unworthy of such an author. It is not in the list which he published himself, but it seems to be generally recognised. (Weidler *in verb.*)

Paris, fifteen-thirty-five. **Orontius Fineus.** 'Arithmetica practica libris quatuor absoluta . . . . . Recens ab Authore castigata . . . . . hoc est in nativum splendorem (quem priorum impressorum amiserat incuria) summa fidelitate restituta.' *Folio in fours.*

This must be at least the third edition; Leslie says the first was in fifteen-twenty-five. Orontius died in 1555.

Strasburg, fifteen-thirty-six. **Hudalrich** or **Huldrich Regius.** 'Utriusque Arithmetices Epitome ex variis authoribus concinnata per Hudalrichum Regium.' *Octavo* (duodecimo size).

The first book is on the Boethian arithmetic: the second on the rules of computation with integers and fractions, and on the use of the abacus. There are but 96 small folios with large print: so that if I wished to bring any person into the closest contact with the middle ages at the least expense of reading, with reference both to their mode of expression and operation, I should certainly prescribe this book.

Paris, fifteen-thirty-eight. **Nicomachus** of Gerasa. 'Νικομαχου Γερασινου Ἀριθμητικης βιβλια δυο Nicomachi Gerasini Arithmeticæ Libri duo. Nunc primum typis excusi, in lucem eduntur. Parisiis in officina Christiani Wecheli.' *Quarto.*

Nicomachus, who lived about the time of Tiberius, was a Pythagorean, and of course devoted to arithmetic. He was in his day, as Lucian intimates, a Cocker, a person whose name passed as an allusion to reckoning and numbering. The work of Boethius is wholly drawn from him: some people call it a mere translation of, others a comment on, Nicomachus. Perhaps a free rendering might be correct. This edition is Greek without Latin: [there is another, also Greek without Latin, at the end of the *Theologumena Arithmeticæ* attributed to Jamblichus, *Leipsic*, eighteen-seventeen, *octavo* (?)]. It is then ultimately to Nicomachus being a Pythagorean that we owe those once never-ending denominations of numbers and ratios, of which it might now be difficult to trace any vestige in modern language, except the common words *multiple* and *sub-multiple* and the *sesquialter* stop of an organ.

Milan, fifteen-thirty-nine. **Jerome Cardan.** 'Practica Arithmetice et mensurandi singularis.' *Octavo* (O).

This was reprinted, under the title of 'Practica Arithmetica, in the fourth volume of his collected works, *Leyden*, sixteen-sixty-three, ten volumes, *folio in twos*. On the arithmetic there is no remark to make, except that, as might be expected from an Italian of that day, Cardan shews more power of computation than the French and German writers. There is a chapter recapitulating the numbers which have mystic properties as he calls them, one use of which is in foretelling future events. These are mostly the numbers mentioned in the Old and New Testaments, but not altogether: 400, for instance, makes its appearance, because it was (but it was not) the number of bishops at the Nicene Council. In the tenth volume of the same collection is an unfinished treatise by Cardan, headed 'Artis Arithmeticæ Tractatus de Integris,' which seems to be the commencement of a more extensive treatise than the former one. It may be cited from it that decidedly, in Cardan's opinion, it was Leonard of Pisa who first introduced the Arabic numerals into Europe.

London. 'This boke sheweth the maner of measurynge of all maner of lande, as well of woodlande, as of lande in the felde, and comptynge the true nombre of acres of the same. +

Newlye invented and compyled by **Syr Rycharde Benese,** Chanon of Marton Abbay besyde London. Prynted in Southwarke in Saynt Thomas hospitall, by me James Nicolson.' *Quarto.*

Again—*London.* 'The Boke of measuryng of Lande as well of Woodland as Plowland, and pasture in the feelde: and to compt the true nombre of Acres of the same. Newly corrected and compiled by **Sir Richarde de Benese.** ¶ Impynted at London, by Thomas Colwell.'

The approximate date of the first we must guess at from the fact that James Nicolson's dated works are from fifteen-thirty-six to thirty-eight, and from the presumption that a printer's undated works come before the dated ones. Thomas Colwell's dated works are from fifteen-fifty-eight to seventy-five. The acre is four roods, each rood is ten *daye-workes,* each daye-worke four perches. So the acre being 40 daye-workes of 4 perches each, and the mark 40 groats of 4 pence each, the aristocracy of money and that of land understood each other easily.* The reprint of the above work wants a set of tables which were in the original, and which I take to be the earliest mathematical tables published in England, except the smaller tables of the same kind which appeared in some acts of parliament. They are of double entry, the entries being made by perches in length and breadth. Thus under 79 opposite 9 we find $\frac{4\cdot1}{7\cdot3}$ meaning that 79 perches by 9 is 4 acres 1 rood, 7 day-works and 3 square perches. Perhaps this sort of table became common: I find it, without the slightest deviation of form, in Arthur Hopton's 'Baculum Geodæticum sive Viaticum, or the Geodeticall Staffe,' *London,* sixteen-ten, *quarto.*

Paris, fifteen-thirty-nine. **Johannes Noviomagus.** 'De Numeris libri duo.' *Octavo* (small) (O).

Professedly on computation, but interspersed with many curious historical remarks. It contains the first hints I have found of the most probable explanation of the origin of the Roman numerals.

Paris, fifteen-forty-four. **Orontius Finæus.** 'Arithmetica Practica . . . . . multisque accessionibus locupletata.' *Octavo* (O).

* More easily than they do at the time when this is written. In the old day, when the aristocracy of money was a Jew, the aristocracy of land used to shut him up, and draw a tooth a day until he protected agriculture to the amount demanded. When Christians became wealthy, a tax upon the use of the teeth seems to have been substituted for their forcible removal. The debates on the abolition of this tax are now proceeding. But, speaking of weights and measures, there is something else which tells a tale about the feasts and Sundays of old England. The sack of wool was 13 tods of 28 pounds each, or 364 pounds. This was arranged, that the tasks of those who spun, &c. might be easily calculated; a pound a day being a tod a month and a sack a year. But where are the Sundays and holidays? Most likely they were made up on the other days.

Paris, fifteen-fifty-five. **Orontius Finæus.** 'De Arithmetica Practica libri quatuor.' *Quarto* (O).

An augmented edition. Had this book been simply headed as arithmetic, I might have let it alone. But as it is called *practical* arithmetic, it must stand as an index of the great advance of the English over their neighbours in computation, as shewn by Tonstall and Recorde. P. 428, 436.

Nuremberg, fifteen-forty-four. **Michael Stifel.** 'Arithmetica integra. Authore Michaele Stifelio. Cum præfatione Philippi Melancthonis.' *Quarto.* (Hutton's Tracts, v. ii. p. 237.)

The two first books are on the properties of numbers, (O), on surds and incommensurables, learnedly treated, and with a full knowledge of what Euclid had done on the subject. The third book is on algebra, and passes for the introduction of algebra into Germany. But Stifel himself, in his preface, acknowledges his obligations to Adam Risen, and professes to have taken all his examples from Christopher Rudolph. Of the latter I can find nothing except a statement that his work was subsequently published: but Kastner (v. i. p. 108) gives some information on the works of Adam and Isaac Riese. P. 424, 436, 474, 450.

Stifel is supposed to have been the inventor of the signs + and – to denote addition and subtraction: and the manner in which he speaks of these signs seems to imply that they were of his own introduction (p. 110). It is "*we place* this sign," &c. and "*we say* that the addition is thus completed;" and so on. The sign +, in the hands of Stifel's printer, has the vertical bar much shorter than the other: and when it is introduced in the woodcuts by the engraver, the disproportion is greater still. One would (supposing the wood-engraver to have imitated the handwriting) have supposed that – was first used, and that + was derived from it by putting a small cross-bar for a distinction. This form is rather against the late learned Professor Rigaud's idea (see Mr. Davies' *Solutions of Hutton's Mathematics*, p. 11) that + is a corruption of P, the initial of *plus*, and also against Mr. Davies' ingenious conjecture that it is a corruption of *et* or &. Stifel does not call the signs *plus* and *minus*, but *signum additorum* and *signum subtractorum*. In no instance do I find him using the first pair of appellatives at all.

I had written the above, and become fully impressed, by repeated examination, with the suspicion that the vertical bar in + was, in Stifel's invention, a small distinction superadded on the sign –, when I happened, in the immediate preparation of these sheets for press, to remember Colebrooke's statement that the Hindoos (our original teachers in algebra) use a dot for subtraction, and nothing but the absence of the dot for addition. It may be likely then, that in the first instance the Hindoo dot was elongated into a bar, to signify subtraction, addition having no sign: and that the first who found it convenient to introduce a sign for addition, merely adopted the sign for subtraction with a difference.

Against the above conjecture it may be argued, that the European algebra comes from India through the Mahometan writers, who use no signs at all. On this I can only remark at present, that I have long suspected, from many circumstances, that there was a more direct communication with India in the introduction of algebra than any one now believes.

Dr. Ritchie suggested that perhaps + was used to denote addition as being two marks joined together, and made significative of two numbers joined together: and that – denoted subtraction, as indicating what is left after one of the marks of the junction is removed. This suggestion is more like what *ought to have been* the derivation than any I have seen: but there is little reason to suppose that any such attempt at significant symbols would have been made in the sixteenth century.

M. Libri attributes the invention of + and – to Leonardo Da Vinci, in whose manuscripts he says he had seen them. But as it happened to me to discover, in a manuscript of that justly celebrated man (if it be not more correct to say *unjustly* celebrated, as not celebrated enough for his merits) now in the British Museum, the sign + used for the figure 4, I think the question must rest open until citations which establish the point without mistake can be produced.

No place, nor date. **Wm. Buckley.** 'Arithmetica memorativa sive compendaria Arithmeticæ tractatio.' Apparently *octavo*.

This book has been described both by Leslie and Peacock, who have severally quoted some of the verses of which it consists. They say it was first printed in fifteen-fifty, and afterwards at the end of Seton's Logic in sixteen-thirty-one. I should probably not have seen it to this day (for after Leslie's full citations, I should hardly have made active search for it), had not an anonymous correspondent (whom I have only this way of thanking) obligingly forwarded to me an imperfect copy which he happened to find among his papers. It is, I suppose, the extract from Seton's Logic: the signature in the title-page is Q 2, after a *libellus ad lectorem*, 'Quisquis Arithmeticam . . . . qui meminisse dedit:' and there is a preface signed T. H. It is worth noting that Buckley's Latin name, *Buclæus*, has been frequently written *Budæus* by the continental bibliographers, so that in all probability the learned author of the book *De Asse* has been taken for the author of these verses. Hylles calls him *Buckle*. P. 437.

[According to Watt, there were editions of Seton's Logic, with 'Huic accessit ob artium ingenuarum inter se cognationem Arithmetica Gulielmi Buclæi,' *London*, fifteen-seventy-two, seventy-four, and seventy-seven, *octavo*.]

Basle, fifteen-fifty. **John Scheubel.** 'Euclidis Megarensis . . . . . Algebræ porro Regulæ . . . . . his libris præmissæ sunt . . .' *Folio in twos*.

The algebra was reprinted, *Paris*, fifteen-fifty-two. *Quarto*. 'Algebræ compendiosa facilisque descriptio,' &c. For a description of the work see Hutton's Tracts, vol. ii. p. 241.

Leyden (*Lugduni*, without further description), fifteen-fifty. **Henry Cornelius Agrippa.** 'De Occulta Philosophia Libri Tres.' *Octavo* (Italic letter).

As the preface is dated *Mechlin*, fifteen-thirty-one, and [an edition appeared there in fifteen-thirty-three], this latter must be the first. This book has as much right in my list as that of Jordano Bruno presently noticed, treating of numbers in much the same way. It is a property of seven that there are seven angels who stand before God; of twelve, that there are twelve apostles, twelve permutations of the letters of the tetragrammaton, &c. A good many of the more formal works on arithmetic of this period abound in enumerations of the same kind, evidently considered as appertaining to the higher uses of the science.

Augsburgh, fifteen-fifty-four. **Psellus.** 'De quatuor scientiis mathematicis.' Edited by Xylander. Gr. Lat. *Octavo* (duodecimo size).

Psellus lived in the ninth century. The part relating to Arithmetic is a mere compendium of arithmetical terms, perhaps from Boethius.

Venice (the first two parts fifteen-fifty-six, the rest fifteen-sixty). **Nicolas Tartaglia.** 'La prima parte dei numeri e misure aritmetica geometria et algebra.' *Folio in threes* (O).

Of this enormous book I may say, as of that of Pacioli, that it wants a volume to describe it. P. 430, 450. It was partly translated into French as follows: *Paris*, sixteen-thirteen. 'L'Arithmetique de Nicolas Tartaglia, Brescian, grand mathematicien et prince de praticiens, recueillie et traduite d'Italien en Francois par Guillaume Gosselin de Caen.' *Octavo*.

London, fifteen-fifty-seven. **Robert Recorde.** 'The whetstone of witte, whiche is the seconde parte of Arithmetike: containyng thextraction of Rootes: The Cossike practise, with the rule of Equation: and the woorkes of Surde Numbers.' *Quarto*.

On this see the *Companion to the Almanac* for 1837, p. 36; Hutton's Tracts, v. ii. p. 243; P. 369. It is rarely remembered that the old name of Algebra, the *cossic* art (from *cosa*, thing), gave this first English work on algebra its punning title the *whetstone* of wit, *Cos ingenii*.

Lione, fifteen-fifty-eight. **Giov. Franc. Peverone.** 'Due brevi e facili trattati. Il primo d'Arithmetica : l'altro di Geometria,' &c. *Quarto.*

A book of the very simplest examples of computation.

Leyden, fifteen-fifty-nine. **Joh. Buteo.** 'Joan. Buteonis Logistica quæ et Arithmetica vulgo dicitur.' *Quarto in fours.* (O).

In the same year was published, by the same author, his *Opera Geometrica,* the first of which is a plan of Noah's ark, with calculations as to its sufficiency for holding all the animals and their provender. P. 437.

Paris, fifteen-sixty. **James Pelletier.** 'Jacobi Peletarii Cenomani, de occulta parte numerorum, quam algebram vocant, libri duo.' *Quarto.* [First edition, fifty-eight.]

This is said to be the first French work on algebra. There is little purely arithmetical in it: at the end is a table of squares and cubes up to those of 140. See Hutton's Tracts, vol. ii. p. 245.

London, fifteen-sixty-one. **Robert Recorde.** 'The Grounde of artes: Teaching the worke and practise of Arithmetike, both in whole numbres and Fractions, after a more easyer and exacter sorte than any like hath hitherto been sette forthe: Made by M. Robert Recorde, Doctor of Physik, and now of late ouerseen and augmented with new and necessarie Additions.

I[ohn] D[ee].

All youth and Elde that reasons Lore
Within your breastes will plant to trade
Of Numbers might the endles store
Fyrst vnderstand, than farther wade'

Octavo (duodecimo size) (O).

See the *Companion to the Almanac* for 1837, p. 32. That this book was published about fifteen-forty there is internal evidence: and Tanner gives it that date. Dr. Peacock has fifteen-forty-two. But I have never seen any edition earlier than this. John Dee published it again, *London,* fifteen-seventy-three, *octavo,* with other verses in the title-page, and the original dedication to Edward VI. restored. The preceding edition, though published under Elizabeth, was probably arranged in the life of her sister, which may account for the omission of the dedication. There have been many editions. Hartwell's edition of Mellis's edition was published thrice at least, in sixteen-forty-eight and sixteen-fifty-four, and once more with the date torn out in our copy. Mellis added a third part, on practice and other things: in Roman letter, the sacred text of Recorde having always its old black letter. The last edition I know of it is by Edw. Hatton, *London,* sixteen-ninety-nine, *quarto;* it has an additional book called 'Decimals made easie.' P. 408-10, 434, 454, 476.

Vienna, fifteen-sixty-one. **Christoffer Rudolff.** 'Kunstliche rechnung mit der ziffer und mit der zalpfenningen,' &c. *Octavo* (small).

This is the man to whom Stifel refers as his instructor in algebra. But there is nothing like either of the signs + or − in his book: so that Stifel does not appear to have got them from him. This book is not algebra: and whether any thing of Rudolph's deserving of that name was published is more than I can now settle.

Antwerp, fifteen-sixty-five. **Valentine Menher de Kempten.** 'Practicque pour brievement apprendre à Ciffrer, et tenir livre de Comptes, avec la regle de Coss, et Geometrie.' *Octavo.* [Date at the end, sixty-four, July.]

Here are books on Arithmetic, Book-keeping, Algebra, Geometry, and Trigonometry: it appears that part of the arithmetic had been published in fifteen-fifty-six, of the geometry in fifteen-sixty-three, of the algebra in fifteen-fifty-six. The arithmetic is an early specimen of the book of mere rules. In this and other French books of the same date, and even much later, the words *septante* and *nonante* are used for seventy and ninety. The book-keeping is a short treatise, or rather example, of double entry. The algebra is more extensive, on the model of Stifelius, with many questions leading to simple equations, each embellished with an illustrative picture in wood. The geometry is mensuration. The spherical trigonometry is from Regiomontanus, and refers to no other tables. This book has some pretension to be called a course of mathematics. Menher is often quoted in England for his tables of rebate or discount.

Antwerp, fifteen-sixty-seven. **Pedro Nunez.** 'Libro de algebra en arithmetica y geometria.' *Octavo* (small).

This book is wholly algebraical. See Hutton's Tracts, vol. ii. p. 250.

London, fifteen-sixty-eight. **Humfrey Baker.** 'The Well Spryng of Sciences, which teacheth the perfecte woorke and practise of Arithmeticke . . . . beautified with moste necessary rules and questions . . . .' *Octavo* (very small size) (O).

The reprints will be presently mentioned.

Paris, fifteen-seventy-one. **Alex. Vandenbussche.** 'Arithmetique militaire.' *Quarto.*

A short work, mostly on the arithmetic of the arrangement of troops, with a large number of military maxims and observations.

The first book ends with a question described as *non moins plaisante que belle*, as follows. A *soldat désespéré* asked of an *arithméticien fantastique* how far it was from Milan to hell. The mathematician, placing it at the centre of the earth, made the distance 1145¼ leagues. This, and *Je prie à Dieu nous garder d'y aller*, is the whole of the joke.

London, fifteen-seventy-one. **Leonard Digges.** 'A Geometricall Practise, named Pantometria . . . . framed by Leonard Digges, Gentleman, lately finished by Thomas Digges his sonne. Who hath also thereunto adjoyned a Mathematicall Treatise of the five regulare Platonicall bodies,' &c. *Quarto.* Reprinted as follows: *London*, fifteen-ninety-one. 'A Geometrical practical treatize named Pantometria. . . . . First published by Thomas Digges Esquire . . . . lately reviewed by the author himself,' &c. *Quarto.*

Almost the whole is application of Arithmetic to Geometry, in measurement both of planes and solids.

In this book I find the earliest printed mention I ever met with of the *theodelite*: a word the derivation of which has long been a puzzle. It is known to be an exclusively English word in all time preceding the middle of the last century. Digges calls it 'the circle called *theodelitus*.' A moveable radius travelling round a circle, for the measurement of angles, had been long denoted by the Arabic word *alhidada* (whence the French word *alidade*). *Theodelite* seems unlikely to be a corruption of this, at first: nor should I have suspected such a thing, if I had not found in Bourne's 'Treasure for travailers,' *London*, fifteen-seventy-eight, *quarto*, the intermediate formation, *athelida*, used for the same thing. I hold it then pretty certain that this is the true origin of the word *theodelite*, which was so spelt, not *theodolite*. (See the *Philosophical Magazine*, April, 1846.)

Venice, fifteen-seventy-five. **Franc. Maurolycus.** 'Arithmeticorum libri duo.' *Quarto in fours.*

On the properties of numbers and the doctrine of incommensurables; a superior work to the mass of those which then treated of similar subjects.

Basle, fifteen-seventy-nine. **Christian Urstisius.** 'Elementa Arithmeticæ, logicis legibus deducta.' *Octavo* (duodecimo size) (O).

It is said that Urstisius was a follower of Copernicus and a teacher of Galileo. This Arithmetic was translated by T. Hood, *London*, fifteen-ninety-six, *octavo* (small). 'The Elements of Arithmeticke most methodically delivered,' &c.

Brescia, fifteen-eighty-one. **Ant. Maria Vicecomes** (Visconti). 'Practica numerorum et mensurarum,' &c. *Quarto.*

A book of mixed arithmetic and algebra, in which the modern method of division is used. It would, I think, repay an examination, particularly on the part which relates to the extraction of roots.

Antwerp, fifteen-eighty-one. **Gemma Frisius.** 'Arithmeticæ practicæ methodus facilis, in eandem J. Steinii et J. Peletarii annotationes,' &c. *Octavo* (small size) (O).

This is a reprint, much augmented, of a compendious work of Frisius, the 'Ar. pract. meth. fac.' *Witteberg*, fifteen-sixty-three, *octavo* (small). But there is an earlier edition of Gemma Frisius, also by Peletarius, being the earliest I have met with, as follows:

Paris, fifteen-sixty-one. 'Arithmeticæ Practicæ Methodus facilis, per Gemmam Frisium . . . . jam recens ab ipso authore emendata, et multis in locis insigniter aucta. Huc accesserunt Jacobi Peletarii Cenomani annotationes . . . . Quibus demum ab eodem Peletario additæ sunt Radicis utriusque demonstrationes.' *Octavo* (small) (O).

These demonstrations of the rules for the square and cube roots Peletarius treats as a great boon to his reader: and states that he was very near contenting himself with a reference to his book on *occult* properties of numbers (his Algebra of 1560, before mentioned).

London, fifteen-eighty-three. **H. Baker.** 'The Well spring of sciences. Which teacheth the perfect worke and practise of Arithmeticke . . . . set forthe by Humfrey Baker, Londoner, 1562. And nowe once agayne perused augmented and amended in all the three partes, by the sayde Aucthour.' *Octavo* (duodecimo size).

Reprinted (same place and size) in sixteen-fifty-five, with a reference at the end to an edition of sixteen-forty-six. It is one of the books which break the fall from the 'grounde of Artes' to the commercial arithmetics of the next century. There are some short rules for particular cases, and great attention to the rule of practice. Among the peculiarities of the book is a notion, apparently, that none but fractions should deal with fractions: for Baker will not double $\frac{3}{5}$, for instance, by multiplying by 2, but only by dividing by $\frac{1}{2}$. P. 452.

London, fifteen-eighty-four. **John Blagrave.** 'The mathematical jewel.' *Folio in twos.*

See the *Companion to the Almanac* for 1837, p. 41. This book is curious, because the woodcuts are all done by the author's own hand, and Walter Venge, the printer, is not known to have printed any other work. [For Blagrave's legacy to the three parishes of Reading, see Ashmole's Berkshire, v. iii. p. 372: for the appearance of his ghost, "credibly reported by many honest and discreet persons," see the *Annus Mirabilis*, 1662, p. 49.] This book is not arithmetical, but the contrary: it is an avowed attempt to drive computation out of astronomy by the introduction of an instrument called the *Jewel*, which is a projection of the sphere. Of the tables of sines he says, 'But now there are in like maner an infinite number of intricate questions more harder far than any yet propounded, which by Regiomont. Copernicus and others doctrine, grow to great toile with their Synes, calculations, and proportions, wherein first they hunt about for one Syne, which they call *inventum primum*, then for another, which they call *inventum secundum*, and commonly *inventum tertium*, and perhaps *quartum*, and so foorth. After all these are found, then they multiplie and devide them, and compare their proportions: and when al is done, that they have founde the Seyne sought for: they yet are faine to goe to their tables for the arch correspondent.' There was a great disposition at this time among the English to try to substitute instruments for computation. Blagrave's Jewel stood in high estimation; and as late as 1658, John Palmer, who published a description of it under the name of the *Catholic Planisphere*, says it was a subject of frequent inquiry where the instrument and the description were to be met with.

Leyden, fifteen-eighty-five. **Simon Stevinus.** 'L'Arithmetique de Simon Stevin de Bruges: Contenant les computations des nombres arithmetiques ou vulgaires: Aussi l'algebre, avec les equations de cinc quantitez. Ensemble les quatre premiers livres d'Algebre de Diophante d'Alexandrie maintenant premierement traduicts in François. Encore un livre particulier de la Pratique d'Arithmetique, contenant entre autres, Les Tables d'Interest, La Disme; Et un traicté des Incommensurables grandeurs: Avec l'Explication du Dixiesme livre d'Euclide.' *Octavo.*

Leyden, sixteen-thirty-four. **Albert Girard.** 'Les Oeuvres Mathematiques de Simon Stevin de Bruges,' &c. *Folio in threes.*

See Hutton, Tracts, vol. ii. p. 257, and Peacock, Enc. Metr. Arithmetic, p. 440. My copy of the first work is imperfect, stopping at the end of the Diophantus; but as Alb. Girard gives every thing in the collected works, just in the order announced in the above title-page, I suppose there can be no doubt that every thing there announced was published in 1585. The *Disme* is the first announcement of the use of decimal fractions. Dr. Peacock supposes it was first published in Dutch about 1590: we here see that it was published in French in 1585. Hutton speaks of a Dutch edition in 1605. I do not know where either date* was got from. On looking at the verses with which Stevinus's friends (according to cus-

* Perhaps 1605 is a mistake for 1626, at which date [R. Soc. Libr. Cat.] De Thiende, leerende alle rekeninghen, &c. was published at Gouda.

tom) have eulogised him, a part of a book to which I am now and then under obligation, I find that the *Disme*, &c., was attached to the edition of 1585, with a new paging. Jagius Tornus, *philomathes*, be he whom he may, quotes and writes as follows, with a side-note:

Non fumum ex fulgore sed ex fumo dare lucem
Cogitat, ut speciosa dehinc miracula promat.
Sume unum è multis. quid non Decarithmia præstat
Divinum scriptoris opus? cui non ego si vel
Aurea mî vox sit, centum linguæ, oráque centum
Omni ætate queam laudes persolvere dignas.

Decarithmia
La Disme
pag. 132.

Accordingly, assuming Girard to have faithfully copied his original in his headings (which he has done as far as my means of comparison go), the method of decimal fractions was announced before 1585, in Dutch.

The characteristic of Stevinus is originality, accompanied by a great want of the respect for authority which prevailed in his time. For example, great names had made the point in geometry to correspond with the unit in arithmetic: Stevinus tells them that 0, and not 1, is the representative of the point. And those who cannot see this, he adds, may the Author of nature have pity upon their unfortunate eyes; for the fault is not in the thing, but in the sight which we are not able to give them. P. 426, 440, 460.

[Dr. Peacock (p. 440) mentions an English translation of this tract by Richard Norton, sixteen-eight, under the title 'Disme, the arte of tenths, or decimal Arithmetike . . . . . invented by the excellent mathematician, Simon Stevin.']

Since writing the above I have examined the account of Stevinus given by M. Quetelet, presently referred to. It is stated that the first work of Stevinus was his table of interest, *Antwerp*, fifteen-eighty-four ——; this was reprinted the next year in the Arithmetic. In the Supplement of the *Penny Cyclopædia*, 'Tables,' I have noted, from the contents of this table and of the *Disme*, the great probability that the table of compound interest *suggested decimal fractions:* the only doubt was, which was written first, the table or the Disme. The statement of M. Quetelet clears this up: and I hold it now next to certain that the same convenience which has always dictated the decimal form for tables of compound interest was the origin of decimal fractions themselves.

London, fifteen-eighty-eight. **John Mellis.** 'A briefe instruction and maner hovv to keepe bookes of Accompts after the order of Debitor and Creditor, and as well for proper Accompts partible, &c. By the three bookes named the Memoriall Journall and Leager, and of other necessaries appertaining to a good and diligent marchant. The which of all other reckoninges is most lawdable: for this treatise well and sufficiently knowen, all other wayes and maners may be the easier and sooner discerned, learned and knowen. Newely augmented and set forth by John Mellis Scholemaister, 1588.

Imprinted at London by John Windet, dwelling at the signe of the white Beare, nigh Baynards Castle. 1588.' *Octavo.*

This is the earliest English book on book-keeping by double entry which has ever been produced. At the end of the book-keeping is a short treatise on arithmetic (O). But Mellis says: 'Truely I am but the renuer and reviver of an auncient old copie printed here in London the 14 of August, 1543. Then collected, published, made and set forth by one Hugh Oldcastle Scholemaster, who, as appeareth by his treatise then taught Arithmetike and this booke, in Saint Ollaves parish in Marke Lane.'

[Watt and others (but not Ames) mention a *quarto* treatise having this date, fifteen-forty-three, and entitled 'A profitable Treatyce called the Instrument or Boke to learne to knowe the good order of the kepyng of the famouse reconynge, called in Latyn, Dare and Habere, and in Englyshe, Debitor and Creditor.' But no author's name is given.]

Beckman quotes Anderson as to a book of James Peele, *London*, fifteen-sixty-nine, *folio*, on book-keeping: but it may be doubted whether this was not a book on single entry.

London, fifteen-ninety. **Cyprian Lucar.** 'A treatise named Lucarsolace,' &c. *Quarto.*

A work on mensuration, interspersed with arithmetical rules. It contains the description of a fire-engine. We find from it that the surveyors used, not black-lead, but 'fine-pointed coles or keelers,' of which three or four might be bought of any painter for a penny. Lucar also translated part of Tartaglia on Artillery, with a long appendix. *London*, fifteen-eighty-eight. *Folio in threes.*

London, fifteen-ninety. **Thomas Digges.** 'An Arithmetical warlike treatise named Stratioticos . . . . first published by Thomas Digges Esquire, anno Salutis, 1579,' &c. *Quarto* (O).

There is here a brief and good treatise on arithmetic, and some algebra of the school of Recorde and Scheubel: but the greater part of the work is on military matters.

Helmstadt, fifteen-ninety. **Heizo Buscher.** 'Arithmeticæ libri duo.' *Octavo* (small) (O).

A short and very backward work.

Bergomi, fifteen-ninety-one. **Peter Bungus.** 'Numerorum Mysteria.' Second edition. *Quarto in fours.*

Dr. Peacock gives some description of this fantastical work, page 424.

Frankfort, fifteen-ninety-one. **Jordano Bruno.** 'De Monade numero et Figura liber Consequens Quinque de Minimo magno et Mensura. Item de Innumerabilibus, immenso, et Infigurabili; seu De Universo et Mundis libri octo.' *Octavo* (duodecimo size).

On Jordano Bruno see Bayle, Drinkwater-Bethune's Life of Galileo, p. 8, and Libri, vol. iv. p. 141, &c. who makes him a vorticist before Des Cartes, an optimist before Leibnitz, a Copernican before Galileo. The end of it was that he was roasted alive at Rome, February 17, 1600, at the age of fifty: ostensibly for heresy (he had been a Dominican), but most probably in revenge for his satirical writings. He does not seem, however, to have been a Protestant martyr; as his opinions would probably have procured his death in almost any state in Europe, as matters were at that time. Bruno, says M. Libri, seems to have become a Copernican by a sort of intuition, for he was any thing rather than a mathematician. His book on monad and number fully confirms this last statement: it is a collection of dissertations on individual numbers. But it has the advantage over Pacioli and Bungus (P. 424), that the Latin is better, and a great part of it is in verse. In the *triads*, for instance:

> Efficiunt totum Casus, Natura, Voluntas,
> Dat triplicem mundum Deitas, Natura, Mathesis,
> Hinc tria principia emanant, Lux, Spiritus, Unda,
> Est animus triplex Vitâ, Sensu, Ratione.

M. Libri has quoted at length (vol. iv. note x.) the Copernican chapter of the work *de Immenso*.

Frankfort, fifteen-ninety-two. **Peter Ramus** and **Lazarus Schoner.** 'Petri Rami Arithmetices Libri Duo, et Algebræ totidem: a Lazaro Schonero emendati et explicati. Ejusdem Schoneri libri duo: alter, De Numeris figuratis; alter, De Logistica Sexagenaria.' *Octavo* (O).

The first edition of Ramus is said to be of fifteen-eighty-four. P. 427.

London, fifteen-ninety-two. **Thomas Masterson,** 'his first booke of Arithmeticke.'

London, fifteen-ninety-two. **Thomas Masterson,** 'his second booke of Arithmeticke.'

London, fifteen-ninety-four. **Thomas Masterson,** 'his Addition to his first booke of Arithmetick.'

London, fifteen-ninety-five. **Thomas Masterson,** 'his thirde booke of Arithmeticke.' All *Quarto* (O).

The first book and the addition are on abstract numbers and fractions: the second book is on commercial arithmetic: the third book is on the extraction of roots, on surds, and on cossic numbers, or algebra. Masterson must have been a valuable help to the student: though he would have been more so if he had used the modern method of division.

London, fifteen-ninety-four. **Blundevile.** 'M. Blundevile His Exercises, containing sixe Treatises,' &c. *Quarto in fours.*

He was the first introducer of a complete trigonometrical canon into English, the first announcer of Wright's discovery of meridional parts, and he exercised much influence on the studies of the first part of the seventeenth century. Nor is he altogether out of date yet; for I lately observed (*Mechanic's Magazine*, vol. xliv. p. 477) that a patent for an improvement in horse-shoes was upset in the Court of Chancery, on proof that Blundevile had described it in one of his books on horses, to which he refers several times in the present work. The first of these exercises is a treatise on arithmetic, in the form of question and answer, written for Elizabeth Bacon, the sister of the great philosopher. It is a dogmatical treatise, sufficiently clear. The seventh edition of these exercises was published in sixteen-thirty-six, and I believe there was none later.

London, fifteen-ninety-six. 'The Pathway to Knowledge. Conteyning certaine briefe Tables of English waights, and Measures .... With the Rules of Cossicke, Surd, Binomicall, and Residuall Numbers, and the Rule of Equation, or of Algebere .... And lastly the order of keeping of a Marchants booke, after the Italian manner, by Debitor and Creditor. ... Written in Dutch, and translated into English, by W. P.' *Quarto* (O).

This work has escaped the notice of Ames, and was perhaps confounded with the geometrical work of the same name by Recorde. The English translator's preface, giving an account of the existing weights and measures, is the most complete thing there is of the day. The arithmetic is not equal to that of Recorde or Masterson, nor the book-keeping to that of Mellis, and the algebra is far behind Recorde's. The translator gives the following verses, the first of which are now well known. I suspect he is the author of them, having never seen them at an earlier date: Mr. Halliwell, who is more likely than myself to have found them if they existed very early, names no version of them earlier than 1635:

> Thirtie daies hath September, Aprill, June, and November.
> Febuarie eight and twentie alone, all the rest thirtie and one.

> Looke how many pence each day thou shalt gaine,
> Just so many pounds, halfe pounds and groates:
> with as many pence in a yeare certaine,
> Thou gettest and takest, as each wise man notes.

> Looke how many farthings in the weeke doe amount.
> In the yeare like shillings, and pence thou shalt count.

To complete this subject: Mr. Davies [Key to Hutton's Course,

p. 17] quotes the following from a manuscript of the date 1570, or near it:

> Multiplication is mie vexation,
> And Division is quite as bad,
> The Golden Rule is mie stumbling stule,
> And Practice drives me mad.

Alcmaar (preface dated from Amsterdam), fifteen-ninety-six. **Nicolaus Petri.** 'Practicque omte Leeren Rekenen Cypheren ende Boeckhouwen met die Reghel Coss . . . .' *Octavo* (small) (O).

Contains arithmetic, algebra, geometry, and an example of book-keeping.

Venice, fifteen-ninety-nine. **Joh. Bapt. Benedictus.** 'Speculationum Liber.' *Folio in twos.*

The first speculation is entitled 'Theoremata Arithmetica,' and is a laboured explanation of the principles of arithmetical rules by reference to geometry, beginning with the old difficulty of the product of two fractions being less than either of the factors.

London, sixteen-hundred. **Thomas Hylles.** 'The arte of vulgar arithmeticke, both in integers and fractions, devided into two bookes . . . . Nomodidactus Numerorum . . . . Portus Proportionum . . . . whereunto is added a third book entituled Musa Mercatorum, comprehending all rules used in the most necessarie and profitable trade of merchandise . . . . Newly collected, digested, and in some part devised, by a wel willer to the Mathematicals.' *Quarto* (O).

This is in dialogue; but whenever any rule or theorem is delivered, it is in verse. It is a big book, heavy with mercantile lore. The following are specimens of the verses:

> Number is first divided as you see
> For number abstract, and number contract
> And numbers abstract are such as stand free
> From every substance so cleare and exact
> That they have no sirname or demonstration
> Save the pure units of their numeration.

> All primes together have no common measure
> Exceeding an ace which is all their treasure.

> Addition of fractions and likewise subtraction
> Requireth that first they all have like basses
> Which by reduction is brought to perfection
> And being once done as ought in like cases,
> Then adde or subtract their tops and no more
> Subscribing the basse made common before.

The partition of a shilling into his aliquot parts.

> A farthing first findes fortie eight
> An halfepeny hopes for twentie foure

Three farthings seekes out 16 streight
A peny puls a dozen lower.
Dicke dandiprat drewe 8 out deade
Twopence tooke 6 and went his way
Tom trip and goe with 4 is fled
But goodman grote on 3 doth stay
A testerne only 2 doth take
Moe parts a shilling cannot make.

In spite of all this trifling, Hylles was a man of learning. He cites both *Lucas* de Burgo and *Peter* de Burgo.

Paris, sixteen-hundred. **John Chambers.** 'Barlaami Monachi Logistica nunc primum Latiné reddita et scholiis illustrata.' *Quarto.*

The Greek is given at the end. Barlaam lived in the fourteenth century, and his work is mostly on fractions and on proportions. See the *Penny Cyclopædia,* 'Barlaam,' for information on the author and on this edition.

Leyden, sixteen-three. **C. Dibuadius.** 'In Geometriam Euclidis prioribus sex Elementorum libris comprehensam Demonstratio Numeralis.' *Quarto.*

The six books of Euclid are, for the most part, here verified by arithmetical or trigonometrical instances. The usual mode, or *linear* demonstration, is made to follow in another book.

Leyden, sixteen-five. **Simon Stevinus.** 'Tomus secundus Mathematicorum hypomnematum de Geometriæ praxi.' *Folio in threes.* (The printing delayed beyond the title-date.)

This work contains the celebrated Latin treatise on book-keeping (Peacock, *Encyl. Metrop.* 'Arithmetic,' p. 464), in which the difficulty of finding Latin renderings of the technical terms is well got over. All writers attribute this translation to the celebrated Willebrord Snell, who passed from fourteen to eighteen years of age while the work was printing. Now the fact is, that not only does Stevinus date his own preface to this treatise on book-keeping in August 1607, but at the end of the volume he excuses himself from giving several matters which had been announced at the beginning, because the printer was tired of waiting, and he had not satisfied himself about them. And in this same final notice he mentions *Will. Sn.* (Snell) as a person who had written a letter for him : styling him *mathematum neque ignarus neque expers;* much too poor a compliment for a youth who, at such an age, had translated, and therefore understood, all the writings of Stevinus. The story about the translation is from Gerard Vossius, who was afterwards Snell's colleague, and who made a mistake and confusion of persons which was copied by Bayle, Beckmann, &c. M. Quetelet has recently published a notice of Stevinus, which states that he was born in 1548, and died in 1620. P. 464.

Augsburgh, sixteen-nine. **Geo. Henischius.** 'Arithmetica perfecta et demonstrata.' *Quarto* (ON).

A laboured work in seven books. Its algebra is that of a former day; its power of computation very fair. Dr. Peacock refers several times to another work of Henischius, 'De numeratione multiplici,' sixteen-five.

Mayence, sixteen-eleven. **Christopher Clavius.** 'Epitome Arithmetices Practicæ.' *Folio in threes* (O).

This is in the second volume of Clavius's collected works. [It is said to have been first published in fifteen-eighty-three.] Perhaps there are more extensive examples of the square root worked by the old method in this treatise than would easily be found elsewhere. P. 426.

Bologna, sixteen-thirteen. **Pietro Antonio Cataldi.** 'Trattato del modo brevissimo di trovare la Radice quadra delli numeri.' *Folio.*

The rule for the square root is exhibited in the modern form, and Cataldi shews himself a most intrepid calculator. But the greatest novelty of the book is the introduction of continued fractions, then, it seems, for the first time presented to the world. Here, with great labour, but still with success, Cataldi reduces the square roots of even numbers to continued fractions of the form $a+\frac{m}{n+}\,\frac{m}{n+}$ &c. He then uses these fractions in approximation, but without the assistance of the modern rule by which each approximation is educed from the preceding two. Thus he reduces, among other examples, the square root of 78 to

$$8\frac{30737638259358854906836 81}{36954898341229855021262 40}$$

London, sixteen-thirteen. **John Tap.** 'The Pathway to Knowledge,' &c. *Octavo* (O).

This professes to be a reprint of the work published under that name (1596), and above described. But it is in fact the substance of the above work thrown into the form of dialogue, with the book-keeping reprinted, and some algebra taken from V. Menher above mentioned.

London, sixteen-thirteen. **Richard Witt.** 'Arithmeticall questions, touching the Buying or Exchange of Annuities; Taking of Leases for Fines, or yearly Rent; Purchase of Fee-Simples; Dealing for present or future Possessions; and other Bargaines and Accounts, wherein allowance for disbursing or forbeareance of money is intended; Briefly resolved by means of certain Breviats.' *Quarto* (octavo size) (O).

There is also in the title-page 'Examined also and corrected at the Presse, by the Author himselfe.' A great many circumstances induce me to think that the general fashion of correcting the press by the author came in with the seventeenth century, or thereabouts.

Breviats are tables. As far as I know, this is the first English book of tables of compound interest. And there are *real* tables of half-yearly and quarterly compound interest. Again, decimal fractions are really used: the tables being constructed for ten million of pounds, seven figures have to be cut off, and the reduction to shillings and pence with a *temporary* decimal separation, is introduced when wanted. For instance, when the quarterly table of amounts of interest at ten per cent is used for three years, the principal being 100*l.* (page 99), in the table stands 137266429, which multiplied by 100 and seven places cut off gives the first line of the following citation:

'The Worke

Facit	l	1372	66429
	sh	13	2858
	d	3	4296.'

Giving 1372*l.* 13*s.* 3*d.* for the answer. And the tables are expressly stated to consist of *numerators*, with 100... for a denominator.

[A reprint of this work, by T. Fisher, *London*, sixteen-thirty-four, *duodecimo*, is in the Royal Society's Library.]

Paris, sixteen-fourteen. **Artabasda.** 'Nic. Smyrnæi Artabasdæ Græci Mathematici ΕΚΦΡΑΣΙΣ Numerorum Notationis per gestum digitorum. Item Venerab. **Bedæ** de Indigitatione et manuali loquelæ lib.' *Octavo in twos* (edited by F. Morell, the first Gr. Lat., the second Lat.).

These tracts are on nothing but the mode of representing numerals by the fingers. Bede's tract is not a separate work, but a chapter from his treatise *de natura rerum*, in which it will be found in the 'Bedæ . . . opuscula plura,' edited by Noviomagus, *Cologne*, fifteen-thirty-seven, *folio in twos*. But the first edition is, *Venice*, fifteen-twenty-five, Joh. Tacuinus (editor and printer), 'hoc in volumine hæc continentur M. Val. Probus . . . . . Ven. Beda de Computo per gestum digitorum . . . . nunc primum edita.' *Quarto in fours*.

Paris, sixteen-fourteen. **Honorat Meynier.** 'L'Arithmetique . . . . enrichie de ce que les plus doctes mathematiciens ont inventés . . . . tant aux formes que nos anciens ont practiquees comme en celles qui se practiquent aujourd'huy en France, en Holande, en Allemagne, en Espagne et autres nations.' *Quarto* (O).

There is a great deal of matter in this book, which runs to 664 pages; and it might be historically useful. Part of it is an attack on Stevinus, who, by the way, is represented as then alive. The following is one of the examples in subtraction: 'L'an 1535, Jean Calvin composa son labirinthe abominable (puisque dans iceluy s'est perdu un grand nombre d'hommes de bien) . . . je demande combien s'est passé d'années depuis qu'il composa le dit Dedale.'

London, sixteen-fourteen. **Thomas Bedwell.** 'De Numeris Geometricis. Of the nature and properties of geometrical numbers, first written by Lazarus Schonerus, and now Englished,' &c. *Quarto*.

On figurate, square, &c. numbers, with applications to mensuration.

Bologna, sixteen-sixteen. **P. A. Cataldi.** 'Quarta parte della pratica Aritmetica Dove si tratta della principalissima, et necessariissima Regola chiamata comunemente del Tre,' &c. *Folio* (quarto size).

A long and wordy treatise on the rule of three, but venturing on examples of a much higher order of difficulty of computation than had been previously attempted.

Edinburgh, sixteen-seventeen. **John Napier.** 'Rabdologiæ seu Numerationis per Virgulas Libri duo: cum Appendice de expeditissimo Multiplicationis Promptuario, Quibus accessit et Arithmeticæ Localis Liber unus. Authore et inventore Johanne Nepero,' &c. *Duodecimo* (O).

A posthumous work. Napier's rods are well known, and this book is the *descriptio princeps* of them, with applications. The use of decimal fractions, expressly attributed to Stevinus, renders it remarkable. It is stated (P. 441) that Napier invented the decimal point, but this is not correct: 1993·273 is written by him 1993$2'7''3'''$.

Oppenheym, sixteen-seventeen and nineteen. **Robert Fludd.** 'Utriusque Cosmi Majoris scilicet et minoris metaphysica, physica atque technica historia.' Two volumes. *Folio in twos*.

The second volume has another title-page, purporting that it gives the supernatural, natural, preternatural, and contranatural history. Of Robert Fludd (though he was great enough in his day to engage the attention of Descartes) I shall here say nothing, except that he gave his mixture of mysticism and science two dedications, one on each side of a leaf. The first, signed *Ego, homo*, was addressed to his Creator: the second, signed 'Robert Fludd' to James I. of England. The first volume contains a treatise on Arithmetic, and on *Cossic* Arithmetic, or Algebra. The arithmetic is rich in the description of numbers, the Boethian divisions of ratios, the musical system, and all that has any connexion with the numerical mysteries of the sixteenth century. The algebra is of the four rules only, referring for equations and other things to Stifel and Recorde. The signs of addition and subtraction are P and M with strokes drawn through them. The notation for powers is the *q*, *c*, *qq*, &c. series of symbols. Perhaps the most remarkable thing about the algebra is, that Fludd, who wrote it in France for the instruction of a Duke of Guise, should have known nothing of Vieta. The second volume is strong upon the hidden theological force of numbers.

London, sixteen-nineteen. **Henry Lyte.** 'The art of tens, or decimall arithmeticke.' *Octavo* (duodecimo size).

One of the earliest English users of decimal fractions. P. 440.

Hamburgi, sixteen-twenty. **Francis Brasser** and **Otto Wesellow.** 'Arithmetica Francisci Brasseri . . . . ab Ottone Wesellow ex Germanico in Latinum sermonem versa, atque in lucem edita.' *Octavo* (O).

Brasser seems to have been a celebrated teacher: it is not clear whether he published or not. The book itself is a mixture of arithmetic and algebra in Scheubel's form, with much of algebraical application to commercial questions. There is extraction of roots as far as the fourth.

Hamburg, sixteen-twenty-one. **Peter Lauremberg** of Rostoch. 'Institutiones Arithmeticæ.' *Octavo* (small).

A demonstrative book.

Rostoch, sixteen-twenty-one. **John Lauremberg.** 'Organum analogicum, sive instrumentum proportionum.' *Quarto.*

This instrument is the sector.

Rostoch, sixteen-twenty-three. **John Lauremberg.** 'Virgularum numeratricum et promptuarii arithmetici descriptio, figuræ et usus.' *Quarto.*

A tract of fourteen pages on Napier's rods.

London, sixteen-twenty-four. **William Ingpen.** 'The Secrets of numbers, according to theologicall, arithmeticall, geometricall, and Harmonicall computation.' *Quarto.*

A worthy follower of Peter Bungus. P. 425.

London, sixteen-twenty-four. **Thomas Clay.** 'Briefe easie and necessary tables of interest and rents forborne,' &c. *Octavo* (very small size).

There is no hint here of decimal fractions. Attached is a 'chorologicall discourse' on the management of estates.

London, sixteen-twenty-eight. **John Speidell.** 'An Arithmeticall Extraction or Collection of divers questions with their answers.' *Octavo* (small).

For John Speidell, see the article *Tables* in the Supplement of the *Penny Cyclopædia.* The book is literally what it professes to be, questions with their *facits* or answers, and nothing else, mostly in reduction, the rule of three, and practice.

Amsterdam, sixteen-twenty-nine. **Albert Girard.** 'Invention nouvelle en l'algebre,' &c. *Quarto.*

In this celebrated work there is a slight treatise on arithmetic, the most remarkable part of which is, that the author gives no examples in division by more than one figure, and seems to decline them as too difficult for his readers: he gives some results only. P. 426, 442. On one occasion he uses the decimal point.

Oughtred's *Clavis* is a book of arithmetic as well as of algebra, and one of great celebrity. The editions that have fallen in my way are:

London, sixteen-thirty-one. 'Arithmeticæ in numeris et speciebus institutio: quæ tum logisticæ, tum analyticæ, atque adeo totius mathematicæ, quasi clavis est.' Signed 'Guilelmus Oughtred,' at the end of the dedication. *Octavo.*

London, sixteen-forty-seven. 'The Key of the Mathematicks New Forged and Filed: Together with A Treatise of the Resolution of all kinde of Affected Æquations in Numbers. With the Rule of Compound Usury; And demonstration of the Rule of false Position. And a most easie Art of delineating all manner of Plaine Sun-Dyalls. Geometrically taught by Will. Oughtred.' *Octavo.*

Oxford, sixteen-fifty-two. 'Guilelmi Oughtred . . . . . . Clavis Mathematicæ denuo limata, sive potius fabricata . . . . Editio tertia auctior et emendatior.' *Octavo.*

Oxford, sixteen-sixty-seven. 'Do. do. . . . . Editio quarta auctior et emendatior.' *Octavo.*

Oxford, sixteen-ninety-three. 'Do. do. . . . . Editio quinta auctior et emendatior.' *Octavo.*

Oxford, sixteen-ninety-eight. 'Do. do. Editio quinta auctior et emendatior. Ex recognitione D. Johannis Wallis, S.T.D. Geometriæ Professoris Saviliani.' *Octavo.*

London, sixteen-ninety-four (again in seventeen-two). 'Mr. William Oughtred's Key of the Mathematics. Newly translated . . . with Notes . . . . . . Recommended by Mr. E. Halley.' *Octavo.*

The third edition was an extended re-translation into Latin; and the preface of it, which was copied into succeeding editions, shews that it was carefully performed under Oughtred's own eye. The treatise on dialling was translated into Latin by a young man of sixteen, of Wadham College, whom Oughtred describes as already an inventor in Astronomy, Gnomonics, Statics, and Mechanics, and of whose future fame he augurs great things: his name was Christopher Wren. There are two fifth editions in the list, the second of them revised, it appears, by Dr. Wallis, who ought, one would think, to have known better the history of the book he edited. But on examining them, I find that they are the same impressions with different title-pages: so that it would seem as if Wallis had allowed his eminent name to appear as a guarantee for a book he had never revised. Wrong again, however: for in the preface to the third edition it appears that Wallis had given all the help of an editor at that stage of the work; so that his name should rather have appeared before.

The editions which follow the first, besides other additions, have the solution of adfected Equations, or Vieta's method (see the Index, 'Horner'). Throughout, the old, or *scratch* method of division, is retained. Dr. Peacock observes of this method that it lasted nearly to the end of the seventeenth century: but it thus appears that it even got into the eighteenth, and that with Halley's name as a recommendation. Throughout all the editions, Oughtred's ancient algebraical notation is retained, as also his way of writing decimal fractions ($12\underline{|3456}$ for our 12·3456). I cannot tell why Dr. Peacock always spells his name Oughtred*e* (p. 441, &c.).

London, sixteen-thirty-two. **William Oughtred.** 'The Circles of Proportion and the Horizontal Instrument. Both invented and the uses of both Written in Latine by Mr. W. O. Translated into English: and set forth for the publique benefit by William Forster.' Published with it, *London*, sixteen-thirty-three, 'An addition unto the use of the instrument . . . . . For the Working of Nauticall Questions . . . . . Hereunto is also annexed the excellent Use of two Rulers for Calculation. And it is to follow after the 111 Page of the first Part.' *Quarto*.

The second edition, *Oxford*, sixteen-sixty, *octavo*, is 'by the Author's consent Revised, Corrected . . . . by A. H. Gent.' [The editor was Arthur Haughton.] The addition above named gives the first description of the *sliding rule*. The circles of proportion are what would now be called circular sliding rules; and the two rulers are the common sliding rule, the rulers being kept together by the hand. Various errors are afloat about the invention of the sliding rule. Some give it to Wingate, some to Gunter, some to Partridge; the official book of the Excise Office gives it to an excise officer whose name I forget. But the truth is that Oughtred invented it. (See Penny Cycl. 'Slide Rule.')

London, sixteen-thirty-three. **Robert Butler.** 'The

Scale of Interest; or Proportionall Tables and Breviats..... Together with the valuation of Annuities,' &c. *Octavo.*

These tables are for discount or present value; and as far as annuities, &c. are concerned, that is, in compound-interest questions, resemble Witt's. But decimals are really used in the applications, the separating notation being as in 12|3456.

London, sixteen-thirty-three. **Nicholas Hunt, M.A.** 'The Hand-Maid to Arithmetick refined: Shewing the variety and facility of working all Rules in whole Numbers and Fractions, after most pleasant and profitable waies. Abounding with Tables above 150. for Monies, Measures and Weights, tale and number of things here and in forraigne parts; verie usefull for all Gentlemen, Captaines, Gunners, Shopkeepers, Artificers, and Negotiators of all sorts: Rules for Commutation and Exchanges for Merchants and their Factors. A Table from 1l. to 100 thousand, for proportionall expences, and to reserve for Purchases.' *Octavo* (ON). P. 442.

The first thing that strikes any one in this book is the slavishness of the dedication to an Earl, and the grotesque appearance which the use of the ambiguous word *rare* gives it. "When I reflect on ancient Nobility, earth's glory, it being found in the way of vertue; this so rare a thing transports my soule with thoughts of a glorious eternitie. And for as much as the Lord hath said (*Dixit dii estis*) that you are titular and tutelar terrestriall gods, powerfully protecting the meaner shrubbes under the spreading branches of your tall Cedars: I wonder not to see almost every Pamphlet to crouch and deject it selfe in all humility and lowlinesse, seeking patronage from persons of eminency." He afterwards calls on his patron to "imitate the propitiousnesse of the divine essence," which will excite him to "a voluntarie prostitution of his humble service," if the patron will protect him from "that squint-ey'd foole, that antipathie to vertue (the envious man)."

The book itself is very full on weights and measures, and commercial matters generally. It does not treat of decimal fractions: what the author calls 'decimall Arithmeticke' is the division of a pound into 10 primes of two shillings each; each shilling into *six* primes of two pence each. There are some verses, perhaps suggested by the republication of Buckley, headed: 'Arithmeticke-Rithmeticall, or the Handmaid's Song of Numbers;' of which the rules of addition and subtraction will give a sufficient specimen:

> Adde thou upright, reserving every tenne,
> And write the Digits downe all with thy pen,
>   The proofe (for truth I say,)
>   Is to cast nine away.
> From the particular summes, and severall
> Reject the Nines; likewise from the totall
> When figures like in both chance to remaine
> Clear light of working right shal be your gain;

Subtract the lesser from the great, noting the rest,
Or ten to borrow, you are ever prest,
To pay what borrowed was thinke it no paine,
But honesty redounding to your gaine.

Paris, sixteen-thirty-four (with another title-page, sixteen-forty-four, but it is all one impression). **Peter Herigone.** 'Cursus Mathematici tomus secundus.' *Octavo* (O).

The volumes of this course, which is said to be the first complete course published, have different dates. It is a polyglott book, Latin and French. This second volume contains practical arithmetic. It introduces the decimal fractions of Stevinus, having a chapter 'des nombres de la dixme.' The mark of the decimal is made by marking the place in which the last figure comes. Thus when 137 livres 16 sous is to be taken for 23 years 7 months, the product of 1378′ and 23583‴ is found to be 32497374⁗ or 3249 liv. 14 sous, 8 deniers. There is a *memoria technica* for numbers in this book, which I subjoin, that those who know the old system of Grey may compare the two. The consonants and vowels which stand for numbers are:

P	B	C	D	T	F	G	L	M	N
1	2	3	4	5	6	7	8	9	0
a	e	i	o	u	ar	er	ir	or	ur }
					ra	re	ri	ro	ru }

So that *r* signifies *five* except before or after *u*, when it destroys the meaning altogether. Grey's system is:

B	D	T	F	L	S	P	K	N	Z
1	2	3	4	5	6	7	8	9	0
a	e	i	o	u	au	oi	ei	ou	y

The vowel-systems of the two might be combined.

London, sixteen-thirty-four. **William Webster.** 'Webster's tables for simple interest direct . . . . Also his Tables for Compound Interest . . .' *Octavo* (small). Third edition.

This work treats decimal arithmetic as a thing known: but all the tables are not decimals. Neither is there yet any recognised decimal point; only a partition-line to be used on occasion. In this book I find the first head-rule for turning a decimal fraction of a pound into shillings, pence, and farthings: though not so perfect a one as was afterwards found. I have seen the second edition, place, title, and form as above, sixteen-twenty-nine. The copy of each edition which I have examined is bound up with King James's list of customs, 'The Rates of Marchandizes' &c. promulgated in 1605.

London, sixteen-thirty-seven (again in forty-seven). **Leonard Digges.** 'A book named Tectonicon . . . . Published by Leonard Digges, Gentleman, in the yeere of our Lord 1556.' *Quarto*.

A book of mensuration: one of the instances in which a book was republished for the name of its author, long after its methods were obsolete. This occurred so frequently in the seventeenth century that we must take for granted the existence of a class of mathematicians to whom the light of Napier, &c. had not penetrated.

London, sixteen-thirty-eight; reprinted in seventeen-forty-eight. **John Penkethman.** 'Artachthos, or A New Book declaring The Assise or Weight of Bread by Troy and Avoirdupois Weights. Containing divers Orders and Articles made and set forth by the Right Honble the Lords and Others of his Majesty's most Honble Privy Council . . . . . . Published by their Lordships Orders.' *Quarto.*

This does not look like a book of arithmetic; but Penkethman has thought it necessary to prefix instruction in numeration, both Roman and Arabic.

'Note that IV signifies IIII, as IX signifies nine which takes as it were by stealth, or pulls back one from foure and ten.' So that, in fact, I stands behind X and picks his pocket.

⌊*Leyden*, sixteen-forty.⌋ **Adrian Metius.** 'Arithmetica Practica.' *Quarto* (N).

There is no prolixity about this book. Sexagesimal fractions are taught, but not decimal ones. The title is torn out in my copy.

Herbornæ Nassoviorum, sixteen-forty-one. **Joh. Henr. Alsted.** 'Methodus Admirandorum Mathematicorum Novem libris exhibens universam Mathesin.' *Duodecimo* (N).

This has a slight treatise on arithmetic, with a few words on algebra: nothing on decimal fractions.

Sora, sixteen-forty-three. **John Lauremberg.** 'Arithmetica . . . . exemplis historicis illustrata, itidem algebræ principia.' *Quarto* (N).

A book which uses the Italian method of division is a rarity among German books of this period. The examples are drawn from historical matters, and from the military art.

Paris, sixteen-forty-four. **Theon** of Smyrna. '*Τῶν κατὰ μαθηματικὴν χρησίμων εἰς τὴν τοῦ Πλάτωνος ἀνάγνωσιν.*' *Quarto.* Edited by Ismael Bullialdus (Bouillaud).

Leyden, eighteen-twenty-seven. **Theon** of Smyrna. The same title. *Octavo in twos.* Edited by J. J. de Gelder.

These are the only editions of Theon of Smyrna, neither having the complete work, if indeed any complete manuscript exist. The first has the arithmetic and music; the second, with various readings, the arithmetic only. The arithmetic is nothing more than that classification and nomenclature of numbers which is also given by Nicomachus (the two were contemporaries), and followed by Boethius. The music, as in other cases, is only discussion of fractions under names derived from the scale. Theon of Smyrna must not be confounded with his greater namesake of Alexandria.

London, sixteen-forty-five. **Edmund Wingate.** 'Use of the Rule of Proportion in Arithmetique and Geometrie.' *Duodecimo.* (A translation from Wingate's own French.)

This book is on the use of Gunter's logarithmic scale, not the sliding rule, which Wingate never wrote upon.

Leyden, sixteen-forty-six. **Francis Vieta.** 'Opera Mathematica.' *Folio in twos.*

In this collection is the 'De numerosa potestatum purarum atque adfectarum ad exegesin resolutione tractatus,' which was first published, *Paris*, sixteen-hundred, *folio*, and is the first exposition of the method by which the general evolution of the roots of equations is connected in operation and principle with the extraction of the square root and common division. See the references presently given under the name of Horner. The first departure from Vieta's form of solution, which amounts to incorporating with it the method afterwards known by the name of Newton, was made by Henry Briggs in the 'Trigonometria Britannica,' *Gouda*, sixteen-thirty-three, *folio in twos and threes.* Accordingly, Briggs has a fair claim to be the first user of that which is called Newton's method of approximation, though he does not explain it, but leaves the explanation to be collected from his examples. The work which introduced Vieta's method into England was the 'Artis Analyticæ Praxis' of Thomas Harriot, *London*, sixteen-thirty-one, *folio.* All these writers would have called their process algebraical, but the progress of the art of computation will make their works essential parts of the history of arithmetic. I believe the most complete account of Vieta is that which I have given in the *Penny Cyclopædia:* with which may be read the account of his works in Hutton's *Tracts*, vol. ii. pp. 260-274.

London, sixteen-forty-eight. **Seth Partridge.** 'Rabdologia, or the Art of numbring by Rods.' *Duodecimo* (O).

The use of Napier's rods or bones explained.

Napier's *Rabdologia* brought the well-known *Napier's rods* into vogue for half a century (P. 411). And the reason seems to have been that this contrivance really was found useful, in the state of the habit of computation as it then existed. P. 432, 440.

Leyden, sixteen-forty-nine. **Joh. Hen. Alsted.** 'Scientiarum omnium Encyclopædia.' Two volumes. *Folio in threes.*

The first edition is said to be sixteen-thirty, and the preface of this one is dated sixteen-twenty-nine, but it refers to a course of philosophy published in sixteen-twenty, of which it seems to be an enlargement. As far as I can learn, this is the first of the works to which moderns attach the idea of a *Cyclopædia*. For though Ringelberg and Martinius had published under that name, and though the *Margarita Philosophica* above mentioned and other works ran through the whole circle of elementary knowledge—yet I cannot establish that any work before Alsted's was so large and comprehensive as to be a Cyclopædia in the sense in which Chamber's Dictionary was afterwards so called. The course of arithmetic (O) in the second volume is rather scanty, but seems less so when the bareness of the professed mathematical courses is considered. There is a table of the squares and cubes of all numbers up to 1000. Logarithms are not mentioned, but Napier's rods are. Moreover, by the date of the first edition, we must call this a course of mathematics prior to Herigone's, usually called the first.

London, sixteen-fifty. **John Wybard.** 'Tactometria. seu, Tetagmenometria. Or, The Geometry of Regulars practically proposed . . . . by J. W.' *Octavo in twos.*

An excellent book of mensuration of solids, full of remarkable information on the subject of weights and measures.

London, sixteen-fifty. **Jonas Moore.** 'Moore's Arithmetick: Discovering the secrets of that Art, in Numbers and Species.' *Octavo* (N).

This is Jonas Moore's first work, and is a very good one. It is very complete in decimals, giving the contracted multiplication and division. It has the use of logarithms (in which the operations with the negative characteristic are fully given), algebra from Oughtred, and squares and cubes of all numbers up to 1000, and the fourth, fifth, and sixth powers up to 200. The second edition is:

London, sixteen-sixty. 'Moor's Arithmetick in two books.' *Octavo.*

It is worthy of remark that the common arithmetic of this edition is in black letter, the second book, or Algebra, being Roman letter, as all the previous edition was. Perhaps Recorde, &c. had given common readers a prejudice for black letter arithmetics. P. 442.

Dantzig, sixteen-fifty-two. **Joh. Broscius.** 'Apologia pro Aristotele et Euclide . . . . . Additæ sunt duæ disceptationes de numeris perfectis.' *Quarto.*

A perfect number is one which is equal to the sum of its less factors: thus 28, being the sum of 1, 2, 4, 7, and 14, is a perfect number. And a number was called defective or redundant, according as the sum of its factors fell short of or exceeded the number. Thus 8 is defective, and 12 is redundant.

London, sixteen-fifty-three. **Noah Bridges.** 'Vulgar Arithmetique, explayning the Secrets of that Art, after a more exact and easie way than ever.' *Octavo* (N).

London, sixteen-sixty-one. **Noah Bridges.** 'Lux Mercatoria, Arithmetick Natural and Decimal,' &c. *Octavo* (N).

Dr. Peacock (p. 452) has given the first book notoriety by his citations from the eulogistic poetry at the beginning, which he has by no means exhausted. The poem of Geo. Wharton, the royalist astrologer, on the existing state of things, is a political satire. Another commends the plainness of the work, hints at decimals and logarithms becoming too common, and pronounces that a merchant

'may fetch home the Indies, and not know
What Napier could, or what Oughtred can doe.'

Dr. Peacock speaks lightly of the work, in mentioning the excessive praises with which it was announced. But a very short treatise, explicit upon the modern mode of division, which those who praised it had perhaps never seen before, and upon the use of practice, was rather an exception to the rule at the time at which this book was published.

The second work is more learned, and has an appendix on decimals. The author disapproves of the use which some would make of decimals, and avers that the rule of practice is more convenient in many cases; which is perfectly true. Bridges seems to have had a very clear view of the capabilities of arithmetic without formal fractions. He mentions some preceding authors, and his list is 'The Merchant's Jewel,' 1628; 'The Handmaid to Arithmetick,' 1633; 'The Map of Commerce,' 1638; Masterman, Johnson, Hill, Leybourn, Recorde, Baker, Val. Menher, Boteler.

London, [sixteen-fifty-four]. **E[dmund] W[ingate].** 'Ludus Mathematicus. Or, The mathematical Game: Explaining the description, construction,, and use of the Numericall Table of Proportion.' *Duodecimo.* (Again in eighty-one.)

This is the description of a logarithmic instrument, which it would be impossible to give a notion of without the instrument itself, or a drawing. The date is cut out in my copy.

London, sixteen-fifty-five. **Thos. Gibson.** 'Syntaxis Mathematica.' *Octavo.*

This work is filled with the construction of equations from Des Cartes; but it gives the arithmetical solution, or Vieta's Exegesis, from Harriot: also interest-tables, partly from *Simon Stevens.*

Oxonii, sixteen-fifty-seven. **John Wallis.** 'Mathesis Universalis: sive Arithmeticum Opus Integrum,' &c. *Quarto.*

This and other works, with their separate title-pages, are collected (apparently so published) as *Opera Mathematica,* and were afterwards republished, in substance, in the folio collection of Wallis's works. The Arithmetic gives both modes of division : but the old notation, as $12|\overline{345}$, is used for decimal fractions. Wallis afterwards adopted the decimal point in his Algebra.

Wurtzburg, sixteen-fifty-seven-and-eight. **Gaspar Schott.** 'Magia Universalis Naturæ et Artis.' *Quarto* (four parts, variously bound in volumes).

This work, and others of a similar character by the same author, are the precursors of the works of mathematical amusement compiled by Ozanam, Montucla, Hutton, &c. In the fourth part he treats of arithmetic, mostly on the wonders of combinations. The gem of the book is his accurate calculation of the degrees of grace and glory of the Virgin Mary, which are exactly

1157920892373161954235709850086879078532
69984665640564039457584007913129639936

not one more nor less. This is the 256th power of 2, and is repeated three times, printed in the same way. Others solved the same problem, and found out the number of stars at the same time, by writing down every possible way in which the words of

Tot tibi sunt dotes, Virgo, quot sidera cœlo

can be arranged in an hexameter line.

As Schott here describes the digit of his geometrical pace as being forty poppy-seeds placed side by side, I thought it possible I might get a fellow to the barley measure (see page 8). But it seems that he has not played fair, or has thought it necessary to bring the degrees of glory of the Virgin up to the 256th power of 2, at any sacrifice. For I find that forty poppy-seeds, side by side, measure an inch and a half (English); so that the geometrical foot of sixteen digits would be two feet English.

Leyden, sixteen-sixty. **Vincent Leotaud.** 'Institutionum arithmeticarum libri quatuor, in quibus omnia quæ ad numeros simplices, fractos, radicales, et proportionales pertinent precepta clarissimis demonstrationibus tum arithmeticis tum geometricis illustrata traduntur.' *Quarto* (O).

This is a clear but heavy work, running to 700 pages.

Herbipoli (*Wurtzburg*), sixteen-sixty-one. **Gaspar Schott.** 'Cursus Mathematicus.' *Folio in threes.*

It has a shabby arithmetic, which, considering the magnitude of the course, reminds us of Falstaff's halfpennyworth of bread. In the frontispiece, a lion and a bear draw a car surmounted by an armillary sphere, and having for its wheels (or rather castors) the celestial and terrestrial globes, along an amphitheatre the pavement of which is studded with diagrams. These diagrams contain a sly little bit of freemasonry. Schott, as a good Jesuit, adheres to the Ptolemaic system, grudgingly, and because his church had condemned the other, as appears from his words. But in the diagrams on the pavement of his frontispiece, he draws the Copernican system, and no other. Dr. Peacock (p. 408) refers to the 'Arithmetica Practica' of Schott [*Herb.* sixteen-sixty-two——].

London, sixteen-sixty-four. **James Hodder.** 'Hodder's Arithmetick: or, That necessary Art made most easie.' The third edition, much enlarged. *Duodecimo* (O).

Had this work given the new mode of division, it must have stood in the place of Cocker. The ninth edition was published in sixteen-seventy-two. [The first edition is said to have had the date 1661.] In Davis's sale catalogue (1686), there are marked as by Hodder, a Vulgar Arithmetic, 1681, and a Decimal Arithmetic, 1671. There are no decimal fractions in the book before us.

Deventer, sixteen-sixty-seven. **Joachim Camerarius.** 'Explicatio in duos libros Nicomachi Geraseni .... et notæ **Samuelis Tennulii** in Arithmeticam Jamblichi ....' *Quarto.* Also, (one publication, though different in place and date,)

Arnhem, sixteen-sixty-eight. **Jamblichus.** 'In Nicomachi Geraseni Arithmeticam introductionem et de Fato.' *Quarto* (Gr. Lat.).

These comments of Jamblichus and Camerarius are the easiest accesses to a knowledge of the matter of the work of Nicomachus, of whom no Latin version exists, of which I can find any mention.

London, sixteen-sixty-eight. **John Newton.** 'The Scale of Interest: Or the Use of Decimal Fractions.' *Octavo.*

This book is further described in 'Tables,' *Penny Cyclop. Suppl.:* it was expressly intended for a school-book, though it is a strange one for the time.

London, sixteen-sixty-eight. **William Leybourn.**

A

Platform { for } Purchasers
Guide { for } Builders
Mate { for } Measurers.

Octavo (O).

The first book is on interest, the second and third on building and mensuration of work.

Toulouse, sixteen-seventy. **Diophantus.** 'Arithmeticorum libri sex.' *Folio in twos.*

This is perhaps the best edition of Diophantus. It has the commentaries of Bachet and of Fermat. The character of the work of Diophantus on the properties of numbers, treated algebraically, is too well known to need description. I do not consider him as belonging to the genuine Greek school. His date is always spoken of as very uncertain, though the preponderance of opinion seems to place him in the second century. But I think I have given sufficient reason for supposing him to have written as late as the beginning of the seventh century. (See Dr. Smith's Dictionary of Biography, article *Hypsicles.*)

Diophantus is usually placed in the second century, because Suidas says that the celebrated Hypatia wrote a comment on the astronomical table of *Diophantus.* This was a common name: Fabricius has collected upwards of twenty writers or philosophers who bore it; and the arithmetician of whom I am speaking shews no appearance* of having attended to astronomy. Diophantus mentions Hypsicles, a mathematician: now (unless this mention make a second) there is only one Hypsicles in Greek literature, namely, the author of the two last books of Euclid's elements. Of Hypsicles, Suidas says that he was a pupil of "Isidore the philosopher." But in another place (and this passage, occurring at the word *Syrianus,* has been overlooked till now) Suidas quotes Damascius for his information upon this Isidore. Now it is certain that the Isidore, whose life John Damascius wrote, must have been a contemporary of the Emperor Justinian, at the earliest: and his pupil, Hypsicles, can hardly have written before the middle of the sixth century. Diophantus, who makes mention of Hypsicles, may have flourished towards the end of the sixth, or the beginning of the seventh century. The silence of Proclus, Pappus, and Theon, as to both Hypsicles and Diophantus, is a strong presumption of the latter two having come after the former three. When Dr. Peacock (p. 397) says that Theon was well acquainted with the writings of Diophantus, I suppose it is upon the presumption that the father and teacher could not have been ignorant of the author on whom the daughter and pupil was said to have published a commentary. But, this apart, I cannot imagine any reason to suppose that Theon had ever seen the work of Diophantus. P. 404.

London, sixteen-seventy-three. **Sam. Morland.** 'The Description and Use of two Arithmetick Instruments, together With a Short Treatise explaining and Demonstrating the Ordinary Operations of Arithmetick, As likewise A Perpetual Almanac And several Useful Tables.' *Octavo* (duodecimo size).

There is also the following second title-page (first in order of time) which is all that some copies have: *London,* sixteen-seventy-two. 'A New, and most useful Instrument for Addition, and Subtraction of Pounds, Shillings, Pence, and Farthings. . . . By S. Morland.'

* For this reason Fabricius wants to alter the words "a commentary upon of Diophantus the astronomical table," into "a commentary upon Diophantus and an astronomical table."

A very miscellaneous book, embodying computation (N), some of Euclid, tables for Easter, description of the calculating machine, &c.

London, sixteen-seventy-three. **John Kersey.** 'Mr Wingate's Arithmetick, containing a plain and familiar method,' &c. *Octavo.*

This is called the sixth edition of Wingate (N), one of the very best of the old writers. This edition has an appendix in augmentation of the commercial part, by Kersey, who says the first edition (which I have never seen) was published about sixteen-twenty-nine.

Amsterdam, sixteen-seventy-three. **William Bartjen.** 'Verniende cyfferinge . . . .' *Octavo* (small) (O).

A book of a decidedly commercial character, and with good force of examples.

London, sixteen-seventy-four. **John Mayne.** 'Socius Mercatoris: or the merchant's companion, in three parts.' *Octavo* (N).

The three parts are on arithmetic, vulgar and decimal, on interest, and on solids and cask-gauging: the second and third have two separate title-pages, dated sixteen-seventy-three. There is a little algebra.

London, sixteen-seventy-four. **John Collins.** 'An Introduction to Merchants-Accompts.' *Folio in twos.*

Book-keeping by double entry. Collins says that his first edition was in 1652, and his second (which was nearly all destroyed by the fire) in 1664-5.

London, sixteen-seventy-six. **A. Forbes.** 'The Whole Body of Arithmetick Made Easie.' *Duodecimo* (O).

Not a very easy system.

London, sixteen-seventy-seven. **Michael Dary.** 'Interest epitomised, both compound and simple . . . . whereunto is added, a Short Appendix For the Solution of Adfected Equations in Numbers by Approachment: Performed by Logarithms.' *Quarto* (octavo size) (N).

Of Dary, who was a gunner, then a tobacco-cutter, then a teacher, &c., and who was the correspondent of Collins, J. Gregory, Newton, &c., a good deal is to be found in the Macclesfield correspondence. This tract contains the distinct announcement and use of a principle which is now well known, and of which Newton's (or Briggs's) method of approximation is only a particular case. If an equation be reduced to the form $x=\phi x$, ϕx being a known function of x, successive formation of $\phi a=b$, $\phi b=c$, $\phi c=d$, &c., gives an approximation to a root of the equation, whenever the results do not increase without limit. Dary saw and applied this, expediting the process occasionally by using the mean of the two last results, instead of the last result alone, to produce the next result. The book itself is quite out of notice; and neither the ingenuity nor the importance of the principle was appreciated by Dary's contemporaries.

London, sixteen-seventy-eight. **William Leybourn.** 'Arithmetick Vulgar, Decimal, Instrumental, Algebraical.' *Octavo.* Fourth edition. (The four parts have separate titles and paging.) (ON).

I have not met with any earlier edition. This work seems to have been substantially inserted in the course of mathematics by Leybourn, presently mentioned.

London, sixteen-eighty. **Thomas Lawson.** 'A Mite into the Treasury, being a Word to Artists, Especially to Heptatechnists.' *Quarto.*

This book I have several times met with in lists, as one of arithmetic, so that I here insert a true account of it. It is written against human learning, and arithmetic among the rest, as being of no use to "open the seals of the book," or to interpret the Bible. It reminds me much, difference of age and manners apart, of some books I have lately seen, of arithmetical examples, in which no name is introduced except from the Bible: so that Judas Iscariot and the devil may be mentioned, but Socrates or Milton may not. Of arithmetic T. Lawson says: "Herein any member of Italian Babilon with his Mass-Book, Mass for the Dead, Fabulous Legend: any Mahumetan with his dreggy Alcoran; any Flint-hearted Jew with his Talmud, a mingle-mangle of Jewish Divine and Humane matters; any Dead, Dry, Unfruitful Formalist, may grow Profound, Exquisit, Nimble, yea, and though involved in the intricate windings of Degeneration, out of the Royal state of Regeneration and Heavenly Transformation may apprehend the Feates, Termes and Parts of this Natural Art, as Digits, Articles mixt Numbers, Cyphers, Terniries, Golden Rule direct, Golden rule reverse, a Cube, Phythagoras's Table, Algorism, &c. yet be Strangers to the Divine Exercise which leads to Christ, the Lion of the Tribe of Judah, who alone opens the Seals of the Book." Attention is then directed to the number of the Beast.

London, sixteen-eighty-one. **Jonas Moore.** 'A new Systeme of the Mathematicks.' Two volumes. *Quarto.*

The first volume contains arithmetic, much abridged from Moore's larger work above noted. This course was written for Christ's Hospital.

London, sixteen-eighty-two. **Gilbert Clark.** 'Oughtredus explicatus, sive Commentarius in Ejus Clavem Mathematicam.' *Octavo.*

This commentary is now of no use. A wise commentator will endeavour to fill his book as much as he can with such little matters as will be historically interesting a couple of centuries after his own time; and this will in many cases cause him to outlive his author. I suppose this commentator is the Gilb. *Clerke* who published, in sixteen-eighty-seven, on what he called the *Spot-dial.*

London, sixteen-eighty-three. **Peter Galtruchius.** 'Mathematicæ totius . . . Clara, Brevis et accurata Institutio.' *Octavo.*

There is a short treatise on arithmetic (O), and logarithms are mentioned as a recent invention, without any examples of their use.

London, sixteen-eighty-four. **Rich. Dafforne.** 'The Merchant's Mirrour.' *Folio in twos.*

A book on book-keeping, translated from the Dutch.

London, sixteen-eighty-four. **Abr. Liset.** 'Amphithalami or the Accountants closet.' *Folio.*

Book-keeping by double entry. This and the two last books on the subject (Dafforne and Collins) I have bound together with the *Lex Mercatoria* of Gerard Malynes and other books. Mr. M'Culloch has precisely the same collection, and gives them all as one publication (*Liter. Polit. Econ.* p. 129-30). But they have different printers, different dates, and different signatures: most being in twos, and the rest simple folios; and from Collins's preface it seems that his book was certainly a separate publication. I merely mention this (having seen the like in other cases) to advert to the probability of it not having been unusual for booksellers to bind parcels of books on the same subject together for sale.

London, sixteen-eighty-four. **Wm. Leybourn.** 'The line of Proportion or [of?] Numbers, Commonly called Gunter's Line made easie,' &c. *Duodecimo.*

On the sliding rule, which invention is attributed to Seth Partridge. But he says that Milbourn (see Sherburne's Manilius, or *Penny Cycl.* 'Horrocks') 'disposed it in a Serpentine or Spiral Line;' the logarithmic spiral, of course (*Penn. Cycl.* 'Slide Rule').

London, sixteen-eighty-four. **Thos. Baker.** 'The Geometrical Key: or the Gate of Equations unlocked:' &c. *Quarto.*

Polyglott: Latin and English. An attempt to do without arithmetic in the solution of equations, by means of construction. See the correspondence of Baker and Collins in the Macclesfield correspondence, beginning of vol. ii.

London, sixteen-eighty-five. **Seth Partridge.** 'The description and use of an instrument called the double scale of proportion.' *Octavo.*

On the sliding rule.

London, sixteen-eighty-five. **John Collins.** 'The doctrine of decimal arithmetick, simple interest, as also of compound interest . . . . made publick by T. D.' *Octavo* (small).

A clear and short work, posthumous.

Naples, sixteen-eighty-seven. **Ægid. Franc. de Gottignies.** 'Logistica Universalis.' *Folio in twos.*

The author is the man mentioned by Montucla (ii. 643) as having claimed some of the discoveries of Cassini. He was a Jesuit, and was professor at Rome. This work is the last of several, of which it seems to contain the substance, amplified (Heilbronner, *in verb.*). I have never met with this book till very lately; and, though I have had no time as yet to examine it fully, I have been much surprised to find it unnoticed by mathematical historians. In his views of algebra, the author seems much before his time. His recognition of its existence at that time as an art only; of its containing principles and definitions which are not arithmetic, but perfectly distinct; its definition as a science in which subtraction is universalised by express convention; the avowed enlargement of meaning of + and − &c. &c. — shew that the author had got more than a glimpse of what was coming. The book is a folio of 400 pages, with much prolixity of expression, but vigorous efforts of thought.

London, sixteen-eighty-seven. **William Hunt.** 'The Gauger's Magazine.'

A treatise on decimal fractions and mensuration, once well known. There is a table of squares in it up to that of 10,000.

London, sixteen-eighty-seven. **D. Abercromby.** 'Academia Scientiarum. . . . . Being a Short & Easie Introduction to the Knowledge Of the Liberal Arts and Sciences.' *Octavo* (small).

A real smatterer's book, published contemporaneously with the *Principia* of Newton. It gives names and general notions, with the names of a few celebrated authors; just enough to set up a man about town. There is another book, which may go with this, out of its place, being the earliest I have met with of the kind.

London, sixteen-forty-eight. **Sir Balthazar Gerbier.** 'The interpreter of the Academie for forrain languages, and all noble sciences, and exercises.' *Quarto.*

A polyglott, English and French, announcing the intention of setting up a College in London. It gives general ideas on the sciences in dialogue; and that which relates to arithmetic passes between the tutor and a page of the court, who begins thus: "If you would favour me now with some particular discourses on all the necessarie sciences (which would make me capable of the Theoricall part), I should esteeme myselfe very much obliged unto you for that it would make me passe for a gallant wit in good companies of men that love to discourse pertinently of things." With this view, my predecessor (in intention) gives two pages and a half of arithmetical terms, and then the young gentleman says: "Ther's enough for my Theory." This book was dated from Paris the day before the funeral of Charles I. of England: but it was dedicated to the Duke of York, with a hope that his turn for literature might increase the comfort of his great parents.

London, sixteen-eighty-eight. **Jonas Moore.** 'Moore's Arithmetick: in Four Books.' The third edition. *Octavo.*

A complete book of Arithmetic (N), with book III. on Logarithms, and book IV. on algebra, both dated sixteen-eighty-seven. This book was edited by John Hawkins, Cocker's editor, who has put his name to the last three parts. See more of this presently, under *Cocker.*

London, sixteen-eighty-eight. **Peter Halliman.** 'The Square and Cube Root compleated and made easie.' *Octavo* (second edition) (O).

That this book should have reached a second edition shews how great an object it was to save a little calculation. The author simply approximates to the fractional part of the root by first interpolation. Thus his general formula would be:

$$\sqrt[n]{(a^n+b)} = a + \frac{b}{(a+1)^n - a^n}$$

to help which he has a small table of squares and cubes, with first differences. But he evidently seems to think he has got the *exact* square and cube roots: and he says:

Now Logarithms lowre your sail,
And Algebra give place,
For here is found, that ne'er doth fail,
A nearer way, to your disgrace.

London, sixteen-ninety. **Henry Coggeshall.** 'The Art of Practical Measuring Easily Performed, By a Two Foot Rule Which slides to a Foot,' &c. *Duodecimo in threes.*

On the sliding rule and the use of logarithms.

London, sixteen-ninety. **William Leybourn.** 'Cursus Mathematicus . . . in nine books.' *Folio in twos and fours.*

The arithmetic (ON) Vulgar, Decimal, Instrumental, and Algebraical, is stated to be the first book: but algebra is afterwards promoted to follow geometry. Leybourn gives both methods of division, but thinks the old one not inferior to the new.

Leyden, sixteen-ninety. **Claude Fr. Milliet Dechales.** 'Cursus seu Mundus Mathematicus.' *Folio in twos.* Four volumes.

The first volume contains a course of Arithmetic (O), on which I may remark, as just now in the case of Schott, that it is very scanty. It is, perhaps, only in a collected course of Mathematics, that we see how difficult a thing computation must have been considered. And I begin to see perfectly, by instance after instance, that it was not the Greeks only who fled to geometry to escape from arithmetic.

Amsterdam, sixteen-ninety-two. **Bernard Lamy.** 'Elemens des Mathematiques,' &c. *Duodecimo.* Third edition (O).

The arithmetic is very poor: p. 410, a friend sends him the mode of using the *carriage* in subtraction; he having previously *borrowed* from the upper line: he prints this as a novelty.

London, sixteen-ninety-three. **John Wing.** 'Heptarchia Mathematica .... Arithmatick ....' *Octavo* (N).

A plain book, containing, among other things, arithmetic, and a larger collection than usual of tables useful in mensuration. It has the licenser's permission dated after the Revolution (Sept. 14, 1689).

Paris, sixteen-ninety-three. **Nicolas Frenicle.** In the 'Divers Ouvrages de Mathématique et de Physique,' *folio*, are contained the following tracts of Frenicle, namely, 'Methode pour trouver la solution des Problemes par les exclusions—Abregé des combinaisons—Des Quarrez magiques—Table generale des Quarrez magiques de quatre de costé.' And in another collection of Frenicle's tracts only, *quarto*, no date nor place, with only fly title-pages (pp. 374, register A—Yy), there are the preceding tracts and also 'Traité des Triangles Rectangles en nombres,' in two parts, [of which last there is said to be an edition *Paris*, sixteen-seventy-six, *octavo*, which may be the one just mentioned, for aught I know].

Frenicle (who died in 1675) was one of those *useless* men, as many appliers of mathematics have called them, who work subjects into shape before the time is come for applying them. In this point we have degenerated from our ancestors in the soundness of our views as to what knowledge is, and how it comes. For instance, at the end of the seventeenth century, Picard, an eminently useful (as well as *useful*) applier of mathematics, who never speculated on any subject, was the careful preserver of Frenicle's manuscript combinations, and exclusions, and magic squares. A century after,

Condorcet, a useful (but *useless*) speculator, who never handled any so-called *practical* subject, finds it necessary, in his *éloge* of Frenicle, to apologise for the *useless* character of his writings, though the doctrine of probabilities,—our power over which depends on the knowledge of combinations, &c., in which Frenicle was a successful workman—was, when the *éloge* was written, almost as far advanced as the theory of probable errors and the method of least squares, on which every observer who knows his business now relies, and to which astronomy in particular is much indebted for its modern accuracy. Condorcet had caught a slang which was current in his day, and has been very popular among us. The time is coming when really learned men will again be ashamed of not seeing the value of all the uses of mind: when nothing but thoughtlessness or impudence, mercurial brain or brazen forehead, will aver that no knowledge is practical, except that which ends in the use of material instruments.

Paris, sixteen-ninety-four. **Jean Prestet.** 'Nouveaux Elemens de Mathematiques.' *Quarto* (two volumes, third edition) (ON).

This work is a complete body of arithmetic and algebra, treated with depth and clearness, and embracing a great extent of subject. The author is much more impressed with the necessity of strict and sustained reasoning in algebra than most of his contemporaries; and in particular, his demonstrations of the processes of arithmetic are ample; a very uncommon thing in his day. Montucla, who does not appear to have known this third edition (for he mentions only the two former ones as of sixteen-seventy-five and eighty-nine), dismisses it with an opinion that it is *ouvrage très estimable*. He speaks of many inferior works with much higher praise.

London, sixteen-ninety-four. **Wm. Leybourn.** 'Pleasure with Profit: Consisting of Recreations of Divers Kinds, viz. Numerical, &c.' *Folio in twos.*

The recreations on arithmetic contain a great many of those which are still common. But some have gone out of vogue. I do not remember ever having had the pleasure or profit of the following:—To put together five odd numbers to make 20. Answered as follows:—Three nines turned upside down, and two units. Honest Leybourn thinks this answer is a *Falacy;* in which I differ from him: I think the question more than answered, viz. in *very* odd numbers.

London (no date, but about sixteen-ninety-five). **Venterus Mandey.** 'Synopsis Mathematica Universalis: or the Universal Mathematical Synopsis of John James Heinlin, Prelate of Bebenhusan.' *Octavo.*

Translated from the third edition, Latin, *Tubingen*, sixteen-ninety. Heinlin's system of arithmetic (O) is somewhat theological. 'In Unity all numbers are virtually, although they are infinite. So all things are in God, in him they live and move. Seven is a Sacred Number, chiefly used in Holy Scripture. It seems to have its Original from the Inscrutable Unity of Divine Essence, and Sextuple in respect of the Divine Persons among themselves: whence also the Description of a Septangled form is impossible, and cannot be known by Human Minds.'

London, sixteen-ninety-six. **Samuel Jeake.** 'Λογιστικηλογία, or Arithmetick Surveighed and Reviewed.' *Folio* (ON).

To see the size and weight of this book, one would have thought arithmetic had been a branch of controversial divinity. But it is now very valuable, from the variety of the information which it contains, particularly on weights, measures, and coins; and from the goodness of the index. There is a good deal of algebra in it, many quaint names, and stories of the same kind. Thus, in stating the famous story of the Delian oracle telling the inquirers to double a cube, in order to rid themselves of the plague, it is stated that the meaning was, that the best method to deliver realms from such contagion, was to abate of their voluptuousness, and apply themselves to literature. But for all this, those who know the value of a large book with a good index, will pick this one up when they can. P. 442.

Paris, sixteen-ninety-seven. **De Lagny.** 'Nouveaux Elemens d'Arithmetique et d'Algebre.' *Octavo* (duodecimo size).

He has the new method of division, but is obliged to write down the divisor afresh every time he wants to use it. As an instance of the effect of change of times and methods, the following trivial circumstance may be worth repeating. De Lagny says that it would take an ordinary computer a month to find the integers in the cube root of 696536483318640035073641037. To shew what would be thought of such an assertion now, I had put down the work for all the integers and a few decimals, for insertion in the *Companion to the Almanac*. When the time came for making up the manuscript, the slip of paper on which the above was written found itself (as the French say, which was more than I could do) in a heap of unarranged papers: and it was better worth my while to repeat De Lagny's month's work, than to sort the papers and pick it out.

Oxford, sixteen-ninety-eight. **E. Wells.** 'Elementa arithmeticæ numerosæ et speciosæ.' *Quarto* (octavo size) (N).

This is an elegant work, in which arithmetic and algebra are exhibited to the university student in a simple form, accompanied by the historical learning which such a student ought to have. The works on arithmetic were taking too commercial a turn to be all-sufficient for the purposes of liberal education: and Wells, without by any means rejecting commercial questions, writes in order that the student may not of necessity be driven to the works in which, as he says, *exempla non aliunde petuntur quam a Butyro et Caseo, Zingibere et Pipere, aliisque consimilibus.*

London, no date (but printed for W. and J. Marshal). **Andrew Tacquet.** 'The Elements of Arithmetick In Three Books, The Seventh, Eighth and Ninth of Euclid: With the Practical Arithmetic In Two Books .... Translated into English by an Eminent Hand.' *Octavo.*

Dr. Peacock mentions Tacquet as a late instance of the old method of division. Both methods are given in this translation, "whereof the first," meaning the Italian method, "is least used but is the best; the other most used, but the difficultest." [According to Dr. Peacock, the original is "Arithmeticæ Theoria et Praxis," *Antwerp,* sixteen-fifty-six ——.] I have described an edition of Tacquet elsewhere (see the Index).

London, seventeen-hundred. **Edward Cocker.** 'Cocker's Arithmetick: Being A plain and familiar Method, suitable to the meanest Capacity for the full understanding of that Incomparable Art, as it is now taught by the ablest School-Masters in City and Country. Composed By Edward Cocker, late Practitioner in the Arts of Writing Arithmetick, and Engraving. Being that so long since promised to the World. Perused and Published by John Hawkins Writing-Master near St. Georges Church in Southwark, by the Authors correct Copy, and commended to the World by many eminent Mathematicians and Writing-Masters in and near London.' *Duodecimo.*

This is called "The *twentieth* edition carefully corrected, with additions:" but I find that it had become usual, when any one edition was augmented, to make the assertion with reference to all succeeding editions. The first edition of this famous work was in sixteen-seventy-seven (I have seen one copy, which appeared in a sale a few years ago), the fourth in sixteen-eighty-two, the twentieth as above, the thirty-third in seventeen-fifteen, the thirty-fifth in seventeen-eighteen, the thirty-seventh in seventeen-twenty. The earliest edition I ever possessed is one of sixteen-eighty-five; what edition it is, is not stated. But there is confusion among the title-pages. For though the above is unmistakeably marked seventeen-hundred, and twentieth edition, I have also compared together two editions, both of seventeen-*twelve,* one called the *fourteenth,* and the other the *thirtieth:* all three by one printer, Eben. Tracy, at the three Bibles on London Bridge. How long it went on in England I do not know: but there was an edition at Edinburgh in seventeen-sixty-five, and another at Glasgow in seventeen-seventy-one, both edited by John Mair, whose preface is dated seventeen-fifty-one. Some account of Cocker is given in the *Penny Cyclopædia* (art.

'Cocker'). Not much more is known of him than this, that he was a skilful writing-master and engraver. At the beginning of the Arithmetic is a recommendation signed "John Collens" (no doubt the famous John Collins, or intended to pass for him) certifying that the deceased author was "knowing and studious in the Mysteries of Numbers and Algebra, of which he had some choice Manuscripts and a great collection of Printed Authors in several Languages." Collins doubts not "but he hath writ his Arithmetick suitable to his own Preface and worthy acceptation:" which means that Collins or Collens had only seen the preface of the forthcoming work at most. Then follows the attestation of fifteen teachers to the merits of the work. All this looks odd; because, according to the editor, the book was that which had been long promised to the world by a celebrated writer. All attestation was unnecessary; and the certificate of a celebrated name, wrong spelt, to the effect that he had no doubt the work, then printed, would be good, may now excite a little curiosity, if not suspicion.

I am perfectly satisfied that Cocker's Arithmetic is a forgery of Hawkins, with some assistance, it may be, from Cocker's papers: that is to say, there has certainly been more or less of forgery, without any evidence being left as to whether it was more or less. I could easily believe that all was forged; and my reasons are as follows:

In both the editions of Hodder which I have seen (1664 and 1672) is the following advertisement: "There is newly printed Mr. Cocker's book called the Tutor to Writing and Arithmetic." It appears then that during his lifetime he had published a book on Arithmetic, which I suspect to have been what would now be called an arithmetical copy-book, with engraved questions and space left for the work. But neither the posthumous work, nor its preface signed by Cocker himself, make the least allusion either to the previous work, or to the promise of another. On the contrary, the language of Cocker's own preface implies that it is the first work he has published on arithmetic, and agrees with many other prefaces (which are usually written last) in speaking of the work *as already published*. To establish these and other contradictions, I first give Hawkins's account in *his* preface (with my own Italics). "I Having the Happiness of an Intimate Acquaintance with Mr. Cocker in his Life-time often solicited him to remember his Promise to the World, of Publishing his Arithmetick, *but* (for Reasons best known to himself) *he refused it;* and (after his Death) the Copy falling *accidentally* into my hands, I thought it not convenient to smother a work of so considerable a moment," &c. But Cocker himself writes, or is made to write, as follows: "By the sacred Influence of Divine Providence, I have been Instrumental to the benefit of many; by vertue of those useful Arts, Writing and Engraving: and do *now* with the same *wonted alacrity* cast this my Arithmetical Mite into the Publick Treasury. . . . For you the pretended Numerists* . . .

* *Numerist* is a word which Hawkins uses in his own professed writings; and it was by no means a common word.

was this Book composed and published. . . ." This is an odd preface for a book which the author never meant to publish, and refused to publish, though pressed to do so. Of course it is possible that though he wrote with an intention of publishing, he afterwards changed his mind. This is one explanation; that Hawkins forged clumsily is another: and which is the most probable must be gathered from a review of all the circumstances.

Next, at the end of the work, Hawkins gives a hint of a book on decimals which would be forthcoming in time. Accordingly, we have:

London, sixteen-eighty-five. 'Cocker's Decimal Arithmetick . . . Whereunto is added his Artificial Arithmetick . . . . also his Algebraical Arithmetick . . . . according to the Method used by Mr. John Kersey in his Incomparable Treatise of Algebra. Composed by Edw. Cocker . . . . Perused, Corrected, and Published by John Hawkins . . . .' *Octavo.*

This book came to its third edition in seventeen-two. The artificial (or logarithmic) arithmetic, and the algebra, have separate title-pages, dated sixteen-eighty-four. Cocker gives no preface here, but Hawkins does, stating that he had in the preceding work given "an account of the speedy publication of his Decimal, Logarithmical, and Algebraical Arithmetick." He has here mended his hand: for, except the words "such Questions being more applicable to Decimals are omitted till we come to acquaint the Learner therewith," the first treatise does not give a hint of the second. Again, Kersey's Algebra, on which part of Cocker's second work is founded, was published in 1673, and the latter had been dead some time before the manuscript of the first work of 1677 "accidentally" fell into Hawkins's hands. This is again singular, on any supposition but that of the forgery. Moreover, at the end of the preface, Hawkins writes a letter to his friend John Perkes, *in cipher* (*Penny Cycl.* 'Cocker') in which he says, "If you peleas to bestow some of your spare houres in perusing the following tereatise, you will then be the better able to judg how I have spent mine." This looks like a confession of authorship. And in 1704, as presently noted, appeared *Cocker's* English Dictionary by *John Hawkins*, who would perhaps, had he lived, have found Cocker's Complete Dancing-Master and Cookery-book among the papers of the deceased.

The famous book itself I take to be a compilation or close imitation in all its parts. Even the Frontispiece, &c. is fashioned upon Hodder. Thus, Hodder begins with his own portrait, and verses of exaggerated praise under it; and so does Cocker. The former begins his title with *Hodder's Arithmetick*, the latter with *Cocker's Arithmetic.* The former speaks of that *necessary art*, the latter of that *incomparable Art.* The former has it 'explained in a way *familiar* to the *capacity*,' the latter a '*familiar* method suitable to the meanest *capacity*:' the two words being then by no means so common in the senses put upon them as they are now. Turning over the title-page we find that each of them 'humbly dedicateth this Manual (manuel in Hodder) of Arithmetick,' the first to a 'most worthily honoured friend,' the second to 'much honoured friends;' the first 'in token of true gratitude for *unmerited* kindnesses,' the second 'as an acknowledgment of *unmerited* favours.' There are too many small coincidences here. And it must be remembered that every resemblance to a work so well known as Hodder's (it is one of the few English works of the century which have found their way into Heilbronner's list) would help the sale.

From all these circumstances I was tolerably sure that there was no dependence to be placed on the famous Cocker being any body but Hawkins, so far as this book is concerned: though I must say I hardly expected to find such confirmation as would arise from catching Hawkins at a similar trick in another quarter. But on looking at the work above described as the third edition of Jonas Moore's Arithmetick, my eye was caught by the following sentence: "You may likewise prove Division by Division, as I have shewed at large in the 7. chap. Page 100, 101, 102. of Mr. Cocker's Arithmetic, printed in the year 1685." Now Jonas Moore was dead before 1685; and moreover, could have shewn nothing in Cocker's Arithmetic: and on looking farther to see who it is that thus speaks in the first person, I find the name of John Hawkins to the second part of the work, as editor. And on looking farther, I find that a good deal from Moore's *own* editions has been introduced *verbatim* into Cocker. For instance, this sentence is in both: "Notation teacheth how to describe any number by certain notes and [or] characters, and to declare the value thereof being so described." And throughout the book, paragraphs are frequently introduced from Moore, with alterations of phrase here and there. So that we have Hawkins arbitrarily altering and adding, in the first person, to the text of a book which had been for thirty years before the world under Moore's name. What are we to suppose he would do with Cocker's papers, if indeed he had any? Moreover, we find in Cocker sentences which had been previously written by Moore.

To see whether much was gained by Cocker's Arithmetic, as well as for the interest of the comparison itself, I will write down the definitions of addition, subtraction, multiplication, and division, from Recorde, Wingate, Johnson (see Additions to this work), Moore, Bridges, Hodder, and Cocker.

ADDITION.

Recorde. Addition is the reduction and bringing of two summes or more into one.

Wingate. Addition is that by which divers Numbers are added together, to the end that their sum, aggregate, or total, may be discovered.

Johnson. Addition serveth to adde or collect divers summes of severall denominations, and to expresse their totall value in one summe.

Moore. Addition is that part of Numbring or Numeration,

whereby two or more numbers are added together, and so the totall or summe of them is formed.

Bridges. Addition is the gathering together and bringing of two numbers or more into one summe.

Hodder. Addition teacheth you to add two or more sums together, to make them one whole or total sum.

Cocker. Addition is the Reduction of two, or more numbers of like kind together into one Sum or Total. Or it is that by which divers numbers are added together, to the end that the Sum or Total value of them all may be discovered.

SUBTRACTION.

Recorde. Subtraction diminisheth a grosse sum by withdrawing of other from it, so that Subtraction or Rebating is nothing els, but an arte to withdrawe and abate one sum from another, that the Remainer may appeare.

Wingate. Subtraction is that by which one number is taken out of another, to the end that the remainder, or difference, between the two numbers given may be known.

Johnson. Subtraction serveth to deduct one summe from another; the lesser from the greater, and to shew the remaines.

Moore. Substraction is that part of Numeration where one number is substracted or taken out of another, and so the Remainder is gotten, which is also called the difference or excesse.

Bridges. Subtraction is the taking of one number from another, whereby the residue, remainder or difference is found.

Hodder. Substraction teacheth to take any lesser number out of a greater, and to know what remains.

Cocker. Subtraction is the taking of a lesser number out of a greater of like kind, whereby to find out a third number, being or declaring the Inequality, excess, or difference between the numbers given, or Subtraction is that by which one number is taken out of another number given, to the end that the residue, or remainder may be known, which remainder is also called the rest, Remainder, or difference of the numbers given.

MULTIPLICATION.

Recorde. Multiplication is such an operation, that by two summes producyth the thirde: whiche thirde summe so manye tymes shall containe the fyrst, as there are unites in the second. And it serveth in the steede of many Additions.

Wingate. Multiplication teacheth how by two numbers given to find a third, which shall contain either of the numbers given, so many times as the other contains 1 or unitie.

Johnson. Multiplication is a number of additions speedily performed.

Moore. Multiplication, is a part of conjunct Numeration, or numbring, whereby the Multiplicand (which is the number to be multiplied) is so often added to it selfe, as an unite is contained in the

Multiplyer (which is the number multiplying) and so the *Factus* (or Product) which is the result of the worke, is had.

Bridges. Multiplication (which serveth for many additions) is that by which we multiply two numbers the one by the other, to the end their product may be discovered.

Hodder. Multiplication serveth instead of many additions, and teacheth of two numbers given to increase the greater as often as there are Unites in the lesser.

Cocker. Multiplication is performed by two numbers of like kind, for the production of a third, which shall have such reason to the one, as the other hath to unite, and in effect is a most breif and artificial compound Addition of many equal numbers of like kind into one sum. Or Multiplication is that by which we multiply two or more numbers, the one into the other, to the end that their Product may come forth, or be discovered. Or, Multiplication is the increasing of any one number by another; so often as there are Units in that number, by which the other is increased, or by having two numbers given to find a third, which shall contain one of the numbers as many times as there are Units in the other.

DIVISION.

Recorde. Division is a partition of a greater summe by a lesser.

Wingate. Division is that by which we discover how often one number is contained in another, or (which is the same) it sheweth how to divide a number propounded into as many equal parts as you please.

Johnson. Apparently considers division not enough of a technical term to need definition: his first example is, "I would divide 65490 pound amongst 5 men."

Moore. Division is that part of conjunct Numeration, wherby one Number is substracted from another, as often as it is contained in it, and by that meanes it is found how many of the one is contained in the other.

Bridges. Division is that by which we discover how often one number is contained in another.

Hodder. Division is that by which we know how many times a lesser sum is contained in a greater.

Cocker. Division is the Separation, or Parting of any Number, or Quantity given, into any parts assigned; Or to find how often one Number is Contained in another; Or from any two Numbers given to find a third that shall consist of so many Units, as the one of those two given Numbers is Comprehended or contained in the other.

The six predecessors of Cocker whom I have chosen, stop when they think enough has been said. But the illustrious discoverer, or at least the first general propagator, of the fact that two and two make four (for his current reputation amounts to this) must have had more in view. He seems to be laying every offence against accuracy in different ways, so that the unfortunate schoolboy who commits it may be sure of a flogging under one count or another of his definition. And the vice of confounding abstract and concrete number, which leads him to imply that five shillings can be multiplied by five shillings, runs through his whole book: as does also the tendency to prolixity and reduplication of things which confuse each other. As to the general notion of what Arithmetic is, Cocker tells his beginners that it is either Natural, Artificial, Analytical, Algebraical, Lineal, or Instrumental. The natural is "that which is performed by the Numbers themselves; and this is either Positive or Negative. Positive, which is wrought by certain infallible numbers propounded, and this either Single or Comparative; Single, which considereth the nature of numbers simply by themselves; and Comparative, which is wrought by numbers as they have Relation one to another. And the Negative part relates to the Rule of False." Artificial Arithmetic is performed by artificial or borrowed numbers invented for that purpose, called Logarithms. Analytical Arithmetic "is that which shews from a thing unknown to find truly that which is sought; always keeping the Species without Change." Algebraical Arithmetic "is an obscure and hidden Art of accompting by numbers in resolving of hard Questions." Lineal Arithmetic "is that which is performed by lines, fitted to proportions, as also Geometrical projections." Instrumental Arithmetic "is that which is Performed by Instruments, fitted with Circular and Right lines of proportions, by the motion of an index or otherwise." So much for Cocker (or Hawkins) as an explainer. As to the actual modes of operation, they are neither better nor worse than those pointed out before by Wingate, Moore, and Bridges. The famous book looks like a patchwork collection, and, I believe, is nothing more. The reason of its reputation I take to be the intrinsic goodness of the processes, in which the book has nothing original; and the systematic puffing with which it was introduced. The long-promised book of the great Mr. Cocker, with Collins and fifteen other teachers to recommend it, pushed aside better productions. I am of opinion that a very great deterioriation in elementary works on arithmetic is to be traced from the time at which the book called after Cocker began to prevail. This same Edward Cocker must have had great reputation, since a bad book under his name pushed out the good ones.

London, seventeen-hundred. **Christopher Sturmius.** 'Mathesis Enucleata, or the Elements of the Mathematicks. . . . . Made English by J. R. M. and R. S. S.' *Octavo*.

Arithmetic and Algebra are here very nearly separated.

Edinburgh, seventeen-one. **George Brown.** 'A Compendious, but a Compleat System of Decimal Arithmetick, Containing more Exact Rules for ordering Infinites, than any hitherto extant . . . . . First Course.' *Quarto*.

The author's knowledge of what was then extant, seems far from complete.

London, seventeen-three. **John Parsons** and **Thos. Wastell.** 'Clavis Arithmeticæ.' *Octavo* (small) (ON).

The old system of division is rather recommended. There is a very neat work on algebra at the end.

Amsterdam, seventeen-four. **Andrew Tacquet.** 'Arithmeticæ Theoria et Praxis.' *Octavo* (small) (ON).

There had been several preceding editions. The theory consists in a version of the seventh, eighth, and ninth books of Euclid : the practice in an ordinary treatise. Both methods of division are given : the Italian as best, but least used.

London, seventeen-six. **William Jones.** 'Synopsis Palmariorum Matheseos ; or a New Introduction to the Mathematics.' *Octavo in twos* (N).

Jones is well known among the contemporaries of Newton, and was the father of his celebrated namesake, the Indian Judge. The book, as its name imports, is a kind of syllabus.

London, seventeen-seven. * * * *, Art. Bac. Trin. Col. Dub. 'Arithmetica absque Algebra aut Euclide demonstrata. Cui accesserunt, Cogitata nonnulla de Radicibus Surdis, de Æstu Aeris, de Ludo Algebraico, &c.' *Octavo in twos* (NO).

There is no doubt that the author was the celebrated Bishop Berkeley, then a youth under twenty. The object of the arithmetic is stated in the title ; and at that time the effort was much wanted. The algebraical game defies brief explanation.

London, seventeen-seven. **John Smart.** 'Tables of simple interest and discount.' *Octavo in twos* (small).

This is the first edition of these celebrated tables, and is little known. The second edition, *London*, seventeen-twenty-six, *quarto*, is the best having Smart's name. But in reality the Tables in Francis Baily's 'Doctrine of Interest and Annuities,' *London*, eighteen-eight, *quarto*, are Smart's, and do not profess to be any thing else. Mr. Baily (who, by the way, did not know this first edition) says, 'I have neither time nor inclination to calculate them anew ; and therefore I give them to the world with all their imperfections on their head. I am happy, however, to observe, that after many years experience, I have not met with any errors but such as might be discovered on inspection.' Brand's edition of seventeen-eighty is said by Mr. Baily to have a good many errors.

There is yet another edition, *London*, seventeen-thirty-six. John Smart. 'Tables of Interest, &c. Abridged for the Use of Schools, in Order to Instruct Young Gentlemen in Decimal Fractions.' *Octavo*. Here is another instance of what I have before remarked, that compound interest has always been considered *the* application of decimal fractions, by those whose arithmetic has been commercial.

Manuscript (in my possession) no place, after seventeen-ten.

I insert this here, as proving that, so late as the date above mentioned, there were French schools in which the decimal point was not introduced, the old method of division was employed, and the Ptolemaic system was taught. It is the collection of notes of lectures made by a young Englishman educated in France, and was sold a few years ago by his descendant.

London, seventeen-fourteen. **Samuel Cunn.** 'A New and Compleat Treatise of the Doctrine of Fractions, Vulgar and Decimal,' &c. *Octavo in twos* (small) (N).

Prefixed is a testimonial from Halley, vouching for the goodness of the work and the novelty of some of its rules.

London, seventeen-fourteen. **Edward Wells.** 'The young Gentleman's Course of Mathematicks.' Three volumes. *Octavo.*

These volumes (in my copy) exhibit what I have more often found at the beginning of the eighteenth century than at any other time; namely, volumes of different editions in one set. The Arithmetic title-page of the first volume, which follows the general title-page, has 'seventeen-twenty-three,' and 'second edition.' The other volumes have seventeen-fourteen and seventeen-eighteen. This Wells is the same whose work I have mentioned above; he qualifies himself D.D. and rector of Cotesbach in Leicestershire. When I quoted his diatribe against butter and cheese, ginger and pepper (which I did before I had seen this work), I sympathised with him, thinking he meant that liberal education had its wants as well as professional. But I was mistaken: it is *gentlemanly* education, as opposed to that of "the meaner part of mankind," that he wants to provide for. Every page is headed on one side 'The young gentleman's,' on the other 'arithmetic,' 'geometry,' or 'mechanics,' as the case may be. The gentlemen are those whom God has relieved from the necessity of working, for which he expects they should exercise the faculties of their minds to his greater glory. But they must not 'be so Brisk and Airy, as to think, that the knowing how to cast Accompt is requisite only for such Underlings as Shop-keepers or Trades-men;' and, for the sake of taking care of themselves, 'no Gentleman ought to think Arithmetick below Him, that do's not think an Estate below Him.' This Wells might be made as useful now as the Spartans used to make their slaves. The Arithmetic is an abridged version of the work of sixteen-ninety-eight above described.

London, seventeen-fourteen. **Joh. Ayres.** 'Arithmetick Made Easie For the Use and Benefit of Trades-Men.' *Duodecimo.* Twelfth edition; with an Appendix on Book-keeping by Chas. Snell.

A work of the immediate school of Cocker.

London, seventeen-fifteen. **John Hawkins.** 'Cocker's English Dictionary.' Second edition. *Octavo in twos* (small).

I have entered this book here only because Hawkins asserts that it was the work of the celebrated arithmetician: which I do not believe, for the reason above given. [The first edition is said to have the date seventeen-four.]

London, seventeen-seventeen. **Wm. Hawney.** 'The Compleat Measurer . . . .' *Duodecimo in threes.*

A full treatise on decimal arithmetic.

No place marked, seventeen-$\frac{\text{seventeen}}{\text{eighteen}}$. **George Brown.** 'Arithmetica Infinita, or the Accurate Accomptant's Best Companion.' *Octavo in twos* (small, oblong).

This is not a work on arithmetic, but a set of tables, which will certainly be reprinted as soon as the decimals of a pound gain their proper footing. The main part of it is the first nine multiples (and the 365th) of the decimals which express each farthing of the pound. Thus under 4*s.* $1\frac{1}{4}$*d.* are given the multiples of ·20520833..... The whole work is copperplate engraving from beginning to end. From several indications, I gather that Geo. Brown of this work is also the author mentioned under seventeen-one.

London, seventeen-seventeen. **Roger Rea.** 'The Sector and Plain Scale, Compared . . . . Unto which is annexed, So much of Decimal Arithmatick and the Extraction of the square Root, as is necessary for the Working of Arithmetical Trigonometry.' *Octavo* (N).

The treatise of an illiterate and confused person. Nothing has been more common than for those who write on application to consider it advisable not to trust the books to teach, nor the readers to know, decimal fractions, and to supply a fresh treatise. Rea says he uses the Italian mode of division (N) as being that which is *most commonly used:* nothing more than this, even in 1717.

London, seventeen-eighteen. **Wm. Bridges.** 'An Essay to facilitate Vulgar Fractions; After a New Method, and to make Arithmetical Operations Very Concise:' &c. *Duodecimo.*

London, seventeen-nineteen. **Good.** 'Measuring made Easy.' *Octavo* (duodecimo size).

A description of Coggeshall's sliding-rule, corrected and enlarged by James Atkinson.

London, seventeen-nineteen. **John Ward.** 'The Young Mathematician's Guide.' The third edition. *Octavo in twos* (N).

This useful course, which commences with arithmetic [was first published about seventeen-six]. It is recommended by Raphson and Ditton. The sixth edition was in seventeen-thirty-four; the eighth edition was in seventeen-forty-seven.

London, seventeen-twenty-one. **Wm. Beverege** (Bp. of St. Asaph). 'Institutionum Chronologicarum Libri Duo. Unà cum totidem Arithmetices Chronologicæ Libellis.' Third edition. *Octavo in fours.*

The date of the preface is sixteen-sixty-eight. The arithmetical part is a treatise on the numerals of different nations, learned, but not always judicious, according to modern views of the history of symbols. It is followed by a brief elementary treatise on arithmetic, with chronological examples.

London, seventeen-twenty-six. **E. Hatton.** 'The Merchant's Magazine: or, Trades-Man's Treasury.' *Quarto.*

This is the eighth edition of a work of some celebrity, but which must not be confounded with Hatton's edition of Recorde. The only guide to the date of the first edition which I have is the statement of the eighth that it was reviewed in the *Ouvrages des Savans** for 1695.

It is interspersed with copperplate pages of flourished writing, containing examples and definitions. There is somewhat more of reason given for rules than was very common, and a vast quantity of mercantile terms, usages, &c. are explained.

Witemberg, seventeen-twenty-seven. **Joh. Fr. Weidler.** 'De Characteribus Numerorum Vulgaribus et eorum Ætatibus . . . . Dissertatio Critico-mathematica.' *Quarto.*

London, seventeen-twenty-eight. **E. Hatton.** 'A Mathematical Manual: or Delightful Associate.' *Octavo.*

Mostly on the use of the globes, but containing some "mysterious curiosities in numbers."

London, seventeen-thirty. **Alexander Malcolm.** 'A New System of Arithmetick, Theorical and Practical.' *Quarto.*

One of the most extensive and erudite books of the last century, having 640 heavy quarto pages of small type; "wherein," to go on with the title-page, "the science of numbers is demonstrated in a regular course from its first principles through all the parts and branches thereof, either known to the ancients or owing to the improvements of the moderns; the practice and application to the affairs of life and commerce being also fully explained: so as to

* Or else the *Acta Eruditorum*. This comes of translating. The phrase is "works of the learned."

make the whole a complete system of theory, for the purposes of men of science; and of practice for men of business." I quote this lengthy title as a true description of the work, at the date of publication. Probably the union of such masses of scientific and commercial arithmetic made the book unusable for either purpose.

London, seventeen-thirty-one. **Edw. Hatton.** 'An Intire System of Arithmetic . . . . containing, I. Vulgar. II. Decimal. III. Duodecimal. IV. Sexagesimal. V. Political. VI. Logarithmical. VII. Lineal. VIII. Instrumental. IX. Algebraical. With the Arithmetic of Negatives and Approximation or [sic] Converging Series,' &c. Second edition. *Quarto.*

A sound, elaborate, unreadable work, of 500 pages, of the same character as Malcolm's.

London, seventeen-thirty-one. **Wm. Hodgkin.** 'A Short New and Easy Method of Working the Rule of Practice in Arithmetick.' *Octavo in twos.*

An author who chooses his own examples can write a short method on any rule: but the first example taken at hazard will probably defy the abbreviations.

London, seventeen-thirty-two. **Joseph Champion.** 'Practical Arithmetick compleat.' *Octavo in twos.*

London, seventeen-thirty-five. **John Kirkby.** 'Arithmetical Institutions, containing a Compleat System of Arithmetic, Natural, Logarithmical, and Algebraical.' *Quarto in ones.*

A system of arithmetic is mixed up with algebra. In the extraction of roots, Halley's formula is applied in such a manner as to make the operation seem continuous, though it is just as difficult as before.

London, seventeen-thirty-five. **James Lostau.** 'The Manual Mercantile, Second Book: Concerning Decimal Arithmetic . . . . .' *Quarto.*

The first book was never published. This work contains a slight treatise on Arithmetic, but the body of it consists of all the various integers and fractions that may be useful in commerce, with the first nine multiples of each. It is 452 pages entirely of copperplate, the figures being rudely worked in, apparently by the author's own hand. It is a posthumous work, and the editor says it took 17 years. This mode of stereotyping was adopted in several instances in the first half of the last century. And it must be observed, that if decimal arithmetic have not thriven in commercial affairs, it has not been for want of a great many attempts to facilitate the use of it, by publishing books of multiples.

London, seventeen-thirty-five. **Benjamin Martin.** 'A new Compleat and Universal System or Body of Decimal Arithmetick.' *Octavo in twos.*

A very full system of decimal arithmetic, applied to all parts of commercial arithmetic.

London, seventeen-thirty-six. **Thomas Weston.** 'A New and Compendious Treatise of Arithmetick.' *Quarto.* Second edition.

A simple and useful treatise.

Edinburgh, seventeen-thirty-six. ——. 'Arithmeticæ et Algebræ Compendium.' *Octavo in twos* (N).

There is a small treatise on arithmetic. The publishers are Thos. and Wal. Ruddiman.

London, seventeen-thirty-eight. **William Pardon.** 'A New and Compendious System of Practical Arithmetick.' *Octavo in twos.*

Four hundred full octavo pages is not a very compendious book on arithmetic, or would not be so now: but it looked small by the side of Hatton, Malcolm, and Kirkby.

London, seventeen-thirty-eight. **Tho. Everard.** 'Stereometry . . . . by the Help of a Sliding-Rule.' Edited by Leadbetter, tenth edition. *Duodecimo.*

A book which once had a great reputation among excise collectors.

London, seventeen-thirty-nine. **Christian Wolff.** 'A treatise of Algebra,' translated from the Latin by J. H. M. A. *Octavo.*

There is very little of arithmetic, and that mostly on the properties of numbers.

Cambridge, seventeen-forty. **Nicholas Saunderson.** 'The Elements of Algebra, in ten books.' *Quarto,* two volumes.

This is a posthumous work of the well-known Professor Saunderson, the blind lecturer on optics. The first volume contains a synopsis of Arithmetic, and the editor's account of the calculating board, by which Saunderson supplied the want of sight.

London, seventeen-forty. **Wm. Webster.** 'Arithmetick in Epitome.' Sixth edition. *Duodecimo* (ON).

The eighth edition of the author's book-keeping is *London,* seventeen-forty-*four*, *octavo in twos* (small); and the eighth edition of his 'Attempt towards rendering the Education of Youth more easy and effectual,' is *London*, seventeen-forty-*three* (with paging continued from that of the last). He says, "When a Man has tried all Shifts, and still failed, if he can but scratch out any thing like a fair *Character*, tho' never so stiff and unnatural, and has got but *Arithmetick* enough in his Head to compute the Minutes in a Year, or the Inches in a Mile, he makes his last Recourse to a Garret, and, with the Painter's Help, sets up for a Teacher of *Writing* and *Arithmetick*; where, by the Bait of low Prices, he perhaps gathers a Number of Scholars."

London, seventeen-forty. ——. 'A small Treatise of the Square and Cube . . . .' Second edition. *Quarto* (O). Also,

London, seventeen-forty. 'A Supplement to the Square and Cube . . . .' *Quarto in twos.*

The author of this treatise (the last, I think, in which I have seen the old method of extracting the square root) is a copier of Peter Halliman, or some similar authority (see sixteen-eighty-eight), only his denominator is less by a unit.

London, seventeen-forty. **Rob. Shirtcliffe.** 'The Theory and Practice of Gauging,' &c.

A work once held in high estimation by the revenue-officers.

Edinburgh, seventeen-forty-one. **John Wilson.** 'An introduction to Arithmetick.' *Octavo in twos.*

A good demonstrative book, in a large type; very full on the *complete* operations with circulating decimals, the *ignes fatui* which have led many an arithmetical writer astray.

London, seventeen-forty-two. **John Marsh.** 'Decimal Arithmetic made perfect.' *Quarto.*

Almost entirely on infinite or circulating decimals. The predecessors whom he cites in his history of the subject are Wallis; Jones, 1706; Ward; Brown, 1708 or 1709 (he has not the work, but it is above at seventeen-one); Malcolm; Cunn; Wright, 1734; Martin, 1735; and Pardon. This subject of circulating decimals was at one time suffered to embarrass books of practical arithmetic, which need have no more to do with them than books on mensuration with the complete quadrature of the circle.

Leipzig, seventeen-forty-two. **Jo. Christoph. Heilbronner.** 'Historia Matheseos Universæ.' *Quarto.*

Though called a history of mathematics, and really a bibliography *raisonée*, yet it is peculiarly devoted to arithmetic, the authors on which have a separate list. There are also dissertations on numerals, on their history, &c. The index of this book is of rare goodness.

Geneva, seventeen-forty-three, forty-six, forty-seven, forty-nine, forty-one. **Christian Wolff.** 'Elementa Matheseos Universæ.' *Quarto.* A second or later edition (N). Five volumes.

The first of the five volumes contains a short treatise on arithmetic. Here, as happens so often in works of this period, a set is made up out of different editions. The first four volumes have *editio novissima*, the fifth has only *nova.* Wolff's course would be better known if it were scarcer. The ordinary reader passes it by as an old book; the collector as one which is very common. But it is replete with pieces of information, which are historical references and suggestions. As far as I can remember, Wolff is much the most learned historian of those who have written extensive courses.

London, seventeen-forty-five. **John Hill.** 'Arithmetick, Both in the Theory and Practice.' *Octavo in twos.*

This is the seventh edition of a work of much celebrity. It seems to have owed its fame partly to a recommendation of Humphrey Ditton, prefixed to the first edition (about 1712), praising it in the strongest terms. Perhaps at this time the only things which would catch the eye are the table of logarithms at the end, and the powers of 2 up to the 144th, very useful for laying up grains of corn on the squares of a chess-board, ruining people by horseshoe bargains, and other approved problems.

London, seventeen-forty-eight. **Charles Leadbetter.** 'The Young Mathematician's Companion.' *Duodecimo in threes.*

Second edition. Begins with an ordinary treatise on arithmetic.

London, seventeen-forty-eight. **William Halfpenny.** 'Arithmetick and Measurement, Improved by Examples.' *Octavo in twos* (N).

This is a surveyor's and artisan's book of application: but it contains decimal fractions.

London, seventeen-forty-nine; second part, seventeen-forty-eight. **Solomon Lowe.** 'Arithmetic in two parts.' *Duodecimo in threes.*

This is a work both learned and foolish: but with the learning and folly so distinct that they can be used separately. The folly consists mostly in an attempt to give the rules of Arithmetic in English hexameters and in alphabetical order; I give a couple of instances.

Barter.

Barter, exchange of commodities: the rule to proportion 'em as follows:
What's to be changd, Value: then, see what That will purchase of T'other.
If an advanc'd price of one, a proportionable find for the other.

Casting out of Nines.

Prove by a careful review: 'tis the safest: the readiest, as follows:
SUB.] right; when -hend and remainder (together) make up the compound.
ADD MULT DIV] add the digits together and cast-out the nines: then
Right; if remainder of Facits agrees with remainder of factors,
Multiplied in Mul: -sor and quotient in Div; to which add the remainder.

The learning consists in a great knowledge of former writers, and a copious account of weights, measures, and coins: together with a list of authors, which I have copied into my own, so far as I could not find the names elsewhere.

London, seventeen-fifty. **James Dodson.** 'The Accountant, or the method of Book-keeping Deduced from Clear Principles.' *Quarto.*

As far as I can find, this is the first book in which double entry is applied to retail trade. James Dodson (my great-grandfather) is best known to mathematicians in general by his *Antilogarithmic Canon, London,* seventeen-forty-two, *folio* (see *Penny Cycl.* 'Tables').

London, seventeen-fifty. **Daniel Fenning.** 'The Young Algebraist's Companion.' *Duodecimo in threes.*

There is a system of fractional arithmetic in this book, which is written in dialogue. The author thought it impossible to understand algebra without some better works on arithmetical fractions than then existed. As it is, says he, it is impossible to understand the *Algorithm* much less the *Algorism,* which he explains by saying that the former means the first principles, and the latter their practice. In this curious confusion of terms we see at its commencement an instance of a process which is always going on (though in this instance it has been arrested), the attachment of different meanings to different spellings of the same word. My curiosity led me to take a little trouble to trace Fenning to his authorities. And I find that of two writers who must have been in his hands, Saunderson and Kirkby, the first uses *Algorithm* for first principles, and the second *Algorism* for practical rules. I think I remember having seen a comparatively recent edition of this work.

London, ———. **R. T. Heath,** assisted by **W. Davidson.** 'The Practical Arithmetician: or Art of Numbers improved.' *Duodecimo.* (Revised by J. Bettesworth.)

Robert Heath was a person who made noise in his day, and in so doing established a claim to be considered a worthless vagabond. He was editor of the *Ladies' Diary* from 1746 to 1753, when the Stationers' Company found it absolutely necessary to strike out some of his scurrility, and dismiss him; appointing Thomas Simpson in his place. But before this, in 1749, Heath had commenced the *Palladium*, an annual publication resembling the *Ladies' Diary*, the very first mathematical question of which is so expressed as to convey an indecent double-meaning, in a manner obviously intended. From 1750 or thereabouts, he began to write against Thomas Simpson. One of his publications against the latter is headed, "Miss Billingsgate in a salivation for a black eye:" and in a letter published in a newspaper, in 1751, he remarks of Simpson and another, that "the best writer against both is one who shall sign the warrant for their execution."

London, seventeen-fifty-one. **T. Smith.** 'Compendious Division. Containing, A Great Variety of Curious and Easy Contractions of Division.' *Octavo in twos.*

London, seventeen-fifty-three. **Sam. Stonehouse.** 'A Compendious Treatise of Arithmetic, By Way of Question and Answer.' *Octavo.* Third edition.

Question and answer is well when the difficulty is in the question and the solution in the answer: but to turn "Decimals are divided like integers" into "Are decimals divided like integers? The division is exactly the same," is trifling. This book gives the abbreviated rule for decimals of a pound, which very few books have given.

London, seventeen-fifty-six. **John Playford.** 'Vade Mecum, or the Necessary Pocket Companion.' Nineteenth edition. *Quarto* (long and narrow size).

This book is a ready reckoner, with miscellaneous tables. I have no information about the origin of these books, which I think are not so ancient as many may suppose. I could almost think from the preface (but such deductions are very deceptive) that the earliest of the books which are now called ready reckoners, meaning those which have totals at given prices ready cast up, was the following: *London*, sixteen-ninety-three. Wm. Leybourn. 'Panarithmologia; Being A Mirror For Merchants, A Breviate For Bankers, A Treasure For Tradesmen, A Mate For Mechanicks, And A Sure Guide for Purchasers, Sellers, Or Mortgagers of Land, Leases, Annuities, Rents, Pensions, &c. In present Possession or Reversion. And A Constant Concomitant Fitted for All Men's Occasions.' *Octavo.*

London, seventeen-fifty-eight. **Benjamin Donn.** 'A new Introduction to the Mathematicks, being Essays on Vulgar and Decimal Arithmetick. Containing, Not only the practical Rules, but also the Reasons and Demonstrations of them.' *Octavo* (ON).

This good book fulfils to a great extent the profession of the title-page as to demonstration. Donn had a good deal of miscellaneous information. He was, if I remember right, one of Humphrey Davy's early teachers.

London, seventeen-fifty-nine and sixty-four. **Benjamin Martin.** 'A New and Comprehensive System of Mathematical Institutions.' Two volumes. *Octavo in twos.*

Old *Ben Martin* (as his admirers called him) was an able, and in this instance a concise writer. He wrote on every mathematical subject (and never otherwise than well, I believe, except on biography), and a complete set of his works is rarely seen. He was a bookseller.

London, no date, but about seventeen-sixty. **William Weston.** 'Specimens of Abbreviated Numbers.' *Octavo in twos.*

Some supposed new rules for formation and use of decimals.

London, seventeen-sixty. **Jacob Welsh.** 'The Schoolmaster's General Assistant.' In two volumes. *Octavo.*

The author claims a hundred curious discoveries; what they are, I cannot find.

London, seventeen-sixty. **James Dodson.** 'A Plain and Familiar Method for Attaining the Knowledge and Practice of Common Arithmetic.' *Octavo.*

An edition of Wingate's Arithmetic (N) called the nineteenth. But Wingate, who first published it about 1629, would not have known his own book, after the various dressings it received from Kersey, Shelley, and Dodson. One of George Shelley's editions of John Kersey the son's edition of John Kersey the father's edition of Wingate, called the fourteenth, is *London,* seventeen-twenty, *octavo in twos.*

Amsterdam, seventeen-sixty-one. **Isaac Newton.** 'Arithmetica Universalis.' *Quarto.*

This is the edition published by Castiglione, in two volumes, and is the best. The original book, published by William Whiston, consisted of the records laid up in the University Archives of the lectures which Newton delivered as Lucasian professor. [S' Gravesande, in the preface to his edition of *Leyden,* seventeen-thirty-two, says it was published without the author's knowledge, and much to his displeasure: but Whiston, in his Memoirs, says it was with Newton's consent: most likely it was both with his consent, and to his displeasure. Though properly a book on Algebra, and its application to Geometry, yet it does contain a system of Arithmetic. Various alterations, both of matter and arrangement, were made in the (so called) second edition [By Machin, 1722], which are supposed to have been approved, if not furnished, by Newton himself.]

This information comes from Castiglione. The other editions I have seen are the original Latin of Whiston, *Cambridge*, seventeen-seven, *octavo*. *London*, seventeen-twenty. 'Universal Arithmetic, ... translated . . . . by the late Mr. Raphson, and revised and corrected by Mr. Cunn.' *Octavo in twos*. The same reprinted, *London*, seventeen-twenty-eight, *octavo in twos*, called second edition; it is advertised as carefully compared with the correct edition that was published in seventeen-twenty-two. There is also Wilder's edition of Raphson's edition, *London*, seventeen-sixty-nine, *octavo*.

London, seventeen-sixty-five. **J. Randall.** 'An Introduction To so much of the Arts and Sciences, More immediately concerned in an Excellent Education for Trade In its lower Scenes and more genteel Professions,' &c. *Duodecimo in threes*.

Mr. Randall was a quaint man, but his book is well done. It contains arithmetic, mensuration, and geography; and ends with a dialogue between the heavenly bodies, upon their mutual arrangements, in which the earth insists upon being allowed to stand still, and quotes Scripture like an anti-Copernican, but is brought to reason by the arguments of the others. This is almost the only writer I have met with who has given the student a few hints upon habits of computation. Thus he will not let him say, three and four are seven, seven and five are twelve, &c.; but only three, seven, twelve, &c. For, says he, (the example being the addition of some rents) "as you have this pretty Income, you must talk like a Gentleman to your Figures."

London, seventeen-sixty-six. **W. Cockin.** 'A Rational and Practical Treatise of Arithmetic.' *Octavo in twos*.

Dublin, seventeen-sixty-eight. **Dan. Dowling.** 'Mercantile Arithmetic.' *Octavo in fours*.

Mostly on exchanges. The author has given the rule for the instantaneous formation of the fourth and fifth and succeeding places of the decimal of a pound, which I never saw till now in any book but my own. The third edition of this book is *Dublin*, seventeen-ninety-five, *duodecimo in threes*. There is another author of the same name presently mentioned.

London, seventeen-seventy-one. **Wm. Rivet.** 'An Attempt to illustrate the Usefulness of Decimal Arithmetic.' Second edition. *Octavo* (small).

This book contains what I thought no one had given before myself, a *complete* head-rule for turning fractions of 1*l.* into decimals to any number of places. This method, which is much wanted in commercial arithmetic, is here lost amidst attempts to compute with interminable fractions; things on which real business will never waste a thought.

Amsterdam, seventeen-seventy-three. **Nicolas Barreme.** 'Comptes-faits, ou Tarif général des Monnoies.' *Duodecimo in threes.*

This is a Dutch reprint of the work of a man who has given his name to ordinary mercantile computation in France, even more than Cocker in England. That he was a real person appears from the *privilège* copied into this book, dated Jan. 26, 1760, whereby Louis XV. grants the usual rights over his book to Nicolas Barreme. I have also seen in a catalogue an edition marked seventeen-forty-four. The name became an institution of France, which even the Revolution did not destroy. The Citoyen Blavier published a 'Nouveau Barême,' *Paris*, seventeen-ninety-eight, *octavo*, which he called a *Barême décimal*, in which there is a well-marked distinction between Barême the person and Barême the thing.

London, seventeen-seventy-three. **Thomas Dilworth.** 'Miscellaneous Arithmetic.' In seven parts. *Duodecimo.*

This work is but little known. Its contents are on the calendar; on logarithms; on the rule of three, when the first term is 1*l.* and all the terms are money; on the weather; a collection of riddles, answered, in the midst of which are seriously set forth Bacon's paradoxes on the characteristics of a Christian, and an essay on the education of children. Dilworth had made his name a selling one, and was determined to make use of it.

Shrewsbury, seventeen-seventy-three. **Thomas Sadler.** 'A Complete System of Practical Arithmetic . . . on an entire new plan.' *Duodecimo in threes.*

The newness of plan seems to consist in putting the rules into unintelligible verse, and beating even the older rule-mongers in puzzling plain questions. Thus, a cargo consists of 84, 61, and 35 tons, of which three-fifths is lost; what must each bear of the loss? This is done by first taking $\frac{3}{5}$ of 180, namely 108; then 108 is divided by 180, producing ·6; then each of the parcels is multiplied by ·6.

London, seventeen-seventy-four. **Anth.** and **Joh. Birks.** 'Arithmetical Collections and Improvements.' Second edition. *Octavo.*

London, seventeen-seventy-four. **Nich. Salomon.** 'The Expeditious Accountant; or, Cyphering rendered *so short*, That Half the Trouble . . . . A VERY CURIOUS WORK, Totally different from all that have preceded it.' *Octavo in twos.*

There is something new, says the author, in almost every rule:

but I cannot find it. The head-rule for the decimals of a pound is introduced.

London, seventeen-seventy-seven. **James Hardy.** 'The Elements or Theory of Arithmetic.' *Duodecimo in threes.*

The author was a teacher at Eton, where, according to common notions, there could have been no such thing at the date above as a teacher of arithmetic. It is true there is an ambiguous comma in "Teacher of Mathematics, and writing-master at Eton College." The book is a very creditable one, of great extent, including logarithms, &c.

London, seventeen-eighty. **John Bonnycastle.** 'The Scholar's Guide to Arithmetic.' *Duodecimo.*

The first edition of this well-known work. It had from the beginning algebraical demonstrations attached.

Birmingham, seventeen-eighty-three. **Wm. Taylor.** 'A Complete System of Practical Arithmetic.' *Octavo in twos.*

An enormous book of 600 pages, with arithmetic, mensuration, geography, astronomy, algebra, book-keeping, &c., in the above order.

London, seventeen-eighty-four. **G[eorge] Anderson.** 'The Arenarius of Archimedes . . . from the Greek . . . . the dissertation of Christopher Clavius on the same subject . . .' *Octavo.*

Oxford, eighteen-thirty-seven. **Steph. Pet. Rigaud.** 'On the Arenarius of Archimedes.' *Octavo.*

I choose this edition to introduce the only purely arithmetical work of Archimedes. Its object is to shew that any number, even a universe full of grains of sand, can be easily expressed: a thing by no means likely to be self-evident to a Greek, whose numerical notation was, till Archimedes and Apollonius shewed how it might be extended, far from sufficient for such a purpose. Professor Rigaud, one of the most learned and accurate of our modern inquirers into mathematical history, has given an account of Anderson, and many valuable remarks on his translation, as well as on the subject of it.

London, seventeen-eighty-four. **Thomas Dilworth.** 'The Schoolmaster's Assistant.' *Duodecimo.*

This is the twenty-second edition. By the dates of the commendations prefixed, it would seem that the first edition was published in seventeen-forty-four or forty-five.

Great-Yarmouth, seventeen-eighty-five. **Thos. Sutton.**

'The Measurer's Best Companion; or, Duodecimals brought to perfection.' *Octavo.*

In truth this is the most elaborate system of duodecimals I have met with.

London, seventeen-eighty-six. **George Atwood.** 'An Essay on the Arithmetic of Factors applied to various Computations which occur in the Practice of Numbers.' *Quarto.*

Edinburgh, seventeen-eighty-six. **John Mair.** 'Arithmetic, Rational and Practical.' *Octavo.* Fourth edition.

The name of Mair, who was rector of the Perth Academy, is highly respected in Scotland. His 'Book-keeping moderniz'd,' of which the *ninth* edition is *Edinburgh,* eighteen-seven, *octavo,* had a great run. Completeness of subjects, and copiousness of examples, characterise both works, which extend to six hundred pages of small print each.

London, seventeen-eighty-seven. **C. G. A. Baselli.** 'An Essay on Mathematical Language; or, an Introduction to the Mathematical Sciences.' *Octavo in twos.*

It cannot be both, the reader will say: and in truth it is only the second, for arithmetic and algebra, with some good points about it.

London, seventeen-eighty-eight. 'Clavis Campanologia, or a Key to the Art of Ringing.' *Duodecimo in threes.*

As large a list as the present ought to have one book at least on bell-ringing, the whole theory of which is arithmetical. No art has had greater enthusiasts for it. The authors of the present treatise, Wm. Jones, Joh. Reeves, and Thos. Blakemore, are of the number. They record several names of inventors to whom they give words of praise which might apply to Newton or Euler; among them is Hardham, who is known to this day by the snuff mixture which he invented and sold in Fleet Street, where his name still remains. I should think that few of those whose noses he has tickled are aware that he may have done the same for their ears.

Oxford, seventeen-eighty-one (one volume); *London,* seventeen-eighty-eight (the other). **James Williamson.** 'The Elements of **Euclid,** with dissertations . . . .' *Quarto.*

The arithmetical books of Euclid are here with the rest. I have chosen this edition by which to introduce the name, because it is the only modern translation of *Euclid.* All the works which go by that name are versions thickly scattered with the views of the editors as to what Euclid *ought to have been,* instead of the rendering of *what he was.* For these "many tamperings with his text," a countryman* of Robert Simson has been the first to call them the "per-

* Sir William Hamilton of Edinburgh, in his notes to Reid, p. 765.

fidious editors and translators of Euclid;" a name which, in a sense, they richly deserve. Williamson was a real disciple of Euclid; and he translated so closely, that such words as, not being in the Greek, English idiom renders necessary, are put in Italics. For many editions of Euclid, the reader may consult my article on that name, in Dr. Smith's Biographical Dictionary: and, for the contents of the tenth book, the article *Irrational Quantities* in the *Penny Cyclopædia.*

If the demonstrative system of Euclid had taken as great a hold in arithmetic as in geometry, we should not have had to complain of one of the best exercises of thought being employed for no other purpose than to make machines.

London, seventeen-eighty-eight. **Thomas Keith.** 'The complete practical Arithmetician.' First edition. *Duodecimo in threes.*

London, seventeen-ninety. **Thomas Keith.** 'A Key to the complete practical Arithmetician.' First edition. *Duodecimo in threes.*

London, no date; about seventeen-ninety (?). **John Duncombe.** 'A new Arithmetical Dictionary.' *Octavo in twos.*

The rules and terms of arithmetic, in alphabetical order.

Calcutta, seventeen-ninety. **Joh. Thos. Hope.** 'A Compendium of Practical Arithmetick.' *Octavo in twos.*

A clear, prolix book, for the Orphan School at Calcutta.

London, seventeen-ninety-one. **William Emerson.** 'Cyclomathesis, or an easy Introduction to the several branches of the Mathematics.' *Octavo.*

These are Emerson's works, collected (not reprinted) in thirteen volumes, with new title-pages. The arithmetic, which is in the first volume, is said by the editor to be of seventeen-sixty-three. Emerson was the writer of many works, which had considerable celebrity: but he was as much overrated as Thomas Simpson was underrated. There is a most amusing life of him prefixed to the collection.

London, seventeen-ninety-one. **Thos. Keith.** 'The New Schoolmaster's Assistant.' *Duodecimo in threes.*

An abridgment of the larger work above mentioned.

Berlin, seventeen-ninety-two. **Leonard Euler.** 'L'Arithmétique raisonnée et démontrée, oeuvres posthumes de Léonard Euler, traduite en François par Bernoulli, Directeur de l'Observatoire de Berlin, &c. &c.' *Octavo in twos* (ON).

The editor calls this the first work of Euler in his preface, and posthumous in the title-page. It is, I suppose, a translation of the work which is set down in Fuss's list as 'Anleitung zur Arithmetic, 2 Th. Petersb. 1738. 8,' the *third* of Euler's separate works. It is mostly on commercial arithmetic, and shews that Euler did not, in 1738, consider the old method of division quite exploded.

London, seventeen-ninety-four. **Rich. Carlile.** 'A Collection of one hundred and twenty . . . . Arithmetical, Mathematical, Algebraical and Paradoxical Questions.' *Octavo.*

What *Paradox* is, as a science, I do not know: but the other distinctions are well known. All who know much of the country schools remember that *mathematics* meant geometry, as opposed to arithmetic and algebra. And it was right it should have been so: for neither the schoolboy's arithmetic nor his algebra were *disciplines.*

London, seventeen-ninety-four. **Henry Clarke.** 'The Rationale of Circulating Numbers.' A new edition. *Octavo.*

Another tract on repeating decimals, with some additions on other subjects.

London, seventeen-ninety-four. **Thomas Molineux.** 'The Scholar's Question-book, or an introduction to Practical Arithmetic. Part the second. For the use of Macclesfield School.' *Duodecimo.*

An ordinary school-book on fractions and commercial arithmetic. I never saw the first part. The school-seal, which is engraved on the title-page, gives the learner to understand the mode adopted of explaining difficulties: it displays a pedagogue with a birch-rod in his right hand and a book in his left; illustrative of primary and secondary method. The fourth edition of this second part is *London,* eighteen-twenty-two, *duodecimo.* As may be supposed from the date, the little hint does not appear.

Paris, seventeen-ninety-five. **Agricol de Fortia.** 'Traité des Progressions . . . . précédé par un Discours sur la nécessité d'un Nouveau Système de Calcul.' Third edition. *Octavo.*

The new system of calculation is a proposal to annex to addition, multiplication, and involution, the next step, as the author takes it to be, in the chain of operations. But he is wrong in his use of the analogy. It appears that he was the author of several works on arithmetic.

Dublin, seventeen-ninety-six. **John Gough** (edited by his son). 'Practical Arithmetick in four books.' *Duodecimo.*

This work, I am told, had such extensive currency in Ireland (where it was first published in seventeen-fifty-eight) that the name of the author became almost synonymous with arithmetic; insomuch that when Professor Thomson's Arithmetic was first published in that country, it went by the name of 'Thomson's Gough.' The second edition appears to have been an augmented and octavo work, afterwards reduced again for schools. It is a book of the ordinary character, and abounds in examples for practice. The last edition I have seen is *Dublin*, eighteen-thirty-one, *duodecimo in threes*, edited by M. Trotter.

New York, seventeen-ninety-seven. **William Milns.** 'The American Accountant.' *Octavo in twos.*

The author seems to have been an emigrant from St. Mary Hall, Oxford. His book has the peculiarity of giving in lieu of answers, the remainders to nine of the answers, for guide in the proof by casting out nines.

Paris, an VI. (seventeen-ninety-seven or ninety-eight). **Condillac.** 'La Langue des Calculs.' *Octavo.*

A posthumous work. The views of a clear-headed mathematician and metaphysician upon the foundation of arithmetic and the formation of its language.

London, seventeen-ninety-eight. **Rich. Chappell.** 'The Universal Arithmetic.' *Octavo in twos* (small).

This book deserves notice for the author's attempt to introduce the practice of subtracting in division, without writing down the subtrahend. He versifies his tables, *ex. gr.*

So 5 times 8 were 40 Scots
Who came from Aberdeen,
And 5 times 9 were 45,*
Which gave them all the spleen.

London, seventeen-ninety-eight. **Francis Walkingame.** 'The Tutor's Assistant; being a Compendium of Arithmetic, and a complete question-book.' *Duodecimo in threes.*

This is the twenty-eighth edition; when the first was published I do not know, any more than what edition is now current. I should be thankful to any one who would tell me who Walkingame was, and when the first edition was published: for this book is by far the most used of all the school-books, and deserves to stand high among them. I have before me John Fraser's 'Walkingame modernized and improved,' *London*, eighteen-thirty-one, called *seventy-first edition;* John Little's edition, *London*, eighteen-thirty-nine; William Birkin's edition, *Derby*, eighteen-forty-three, called the *fifty-first*, and bearing proof that at least seven *Birkins* had appeared; and Samuel Maynard's edition of F. Crosby's edition, *London*, eighteen-forty-four. All these are *duodecimo.* When editors do not agree

* The North Briton, No. 45.

within twenty as to the number of editions of their author which have been published, that author is surely a man of note.

London, seventeen-ninety-eight. **Wm. Playfair.** 'Lineal Arithmetic, applied to shew the Progress of the Commerce and Revenue of England during the present century.' *Octavo in fours.*

Not arithmetic, but plates arranging the several matters in curves, in the manner now much more familiar than it was then.

London, seventeen-ninety-nine. **Charles Vyse.** 'The Tutor's Guide, being a complete system of Arithmetic.' *Duodecimo in sixes.* (Tenth edition, edited by J. Warburton; eleventh in eighteen-one.)

In the same place, year, form, and by the same editor, was published a new edition of the Key. It appears that the first edition was reviewed in the *Monthly Review* for 1771, so that it is probably of the year before. Vyse is one of the most celebrated of the illustrious band who used to adorn the shelves of a country schoolmaster at the beginning of this century; Vyse, Dilworth, Walkingame, Keith, Joyce, Hutton, Bonnycastle (with Cocker for a lost Pleiad). He is also the poet of the lot: and some of his examples have gone through many other books. The following specimen of the muse of arithmetic should be preserved, as the best known in its day, and the most classical of its kind:

When first the Marriage-Knot was tied
  Between my Wife and me,
My Age did her's as far exceed
  As three Times three does three;
But when ten Years, and Half ten Years,
  We Man and Wife had been,
Her Age came up as near to mine
  As eight is to sixteen.
Now, tell me, I pray,
What were our Ages on the Wedding Day?

The book (this tenth edition at least) is crowded with examples, which circumstance makes the Key very large. On the execution there is no remark to make. If a new edition were published, some of the examples must be omitted, as rather opposed to modern ideas of decency.

Paris, eighteen-hundred. 'Séances des Ecoles Normales, recueillies par des sténographes, et revues par les Professeurs. Nouvelle édition.' Thirteen volumes. *Octavo.*

When the Normal School was founded at Paris, in 1794, the professors engaged "pledged themselves to the representatives of the people and to each other" neither to read nor to repeat from memory. Their lectures were taken down in short-hand, and these volumes contain some of them. The professors of mathematics were Lagrange and Laplace: and few persons are aware that the mode in which the two first mathematicians in Europe taught the humblest elements of arithmetic and algebra can thus be judged of. The contents are, so far as these subjects are concerned; vol. I. p. 16, programme and Laplace, arithmetic; 268, Laplace, arithmetic; 381, Laplace, algebra: II. 116, Laplace, algebra; 302, do. do.: III. 24, Laplace, algebra; 227, Lagrange, arithmetic; 276, Lagrange, algebra; 463, do. do.: IV. 41, Laplace, geometry; 223, Laplace, algebraic geometry; 401, Lagrange, algebraic geometry: V. 201, Laplace, new system of measures: VI. 32, Laplace, probability: VII. 1, Biot, account of the 'Mécanique Céleste.'

The last three volumes contain the *debates*, or conferences, between the teachers and their pupils, of which there are three in the first of them, on arithmetic and algebra, not at all worth reading.

Taunton, (no date, perhaps before eighteen-hundred). **William Wallis.** 'An Essay on Arithmetic . . . Briefly, Shewing, First, The Usefulness; Secondly, It's extensiveness; Thirdly, The Methods of it.'

This is a remonstrance by the author, a teacher at Bridgwater in Somersetshire, against the prevailing modes of teaching arithmetic. The following is an extract: "And I have seen a *Fair-Book* (as 'tis call'd) of a young Man's, about 17 *Years* of Age, who had been 6 *Years* at *School*, but never went through that Rule [of three]: In the same Book I found 132 Questions in *Reduction*, in the working of them were 2680 Figures, which might have been better done in 500, so that there were 2180 *superfluous Ones*. In another Rule I saw an Example, in which were 174 Figures, but might have been done in 23; and one of 80 that might have been done in 12: In general, I have found in the Boys Books, 3, or 4 Times as many Figures as need be. These *Methods* have so far hindred their Advances in Learning, that amongst 30 Scholars, since I came hither, I have not found one that understood a Rule beyond *Division*, tho' some of them were 14 or 15 *Years* of Age, and had been kept at School, ever since they were capable of being taught."

Paris, eighteen-hundred. **Condorcet.** 'Moyen d'apprendre à compter surement et avec facilité.' Second edition. *Octavo* (duodecimo size).

One of the simplest explanations of the most elementary arithmetic which has ever appeared. It was written in the last days of the author, while hiding from the fate which he only finally avoided by suicide: and the last sheet was hardly finished when his retreat was discovered.

Madrid, eighteen-one. 'Aritmetica y Geometría práctica de la real Academia de San Fernando.' *Octavo*.

A very clearly written and printed work.

Buckingham, eighteen-two; second edition, eighteen-eight. **John King.** 'An Essay intended to establish a new universal System of Arithmetic . . . .' *Octavo.*

The title of the second edition is more modest: it is 'An Essay, or attempt towards establishing . . . .' The system is the *octonary* system, in which 10 means eight, 100 is sixty-four, &c.

London, eighteen-three. **Rob. Goodacre.** 'Arithmetic adapted to different classes of learners.' *Duodecimo in sixes.*

The ninth edition of this work, by Samuel Maynard, is *London*, eighteen-thirty-nine, *duodecimo.*

Dublin, eighteen-four. **P. Deighan.** 'A Complete Treatise on Arithmetic, rational and practical.' Two volumes. *Octavo in twos.*

This treatise has a list of a thousand subscribers, and has amused me very much. The old notions of the style of a book were, it seems, not extinct in Ireland a hundred years after they had been exploded in England. The author, who handles his subject ably, puts *philomath* after his name, and is perhaps the last of those who rejoiced in a title which, though self-conferred, its owners would not have changed for F.R.S. It is dedicated "to all those who think that a knowledge of accounts is useful to mankind, from the king on the throne to the lowest subject." It has the praises of the author's friends in prose and poetry, duly prefixed. I quote a few lines from one of the poets, desiring the reader to observe where sophs come from, unsought, and whence Irish authors got their stationery.

"How many sophs, to sense and science blind,
Range through the realms of nonsense unconfin'd,
Unaw'd by shame, and unrestrain'd by law,
Their labour chaff, and their reward a straw;
Neglected and despis'd, they sink in shame
To that *oblivion* whence, *unsought*, they came.
The muse, indignant, oft with grief has seen
An author led by ignorance and spleen,
With snail-paced speed, but unremitting toil,
In attic chamber waste the midnight oil,
With waste of paper, loss of ink combined,
*And pens from public offices purloined.**—
But DEIGHAN of a more enlightened mind,
More innate genius, talents more refined," &c.

London, eighteen-five. **Christ. Dubost.** 'Commercial Arithmetic, with an Appendix upon Algebraical Equations.' *Duodecimo in threes.*

Of all works I know professing to be strictly commercial, this has the fullest explanations in words of the rules and processes.

* Really I am afraid that there must have been some truth in this. Mr. Thomas Moore gives a translation of the *Pennis non homini datis* of Horace, which shews that he had heard of the thing at least. Of course he can clear himself: at any rate Lalla Rookh has not much the air of having been written with a pen from a public office.

Madras, eighteen-six. **James Brown.** 'A course of Military and Commercial Arithmetic.' *Small octavo size:* no signatures.

As might be expected, this is full on Indian exchanges, weights, and measures.

London, eighteen-six. **William Frend.** 'Tangible Arithmetic; or the Art of Numbering made easy, by means of an Arithmetical Toy which will express every number up to 16,666,665.' *Octavo.* Second edition.

The toy is the Chinese instrument or abacus, called the *Schwan-pan*, for a description of which see Peacock, p. 408.

Paris, eighteen-seven. **J. B. V.** 'L'Arithmétique enseignée par des moyens clairs et simples.' *Octavo.*

This is in dialogue between a mother and her boy: the author assures us that they are from real life; and it must have been so; for the *ancien officier du génie*, as he calls himself, could no more have written these dialogues than the mother and child could have constructed a sap. The lady has the awkward name of Madame Épinogy; and as the object of the dialogues is to make the child invent, I can find no origin for this name, except the supposition that it is a blundering derivative from *ἐπίνοια*. Nevertheless the dialogues are exceedingly good.

Paris, eighteen-eight. **F. Peyrard.** 'Oeuvres d'Archimède, traduites littéralement, avec un commentaire.' Second edition, two volumes. *Octavo.*

I mention this work, not only for the *Arenarius* already noticed, but for the disquisition by Delambre on the arithmetic of the Greeks, which afterwards appeared in the 'Histoire de l'Astronomie Ancienne,' *Paris*, eighteen-seventeen, *quarto*, two volumes. Delambre was a real reader of the works he cites. He collected his materials from Nicomachus, the *Theologumena*, Barlaam, the two Theons, Ptolemy, Eutocius, Pappus, and Archimedes. I may as well say here what I have to say on those of the above who have not been mentioned elsewhere.

The best edition of Ptolemy is that of Halma, which is a collection of Ptolemy and his commentators, published at different times, and separately,—the whole making distinct works, as well as a set. Of this I know only four volumes, of which the two to the present purpose are *Paris*, eighteen-thirteen and sixteen. 'Κλαυδιου Πτολεμαιου μαθηματικη Συνταξις . . . .' (Gr. Fr.) *quarto*, two volumes. Brunet says that the commentaries of Theon, and the Κανονες Προχειροι of Ptolemy, are in five more volumes, *Paris*, eighteen-twenty-one, twenty-two, twenty-two, twenty-three, twenty-five, *quarto*.

The commentaries of Eutocius on the works of Archimedes are

to be found in several editions, but best in that of Joseph Torelli, *Oxford*, seventeen-ninety-two, ''Αρχιμηδους τα σωζομενα μετα των Εὐτοκιου 'Ασκαλωνιτου ὑπομνηματων' (Gr. Lat.). *Folio.*

The fragment of the second book of Pappus (the only part of the first two books published as yet, if, indeed, any more exist) is to be found (Gr. Lat.) in the third volume of John Wallis's *Opera Mathematica*, Oxford, sixteen-ninety-nine, *folio in twos.*

London, eighteen-ten. **W. Tate.** 'A System of Commercial Arithmetic.' *Duodecimo.*

A work approximating more nearly to modern business than most of those then in use, in its *additions;* but, like most attempts to improve real commercial arithmetic, wanting the corresponding *omissions.*

Hawick, eighteen-eleven. **Chas. Hutton,** edited by Alex. Ingram. 'A Complete Treatise on Practical Arithmetic and Book-keeping.' *Duodecimo in threes.*

According to Hutton's Catalogue, the fifth edition was in seventeen-seventy-eight, and the twelfth in eighteen-six: and at his death he possessed no edition previous to the fifth. The late Dr. Olinthus Gregory published what he called the eighteenth edition, enlarged, &c. *London,* eighteen-thirty-four, *duodecimo:* and a new edition of Ingram's Hutton, by James Trotter, appeared *Edinburgh,* eighteen-thirty-seven, *duodecimo.*

London, eighteen-twelve. **Thomas Clark.** 'A New System of Arithmetic; including Specimens of a Method by which most Arithmetical Operations may be performed without a Knowledge of the Rule of Three; and followed by Strictures on the Nature of the Elementary Instruction contained in English Treatises on that Science.' *Octavo.*

This is an able attempt to draw public attention to the state of instruction in arithmetic. The author asserts, 1. There is not in the English language, a work of any repute whatever, employed in school education, in which the four fundamental rules of arithmetic are clearly and comprehensively laid down. 2. Not one in which the rules laid down are accompanied by examples so detailed as to remove the difficulties which these rules must present to beginners. 3. None in which the rules and examples for abstract and concrete numbers are kept distinct from each other. 4. There is not a work of this description in which ordinary and decimal fractions are properly arranged. 5. Or in which the rationale of arithmetical operations seems of sufficient importance to the instructor to induce him to incorporate it with his work. 6. Or in which the principles and algebraical signs used in arithmetic are given and explained at the time when the science requires their introduction.

Dublin, eighteen-twelve. **John Walker.** 'The Philosophy of Arithmetic . . . and the Elements of Algebra.' *Octavo.*

Mr. Walker was a good scholar, an excellent mathematician, and a most original thinker. Both this work and that which he published on geometry shew great power.

Sheffield, eighteen-thirteen. **Joseph Youle.** 'The Arithmetical Preceptor . . . . to which is added a Treatise on Magic Squares.' *Duodecimo in threes.*

London, eighteen-thirteen. **Edward Strachey.** 'Biga Ganita; or the Algebra of the Hindus.' *Quarto.*

Bombay, eighteen-sixteen. **John Taylor.** 'Lilawati; or a Treatise on Arithmetic and Geometry by **Bhascara Acharya.**' *Quarto size* (no signatures).

London, eighteen-seventeen. **Henry Thomas Colebrooke.** 'Algebra, with Arithmetic and Mensuration, from the Sanscrit of Brahmegupta and Bháscara.' *Quarto.*

The first work has notes by S. Davis, and is from a Persian version of Bhascara's Sanscrit. The second work is also from the Persian. The third, which contains not only the two works of Bhascara, but also an arithmetical chapter from Brahmegupta, is all from the Sanscrit. It is also pretty copious in selections from the Commentators, and has a large body of dissertation by Colebrooke himself. But it does not entirely supersede the former two, which have likewise valuable annotations.

Edinburgh, eighteen-thirteen. **Elias Johnston.** 'A sure and easy Method of learning to Calculate.' *Duodecimo in sixes.*

This is a translation of the work of Condorcet mentioned under the date 1800.

Paris, eighteen-thirteen. **F. Peyrard.** 'Les Principes fondamentaux de l'Arithmétique.' *Octavo in twos.*

An elegant mixture of arithmetic and algebra, by the editor and translator of Euclid, and the translator of Archimedes.

Lille, eighteen-fourteen. ———. 'Manuel d'Arithmétique ancienne et décimale.' *Duodecimo in threes.*

A small book, in question and answer: a transition book from the old system to the new, containing both, and intended for commercial purposes. At the end are some forms for letters of ceremony and business, and for petitions: and it seems rather strange to English eyes to see that a petition for a son condemned to death for homicide, ranks among the matters which are considered near enough to the ordinary course of business to find a place; and a place which, when opened, gives the option of reading the way of turning francs into roubles.

London, eighteen-fourteen. **S. F. Lacroix.** 'Traité Élémentaire d'Arithmétique à l'usage de l'école centrale des quatre-nations.' Tenth edition. *Octavo.*

A well-known work, by one of the most systematic and most widely circulated of elementary writers. The sixteenth edition was *Paris*, eighteen-twenty-three, *octavo.* There was an English translation, *London*, eighteen-twenty-three, *octavo*, anonymous,—an attempt to introduce demonstrative arithmetic into our schools.

The third edition of an American translation, by John Farrar, appeared *Cambridge* (*U. S.*), eighteen-twenty-five, *octavo in twos.*

Dublin, eighteen-fourteen. **R. F. Purdon.** 'Theory of some of the Elementary Operations in Arithmetic and Algebra.' *Octavo.*

Oxford, eighteen-fourteen. **Charles Butler.** 'An Easy Introduction to the Mathematics.' *Octavo.*

This book fulfils the promise of the title-page well, and has been frequently cited for the historical introductions to the several subjects, which are very good, and, as parts of a learner's course, unexampled.

London, eighteen-fifteen. **J. Carver.** 'The Master's and Pupil's Assistant.' *Duodecimo in threes.*

The author of this work, dependent as the sale of it was on teachers, has had the sense and courage to say, that questions with answers are *for the benefit of the masters* and the *injury of the pupils.* It is dated from *Belgrave House*, *Pimlico*,—a name and site which might puzzle an antiquary a century hence.

Paris, eighteen-sixteen. **Bezout.** 'Traité d'Arithmétique.' Eighth edition. *Octavo.*

A work of a somewhat older stamp than those of Lacroix and Bourdon. The eleventh edition was *Paris*, eighteen-twenty-three, *octavo*, edited by A. A. L. Reynaud, with notes and a table of logarithms; the notes a separate work, with another title-page. The next year appeared the twelfth edition of Reynaud's own work, *Paris*, eighteen-twenty-four, 'Traité d'Arithmétique,' *octavo*, also augmented by a table of logarithms. This work enters rather more on the theory of numbers.

London, eighteen-sixteen. **Thos. Taylor.** 'Theoretic Arithmetic, in three books; containing the substance of all that has been written on this subject by Theo of Smyrna, Nicomachus, Jamblichus, and Boetius. Together with some remarkable particulars respecting perfect, amicable, and other numbers, which are not to be found in the writings of any ancient or modern mathematicians. Likewise, a specimen of the manner in which the Pythagoreans philosophised about numbers; and a development of their mystical and theological arithmetic.' *Octavo.*

Edinburgh, circa eighteen-sixteen. **A. Melrose** (edited by A. Ingram). 'A Concise System of Practical Arithmetic.' Second edition. *Duodecimo in threes.*

Horner refers to this book as containing what is nearly an anticipation of his method, in the case of the simple cube root.

London, eighteen-seventeen. **Thos. Preston.** 'A New System of Commercial Arithmetic .... a perfect, a permanent and universal ready reckoner.' *Duodecimo in threes.*

Perhaps the plan of this book is partly taken from one published by Girtanner in seventeen-ninety-four (*Penny Cycl. Suppl.* Tables), in which the logarithms of numbers and certain intermediate fractions are given. But, in the main, it is an application of the same principle as that which has long been used in astronomical logarithms, namely, giving the logarithms of integers, with a column in which those integers, considered as seconds, are turned into degrees, minutes, and seconds. In this way are given the logarithms of pence up to 130*l.*; of pounds up to 7 cwt. 16lbs.; of twelfths up to $333\frac{1}{3}$; of sixteenths up to $312\frac{1}{2}$; of sixteenths of gallons, considered as fractions of a tun of 236 gallons, for seed-oils; the same for a tun of 256 gallons, for fish-oils and wines; of pounds of 120 to the cwt. up to five tons; of grains, up to 25 oz. troy; and two for days and for pounds at 5 per cent, by which one operation gives the interest for days on any number of pounds up to 2600*l.*

Leipsic, eighteen-seventeen. **Fred. Astius.** 'Theologumena Arithmeticæ .... Accedit **Nicomachi** Geraseni Institutio Arithmetica.' *Octavo.* (See p. 17.)

These *θεολογούμενα* have been attributed to Jamblichus and to Nicomachus: but they seem rather to consist of extracts from Nicomachus, Anatolius, and others. They are explanations of the Pythagorean and Platonic opinions on numbers; and form a very good accompaniment for the works of Nicomachus. The notes are full and good.

London, eighteen-eighteen. **George G. Carey.** 'A Complete System of Theoretical and Mercantile Arithmetic.' *Octavo.*

A commercial book with a table of logarithms in it is rare in the nineteenth century.

Edinburgh, eighteen-eighteen. **William Ritchie.** 'A System of Arithmetic .... and a Course of Book-keeping.' *Duodecimo in sixes.*

A book of much greater merit than could be guessed from its pretensions or its notoriety: and, for its size, one of the most comprehensive I have met with. Its author was a little (about twenty years) in advance of his age, and the greater part of the edition was sold as waste paper.

London, eighteen-nineteen. 'Philosophical Transactions.' *Quarto*. **W. G. Horner.** 'A New Method of Solving Numerical Equations of all orders by Continuous Approximation.'

London, eighteen-thirty. **Thos. Leybourn.** 'New Series of the Mathematical Repository.' Volume five. *Octavo in twos.* **W. G. Horner.** 'Horæ Arithmeticæ.'

The first-mentioned paper contains the most remarkable addition made to arithmetic in modern times, the value of which is gradually becoming known. On this subject I may refer to Mr. Horner's paper on Algebraic Transformation in the Mathematician (vols. i. and ii. various numbers); to J. R. Young, 'An elementary treatise on Algebra,' *London*, eighteen-twenty-six, *octavo*, as the first elementary writer who saw the value of Horner's method; J. R. Young, 'Theory and solution of Algebraical Equations of the higher orders,' *London*, eighteen-forty-three, *octavo;* Thos. Stephens Davies, 'A Course of Mathematics . . . . by Charles Hutton,' twelfth edition, *London*, eighteen-forty-three, *octavo;* T. S. Davies, 'Solutions of the principal questions in Dr. Hutton's Course of Mathematics,' *London*, eighteen-forty, *octavo;* Peter Gray, four papers in the Mechanic's Magazine for March, eighteen-forty-four; A. De Morgan, 'Notices of the progress of the problem of Evolution,' in the *Companion to the Almanac* for eighteen-thirty-nine, with the two articles headed 'Involution and Evolution' in the Penny Cyclopædia, and in the Supplement (eighteen-thirty-eight and forty-five), and a Letter to the Editor of the Mechanic's Magazine, published in that work for February, eighteen-forty-six.

In connexion with this subject I ought to mention Mr. Thomas Weddle's 'New simple and general method of solving numerical equations of all orders,' *London*, eighteen-forty-two, *quarto*. This is an organised process, in which the principle of each step is the correction of the preceding result by multiplication, not by addition.

Edinburgh, eighteen-twenty. **John Leslie.** 'The Philosophy of Arithmetic.' *Octavo.*

I have spoken of this work in the Introduction. P. 373, 405, 411, 477.

Vienna, eighteen-twenty-one. **Geo. Fred. Vega.** 'Vorlesungen über die Rechenkunst und Algebra.' *Octavo.*

The fourth edition (with preface dated seventeen-eighty-two, which I presume to be the date of the first) of the Arithmetic of the celebrated editor of the greatest modern table of logarithms.

Leeds, eighteen-twenty-three. [**Walker ?**] 'Elements of Arithmetic for the use of the Grammar School . . . .' *Duodecimo in threes.*

An excellent little work, which I suppose I am right in attributing to Mr. Walker, and calling it the first edition of the work presently mentioned.

London, eighteen-twenty-three. **Thomas Taylor.** 'The Elements of a New Arithmetical Notation, and of a New Arithmetic of Infinites.' *Octavo in twos.*

A curious attempt at establishing a theory of infinites, the unit of which is 1 + 1 + 1 + &c. ad inf. Those who know nothing of Taylor *the Platonist,* should read his life in the *Penny Cyclopædia.* To re-establish Plato and Aristotle (in some sense even their very mythology) was the uniform endeavour of a long life and a most voluminous course of authorship. P. 424.

London, eighteen-twenty-four. **Jas. Darnell.** 'Essentials of Arithmetic, or Universal Chain.' *Duodecimo.*

This treatise reduces most questions to the form known as the *chain-rule.* Several writers have since advocated this plan.

London, eighteen-twenty-five. **J. Joyce.** 'A System of Practical Arithmetic.' *Duodecimo.*

A well-known book. The preface is dated eighteen-sixteen, which I suppose to be the date of the first edition.

London, eighteen-twenty-six. **J. R. Young.** 'An Elementary Treatise on Algebra.' *Octavo.*

I enter this work here as that of the first elementary writer who saw the value of Horner's (or, as he then called it, Holdred's) method of solving equations. This is, I believe, the first of the series of widely-known and much-used works by the same author.

London, eighteen-twenty-six. **Chas. Pritchard.** 'Illustrations of Theoretical Arithmetic.' *Duodecimo in sixes.*

A demonstrated system of Arithmetic, at a time when there were few such things in English.

London, eighteen-twenty-seven. **H.** and **J. Grey.** 'Practical Arithmetic.' Eighth edition. *Duodecimo.*

A book of concise rules for special cases.

London, eighteen-twenty-seven. **George Walker** (Master of the Grammar School at Leeds). 'Elements of Arithmetic, theoretical and practical, for the use of the Grammar School, Leeds.' Third edition. *Duodecimo.*

A clear and excellent work, written by a man of real science. I doubt whether the peculiarity of the work, the introduction of the distinction of integers and decimal fractions at the very outset, be judicious: but it has had advocates of powerful name.

Paris, eighteen-twenty-eight. **Bourdon.** 'Élémens d'Arithmétique.' Sixth edition. *Octavo.*

More complete than Lacroix in details. Bourdon was an excellent elementary writer, both on arithmetic and on algebra. I began my career as an author, by a translation of part of his work on algebra.

London, eighteen-twenty-eight. **John Bonnycastle.** 'The Scholar's Guide to Arithmetic . . . . enlarged and improved by the Rev. E. C. Tyson.' *Duodecimo.*

What edition this is I do not know.

London, eighteen-twenty-eight. 'An Abridgment of the Arithmetical Grammar . . . . By Catechetical Scrutiny.'

I have never met with the larger work.

Newcastle, eighteen-twenty-eight. **William Tinwell** (edited by James Charlton). 'Treatise on Practical Arithmetic, with Book-keeping, by single and double entry.' Twelfth edition. *Duodecimo in threes.*

Dublin, eighteen-twenty-eight. **John Garrett.** 'An Essay on Proportion.' *Octavo.*

Mostly arithmetical, with something on the connexion of number and magnitude.

London, —— **George Peacock.** 'Arithmetic.'

This is from the *Encyclopædia Metropolitana.* As part of the first volume of the *Pure Sciences,* the date is eighteen-twenty-nine: but it was separately published, in the parts, in eighteen-twenty-five or twenty-six. This is the article mentioned in the Introduction. I subjoin a few remarks, either in correction of some slips of the pen, or in addition to what has been said. The paging is that of the *Encyclopædia Metropolitana.*

P. 402, note. A sexagesimal table is sometimes pasted into a work to which it does not belong, by some old owner.

P. 404. For 'Regiomontanus, in his Opus *Palatinum* de Triangulis,' read 'in his work *de Triangulis;*' and for 'as we learn from the relation of Valentine Otho, *in his preface to that work,*' read 'in his preface to the *Opus Palatinum.*' But further observe, that though Regiomontanus did change the usual sexagesimal radius into a decimal one, no decimal tables of his were *published* until some time after Apian had published decimal tables of his own. At least, after much research, I can find none even mentioned. It is a mistake (a universal one) to say that decimal sines were published with the work *de Triangulis*, in 1533. It arises from such tables being in the *second* edition of that work, in 1561.

P. 411. Add that Chaucer, in his work on the *Astrolabe*, makes *augrime* figures to be exclusively the Arabic numerals.

P. 414. The work of *Pacioli* (not Paccioli, though very often so spelt) was published in 1494 (not 1484). Canacci, not Caracci.

P. 419 (note). Besides the *Fasciculus Temporum* (of which, by the way, there are editions dated 1474, and two or three without date, probably earlier, according to Hain), there are the *Almanac* and *Ephemerides* of Regiomontanus, described in the *Companion to the Almanac* for 1846, with extensive masses of Arabic numerals.

P. 425. Note that there is an express commentary by Jamlichus on Nicomachus, and that the former probably did not write the *Theologumena* at all.

P. 434. I have no doubt that Dr. Peacock has authority for saying that the old English Arithmeticians called the then common mode of dividing the *scratch* way (as indeed it was); but I have never met with the phrase.

P. 438 (note). It was quite common, even before the invention of printing, to speak of *Algus* as the inventor of decimal notation, or *Algorithm:* and several of the ornamented title-pages of Simon de Colines, the successor of Henry Stephens, have figures of Ptolemy, Orpheus, Euclid, and *Algus*, on one side, opposite to the Muses of Astronomy, Music, Geometry, and *Arithmetic*, on the other: as if paired for a dance.

P. 440. See what I have said on the *Disme* of Stevinus (1585); and note that Simon of Bruges is Stevinus himself.

P. 442. There is nothing about decimal fractions in Hunt's Handmaid.

P. 444. I suggest as the derivation of *furlong*, a corruption of *forty-long*. The ordinary derivation, *furrow-long*, can hardly have a foundation in fact; for the length of a furrow depends upon that of a field.

P. 454. For 'Wingrave' read 'Wingate.' As to the note, I differ greatly (as appears elsewhere) from Dr. Peacock's opinion of Cocker. With respect to the last sentence: 'It may be worth while observing that this modest and useful book is not honoured with poetical recommendations;' it would seem that the copy consulted wanted the frontispiece, on which, under the portrait of Cocker, are these lines:

> "Ingenious COCKER! (now to Rest thou'rt Gone)
> Noe Art can Show thee fully but thine own,
> Thy rare *Arithmetick* alone can show
> Th' vast *Sums* of Thanks wee for thy Laboure owe."

P. 464. For 'James Peele' read 'John Mellis,' and alter the date to 1588, as above. See on Stevinus what I have said above.

P. 471. The *leuca* was originally a measure of the Gauls; and it may be much doubted whether the measure was ever out of France since the time of Cæsar. See the article *League* in the *Penny Cyclopædia*. Surely Dr. Peacock must have written 'France' meaning to have written 'England.' This alteration of one word makes every thing right.

The small number of inaccuracies noted above are all I have been able to lay my hands on, in going through Dr. Peacock's most valuable article, with the materials for this list about me. I may add, that in two, or perhaps three instances, I remember to have found an edition mentioned as the earliest which is not so. But to this every writer is subject who has the courage to attempt history upon such bibliographical guides to the sources as now exist.

London, [eighteen-twenty-nine]. **James Parker.** 'Arithmetic and Algebra.' *Octavo.*

An early work of the Society for the Diffusion of Useful Knowledge. Arithmetical and algebraical demonstrations are connected.

London, eighteen-thirty. **Joh. White.** 'An Elucidation of the Tutor's Expeditious Assistant.' *Duodecimo.*

The author claims as a new discovery in arithmetic, the notion of forming examples in which the figures of the answer have some perceptible relation, unknown to the pupil, but known to the teacher, who can thereby see whether it is right or wrong. All the questions set, for instance, in multiplication have digits which descend by twos, as in 86420, 20864, 75319, &c. So that it is gravely proposed that the pupil shall be trained upon selected questions in which the answers have all the figures odd, or all even. A pupil so trained would be in danger of being led to think that 3728 could not be the product of any numbers.

Edinburgh, eighteen-thirty. **Adam Anderson.** 'Arithmetic.' *Quarto.*

This is the article so headed in Brewster's Cyclopædia, vol. ii. part 1. It has a fair amount of arithmetical history, though hardly enough for the extent of the work of which it forms a part.

London, eighteen-thirty-one. **Frederic Rosen.** 'The Algebra of **Mohammed Ben Musa.**' *Octavo in twos.* English and Persian.

This algebra is very arithmetical. It was published by the Oriental Translation Fund. M. Libri, in the work presently mentioned (vol. i. pp. 253-297), has given a long extract from old Latin translations now in manuscript in the Royal Library at Paris. This book passes for the one by means of which Leonard of Pisa was the first European who learnt algebra, or at least who wrote upon it.

Derby, eighteen-thirty-three. **Samuel Young.** 'A System of Practical Arithmetic.' *Duodecimo.*

A work with great force of rules, and examples in machinery, manufactures, &c.

Springfield (*U. S.*), eighteen-thirty-three. **Zerah Colburn.** 'A Memoir of Zerah Colburn, written by himself . . . with his peculiar methods of calculation.' *Duodecimo in threes.*

A great many will remember that in 1812 and 1815 two young boys, Zerah Colburn and George Bidder, astonished every one by a power of rapid mental calculation to which the most practised arithmeticians could not make the least approach. Mr. Colburn was, in 1833, a minister among the Methodists in the United States; Mr. Bidder, there is little occasion to say, is a civil engineer in England. The peculiarity of Colburn was, that he could extract roots and find the factors of numbers, to an extent which the mathematician himself had no organised rule for doing. Speaking in the third person, Mr. Colburn says: "Some time in 1818, Zerah was invited to a certain place, where he found a number of persons questioning the Devonshire boy (Mr. Bidder). He displayed great strength and power of mind in the higher branches of arithmetic; he could answer some questions that the American would not like to undertake; but he was unable to extract the roots and find the factors of numbers." This treatise contains an account of Colburn's methods.

London, eighteen-thirty-seven. **Daniel Harrison.** 'A New System of Mental Arithmetic, by the acquirement of which all numerical questions may be promptly answered without recourse to pen or pencil . . . . an entirely novel method of reducing the largest sums of money . . . . to their lowest denominations by means of original quadrantal rationale' . . . . *Duodecimo in threes.* Second edition.

The pretensions of this work are manifestly exaggerated; but the methods are ingenious. Books of mental arithmetic choose their own examples, and thus make their own rules work well. The *quadrantal* method (why so called I cannot guess) is making use of a rule for turning pounds, &c. into farthings, which requires the learning of a table; and then making such applications as the following:—A pound troy contains six times as many grains as there are farthings in the pound; therefore the grains in 4 pounds troy are the farthings in six times as many pounds sterling. Some of these rules would really become very effective, if the method of decimalising the parts of a pound were used. Multiply the pounds by a thousand, and subtract 4 per cent, and the result is the number of farthings. Thus:

£1763 . 17 . $9\frac{3}{4}$ is 1763·89062

	1763890·62
4 per cent	70555·62
Subtract	1693335 = No. of farthings in £1763 . 17 . $9\frac{3}{4}$

Berwick-upon-Tweed, eighteen-thirty-eight. **James Gray** (edited by Wm. Rutherford). 'An Introduction to Arithmetic.' *Duodecimo in threes.*

Mr. Rutherford says this neat little work went through more than forty editions in the half century preceding this publication. How many works on arithmetic there must be which I have never seen! I never met with any one of the forty.

Paris, eighteen-thirty-eight, thirty-eight, forty, forty-one. **Guillaume Libri.** 'Histoire des Sciences Mathématiques en Italie, depuis la renaissance des lettres jusqu'à la fin du xviie siècle.' *Octavo.*

The first volume of this work was printed *Paris*, eighteen-thirty-five, *octavo*, but it was never sold; the whole impression (except a few copies distributed as presents) was burnt. The work is yet unfinished; four volumes, dated as above, being all that have appeared. It is a valuable history as concerns arithmetic, both in text and notes; the latter contain—the *liber augmenti et diminutionis* compiled by Abraham (supposed to be Aben Ezra) *secundum librum qui Indorum dictus est* (vol. i. pp. 304-376)—Account of an extract from the *Liber Abbaci* of Leonard of Pisa, written in twelve-hundred-and-two (vol. ii. pp. 287-304) — the *practica Geometriæ* of the same author, being the whole of his algebra, written in twelve-hundred-and-twenty (vol. ii. pp. 305-476) — extracts from Pacioli (vol. iii. pp. 277-294): besides many other matters not relating to arithmetic. As this work stands, it can be little used for want of an index.

London, eighteen-thirty-eight. **Thos. Keith.** 'The Complete Practical Arithmetician.' *Duodecimo.*

The editor (Mr. Maynard, the mathematical bookseller, who ought to know) calls this the 12th edition: it has a deservedly high character among books of rules only, for precision and completeness. Keith says that the bare names of those who have written on arithmetic in England, from the time of Wingate, would fill a moderate volume. I suspect not: after having examined every source within my reach, and got only 1500 names, out of all times and countries, I should think it impossible that Keith could have known of 300 Englishmen, within the limits mentioned.

London, eighteen-forty. **Thomas Stephens Davies.** 'Solutions of the Principal Questions of Dr. Hutton's Course of Mathematics.' *Octavo.*

This work is also a running comment on the original, and it has a specific right to be in this list from its abounding with instances of Mr. Horner's various methods of manipulating arithmetical and more particularly algebraic questions, independently of his celebrated method of solving equations. In eighteen-forty-one the same author published an edition of Hutton's Course itself.

London, eighteen-forty-two. 'Encyclopædia Britannica.' Vol. III. *Quarto.*

The article on arithmetic was written, I suppose, by Leslie; or if not, by a follower of his views. It is meagre in its history.

London, eighteen-forty-three. **Alfr. Crowquill** (as he calls himself). 'The Tutor's Assistant, or Comic Figures of Arithmetic.' *Duodecimo in sixes.*

Walkingame's arithmetic, with comic woodcuts: but the subject will not furnish good materials. Under 'Subtraction of time,' for instance, is represented the stealing of a watch; as a heading for *Troy weight,* the *wooden horse;* and for *measure of capacity,* a phrenologised head. The worst of it is, that the joke will remain on hand too long for the learner: a picture of a gamekeeper producing a hare from a poacher's pocket must stare him in the face, as an illustration of 'proof,' through twenty-two mortal questions of addition. A comic arithmetic, with the cuts in illustration of the examples for exercise, would give the artist much fairer play.

Liège, eighteen-forty-four. **H. Forir.** 'Essai d'un cours de Mathématiques à l'usage des élèves du collége communal de Liége . . . . . Arithmétique.' Eighth edition. *Octavo in fours.*

A good and well-printed treatise, with many examples.

London, eighteen-forty-five. **Thomas Tate.** 'A Treatise on the First Principles of Arithmetic.' *Duodecimo.*

Two pages at the end, on the use of the properties of nine in constructing questions for pupils, are well worthy the attention of teachers.

London, eighteen-forty-six. **A. De Morgan.** 'The Elements of Arithmetic.' Fifth edition, with Appendixes. *Duodecimo in threes.*

The previous editions were eighteen-thirty, thirty-two, and thirty-five, *duodecimo,* and eighteen-forty *in threes.*

Books of bibliography last longer than elementary works; so that I have a chance of standing in a list to be made two centuries hence, which the book itself would certainly not procure me.

The following are some additional works more briefly described. Here 2+2, 4+4, 6+6, mean quarto in ones, octavo in twos, duodecimo in threes.

Date.	Place.	Author.	Leading Word of Title.	Form.
Seventeen-				
-thirty-four .	London	Alex. Wright . .	Fractions	6+6.
-fifty-four . .	Exon.	Joseph Thorpe . .	Treatise	4+4.
-fifty-seven .	London	R. Gadesby . . .	Decimal	4+4.
-sixty-one . .	London	John Dean. . . .	Rule of Practice	8 (2d ed.)
-sixty-two . .	London	Rich. Ramsbottom.	Fractions anatomised . . .	6+6.
-sixty-six . .	Sheffield	John Eadon . . .	Guide	6+6.
-sixty-six . .	Birmingham . . .	Wm. Crumpton . .	Decimals and Mensuration .	6+6.
-seventy . .	Exeter	G. Dyer	School-Assistant	12.
-seventy-one.	London	Wm. Scott	New System	8.
-eighty-eight	Birmingham . . .	Wm. Taylor . . .	Guide	6+6.
-eighty-nine .	Edinburgh . . .	Wm. Gordon . . .	Institutes	4+4.
Eighteen-				
-four . . .	Bingham	W. Butterman . .	Dialogue	6+6.
-five . . .	Shrewsbury . . .	Geo. Bagley . . .	Assistant	2+2.
-eight . . .	Paris	Edm. Degrange. .	Pratique	8.
-eleven . .	London	Joh. Harris Wicks	Merchant's Companion . .	6+6.
-eleven . .	Birmingham . . .	J. Richards . . .	Practical	6+6.
-twelve . .	Paris	J.Cl.OuvrierDelile	Appliquée au Commerce . .	8.
-sixteen . .	Edinburgh . . .	R. Hay	Beauties	8.
-seventeen .	London	James Morrison .	Concise System	12.
-eighteen . .	London	Joh. Matheson . .	Theory and Practice . . .	12.
-eighteen . .	Dundee	James Nicolson . .	Modern Accountant . . .	6+6.
-twenty . .	Glasgow	C. Morrison . . .	Young Lady's Guide . . .	6+6.
-twenty . .	Paris	J. B. Juvigny . . .	Application au Commerce .	8.
-twenty-two .	Bedford	W. H. White . .	Complete Course	6+6.
-twenty-three	Canterbury . .	H. Marleen and W. Smeeth . . .	Assistant	6+6.
-twenty-three	Lausanne	Em. Develey . . .	D'Emile	8.
-twenty-four	Cork	Joh. Penrose . . .	Treatise	6+6.
-twenty-six .	London	Anth. Peacock . .	Mental	12.
-twenty-six .	London	J. W. Hoar. . . .	Examinations	18.
-twenty-seven	London	Wm. Phillips . .	New and concise System .	12.
-twenty-eight	London	S. P. Reynolds . .	Practical	4+4 (small).
-twenty-eight	London	Rob. Fraiter. . .	Short System	6+6 (2d ed.)
-twenty-eight	[London]	R. W. Woodward .	Guide	12.
-twenty-eight [*circa*]	London	Anonymous . . .	Intellectual (Pestalozzian) .	6+6.
-twenty-nine	Paris	H. L. D. Rivail . .	D'après Pestalozzi . . .	12.

Date.	Place.	Author.	Leading Word of Title.	Form.
EIGHTEEN-				
-twenty-nine	Paris	A. SAVARY	Arithmétique	8.
-twenty-nine	London	ANONYMOUS . . .	Concise Arithmetician . .	6+6.
-twenty-nine	London	DANIEL DOWLING .	Improved System	12.
-twenty-nine	London	JAMES PERRY . . .	Middle Stage and Key . .	6+6.
-thirty . . .	London	W. PUTSEY	Practical	6+6.
-thirty-one .	London	GEO. HUTTON . . .	Theory and Practice . . .	12.
-thirty-two .	Edinburgh . . .	ROB. CUNNINGHAM .	Text-Book	6+6.
-thirty-four .	London	RICH. CHAMBERS . .	Introduction	6+6.
-thirty-four .	Newcastle-on-Tyne	THOS. THOMPSON . .	Business Book	4+4.
-thirty-four .	London	CHRIST. KNOWL. SOC.	Taught by Questions . . .	8 (square).
-thirty-six .	London	RICH. MOSLEY . . .	Elements	12.
-thirty-six .	Edinburgh . . .	JOH. DAVIDSON . .	Guide	6+6.
-thirty-seven	Boston (U. S.) . .	JOS. BARTRUM . . .	Arithmetic	6+6.
-thirty-nine .	London	J. FELTON	Improved Method	6+6.
-forty . . .	Cambridge . . .	W. C. HOTSON . . .	Principles	6+6 (2d ed.)
-forty-three .	London	MERCATOR	Expeditious Calculation . .	4+4.
-forty-five -	London	W. WATSON	School	6+6(4th ed.)
Without date	London	ADAM TAYLOR . .	[Useful Arithmetic] & Sequel	12.
———	London	HENRY AULT . . .	Requisites of Business . .	6+6 (pt. 1).
———	Dartmouth . . .	WM. JARVIS . . .	Catechism	8 (small).

ADDITIONS

No place, date, nor author's nor printer's name. 'Algorithmus Integrorum Cum Probis annexis.' *Quarto in threes.*

I think this is the oldest book in my list. The type is of that manuscript appearance which is so common in books printed before 1480: and this one can hardly be later than 1475, and may be ten years earlier. The matter is that of a writer prior to the school of Borgi and Pacioli. Though the nine digits are given, they are not used. The rules only are given, and they are Sacrobosco's set; for which see p. 15.

No place nor date. 'Algorismus novus de integris compendiose sine figurarum (more Italorum) deletione compilatus. artem numerandi omnemque viam calculandi enucleatim brevissimè edocens. una cum algorismis de minuciis vulgaribus videlicet et phisicalibus. Addita regula proportionum tam de integris quam fractis que vulgo mercatorum regula dicitur. Quibus habitis quivis modica adhibita diligentia omnem calculandi modum facillime adipisci potest.' *Quarto in fours.*

Three different editions of this work are mentioned by bibliographers, all without dates. Numeration is called *prima species;* addition, *secunda species;* and so on. There are slight examples, except upon the square root, in which only results are given. It is something between such a work as the last, with rules only, and that of Borgi, to which I now come, in which there is decided force of examples.

Venice, fourteen-eighty-eight. **Piero Borgi.** Work on Arithmetic without title, as follows. *Quarto in fours.*

This is one of the editions alluded to in page 2, and for the power of mentioning it from inspection (as well as the works of Ghaligai and Texeda following) I am indebted to Dr. Peacock. It begins with the following verses:

Chi de arte matematiche ha piacere
Che tengon di certeza el primo grado
Avanti che di quelle tenti el vado
Vogli la presente opera vedere
Per questa lui potra certo sapere
Se error sara nel calculo notado

Per questa esser potra certificado
A formar conti di tutto [*sic*] maniere
A merchadanti molta utilitade
Fara la presente opera e afatori
Dara in far conti gran facilitade
Per questa vederan tutti li errori
Ede iquaterni soi la veritade
Danari acquisterano e grandi honori
In la patria e de fuori
Sapran far le rason de tutte gente
Per le figure che son qui depente.

After a page of description of symbols, comes 'Qui comenza la nobel opera de arithmeticha ne laqual se tracta tute cosse amercantia pertinente facta et compilata per Piero borgi da Veniesia.' At the end is 'Stampito in Veniexia per zovanne de Hallis 1488.'

The author begins by saying that there are plenty of sufficiently good masters, and not less abundance of most excellent authors. His arrangement of the five *acts* of arithmetic is strange; they are numeration, multiplication, partition, summation, and subtraction. He uses the word *million*, and his numeration goes to millions of millions of millions. He then proceeds to multiplication, and forms a table of the nine multiples of 1, 2, ... 10, with those of 12, 16, 20, 24, 32, and 36. He then points out how to prove processes by casting out sevens and nines; in doing which he shews how to divide by 7, and uses the sum of the digits to find the remainder to 9. Then follow multiplications done each in one line, even when the multiplier has three or four figures. After this follows the common method, in which the final process is addition, done without any formal rule (in fact, addition is not made a formal rule any where). Division is said, at the head of a chapter, to be done in three ways; but only two are given: that *per cholona*, where the whole is done in one line, and that *per batelo*, which is what I have always designated by (O). Subtraction is then explained; and after some applications, the rules for fractions (which are expressed in the same way as now) are introduced. The rule of three (*riegola del tre*) is given, and a large number of money applications.

Paris, fourteen-ninety-six. 'In hoc opere contenta. Arithmetica decem libris demonstrata Musica libris demonstrata quatuor Epitome in libros arithmeticos divi Severini Boetii. Rithmimachie ludus qui et pugna numerorum dicitur.' *Folio in fours.*

This is the first edition of the work mentioned in page 10: so that I was wrong in page 4. I was misled by bibliographers mentioning this as an edition of Jordanus only. The copy before me comes to me in two different parts, though the register at the end (which has the first words of all the sheets) proves that these parts belong to one work. It would be odd if it should have been customary to divide it in this manner, so as to lead to the above imperfect description. The corrector of the press was David Lauxius, and the printers John Higman and Wolfgang Hopilius.

Venice, no date, but apparently before fifteen-hundred. 'Dabaco che insegna a fare ogni ragione mercadantile: et a pertegare le terre con larte di la geometria: et altre nobilissime ragione straordinarie con la Tariffa come respondeno li pesi et monede de molte terre del mondo con la inclita citta de Vinegia. El qual Libbro se chiama Thesauro universale.' *Octavo*.

The order of the operations is as in P. Borgi. The book is very full of florid ornaments and woodcuts.

Among books of the fifteenth century which I have had no opportunity of seeing are the following, described by Hain and others:

Without date. 'Arithmeticæ Textus communis, qui pro Magisterio fere cunctis in gymnasiis ordinarie solet legi, correctus corroboratusque perlucida quadam ac prius non habita commentatione a Conrado Norico. Lipsiæ per Martinum Herbipolensem.' *Folio*.

Without date, *quarto*, with the mark of Martin of Wurtzburg (Herbipolis). 'Algorithmus linealis,' beginning 'Ad evitandum multiplices Mercatorum errores &c.'

Treviso, fourteen-seventy-eight, *quarto*. 'L'arte del Abbacho, Practica molto bona et utile a chiachaduno qui vuole uxare l'arte merchandantia.'

Watt mentions an edition, *Rome*, fourteen-eighty-two, *quarto*, of the 'Ludus Arithmomachiæ,' attributed to John (he should have said William) Shirewood, bishop of Durham. Hain does not mention this.

Babenberg, fourteen-eighty-three, *duodecimo*. 'Rechnungsbüchlein.'

Cologne, fourteen-eighty-five, *quarto*. 'Ars numerandi, seu compendiosus tractatus de dictionibus numerabilibus.'

Strasburg, fourteen-eighty-eight, *quarto*. Anianus. 'Compotus manualis magistri aniani. metricus cum commento Et algorismus.'

Leipzig, fourteen-eighty-nine, *octavo*. 'Rechnung auf alle Kaufmannschaft.' Murhard and Kloss (who cites Panzer) give this book to Joh. Widman, whose name appears in an address to the reader; but Hain, though naming Widman, puts the work down as anonymous.

Two collections of the works of Boethius, with the Arithmetic among them, *Venice*, fourteen-ninety-one and ninety-seven, *folio*.

Deventer (Daventrie), fourteen-ninety-nine, *quarto*. 'Enchiridiom Algorismi sive tractatus de numeris integris.'

Without date, but with a letter from Balth. Licht dated fifteen-hundred, *Leipzig*, printed by Melchiar Lotter, *quarto*. 'Algorithmus linealis cum pulchris conditionibus Regule detri: septem fractionum: regulis socialibus,' &c. And another, seemingly of the same date, by the same printer, 'Algoritmus linealis.'

No doubt the first English print on Arithmetic is cap. x. of 'The Mirrour of the World or Thymage of the same,' headed 'And after of Arsmetrike and whereof it proceedeth,' printed by Caxton in fourteen-eighty. (Peacock, p. 419, and Ames.)

Basle, fifteen-three. **Nicolas de Orbelli.** 'Cursus librorum philosophie naturalis venerabilis magistri Nicolai de Orbelli ordinis minorum secundum viam doctoris subtilis Scoti.' *Quarto in fours.*

This course of natural philosophy requires two pages of explanation of arithmetical, and, in particular of Boethian, terms, and nothing more. It purports to have been intended for those who were studying theology. The *Margarita Philosophica* (page 5,) seems to give the highest arithmetical limit in liberal education, and the work before us the lowest.

See page 12. John de Lortse, &c. I was perfectly justified in writing Lortse, for the name is so spelt in my copy in the clearest black letter. Nevertheless, it is John de L'Ortie, and the book is a French translation of the work of Juan de l'Ortega, mentioned by Dr. Peacock, pp. 426 and 436. Seeing de L'Ortie [sic] mentioned by Meynier, and the word thus written being nothing but the translation of L'Ortega (nettle), I compared the French work with Dr. Peacock's description of the Spanish, and found that the two must necessarily be the same. On looking very narrowly at the f which caused the mistake, I found in it a small hook at the bottom, which the letter f never has elsewhere in the work, and the letter i always has. The f then in Lortse is an i in which the dot (or rather small line, for so it is in the black letter) has been accidentally battered into the top of an f. But the alteration is so perfect that every one reads it for an f. This is the second paragraph which one battered letter has caused. In the new edition of Brunet it is noted that Heber's catalogue spells the word Lort*s*e. My copy is the one which was in Heber's library.

This French translation gives Juan de l'Ortega an earlier date than the Spanish bibliographers, or those who have devoted themselves to the Dominicans, are aware of. Antonio, in his Spanish lists, and Quetif, in his Dominican one, both give Ortega no earlier date than 1534. But it appears that he was translated into French in 1515. Panzer (xi. 465, according to Brunet) mentions a work of L'Ortega, *Messina,* fifteen-twenty-two, beginning 'Sequitur la quarta opera di arithmetica et geometria.'

Florence (?), no date. **Francisco di Lionardo Ghaligaio.** 'Summa De Arithmetica.' *Quarto in threes* (N).

Florence, fifteen-sixty-two. **Francesco di Lionardo Ghaligai.** 'Pratica D'Arithmetica . . . . Nuovamente Rivista.' *Quarto in fours.*

Two editions of the same book. The first must have been published before fifteen-twenty-three, the year in which Julius de Medicis, to whom it is dedicated as Cardinal, would have been addressed as Pope. It is a complete and advanced treatise of arithmetic, in thirteen books. The tenth book begins algebra, or *Arcibra*, the introduction of which from the Arabic seems to be attributed to Guglielmo de Lunis; Leonard of Pisa and Giovanni del Sodo are also mentioned as writers.

Venice, fifteen-twenty-six. **Joh. Fr. Dal Sole.** 'Libretto di Abaco novamente Stampato: Composto per lo excellente maestro Joanne Francisco dal sole Ingegnero: utilissimo a cadauno per imparare per se stesso senza maestro. Dimandato breve introductione.' *Octavo in twos* (O).

A very elementary work, from the school of Borgi and Pacioli, proceeding up to the square root.

——— fifteen-forty-five. **Gaspar de' Texeda.** 'Suma De Arithmetica pratica y de' todas Mercaderias Con la 5 orden de' contadores. Hecho por Gaspar de' Texeda, Con Privilegio Imperial.' *Quarto in fours* (ON).

The colophon, which probably contains the place, is torn out. In numeration the places are distinguished throughout the first part of the book. Thus, 950777200000 is 950||777q°200||000, where || (the two lines should be joined at the bottom) stands for *mill*, and q° for *cuento*, a million. The proof by sevens and nines appears, as in Borgi. The work is rather more advanced than that of the latter, and proceeds to the extraction of roots. The greater part is commercial. Rules for fractions are given, not complete. The rule of three opens with an invocation of the Trinity: Sfortunati goes further, for he connects the rule with the doctrine.

London, fifteen-forty-six. 'An introduction for to lerne to recken with the pen, or with the counters accordyng to the trewe cast of Algorisme, in hole numbers or in broken newly corrected. And certayne notable and goodly rules of false positions thereunto added, not before sene in oure Englyshe tonge by the which all maner of difficile questions may easely be dissolved and assoyled Anno. 1546.' *Octavo*. At the end is: 'Imprynted at London in Aldersgate Strete by Jhon Herford.'

London, fifteen-seventy-four (printed by John Awdeley). 'An introduc- [sic] of Algorisme, to learn to recken wyth the Pen or wyth the Counters, in whole numbers or in broken. Newly overseen and corrected . . .' *Octavo* (small) (O).

These are, no doubt, reprints of the work of nearly the same title, mentioned by Dr. Peacock (p. 419) as having been printed at St. Albans in 1537. It was a good predecessor of Recorde's *Grounde of Artes*, but it makes only a poor follower.

See page 19. In speaking of Rudolph, I had overlooked that Dr. Peacock (p. 424, note) describes his *Algebra* from inspection, as having been edited by Stifel himself in 1571.

Rome, fifteen-eighty-three. **Christ. Clavius.** 'Epitome Arithmeticæ Practicæ.' *Octavo* (small) (O). See page 33.

Witeberg, sixteen-four. **Gemma Frisius.** 'Arithmeticæ Practicæ Methodus Facilis.' *Octavo* (small). See page 25.

See page 26. Macpherson (Annals of Commerce, eighteen-five, quarto, vol. i. p. 146) gives (from Anderson, i. 409) a French edition of Stevinus on book-keeping, *Leyden*, sixteen-two, *folio*, as quoted by Anderson from a copy in his own possession. The title he gives is 'Livre de compte de prince à la manière d'Italie en domaine et finance ordinaire : contenant ce en quoi s'exerce le très illustre et très excellent Prince et Seigneur Maurice Prince d'Orange, &c.'

Frankfort, sixteen-thirteen. **Joh. Faulhaber.** 'Ansa inauditæ et mirabilis novæ artis . . . .' *Quarto*. Same place, author, and form, sixteen-fourteen. 'Numerus figuratus . . . .' and sixteen-fifteen, 'Mysterium Arithmeticum . . . .'

These singular medleys of arithmetic, algebra, prophecy, and nonsense, seem to be all one publication. The date of the first is JVDICIVM, written at the bottom of the title-page. Alter the order of the letters into M.DC.VVJII., and it becomes 1613.

London, sixteen-fourteen. **James Dowson.** 'De Numerorum Figuratorum Resolutione.' *Octavo*.

This work gives Pascal's table.

No place nor date. **Joh. Harpur.** 'The Jewell of Arithmetick : or, the explanation of a new invented arithmeticall Table . . . .' *Quarto*.

Perhaps this work is the *Merchant's Jewel* mentioned by Bridges (p. 44) as of 1628, a date which, from the character of the work, I should be disposed to give it. I can find no mention nor use of decimal fractions in it. The operations are performed on a kind of abacus.

Johnson, presently mentioned, describes a 'most excellent instrument invented by Mr. William Pratt, called, The Jewell of Arithmatick ;' which is perhaps the one described in the work above, though with a different inventor's name.

London, sixteen-thirty-three. **Joh. Johnson** (Survaighor). 'Johnson's Arithmatick In 2 Bookes. The first, of vulgare Arithma : with divers Briefe and Easye rules : to worke all the first 4. partes of Arithmatick in whole numbers and fractions by the Author newly Invented. The Second, of Decimall Arithmatick wherby all fractionall operations are wrought, in whole Numbers, in Marchants accomptes without reduction; with Interest, and Annuityes.' Second edition. *Duodecimo* (O).

In his decimal fractions, Johnson has the rudest form of notation; for he generally writes the places of decimals over the figures, thus—

$$146{\cdot}03817 \text{ would be } 146|\overset{1.2.3.4.5.}{03817}$$

Otherwise, his system is tolerably complete.

London, sixteen-forty-nine. **R. B.** (Mr of Arts). 'Arithmetick Symbolical In one Book. In which the Mystery of Numeration by Symbols is revealed.' *Octavo.*

A short and easy treatise on Algebra.

London, sixteen-fifty-five. **R. B.** 'An Idea of Arithmetick at first Designed for the use of the Free-Schoole at Thurlow in Suffolk. By R. B. Schoolmaster there.' *Octavo* (O).

Decimals and algebra, all on the plan of Oughtred.

Antwerp, sixteen-sixty-two. **Jean Raeymaker.** 'Traicte d'Arithmetique Contenant les quatres especes, avec la regle de trois, et la Practique.' *Octavo.*

Not a first edition. Nothing but questions and their answers, as in John Speidell's work of sixteen-twenty-eight.

London, sixteen-seventy-five. **C. H.** 'The Golden Rule made Plain and Easie.' *Octavo.* Second edition.

No date nor place (I think about sixteen-eighty). **J. W.** (of Brandon). 'A progression in Arithmetical Progression . . . . . with Some things very remarkable in that Mystical Number 666.' *Octavo in twos.*

The author shews how to form squares and cubes by differences. He is very charitable, for his day, to the Papists: for, setting down twelve as the divine number, and that of the apostles, he seems to imply that the Roman church, as having a number made up of sixes, may be half Christian and half apostolic.

London, sixteen-eighty-four. —— 'Enneades Arithmeticæ; the Numbering Nines. Or, Pythagoras His Table extended to All Whole Numbers under 10000.' *Octavo in twos.*

The plan is, to have 99 rods instead of nine, in Napier's set; so as to have the first nine multiples of all numbers short of 100.

London, sixteen-eighty-five. **John Wallis.** 'A treatise of Algebra, both Historical and Practical.' *Folio in twos.*

London, sixteen-ninety-three. **John Wallis.** 'De Algebra Tractatus; Historicus et Practicus. Anno 1685 Anglice editus; Nunc Auctus Latine.' *Folio in twos.*

The Latin, or augmented edition, is the second volume of Wallis's collected works. See the Introduction to this Catalogue.

London, sixteen-ninety-seven ——. 'Computatio Universalis, seu Logica Rerum. Being an Essay attempting in a Geometrical Method, to Demonstrate an Universal Standard, whereby one may judge of the true Value of every thing in the World, relatively to the Person.' *Octavo.*

The author, cutting down life to its half, to allow for childhood, sickness, &c., proceeds to a calculation how the true value of every thing is to be estimated numerically. This, and Craig's more famous book (see Useful Knowledge Library, *Probability*,*) are consequences of the Principia. It is curious that as soon as *force* was widely made known as an object of numerical measurement, some people began to fancy prudence, pleasure, and pain could be submitted to the same process.

Breslau, seventeen-seventy-nine. **Joh. Ephr. Scheibel.** 'Einleitung zur mathematischen Bücherkentnis. Eilftes Stück.' *Octavo* (small).

See the Introduction. This part contains the arithmetical bibliography; but the paging does not begin after that of the 'Zehntes Stück.' In fact, the bibliography of this book of bibliography is a study of itself.

Göttingen, seventeen-ninety-six, ninety-seven, ninety-nine, eighteen-hundred. **Abr. Gotthelf Kastner.** 'Geschichte der Mathematik . . . .' Four volumes *octavo.*

See the Introduction, as to Kästner.

* I have seen this book sold with my name upon the cover as the author. To this I have no objection, except my knowledge of the fact that it is the joint production of Sir John Lubbock and Mr. Drinkwater-Bethune.

LIST OF 1580 NAMES

OF REPORTED AUTHORS, EDITORS, &c. OF WORKS ON ARITHMETIC, INCLUDING THE INDEX TO THIS CATALOGUE.

The numbers refer to pages of the present work; and the asterisk prefixed shews that the author has been referred to by Dr. Peacock in his History of Arithmetic.

INDEXES

INDEX OF DATES

DATES OF PRINTED BOOKS

DATES OF MANUSCRIPTS

INDEX OF NAMES, PLACES, AND SUBJECTS